THE RICHNESS OF LIFE

Stephen Jay Gould was the Alexander Agassiz Professor of Zoology and Professor of Geology at Harvard and the Curator for Invertebrate Palaeontology in the university's Museum of Comparative Zoology. He died in May 2002.

Steven Rose is Professor of Biology and Director of the Brain and Behaviour Research Group at the Open University, Visiting Professor in the Department of Anatomy and Developmental Biology at University College London, and, jointly with sociologist Hilary Rose, Professor of Physic (genetics and society) at Gresham College, London.

Paul McGarr is a mathematics teacher and a leading member of the Respect coalition in Tower Hamlets, London.

ALSO BY STEPHEN JAY GOULD

Ontology and Phylogeny

The Structure of Evolutionary Theory

The Mismeasure of Man

Time's Arrow, Time's Cycle

Life's Grandeur: The Spread of Excellence From Plato to Darwin

The Hedgehog, The Fox and The Magister's Pox: Mending the Gap Between Science and the Humanities

Rocks of Ages: Science and Religion in the Fullness of Life

Wonderful Life: The Burgess Shale and the Nature of History

Questioning The Millennium: A Rationalists Guide to a Precisely Arbitrary Countdown

Collected Essays

Ever Since Darwin: Reflections in Natural History

The Panda's Thumb: More Reflections in Natural History

Hen's Teeth and Horses Toes: Further Reflections in Natural History

The Flamingo's Smile

Dinosaur in a Haystack

Leonardo's Mountain of Clams and the Diet of Worms

Lying Stones of Marrakech: Penultimate Reflections in Natural History

Triumph and Tragedy in Mudville: A Lifelong Passion for Baseball

I Have Landed: Splashes and Reflections in Natural History

An Urchin in the Storm

Bully for Brontosaurus

Eight Little Piggies

STEPHEN JAY GOULD

The Richness of Life

The Essential Stephen Jay Gould

EDITED BY

Paul McGarr and Steven Rose

WITH AN INTRODUCTION BY

Steven Rose

VINTAGE BOOKS
London

Published by Vintage 2007

10 9

First published in Great Britain in 2006 by
Jonathan Cape

Vintage
Random House, 20 Vauxhall Bridge Road,
London SW1V 2SA

www.vintage-books.co.uk

Addresses for companies within The Random House Group Limited can be found at: www.randomhouse.co.uk/offices.htm

The Random House Group Limited Reg. No. 954009

A CIP catalogue record for this book
is available from the British Library

ISBN 9780099488675

Printed and bound by CPI Group (UK) Ltd, Croydon, CR0 4YY

CONTENTS

LIST OF FIGURES

NOTE ON THE TEXT

FOR this anthology of Stephen Jay Gould's writing, many of the illustrations that originally accompanied the articles and extracts collected here have been omitted. In some cases, the text has also been lightly edited and source references have been deleted. Readers who would like to read an essay or extract in its original form, or a book in its entirety, are referred to the bibliographical information provided in the Sources and Acknowledgements.

INTRODUCTION

THERE are few scientists famous enough in their lifetime to be canonized by the United States Congress as one of that country's 'living legends'. It is still more unlikely that the title should have been conferred on a man regarded by many Americans as a notorious radical, a sometime Marxist who was controversial throughout his life as a theorist and polemicist, even amongst colleagues in his own chosen fields of paleontology and evolutionary theory. Yet few would have grudged this accolade to Stephen Jay Gould, whose writings on history – both of the natural world and of the study of that natural world – had made him a household name by the time of his death in 2002. And not just in the Anglophone world, for his books and articles have been widely translated and read in their hundreds of thousands in every society in which debate about evolution and the human condition are the stuff of intellectual life.

Even a select bibliography of Steve Gould's writings – books, academic papers and major reviews – runs to over a dozen pages. Words – learned, ironic, argumentative, whimsical – poured from his pen, and later his keyboard. To the public he is best known for the unbroken sequence of 300 essays published in the magazine *Natural History*, all later reprinted in book form, which began in 1974 and ceased only a year before his death. A stylish writer, he generated each essay by deriving a seemingly abstruse point in natural history or paleontology via a sideways look at a novel, a building, or a reference – esoteric to readers on this side of the Atlantic – to his lifelong love of baseball. He once illuminated the peculiar evolutionary phenomenon in which more recently evolved species within a family group steadily decrease in size by comparing the phenomenon to how the manufacturers of Hershey bars avoided price rises by making the bars smaller whilst keeping the costs the same (see 'Phyletic Size Decrease in Hershey Bars').

It is from this rich stock of writings that we have made a selection for the present anthology. The choice has not been easy. As a scientific essayist Gould's only peers were 'Darwin's bulldog' Thomas Huxley in the nineteenth century and J. B. S. Haldane in the 1930s and '40s. Like them he was also a major public intellectual and powerful public speaker to audiences at meetings ranging from those held in the venerable (if spartan) premises of the Papal Academy in Venice or the Sheldonian in Oxford to a local picket line. The comparison with Haldane is apt in two further ways. Both made fundamental contributions to evolutionary theory and both were politically engaged within science and in the broader political arena. In Gould's case these more fundamental contributions appeared in their turn in a number of major books, beginning in 1977 with *Ontogeny and Phylogeny* and culminating in his last great work, the 1,400-page *Structure of Evolutionary Theory*, published only months before his death. Any one of these major works should have assured Gould an honoured place amongst evolutionary theorists. Yet, unlike Haldane, whose contribution to what became known as 'the modern synthesis' of Darwinism with Mendelian genetics, is uncontested, Gould's remains controversial. His innovations, in collaboration with co-thinkers such as Niles Eldredge, Elizabeth Vrba and Richard Lewontin, of punctuated equilibrium, exaptation and multilevel selection, have not found easy acceptance. What some saw as his Darwinian revisionism, but he would have regarded as extending evolutionary theory in a direction that the great pluralist Darwin himself would have approved of, earned him obloquy amongst more orthodox ultra-Darwinists. Leading evolutionists such as John Maynard Smith and Ernst Mayr remained hostile to Gould's thinking. And in some ways Gould's own felicity with words, his hugely popular essays and – to put it positively – self-assurance, won him enemies as well as friends.

This introduction is not a substitute for a biography, but some bare outline of his life will help put his writings in focus, and indeed his own reflections on his life occupy the first (Autobiography) section of this anthology. Gould's family originated in Eastern Europe. One of his great-grandparents was living in New York in the late nineteenth century; on the other side of the family, his grandparents landed in 1901. Like my own, who arrived at about the same time in London, they worked in the classic Jewish trades of cabinet-making and garment manufacture. Steve was born in 1941 in the New York district of Queens and educated through the city's superb public school system. He trained as a geologist at Antioch College, Ohio and took a doctorate in paleontology at Columbia University in New York in 1967 before moving to Harvard, where in 1982 he was appointed to the Alexander Agassiz chair in zoology, a position he retained for the rest of his life, coupling it to a professorship of geology and curatorship of invertebrate paleontology at the University Museum of Comparative Zoology, and, subsequently, the Vincent Astor visiting professorship of biology at New York University. He accumulated a glittering stock of academic prizes and awards. Though you would be hard put to learn about it from his writings, he was married twice. With Deborah, his first wife, he had two sons, Jesse and Ethan, and with his second wife, Rhonda, two stepchildren, Jade and London. He was much more open about a key turning point in his life: the discovery, in 1982, that he was suffering from mesothelioma, a savage form of cancer with a devastatingly high and fast mortality rate. His account of how he survived is also reprinted in the Autobiography section, where, characteristically, it serves as a way of making a more general statistical point – also demonstrated in his several references to baseball in the essays included in the selection. But it was finally cancer that killed him, sadly prematurely.

Steve's radical and activist anti-racist political perspective was formed early. He recalled, for instance, how during his time as a PhD student he spent a period at the University of Leeds, and there shared in the picketing of an ice rink in neighbouring Bradford which operated a colour bar. The late sixties and seventies were a time of radicalization amongst students and young academics, notably around opposition to the Vietnam war. The 'abuses' of science that became apparent in the United States' conduct of that war led to the development of a general critique of the role of science in capitalist societies and to the formation there of such groups as Scientists and Engineers for Social and Political Action and Science for the People. An early focus for these campaigning groups was the notorious review by the psychologist Arthur Jensen, published in 1969, which reopened an old debate by claiming that American blacks were inferior to whites in intelligence, and that no amount of educational manipulation would ever obviate this genetically determined difference.

This biological determinist assertion heralded a sequence of high-profile claims culminating in the publication in 1975 of E. O. Wilson's *Sociobiology*, whose basic argument as applied to humans was that the inequalities in wealth, status and power between genders, classes and races in human societies were evolved properties, the results of Darwinian selection processes and hence largely beyond the reach of social change. Science for the People established a sociobiology study group, mainly centred in Cambridge, of which Gould was a member, and in 1977 published an initial response in their collective book *Biology as a Social Weapon*.

Some saw this as a specifically Harvard-based battle, as Gould occupied the ground floor of the MCZ and his colleague and sometime co-author, the evolutionary geneticist Richard Lewontin the third, with Wilson one floor above. As I discovered in the quarter I spent at the MCZ in the 1980s working with Dick Lewontin on our own book *Not in Our Genes*, this proximity made Wilson distinctly uncomfortable about entering the elevator. A later essay in the *New York Review of Books* by Lewontin discussing a book critical of biological determinism made sideways reference to this discomfort, by entitling it 'The Corpse in the Elevator'. Incidentally, in the first draft to this Introduction, I referred to Gould as a 'key' member of Science for the People, but on sending my draft to Dick Lewontin, was properly chastised for doing so. SftP was a collective, and as such, and befitting the politics of the time, eschewed hierarchy. However, despite this, I suspect that Steve, along with Dick, was a major influence on the direction of the group in that particular struggle. But the tension in these relationships was not confined to those on the opposite side of the ideological divide, as became apparent when Gould, reviewing Lewontin, Leon Kamin's and my *Not in Our Genes* in the *New York Review*, described it as reading as if written by a committee – not a charge we would willingly plead guilty to. But Steve was never one to show undue modesty, in either print or speech.

Although Gould's more collectivist and activist phase was relatively brief, the critiques of racist and determinist thinking shaped a great deal of his later writing, from *The Mismeasure of Man* (1981) through to the later debates of the 1990s with zoologist Richard Dawkins and philosopher Daniel Dennett as sociobiology became 'rebranded' (to use Dawkins's term) as evolutionary psychology, as well as a famous collaboration with Lewontin in a full-blooded assault on what they called the 'Panglossian' adaptationist programme, after Voltaire's famous character who insisted in the face of endless catastrophes that we humans live in the best of all possible worlds.

However, it would be a mistake to view these conflicts, as did the sociobiologists and their acolytes amongst certain philosophers and sociologists, as merely reflective of prior differences in political orientation between Gouldites and Dawkinsites, as they came to be termed, much to many people's irritation. It isn't that the politics were unimportant, but that the disputes also derive from important differences in the understanding of the mechanisms and processes of evolution. Gould's distancing himself from orthodox neo-Darwinism began early. Like sculptors, paleontologists are fascinated by shapes and patterns, their evolution and development. Organisms (phenotypes) rather than genes are the focus of their attention. Steve's first research subjects and abiding love (along with baseball) were the snails, live or fossilized, of the Bahamas (see 'Opus 100'). Snail shells are coiled helices, and their outer surfaces are characteristically banded. What determines their colours and forms? Are they accidental? Are they adaptive – that is, do they serve some function which enhances the lifetime reproductive success of the snail which bears them, and hence are they subject to evolutionary pressures? Just as Darwin learned a great deal about evolution from the study of earthworms and barnacles, so one of his outstanding followers did from observing snails.

Strict adaptationists are convinced that all such differences are themselves, or at least reflect, some functional advantage to the organism. Their opponents, following theories developed by the Japanese geneticist Motoo Kimura (see 'Betting on Chance – and No Fair Peeking') see many differences as neutral, arising by chance and neither improving nor diminishing genetic fitness. The clash of these two positions was never more forcefully evoked than at a packed meeting at The Royal Society in 1979. Gould and Lewontin's paper 'The Spandrels of San Marco and the Adaptationist Programme' (here reprinted as 'The Spandrels of San Marco and the Panglossian Paradigm') was given at the meeting by Gould. To the mounting incomprehension of the audience of evolutionists, it opens with an account of the glories of the mosaics on the dome of Venice's famous cathedral. Looking at them, the paper argues, you would assume that the arches that support the dome had been constructed to provide the spaces (spandrels or, perhaps more accurately, pendentives) that the mosaics cover. In fact, Gould and

Lewontin continue, the spandrels are merely infillers. The architectural need is for the load-bearing arches to take the form they do, and having constructed them you might as well make the best use of the infillers by decorating them with appropriate devotional scenes. Many observable features of living organisms are, they argued, such spandrels.

The paper was given just before the tea break, and I still remember the quivering rage with which one of Britain's leading snail geneticists harangued me, accusing the paper of being a Marxist plot and assuring me that he could prove that every single variation that he found amongst the coloured bands on the snails he studied had a functional significance. The spandrel row indeed rumbled on, such that the philosopher Daniel Dennett, in his book *Darwin's Dangerous Idea*, took an entire chapter to attempt to refute it both on architectural and on biological grounds (see 'More Things in Heaven and Earth' for Gould's response). Steve's enthusiasm for architectural and indeed cathedral metaphors was undiminished, however, and another cathedral, this time the monstrous *duomo* in Milan, plays a similar and to my mind much less successful didactic role in the vast edifice of *The Structure of Evolutionary Theory*.

This huge book is the most comprehensive statement of Gould's Darwinian revisionism – or the pleasures of pluralism, as he referred to it. This revisionism began in graduate school, when he and fellow student Niles Eldredge developed their critique of one of Darwin's central theses, that of gradual evolutionary change. To the concern of many of his friends and supporters, who had argued that speciation was likely to occur by abrupt transitions, Darwin always insisted that 'nature does not make leaps.' Gould and Eldredge readdressed this question, pointing out that the fossil record was one of millions of years of stasis interspersed with relatively brief periods of rapid change – hence punctuated equilibrium. As I have hinted, punctuated equilibrium, perhaps Gould's most significant contribution to evolutionary debate, and the most consistently misunderstood (see Part III, especially 'The Episodic Nature of Evolutionary Change'), made many traditional evolutionists unhappy, as they saw it as one more example of Gould's alleged Marxism – revolution rather than evolution. 'Evolution by jerks,' they termed it contemptuously, to which the punctuationists responded by calling their opponents' theory 'evolution by creeps'.

A further example of Gould's Darwinian revisionism occurs in his invocation of 'exaptations' – a term he coined with another long-term collaborator, Elizabeth Vrba, to describe features arising in one context but subsequently put to a different use. Feathers, originally evolved as a heat regulation device amongst the reptilian ancestors of today's birds, are a good example. But again, to orthodox neo-Darwinists, who believe that every feature of an organism has been honed by what Darwin called 'nature's continuous scrutiny', this claim too is heretical, charging, once again erroneously in my view, that it harks back to the much derided theories of Robert Goldschmidt in the 1930s, who hypothesised that species evolved by way of large mutations that produced 'hopeful monsters', which then awaited the right environmental chance to emerge.

The intellectual's development from radical young Turk to mature senior academic is traditionally that from iconoclasm to conventional wisdom. Not so with Steve Gould. *The Structure of Evolutionary Theory* is a robust and formidable defence of his key contributions to Darwinian pluralism. Evolution, he argues, is not à la carte but structurally constrained; not all phenotypic features are adaptive but they may instead be spandrels or exaptations – or even contingent accidents, like the asteroid collision that is believed by some to have resulted in the extinction of the dinosaurs, thus making space for mammals and ultimately humans. Wind the tape of history back, Gould insists, notably in his book *Wonderful Life* (see 'The Ladder and the Cone: Iconographies of Progress'), allow it to run forward again, and it is in the highest degree unlikely that the same species will evolve. Chance is crucial, and there is nothing inherently progressive about evolution – no drive to perfection, complexity or intelligent life (see 'Up Against a Wall'). Once again, this insistence on chance and contingency won Gould few friends among the orthodox. For 'Dawkinsites' – and most notably Simon Conway Morris, one of the central figures of the story of the fossils of the Burgess Shale that occupies much of *Wonderful Life* – adaptation implies that certain features of living systems, including perhaps intelligent life, are bound to evolve. This debate is going to run and run.

Above all, Gould insists, natural selection works at many levels. Because genetics has come to dominate much of the life sciences, for many biologists organisms have become almost irrelevant, save as instruments serving the purposes of their genes – as is splendidly encapsulated in Dawkins's famous description of humans as 'lumbering robots' – the gene's way of making copies of itself. Evolution itself came to be defined as a change in gene frequency within a population. By contrast Gould argued for a hierarchical view: that evolution works on genes, genomes, cell lineages and, especially, on species. Ignoring speciation, he says, is like playing *Hamlet* without the prince. This is the central theoretical issue underlying all the polemics that characterize what have become known as 'the Darwin wars', pitting Gould against Dawkins as his principal adversary, although in reality – and to the chagrin of creationists – the two agree on far more than they disagree.

The claim by some of Gould's critics in the United States that the theory of punctuated equilibrium gave succour to creationists angered Steve. His was one of the strongest of voices raised against the pretence that creationism – or its modern avatar, so-called 'intelligent design' (ID) – is a sort of scientific alternative to a disputed 'theory of evolution' (see, for example, 'Trouble in Our Own House' and the last two essays of this collection). Here at least lies the strongest of common grounds between the warring tribes of the inheritors of Darwin. Indeed, as a post-mortem tribute to his rebuttal of creationist myths an array of biologist Steves, Stephens, Stefans and Stephanies put their names to a robust defence of evolution as no more a 'hypothesis' than is gravitation, however much biologists may dispute the subtleties of the mechanisms whereby evolutionary change occurs.

One of the persistent irritations to biologists is that religious arguments for creationism, even when they masquerade as scientific, as does ID, distract from the real and important controversies within evolutionary biology. But it was perhaps his attempts to find allies among other Christian voices that led Gould, this Jewish atheist, towards the end of his life, to seek to make principled alliances with the Catholic Church, via his doctrine of NOMA – 'non-overlapping magisteria' (see the essay of that title in Part VIII). Granted the strength of anti-scientific religious fundamentalism in the United States, Gould's response is understandable, although I cannot say it is one that from my side of the Atlantic I would willingly share.

There is one final feature of Steve's extraordinary range which is displayed to the fullest in his essays, though also running like a thread through all the books. Cutting-edge researchers are often ignorant of their own science's history. Perhaps it was because he was a paleontologist that Gould returned so often in his writings to the history of his subject. His was not the sort of Whiggish approach by which senior scientists tend to beatify the progress of knowledge as a triumphal march from past obscurity to present clarity (indeed Steve was very hostile to any type of progressivist account of history – see Part V). Instead his intent was deeper: to understand the twists and turns of experiments and theories made by past researchers in the context of their own times and problematics. Such an understanding helps ensure our recognition that even our present-day knowledge is provisional and, like life itself, historically constrained.

It has been Paul McGarr's and my pleasure and problem to select, from amongst the large corpus of Stephen Jay Gould's writings, a range of essays, papers and book extracts that constitute perhaps not 'the best of Gould' – for we have had to omit much that could well fall within this rubric – but a range that we feel best illuminates the themes touched upon in this Introduction. To help in this we have arranged our forty-four choices into eight parts, each with a brief introduction that I will not recapitulate here, although the segmentation is in some cases a little arbitrary. In some sections, more academic papers are intermingled with the essays and book extracts. We have decided against 'unpacking' the more formal language of these papers or defining unfamiliar terms, as the large majority of our choices are drawn from the essays and readily accessible to non-biologists. We begin with a few illuminating autobiographical reflections in Part I. Part II shows Steve at his essayist best, evoking past episodes in the study of earth history and evolution through brief biographies of significant – and sometimes obscure – figures. Part III is intended to provide an introduction to Steve's major theoretical contributions to evolutionary theory, including a substantial extract from *The Structure of Evolutionary Theory*. In Part IV we turn to morphology and the paleontological fascination with size, form and shape. Part V illustrates both his interest in iconography and his rejection of fashionable stage and progressivist theories of change. With Parts VI and VII we turn to some of the issues in which Steve was at his polemical best: the fights over sociobiology and evolutionary psychology and, of course, the racism of much biological determinist writing. Finally, in Part VIII, we have his fights with creationism and his doctrine of NOMA.

Enough. This Introduction has kept the reader for too long from the man himself and his writings. Now, as they say, read on.

Thanks to Dick Lewontin for his comments on the draft of this Introduction; to him and David Pilbeam for agreeing to the inclusion of their joint publications; to Paul McGarr for always helpful comments; and to Jörg Hensgen for his as always meticulous editing.

Steven Rose
October 2005

PART I

AUTOBIOGRAPHY

STEVE Gould was neither overly forthcoming in writing about his own life nor especially reticent. There is no formal autobiography, but his essays are peppered almost parenthetically with personal references and anecdotes. Reading them, one learns of his parents and grandparents, his New York Jewish upbringing, his student mentors, and his direct engagement in some of the great biological and social controversies of the day, from scientific racism to creationism. There is no mistaking his passion for baseball, enigmatic though it may seem to readers on this side of the Atlantic. His reaction to being diagnosed with cancer resulted in one of his most memorable essays. Yet on the loves of his life, and on his children, he is silent.

In making 'Autobiography' the first of our series of selections we are constrained by these reticences. Yet it seems appropriate to begin this book with some of the more personal of Gouldian essays. So, for a start, we give his account of his own family history, the introductory chapter to the last of the series of collected essays from *Natural History, I Have Landed*, which, although there were other books to come, even posthumously, in some sense also provides his own funeral oration. There follows the famous account of the diagnosis of his own cancer, mesothelioma, a devastating and usually rapidly lethal condition associated, as he says, with exposure to asbestos. What he does not add is the belief amongst many of his friends that it was the asbestos in the basement of the Harvard Museum of Comparative Zoology, in which his office was located for many years, that contributed to the disease. But Gould uses his own diagnosis to make a different, statistical point, that 'The Median Isn't the Message' – an elegant example of experience-based teaching, made without either self-pity or heroics.

Almost any event can it seems provide the starting point for a statistics lesson, so the next two selections come from Steve's lifelong passion for baseball. 'The Streak of Streaks' provides a safe way into this passion for those unfamiliar with the game, whilst its strongest expression, which also includes a vignette of the young Gould's sporting activities, and his relations with his father, comes in 'Seventh Inning Stretch'.

Finally in this section, reflections on two of the concerns that came increasingly to occupy Steve in his later years: the relationship between science and religion, and that between art and science. The extract from *Rocks of Ages: Science and Religion in the Fullness of Life*, recounts his own part in the current fights over creationism and his favoured final doctrine, NOMA – non-overlapping magisteria – intended to resolve any potential conflict between science and religion. 'Of Two Minds and One Nature' not only explores the relationship between art and science but does so in an article written jointly with Rhonda Shearer, Steve's second wife, and to our knowledge the only example of this collaboration.

I HAVE LANDED

AS a young child, thinking as big as big can be and getting absolutely nowhere for the effort, I would often lie awake at night, pondering the mysteries of infinity and eternity – and feeling pure awe (in an inchoate, but intense, boyish way) at my utter inability to comprehend. How could time begin? For even if a God created matter at a definite moment, then who made God? An eternity of spirit seemed just as incomprehensible as a temporal sequence of matter with no beginning. And how could space end? For even if a group of intrepid astronauts encountered a brick wall at the end of the universe, what lay beyond the wall? An infinity of wall seemed just as inconceivable as a never-ending extension of stars and galaxies.

I will not defend these naïve formulations today, but I doubt that I have come one iota closer to a personal solution since those boyhood ruminations so long ago. In my philosophical moments – and not only as an excuse for personal failure, for I see no sign that others have succeeded – I rather suspect that the evolved powers of the human mind may not include the wherewithal for posing such questions in answerable ways (not that we ever would, should, or could halt our inquiries into these ultimates).

However, I confess that in my mature years I have embraced the Dorothean dictum: yea, though I might roam through the pleasures of eternity and the palaces of infinity (not to mention the valley of the shadow of death), when a body craves contact with the brass tacks of a potentially comprehensible reality, I guess there's no place like home. And within the smaller, but still tolerably ample, compass of our planetary home, I would nominate as most worthy of pure awe – a metaphorical miracle, if you will – an aspect of life that most people have never considered, but that strikes me as equal in majesty to our most spiritual projections of infinity and eternity, while falling entirely within the domain of our conceptual understanding and empirical grasp: the continuity of *etz chayim,* the tree of earthly life, for at least 3.5 billion years, without a single microsecond of disruption.

Consider the improbability of such continuity in conventional terms of ordinary probability: Take any phenomenon that begins with a positive value at its inception 3.5 billion years ago, and let the process regulating its existence proceed through time. A line marked zero runs along below the current value. The probability of the phenomenon's descent to zero may be almost incalculably low, but throw the dice of the relevant process billions of times, and the phenomenon just has to hit the zero line eventually.

For most processes, the prospect of such an improbable crossing bodes no permanent ill, because an unlikely crash (a year, for example, when a healthy Mark McGwire hits no home runs at all) will quickly be reversed, and ordinary residence well above the zero line reestablished. But life represents a different kind of ultimately fragile system, utterly dependent upon unbroken continuity. For life, the zero line designates a permanent end, not a temporary embarrassment. If life ever touched that line, for one fleeting moment at any time during 3.5 billion years of sustained history, neither we nor a million species of beetles would grace this planet today. The merest momentary brush with voracious zero dooms all that might have been, forever after.

When we consider the magnitude and complexity of the circumstances required to sustain this continuity for so long, and without exception or forgiveness in each of so many components – well, I may be a rationalist at heart, but if anything in the natural world merits a designation as 'awesome,' I nominate the continuity of the tree of life for 3.5 billion years. The earth experienced severe ice ages, but never froze completely, not for a single day. Life fluctuated through episodes of global extinction, but never crossed the zero line, not for one millisecond. DNA has been working all this time, without an hour of vacation or even a moment of pause to remember the extinct brethren of a billion dead branches shed from an evergrowing tree of life.

When Protagoras, speaking inclusively despite the standard translation, defined 'man' as 'the measure of all things,' he captured the ambiguity of our feelings and intellect in his implied contrast of diametrically opposite interpretations: the expansion of humanism versus the parochiality of limitation. Eternity and infinity lie too far from the unavoidable standard of our own bodies to secure our comprehension; but life's continuity stands right at the outer border of ultimate fascination: just close enough for intelligibility by the measure of our bodily size and earthly time, but sufficiently far away to inspire maximal awe.

Moreover, we can bring this largest knowable scale further into the circle of our comprehension by comparing the macrocosm of life's tree to the microcosm of our family's genealogy. Our affinity for evolution must originate from the same internal chords of emotion and fascination that drive so many people to trace their bloodlines with such diligence and detail. I do not pretend to know why the documentation of unbroken heredity through generations of forebears brings us so swiftly to tears, and to such a secure sense of rightness, definition, membership, and meaning. I simply accept the primal emotional power we feel when we manage to embed ourselves into something so much larger.

Thus, we may grasp one major reason for evolution's enduring popularity among scientific subjects: our minds must combine the subject's sheer intellectual fascination with an even stronger emotional affinity rooted in a legitimate comparison between the sense of belonging gained from contemplating family genealogies, and the feeling of understanding achieved by locating our tiny little twig on the great tree of life. Evolution, in this sense, is 'roots' writ large.

To close this series of three hundred essays in *Natural History,* I therefore offer two microcosmal stories of continuity – two analogs or metaphors for this grandest evolutionary theme of absolutely unbroken continuity, the intellectual and emotional center of 'this view of life.'* My stories descend in range and importance from a tale about a leader in the founding generation of Darwinism to a story about my grandfather, a Hungarian immigrant who rose from poverty to solvency as a garment worker on the streets of New York City.

Our military services now use the blandishments of commercial jingles to secure a 'few good men' (and women), or to entice an unfulfilled soul to 'be all that you can be in the army.' In a slight

*This essay, number 300 of a monthly entry written without a break from January 1974 to January 2001 and appearing in *Natural History* magazine, terminates a series titled 'This View of Life.' The title comes from Darwin's poetic statement about evolution in the last paragraph of *The Origin of Species*: 'There is grandeur in this view of life . . .'

variation, another branch emphasizes external breadth over internal growth: join the navy and see the world.

In days of yore, when reality trumped advertisement, this motto often did propel young men to growth and excitement. In particular, budding naturalists without means could attach themselves to scientific naval surveys by signing on as surgeons, or just as general gofers and bottle washers. Darwin himself had fledged on the *Beagle*, largely in South America, between 1831 and 1836, though he sailed (at least initially) as the captain's gentleman companion rather than as the ship's official naturalist. Thomas Henry Huxley, a man of similar passions but lesser means, decided to emulate his slightly older mentor (Darwin was born in 1809, Huxley in 1825) by signing up as assistant surgeon aboard HMS *Rattlesnake* for a similar circumnavigation, centered mostly on Australian waters, and lasting from 1846 to 1850.

Huxley filled these scientific *Wanderjahre* with the usual minutiae of technical studies on jellyfishes and grand adventures with the aboriginal peoples of Australia and several Pacific islands. But he also trumped Darwin in one aspect of discovery with extremely happy and lifelong consequences: he met his future wife in Australia, a brewer's daughter (a lucrative profession in this wild and distant outpost) named Henrietta Anne Heathorn, or Nettie to the young Hal. They met at a dance. He loved her silky hair, and she reveled in his dark eyes that 'had an extraordinary way of flashing when they seemed to be burning – his manner was most fascinating' (as she wrote in her diary).

Huxley wrote to his sister in February 1849, 'I never met with so sweet a temper, so self-sacrificing and affectionate a disposition.' As Nettie's only dubious trait, Hal mentioned her potential naïveté in leaving 'her happiness in the hands of a man like myself, struggling upwards and certain of nothing.' Nettie waited five years after Hal left in 1850. Then she sailed to London, wed her dashing surgeon and vigorously budding scientist, and enjoyed, by Victorian standards, an especially happy and successful marriage with an unusually decent and extraordinarily talented man. (Six of their seven children lived into reasonably prosperous maturity, a rarity in those times, even among the elite.) Hal and Nettie, looking back in their old age (Hal died in 1895, Nettie in 1914), might well have epitomized their life together in the words of a later song: 'We had a lot of kids, a lot of trouble and pain, but then, oh Lord, we'd do it again.'

The young and intellectually restless Huxley, having mastered

German, decided to learn Italian during his long hours of boredom at sea. (He read Dante's *Inferno* in the original *terza rima* during a year's jaunt, centered upon New Guinea.) Thus, as Huxley prepared to leave his fiancée in April 1849 (he would return for a spell in 1850, before the long five-year drought that preceded Nettie's antipodal journey to their wedding), Nettie decided to give him a parting gift of remembrance and utility: a five-volume edition, in the original Italian of course, of *Gerusalemme liberata* by the great Renaissance poet Torquato Tasso. (This epic, largely describing the conquest of Jerusalem by the First Crusade in 1099, might not be deemed politically correct today, but the power of Tasso's verse and narrative remains undiminished.)

Nettie presented her gift to Hal as a joint offering from herself, her half-sister Oriana, and Oriana's husband, her brother-in-law William Fanning. She inscribed the first volume in a young person's hand: 'T. H. Huxley. A birthday and parting gift in remembrance of three dear friends. May 4th 1849.' And now we come to the point of this tale. For some reason that I cannot fathom but will not question, this set of books sold (to lucky me) for an affordable pittance at a recent auction. (Tasso isn't big these days, and folks may have missed the catalog entry describing the provenance and context.)

So Nettie Heathorn came to England, married her Hal, raised a large family, and lived out her long and fulfilling life well into the twentieth century. As she had been blessed with accomplished children, she also enjoyed, in later life, the promise of two even more brilliant grandchildren: the writer Aldous Huxley and the biologist Julian Huxley. In 1911, more than sixty years after she had presented the five volumes of Tasso to Hal, Nettie Heathorn, then Henrietta Anne Huxley, and now Granmoo to her grandson Julian, removed the books from such long residence on her shelf, and passed them on to a young man who would later carry the family's intellectual torch with such distinction. She wrote below her original inscription, now in the clear but shaky hand of an old woman, the missing who and where of the original gift: 'Holmwood. Sydney, N. S. Wales. Nettie Heathorn, Oriana Fanning, William Fanning.'

Above her original words, penned sixty years before in youth's flower, she then wrote, in a simple statement that needs no explication in its eloquent invocation of life's persistence: 'Julian Sorel Huxley from his grandmother Henrietta Anne Huxley née Heathorn "Granmoo."' She then emphasized the sacred theme of continuity by closing her rededication with the same words she had written to Hal so many years before: '"In remembrance" 28 July 1911. Hodeslea, Eastbourne.'

If this tale of three generations, watched over by a great woman as she follows life's passages from dashing bride to doting grandmother, doesn't epitomize the best of humanity, as symbolized by our continuity, then what greater love or beauty can sustain our lives in this vale of tears and fascination? Bless all the women of this world who nurture our heritage while too many men rush off to kill for ideals that might now be deeply and personally held, but will often be viewed as repugnant by later generations.

My maternal grandparents – Irene and Joseph Rosenberg, or Grammy and Papa Joe to me – loved to read in their adopted language of English. My grandfather even bought a set of Harvard Classics (the famous 'five-foot shelf' of Western wisdom) to facilitate his assimilation to American life. I inherited only two of Papa Joe's books, and nothing of a material nature could be more precious to me. The first bears a stamp of sale: 'Carroll's book store. Old, rare and curious books. Fulton and Pearl Sts. Brooklyn, N. Y.' Perhaps my grandfather obtained this volume from a *landsman*, for I can discern, through erasures on three pages of the book, the common Hungarian name 'Imre.' On the front page of this 1892 edition of J. M. Greenwood's *Studies in English Grammar*, my grandfather wrote in ink, in an obviously European hand, 'Prop. of Joseph A. Rosenberg, New York.' To the side, in pencil, he added the presumed date of his acquisition: '1901. Oct. 25th.' Just below, also in pencil, he appended the most eloquent of all conceivable words for this context – even though one might argue that he used the wrong tense, confusing the compound past of continuous action with an intended simple past to designate a definite and completed event (not bad for a barely fourteen-year-old boy just a month or two off the boat): 'I have landed. Sept. 11, 1901.'

Of all that I shall miss in ending this series of essays, I shall feel most keenly the loss of fellowship and interaction with readers. Early in the series, I began – more as a rhetorical device to highlight a spirit of interaction than as a practical tactic for gaining information – to pose questions to readers when my research failed to resolve a textual byway. (As a longtime worshiper at the altar of detail, nothing niggles me more than a dangling little fact – partly, I confess, from a sense of order, but mostly because big oaks do grow from tiny acorns, and one can never know in advance which acorn will reach heaven.)

As the series proceeded, I developed complete faith – not from hope, but from the solid pleasure of invariant success – that any posted question would elicit a host of interesting responses, including the desired factual resolution. How did the Italian word *segue* pass from a technical term in the rarefied world of classical music into common speech as a synonym for 'transition' (resolved by personal testimony of several early radio men who informed me that in the 1920s they had transferred this term from their musical training to their new gigs as disc jockeys and producers of radio plays). Why did seventeenth-century engravers of scientific illustrations usually fail to draw snail shells in reverse on their plates (so that the final product, when printed on paper, would depict the snail's actual direction of coiling), when they obviously understood the principle of inversion and always etched their verbal texts 'backwards' to ensure printed readability? Who were Mary Roberts, Isabelle Duncan, and several other 'invisible' women of Victorian science writing who didn't even win a line in such standard sources as the *Encyclopaedia Britannica* or the *Dictionary of National Biography*?

Thus, when I cited my grandfather's text *en passant* in an earlier essay, I may have wept for joy, but could not feign complete surprise, when I received the most wonderful of all letters from a reader:

> For years now I have been reading your books, and I think I should really thank you for the pleasure and intellectual stimulation I have received from you. But how to make even a small return for your essays? The answer came to me this week. I am a genealogist who specializes in passenger list work. Last Sunday I was rereading that touching essay that features your grandfather, Joseph A. Rosenberg who wrote 'I have landed. Sept. 11, 1901.' It occurred to me that you might like to see his name on the passenger list of the ship on which he came.

I think I always knew that I might be able to find the manifest of Papa Joe's arrival at Ellis Island. I even half intended to make the effort 'some day.' But, in honest moments of obeisance to the Socratic dictum 'know thyself,' I'm pretty sure that, absent this greatest of all gifts from a reader, I never would have found the time or made the move. (Moreover, I certainly hadn't cited Papa Joe's inscription in a lazy and intentional 'fishing expedition' for concrete information. I therefore received the letter of resolution with pure exhilaration – as a precious item beyond price, freely given in fellowship, and so gratefully received without any conscious anticipation on my part.)

My grandfather traveled with his mother and two younger sisters on the SS *Kensington*, an American Line ship, launched in 1894 and scrapped in 1910, that could carry sixty passengers first class, and one thousand more in steerage – a good indication of the economics of travel and transport in those days of easy immigration for European workers, then so badly needed for the factories and sweatshops of a booming American economy based on manual labor. The *Kensington* had sailed from Antwerp on August 31, 1901, and arrived in New York, as Papa Joe accurately recorded, on September 11. My page of the 'list or manifest of alien immigrants' includes thirty names, Jewish or Catholic by inference, and hailing from Hungary, Russia, Romania, and Croatia. Papa Joe's mother, Leni, listed as illiterate and thirty-five years of age, appears on line twenty-two with her three children just below: my grandfather, recorded as Josef and literate at age fourteen, and my dear aunts Regina and Gus, cited as Regine and Gisella (I never knew her real name) at five years and nine months old, respectively. Leni carried $6.50 to start her new life.

I had not previously known that my great-grandfather Farkas Rosenberg (accented on the first syllable, pronounced *farkash*, and meaning 'wolf' in Hungarian) had preceded the rest of his family, and now appeared on the manifest as their sponsor, 'Wolf Rosenberg of 644 East 6th Street.' I do not remember Farkas, who died during my third year of life – but I greatly value the touching tidbit of information that, for whatever reason in his initial flurry of assimilation, Farkas had learned, and begun to use, the English translation of a name that strikes many Americans as curious, or even amusing, in sound – for he later reverted to Farkas, and no one in my family knew him by any other name.

My kind and diligent reader then bestowed an additional gift upon me by locating Farkas's manifest as well. He had arrived, along with eight hundred passengers in steerage, aboard the sister ship SS *Southwark* on June 13, 1900, listed as Farkas Rosenberg, illiterate at age thirty-four (although I am fairly sure that he could at least read and probably write Hebrew) and sponsored by a cousin named Jos. Weiss (but unknown to my family, and perhaps an enabling fiction). Farkas, a carpenter by trade, arrived alone with one dollar in his pocket.

Papa Joe's later story mirrors the tale of several million poor immigrants to a great land that did not welcome them with open arms (despite Lady Liberty's famous words), but also did not foreclose the possibility of success if they could prevail by their own wits and unrelenting hard work. And who could, or should, have asked for more in those times? Papa Joe received no further schooling in America, save what experience provided and internal drive secured. As a young man, he went west for a time, working in the steel mills of Pittsburgh and on a ranch somewhere in the Midwest (not, as I later found out, as the cowboy of my dreams, but as an accountant in the front office). His mother, Leni, died young (my mother, Eleanor, bears her name in remembrance), as my second book of his legacy testifies. Papa Joe ended up, along with so many Jewish immigrants, in the garment district of New York City, where, after severing his middle finger in an accident as a cloth cutter, he eventually figured out how to parlay his remarkable, albeit entirely untrained, artistic talents into a better job that provided eventual access to middle-class life (and afforded much titillation to his grandchildren) – as a designer of brassieres and corsets.

He met Irene, also a garment worker, when he lived as a boarder at the home of Irene's aunt – for she had emigrated alone, at age fourteen in 1910 after a falling-out with her father, and under her aunt's sponsorship. What else can one say for the objective record (and what richness and depth could one *not* expose, at least in principle, and for all people, at the subjective level of human life, passion, and pure perseverance)? Grammy and Papa Joe married young, and their only early portrait together radiates both hope and uncertainty. They raised three sons and a daughter; my mother alone survives. Two of their children finished college.

Somehow I always knew, although no one ever pressured me directly, that the third generation, with me as the first member thereof, would fulfill the deferred dream of a century by obtaining an advanced education and entering professional life. (My grandmother spoke Hungarian, Yiddish, German, and English, but could only write her adopted language phonetically. I will never forget her embarrassment when I inadvertently read a shopping list that she had written, and realized that she could not spell. I also remember her joy when, invoking her infallible memory and recalling some old information acquired in her study for citizenship, she won ten dollars on a Yiddish radio quiz for correctly identifying William Howard Taft as our fattest president.)

I loved Grammy and Papa Joe separately. Divorce, however sanctioned by their broader culture, did not represent an option in their particular world. Unlike Hal and Nettie Huxley, I'm not at all sure that they would have done it again. But they stuck together and prevailed, at least in peace, respect, and toleration, perhaps even in fondness. Had they not done so, I would not be here – and for this particular twig of evolutionary continuity I could not be more profoundly grateful in the most ultimate of all conceivable ways. I also loved them fiercely, and I reveled in the absolute certainty of their unquestioned blessing and unvarying support (not always deserved of course, for I really did throw that rock at Harvey, even though Grammy slammed our front door on Harvey's father, after delivering a volley of Yiddish curses amid proclamations that her Stevele would never do such a thing – while knowing perfectly well that I surely could).

The tree of all life and the genealogy of each family share the same topology and the same secret of success in blending two apparently contradictory themes of continuity without a single hair's breadth of breakage, and change without a moment's loss of a potential that need not be exploited in all episodes but must remain forever at the ready. These properties may seem minimal, even derisory, in a universe of such stunning complexity (whatever its inexplicable eternity or infinity). But this very complexity exalts pure staying power (and the lability that facilitates such continuity). Showy statues of Ozymandias quickly become lifeless legs in the desert; bacteria have scuttled around all the slings and arrows of outrageous planetary fortune for 3.5 billion years and running.

I believe in the grandeur of this view of life, the continuity of family lines, and the poignancy of our stories: Nettie Heathorn passing Tasso's torch as Granmoo two generations later; Papa Joe's ungrammatical landing as a stranger in a strange land and my prayer that, in some sense, he might see my work as a worthy continuation, also two generations later, of a hope that he fulfilled in a different way during his own lifetime. I suspect that we feel the poignancy in such continuity because we know that our little realization of an unstated family promise somehow mirrors the larger way of all life, and therefore becomes 'right' in some sense too deep for words or tears. I can therefore write the final essay in this particular series because I know that I will never run out of unkept promises, or miles to walk; and that I may even continue to sprinkle the journey remaining before sleep with a new idea or two. This view of life continues, flowing ever forward, while the current patriarch of one tiny and insignificant twig pauses to honor the inception of the twig's centennial in a new land, by commemorating the first recorded words of a fourteen-year-old forebear.

Dear Papa Joe, I have been faithful to your dream of persistence, and attentive to a hope that the increments of each worthy generation may buttress the continuity of evolution. You could speak those wondrous words right at the joy and terror of inception. I dared not repeat them until I could fulfill my own childhood dream – something that once seemed so mysteriously beyond any hope of realization to an insecure little boy in a garden apartment in Queens – to become a scientist and to make, by my own effort, even the tiniest addition to human knowledge of evolution and the history of life. But now, with my step 300, so fortuitously coincident with the world's new 1,000 and your own 100,* perhaps I have finally won the right to restate your noble words, and to tell you that their inspiration still lights my journey: I have landed. But I also can't help wondering what comes next.

*This essay, the three-hundredth and last of my series, appeared in January 2001 – the inception of the millennium by a less popular, but more mathematically sanctioned, mode of reckoning. My grandfather also began the odyssey of my family in America when he arrived from Europe in 1901.

THE MEDIAN ISN'T THE MESSAGE

My life has recently intersected, in a most personal way, two of Mark Twain's famous quips. One I shall defer to the end of this essay. The other (sometimes attributed to Disraeli) identifies three species of mendacity, each worse than the one before – lies, damned lies, and statistics.

Consider the standard example of stretching truth with numbers – a case quite relevant to my story. Statistics recognizes different measures of an 'average,' or central tendency. The *mean* represents our usual concept of an overall average – add up the items and divide them by the number of sharers (100 candy bars collected for five kids next Halloween will yield twenty for each in a fair world). The *median*, a different measure of central tendency, is the halfway point. If I line up five kids by height, the median child is shorter than two and taller than the other two (who might have trouble getting their mean share of the candy). A politician in power might say with pride, 'The mean income of our citizens is $15,000 per year.' The leader of the opposition might retort, 'But half our citizens make less than $10,000 per year.' Both are right, but neither cites a statistic with impassive objectivity. The first invokes a mean, the second a median. (Means are higher than medians in such cases because one millionaire may outweigh hundreds of poor people in setting a mean, but can balance only one mendicant in calculating a median.)

The larger issue that creates a common distrust or contempt for statistics is more troubling. Many people make an unfortunate and invalid separation between heart and mind, or feeling and intellect. In some contemporary traditions, abetted by attitudes stereotypically centered upon Southern California, feelings are exalted as more 'real' and the only proper basis for action, while intellect gets short shrift as a hang-up of outmoded elitism. Statistics, in this absurd dichotomy, often becomes the symbol of the enemy. As Hilaire Belloc wrote, 'Statistics are the triumph of the quantitative method, and the quantitative method is the victory of sterility and death.'

This is a personal story of statistics, properly interpreted, as profoundly nurturant and life-giving. It declares holy war on the downgrading of intellect by telling a small story to illustrate the utility of dry, academic knowledge about science. Heart and head are focal points of one body, one personality.

In July 1982, I learned that I was suffering from abdominal mesothelioma, a rare and serious cancer usually associated with exposure to asbestos. When I revived after surgery, I asked my first question of my doctor and chemotherapist: 'What is the best technical literature about mesothelioma?' She replied, with a touch of diplomacy (the only departure she has ever made from direct frankness), that the medical literature contained nothing really worth reading.

Of course, trying to keep an intellectual away from literature works about as well as recommending chastity to *Homo sapiens*, the sexiest primate of all. As soon as I could walk, I made a beeline for Harvard's Countway medical library and punched mesothelioma into the computer's bibliographic search program. An hour later, surrounded by the latest literature on abdominal mesothelioma, I realized with a gulp why my doctor had offered that humane advice. The literature couldn't have been more brutally clear: Mesothelioma is incurable, with a median mortality of only eight months after discovery. I sat stunned for about fifteen minutes, then smiled and said to myself: So that's why they didn't give me anything to read. Then my mind started to work again, thank goodness.

If a little learning could ever be a dangerous thing, I had encountered a classic example. Attitude clearly matters in fighting cancer. We don't know why (from my old-style materialistic perspective, I suspect that mental states feed back upon the immune system). But match people with the same cancer for age, class, health, and socioeconomic status, and, in general, those with positive attitudes, with a strong will and purpose for living, with commitment to struggle, and with an active response to aiding their own treatment and not just a passive acceptance of anything doctors say tend to live longer. A few months later I asked Sir Peter Medawar, my personal scientific guru and a Nobelist in immunology, what the best prescription for success against cancer might be. 'A sanguine personality,' he replied. Fortunately (since one can't reconstruct oneself at short notice and for a definite purpose), I am, if anything, even-tempered and confident in just this manner.

Hence the dilemma for humane doctors: Since attitude matters so critically, should such a somber conclusion be advertised, especially since few people have sufficient understanding of statistics to evaluate what the statements really mean? From years of experience with the small-scale evolution of Bahamian land snails treated quantitatively, I have developed this technical knowledge – and I am convinced that it played a major role in saving my life. Knowledge is indeed power, as Francis Bacon proclaimed.

The problem may be briefly stated: What does 'median mortality of eight months' signify in our vernacular? I suspect that most people, without training in statistics, would read such a statement as 'I will probably be dead in eight months' – the very conclusion that must be avoided, both because this formulation is false, and because attitude matters so much.

I was not, of course, overjoyed, but I didn't read the statement in this vernacular way either. My technical training enjoined a different perspective on 'eight months median mortality.' The point may seem subtle, but the consequences can be profound. Moreover, this perspective embodies the distinctive way of thinking in my own field of evolutionary biology and natural history.

We still carry the historical baggage of a Platonic heritage that seeks sharp essences and definite boundaries. (Thus we hope to find an unambiguous 'beginning of life' or 'definition of death,' although nature often comes to us as irreducible continua.) This Platonic heritage, with its emphasis on clear distinctions and separated immutable entities, leads us to view statistical measures of central tendency wrongly, indeed opposite to the appropriate interpretation in our actual world of variation, shadings, and continua. In short, we view means and medians as hard 'realities,' and the variation that permits their calculation as a set of transient and imperfect measurements of this hidden essence. If the median is the reality and variation around the median just a device for calculation, then 'I will probably be dead in eight months' may pass as a reasonable interpretation.

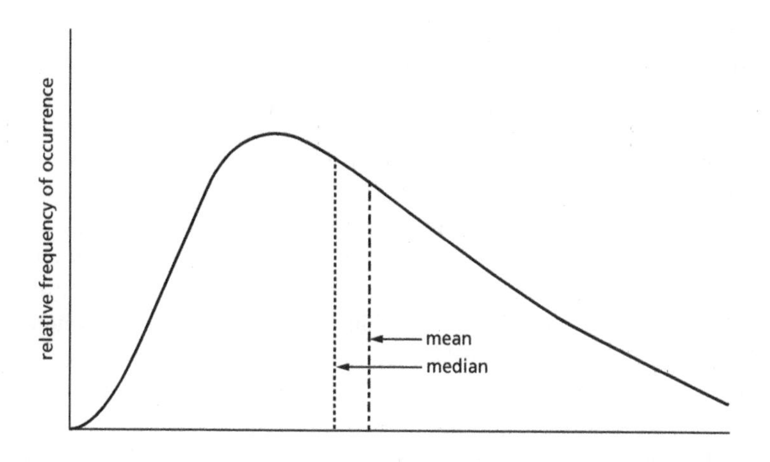

Figure 1. A right-skewed distribution showing that means must be higher than medians, and that the right side of the distribution extends out into a long tail.

But all evolutionary biologists know that variation itself is nature's only irreducible essence. Variation is the hard reality, not a set of imperfect measures for a central tendency. Means and medians are the abstractions. Therefore, I looked at the mesothelioma statistics quite differently – and not only because I am an optimist who tends to see the doughnut instead of the hole, but primarily because I know that variation itself is the reality. I had to place myself amidst the variation.

When I learned about the eight-month median, my first intellectual reaction was: Fine, half the people will live longer; now what are my chances of being in that half? I read for a furious and nervous hour and concluded, with relief: damned good. I possessed every one of the characteristics conferring a probability of longer life: I was young; my disease had been recognized in a relatively early stage; I would receive the nation's best medical treatment; I had the world to live for; I knew how to read the data properly and not despair.

Another technical point then added even more solace. I immediately recognized that the distribution of variation about the eight-month median would almost surely be what statisticians call 'right skewed.' (In a symmetrical distribution, the profile of variation to the left of the central tendency is a mirror image of variation to the right. Skewed distributions are asymmetrical, with variation stretching out more in one direction than the other – left skewed if extended to the left, right skewed if stretched out to the right.) The distribution of variation had to be right skewed, I reasoned. After all, the left of the distribution contains an irrevocable lower boundary of zero (since mesothelioma can only be identified at death or before). Thus, little space exists for the distribution's lower (or left) half – it must be scrunched up between zero and eight months. But the upper (or right) half can extend out for years and years, even if nobody ultimately survives. The distribution must be right skewed, and I needed to know how long the extended tail ran – for I had already concluded that my favorable profile made me a good candidate for the right half of the curve.

The distribution was, indeed, strongly right skewed, with a long tail (however small) that extended for several years above the eight-month median. I saw no reason why I shouldn't be in that small tail, and I breathed a very long sigh of relief. My technical knowledge had helped. I had read the graph correctly. I had asked the right question and found the answers. I had obtained, in all probability, that most precious of all possible gifts in the circumstances – substantial time. I didn't have to stop and immediately follow Isaiah's injunction to Hezekiah – set thine house in order: for thou shalt die, and not live. I would have time to think, to plan, and to fight.

One final point about statistical distributions. They apply only to a prescribed set of circumstances – in this case to survival with mesothelioma under conventional modes of treatment. If circumstances change, the distribution may alter. I was placed on an experimental protocol of treatment and, if fortune holds, will be in the first cohort of a new distribution with high median and a right tail extending to death by natural causes at advanced old age.*

It has become, in my view, a bit too trendy to regard the acceptance of death as something tantamount to intrinsic dignity. Of course I agree with the preacher of Ecclesiastes that there is a time to love and a time to die – and when my skein runs out I hope to face the end calmly and in my own way. For most situations, however, I prefer the more martial view that death is the ultimate enemy – and I find nothing reproachable in those who rage mightily against the dying of the light.

*So far so good.

The swords of battle are numerous, and none more effective than humor. My death was announced at a meeting of my colleagues in Scotland, and I almost experienced the delicious pleasure of reading my obituary penned by one of my best friends (the so-and-so got suspicious and checked; he too is a statistician, and didn't expect to find me so far out on the left tail). Still, the incident provided my first good laugh after the diagnosis. Just think, I almost got to repeat Mark Twain's most famous line of all: The reports of my death are greatly exaggerated.*

*Since writing this, my death has actually been reported in two European magazines, five years apart. *Fama volat* (and lasts a long time). I squawked very loudly both times and demanded a retraction; guess I just don't have Mr Clemens's *savoir faire*.

THE STREAK OF STREAKS[1]

My father was a court stenographer. At his less than princely salary, we watched Yankee games from the bleachers or high in the third deck. But one of the judges had season tickets, so we occasionally sat in the lower boxes when hizzoner couldn't attend. One afternoon, while DiMaggio was going 0 for 4 against, of all people, the lowly St Louis Browns, the great man fouled one in our direction. 'Catch it, Dad,' I screamed. 'You never get them,' he replied, but stuck up his hand like the Statue of Liberty – and the ball fell right in. I mailed it to DiMaggio, and, bless him, he actually sent the ball back, signed and in a box marked 'insured.' Insured, that is, to make me the envy of the neighborhood, and DiMaggio the model and hero of my life.

I met DiMaggio a few years ago on a small playing field at the Presidio of San Francisco. My son, wearing DiMaggio's old number 5 on his Little League jersey, accompanied me, exactly one generation after my father caught that ball. DiMaggio gave him a pointer or two on batting and then signed a baseball for him. One generation passeth away, and another generation cometh: But the earth abideth forever.

My son, uncoached by Dad, and given the chance that comes but once in a lifetime, asked DiMaggio as his only query about life and career: 'Suppose you had walked every time up during one game of your 56-game hitting streak? Would the streak have been over?' DiMaggio replied that, under 1941 rules, the streak would have ended, but that this unfair statute has since been revised, and such a game would not count today.

My son's choice for a single question tells us something vital about the nature of legend. A man may labor for a professional lifetime, especially in sport or in battle, but posterity needs a single transcendant event to fix him in permanent memory. Every hero must be a Wellington on the right side of his personal Waterloo; generality of excellence is too diffuse. The unambiguous factuality of a single achievement is adamantine. Detractors can argue forever about the general tenor of your life and works, but they can never erase a great event.

In 1941, as I gestated in my mother's womb, Joe DiMaggio got at least one hit in each of fifty-six successive games. Most records are only incrementally superior to runners-up; Roger Maris hit 61 homers in 1961, but Babe Ruth hit 60 in 1927 and 59 in 1921, while Hank Greenberg (1938) and Jimmy Foxx (1932) both hit 58. But DiMaggio's fifty-six-game hitting streak is ridiculously, almost unreachably far from all challengers (Wee Willie Keeler and Pete Rose, both with 44, come second). Among sabermetricians (a happy neologism based on an acronym for members of the Society for American Baseball Research, and referring to the statistical mavens of the sport) – a contentious lot not known for agreement about anything – we find virtual consensus that DiMaggio's fifty-six-game hitting streak is the greatest accomplishment in the history of baseball, if not all modern sport.

The reasons for this respect are not far to seek. Single moments of unexpected supremacy – Johnny Vander Meer's back-to-back no-hitters in 1938, Don Larsen's perfect game in the 1956 World Series – can occur at any time to almost anybody, and have an irreducibly capricious character. Achievements of a full season – such as Maris's 61 homers in 1961 and Ted Williams's batting average of .406, also posted in 1941 and not equaled since – have a certain overall majesty, but they don't demand unfailing consistency every single day; you can slump for a while, so long as your average holds. But a streak must be absolutely exceptionless; you are not allowed a single day of subpar play, or even bad luck. You bat only four or five times in an average game. Sometimes two or three of these efforts yield walks, and you get only one or two shots at a hit. Moreover, as tension mounts and notice increases, your life becomes unbearable. Reporters dog your every step; fans are even more intrusive than usual (one stole DiMaggio's favorite bat right in the middle of his streak). You cannot make a single mistake.

Thus Joe DiMaggio's fifty-six-game hitting streak is both the greatest factual achievement in the history of baseball and a principal icon of American mythology. What shall we do with such a central item of our cultural history?

Statistics and mythology may strike us as the most unlikely of bedfellows. How can we quantify Caruso or measure *Middlemarch*? But if God could mete out heaven with the span (Isaiah 40:12), perhaps we can say something useful about hitting streaks. The statistics of 'runs,' defined as continuous series of good or bad results (including baseball's streaks and slumps), is a well-developed branch of the profession, and can yield clear – but wildly counterintuitive – results. (The fact that we find these conclusions so surprising is the key to appreciating DiMaggio's achievement, the point of this article, and the gateway to an important insight about the human mind.)

Start with a phenomenon that nearly everyone both accepts and considers well understood – 'hot hands' in basketball. Now and then, someone just gets hot, and can't be stopped. Basket after basket falls in – or out as with 'cold hands,' when a man can't buy a bucket for love or money (choose your cliché). The reason for this phenomenon is clear enough: It lies embodied in the maxim, 'When you're hot, you're hot; and when you're not, you're not.' You get that touch, build confidence; all nervousness fades, you find your rhythm; *swish, swish, swish*. Or you miss a few, get rattled, endure the booing, experience despair; hands start shaking and you realize that you shoulda stood in bed.

Everybody knows about hot hands. The only problem is that no such phenomenon exists. Stanford psychologist Amos Tversky studied every basket made by the Philadelphia 76ers for more than a season. He found, first of all, that the probability of making a second basket did not rise following a successful shot. Moreover, the number of 'runs,' or baskets in succession, was no greater than what a standard random, or coin-tossing, model would predict. (If the chance of making each basket is 0.5, for example, a reasonable value for good shooters, five hits in a row will occur, on average, once in thirty-two sequences – just as you can expect to toss five successive heads about once in thirty-two times, or 0.5^5.)

Of course Larry Bird, the great forward of the Boston Celtics, will have more sequences of five than Joe Airball – but not because he has greater will or gets in that magic rhythm more often. Larry has longer runs because his average success rate is so much higher, and random models predict more frequent and longer sequences. If Larry shoots field goals at 0.6 probability of success, he will get five in a row about once every thirteen sequences (0.6^5). If Joe, by contrast, shoots only 0.3, he will get his five straight only about once in 412 times. In other words, we need no special explanation for the apparent pattern of long runs. There is no ineffable 'causality of circumstance' (to coin a phrase), no definite reason born of the particulars that make for heroic myths – courage in the clinch, strength in adversity, etc. You only have to know a person's ordinary play in order to predict his sequences. (I rather suspect that we are convinced of the contrary not only because we need myths so badly, but also because we remember the successes and simply allow the failures to fade from memory. More on this later.) But how does this revisionist pessimism work for baseball?

My colleague Ed Purcell, Nobel laureate in physics but, for purposes of this subject, just another baseball fan, has done a comprehensive study of all baseball streak and slump records. His firm conclusion is easily and swiftly summarized. Nothing ever happened in baseball above and beyond the frequency predicted by coin-tossing models. The longest runs of wins or losses are as long as they should be, and occur about as often as they ought to. Even the hapless Orioles, at 0 and 21 to start the 1988 season, only fell victim to the laws of probability (and not to the vengeful god of racism, out to punish major league baseball's only black manager).[2]

But 'treasure your exceptions,' as the old motto goes. Purcell's rule has but one major exception, one sequence so many standard deviations above the expected distribution that it should never have occurred at all: Joe DiMaggio's fifty-six-game hitting streak in 1941. The intuition of baseball aficionados has been vindicated. Purcell calculated that to make it likely (probability greater than 50 percent) that a run of even fifty games will occur once in the history of baseball up to now (and fifty-six is a lot more than fifty in this kind of league), baseball's rosters would have to include either four lifetime .400 batters or fifty-two lifetime .350 batters over careers of 1,000 games. In actuality, only three men have lifetime batting averages in excess of .350, and no one is anywhere near .400 (Ty Cobb at .367, Rogers Hornsby at .358, and Shoeless Joe Jackson at .356). DiMaggio's streak is the most extraordinary thing that ever happened in American sports. He sits on the shoulders of two bearers – mythology and science. For Joe DiMaggio accomplished what no other ballplayer has done. He beat the hardest taskmaster of all, a woman who makes Nolan Ryan's fastball look like a cantaloupe in slow motion – Lady Luck.

A larger issue lies behind basic documentation and simple appreciation. For we don't understand the truly special character of DiMaggio's record because we are so poorly equipped, whether by habits of culture or by our modes of cognition, to grasp the workings of random processes and patterning in nature.

Omar Khayyám, the old Persian tentmaker, understood the quandary of our lives (*Rubaiyat of Omar Khayyám*, Edward Fitzgerald, trans.):

> Into this Universe, and Why not knowing,
> Nor Whence, like Water willy-nilly flowing;
> And out of it, as Wind along the Waste,
> I know not Whither, willy-nilly blowing.

But we cannot bear it. We must have comforting answers. We see pattern, for pattern surely exists, even in a purely random world. (Only a highly nonrandom universe could possibly cancel out the clumping that we perceive as pattern. We think we see constellations because stars are dispersed at random in the heavens, and therefore clump in our sight. Our error lies not in the perception of pattern but in automatically imbuing pattern with meaning, especially with meaning that can bring us comfort, or dispel confusion. Again, Omar took the more honest approach:

> Ah, love! could you and I with Fate conspire
> To grasp this sorry Scheme of Things entire.
> Would not we shatter it to bits – and then
> Re-mould it nearer to the Heart's Desire!

We, instead, have tried to impose that 'heart's desire' upon the actual earth and its largely random patterns (Alexander Pope, *Essay on Man*, end of Epistle 1):

> All Nature is but Art, unknown to thee:
> All Chance, Direction, which thou canst not see:
> All Discord, Harmony not understood:
> All partial Evil, universal Good.

Sorry to wax so poetic and tendentious about something that leads back to DiMaggio's hitting streak, but this broader setting forms the source of our misinterpretation. We believe in 'hot hands' because we must impart meaning to a pattern – and we like meanings that tell stories about heroism, valor, and excellence. We believe that long streaks and slumps must have direct causes internal to the sequence itself, and we have no feel for the frequency and length of sequences in random data. Thus, while we understand that DiMaggio's hitting streak was the longest ever, we don't appreciate its truly special character because we view all the others as equally patterned by cause, only a little shorter. We distinguish DiMaggio's feat merely by quantity along a continuum of courage; we should, instead, view his fifty-six-game hitting streak as a unique assault upon the otherwise unblemished record of Dame Probability.

Amos Tversky, who studied 'hot hands,' has performed, with Daniel Kahneman, a series of elegant psychological experiments. These long-term studies have provided our finest insight into 'natural reasoning' and its curious departure from logical truth. To cite an example, they construct a fictional description of a young woman: 'Linda is thirty-one years old, single, outspoken, and very bright. She majored in philosophy. As a student, she was deeply concerned with issues of discrimination and social justice, and also participated in anti-nuclear demonstrations.' Subjects are then given a list of hypothetical statements about Linda: They must rank these in order of presumed likelihood, most to least probable. Tversky and Kahneman list eight statements, but five are a blind, and only three make up the true experiment:

Linda is active in the feminist movement;
Linda is a bank teller;
Linda is a bank teller and is active in the feminist movement.

Now it simply must be true that the third statement is least likely, since any conjunction has to be less probable than either of its parts considered separately. Everybody can understand this when the principle is explained explicitly and patiently. But all groups of subjects, sophisticated students who have pondered logic and probability as well as folks off the street corner, rank the last statement as more probable than the second. (I am particularly fond of this example because I know that the third statement is least probable, yet a little homunculus in my head continues to jump up and down, shouting at me – 'but she can't just be a bank teller; read the description.')

Why do we so consistently make this simple logical error? Tversky and Kahneman argue, correctly I think, that our minds are not built (for whatever reason) to work by the rules of probability, though these rules clearly govern our universe. We do something else that usually serves us well, but fails in crucial instances: We 'match to type.' We abstract what we consider the 'essence' of an entity, and then arrange our judgments by their degree of similarity to this assumed type. Since we are given a 'type' for Linda that implies feminism, but definitely not a bank job, we rank any statement matching the type as more probable than another that only contains material contrary to the type. This propensity may help us to understand an entire range of human preferences, from Plato's theory of form to modern stereotyping of race or gender.

We might also understand the world better, and free ourselves of unseemly prejudice, if we properly grasped the workings of probability and its inexorable hold, through laws of logic, upon much of nature's pattern. 'Matching to type' is one common error; failure to understand random patterning in streaks and slumps is another – hence Tversky's study of both the fictional Linda and the 76ers' baskets. Our failure to appreciate the uniqueness of DiMaggio's streak derives from the same unnatural and uncomfortable relationship that we maintain with probability. (If we knew Lady Luck better, Las Vegas might still be a roadstop in the desert.)

My favorite illustration of this basic misunderstanding, as applied to DiMaggio's hitting streak, appeared in a recent article by baseball writer John Holway, 'A Little Help from His Friends,' and subtitled 'Hits or Hype in '41' (*Sports Heritage*, 1987). Holway points out that five of DiMaggio's successes were narrow escapes and lucky breaks. He received two benefits-of-the-doubt from official scorers on plays that might have been judged as errors. In each of two games, his only hit was a cheapie. In game sixteen, a ball dropped untouched in the outfield and had to be called a hit, even though the ball had been misjudged and could have been caught; in game fifty-four, DiMaggio dribbled one down the third-base line, easily beating the throw because the third baseman, expecting the usual, was playing far back. The fifth incident is an oft-told tale, perhaps the most interesting story of the streak. In game thirty-eight, DiMaggio was 0 for 3 going into the last inning. Scheduled to bat fourth, he might have been denied a chance to hit at all. Johnny Sturm popped up to begin the inning, but Red Rolfe then walked. Slugger Tommy Henrich, up next, was suddenly swept with a premonitory fear: Suppose I ground into a double play and end the inning? An elegant solution immediately occurred to him: Why not bunt (an odd strategy for a power hitter). Henrich laid down a beauty; DiMaggio, up next, promptly drilled a double to left.

I enjoyed Holway's account, but his premise is entirely, almost preciously, wrong. First of all, none of the five incidents represents an egregious miscall. The two hits were less than elegant, but undoubtedly legitimate; the two boosts from official scorers were close calls on judgment plays, not gifts. As for Henrich, I can only repeat manager Joe McCarthy's comment when Tommy asked him for permission to bunt: 'Yeah, that's a good idea.' Not a terrible strategy either – to put a man into scoring position for an insurance run when you're up 3–1.

But these details do not touch the main point: Holway's premise is false because he accepts the conventional mythology about long sequences. He believes that streaks are unbroken runs of causal courage – so that any prolongation by hook-or-crook becomes an outrage against the deep meaning of the phenomenon. But extended sequences are not pure exercises in valor. Long streaks always are, and must be, a matter of extraordinary luck imposed upon great skill. Please don't make the vulgar mistake of thinking that Purcell or Tversky or I or anyone else would attribute a long streak to 'just luck' – as though everyone's chances are exactly the same, and streaks represent nothing more than the lucky atom that kept moving in one direction. Long hitting streaks happen to the greatest players – Sisler, Keeler, DiMaggio. Rose – because their general chance of getting a hit is so much higher than average. Just as Joe Airball cannot match Larry Bird for runs of baskets, Joe's cousin Bill Ofer, with a lifetime batting average of .184, will never have a streak to match DiMaggio's with a lifetime average of .325. The statistics show something else, and something fascinating: There is no 'causality of circumstance,' no 'extra' that the great can draw from the soul of their valor to extend a streak beyond the ordinary expectation of coin-tossing models for a series of unconnected events, each occurring with a characteristic probability for that particular player. Good players have higher characteristic probabilities, hence longer streaks.

Of course DiMaggio had a little luck during his streak. That's what streaks are all about. No long sequence has ever been entirely sustained in any other way (the Orioles almost won several of those twenty-one games). DiMaggio's remarkable achievement – its uniqueness, in the unvarnished literal sense of that word – lies in whatever he did to extend his success well beyond the reasonable expectations of random models that have governed every other streak or slump in the history of baseball.

Probability does pervade the universe – and in this sense, the old chestnut about baseball imitating life really has validity. The statistics of streaks and slumps, properly understood, do teach an important lesson about epistemology, and life in general. The history of a species, or any natural phenomenon that requires unbroken continuity in a world of trouble, works like a batting streak. All are games of a gambler playing with a limited stake against a house with infinite resources. The gambler must eventually go bust. His aim can only be to stick around as long as possible, to have some fun while he's at it, and, if he happens to be a moral agent as well, to worry about staying the course with honor. The best of us will try to live by a few simple rules: Do justly, love mercy, walk humbly with thy God, and never draw to an inside straight.

DiMaggio's hitting streak is the finest of legitimate legends because it embodies the essence of the battle that truly defines our lives. DiMaggio activated the greatest and most unattainable dream of all humanity, the hope and chimera of all sages and shamans: He cheated death, at least for a while.

Notes

1. This essay originally appeared in the *New York Review of Books* as a review of Michael Seidel's *Streak: Joe DiMaggio and the Summer of 1941* (New York: McGraw-Hill, 1988). I have excised the references to Seidel's book in order to forge a more general essay, but I thank him both for the impetus and for writing such a fine book.
2. When I wrote this essay, Frank Robinson, the Baltimore skipper, was the only black man at the helm of a major league team. For more on the stats of Baltimore's slump, see my article 'Winning and Losing: It's All in the Game,' *Rotunda*, Spring 1989.

SEVENTH INNING STRETCH: BASEBALL, FATHER, AND ME

THIS piece exists to fulfill a promise made to one of my best baseball friends, author Steve King, who wrote to me in late October 1992: 'I think you should set aside a small but not inconsequential block of time – three weeks, maybe a month – to write a long (20,000 to 30,000 words) "linchpin essay" that would place your love and knowledge of baseball among the other landmarks of your rather remarkable life.'

Obviously, I appreciated the implied compliments in Steve's remarks. But I quickly resolved to follow his suggestion for a set of more literary and personal reasons. I have published eleven volumes of essays, ten from my monthly series in *Natural History* magazine that ran to three hundred successive pieces, without a break, from January 1974 until January 2001, and an eleventh (*Urchin in the Storm*) based primarily on essays originally published in the *New York Review of Books*. I have been writing about my serious dedication to baseball (in a wide variety of formats, from short op-ed statements to fairly lengthy articles) since the early 1980s; and I suppose that, for at least fifteen years, I have been intending to collect these baseball scribblings into a volume once enough material had accumulated. Steve King's suggestion of a decade ago (I began this 'novella' in mid April 2002) made my resolve firm, but the press of other commitments and the need to accumulate more material made this palindromic year of the new millennium an appropriate time to begin in earnest.

As much as I have loved and followed baseball all my conscious life, I never thought, before deciding to 'cash out' Steve's suggestion in an opening novella, that I would ever try to discourse (at any length or seriousness) about why the game continues to hold me so tight after more than half a century of serious and continuous rooting. After all, one loves what one loves, and unless the activity causes clear and intrinsic harm to others, no explicit defense need be provided for following one's bliss.

Yet I have developed personal answers to the two major questions so often thrown at academics and other professional intellectuals as challenges to their baseball commitments, and my responses might be worth sharing, especially since discussion on these issues never seems to abate. First, why are so many American intellectuals so serious about their baseball commitments? Or (to put the matter more specifically, as this form of inquiry so often does), why has baseball alone among major American sports, with boxing as the only other possible contender, generated so much writing of not insubstantial literary quality? Second, does this favoritism toward baseball arise in any way from the plethora of common claims about baseball's imitation of the central rhythms and patterns of our lives? Does life imitate the World Series in any way that might transcend lame and meaningless metaphor? Does time begin on opening day in any sense that might help us, at least by analogy, to accept the loud evenings of July 4 or the silly costumes of neighborhood kids on October 31?

I have devised, over the years, a definite way of treating general questions of this sort, including these two particular inquiries – and though my resolutions satisfy me both rationally and emotionally, I cannot claim for them any abiding status as provable or general truths. These resolutions do, however, set a groundwork for permitting me to begin with an autobiographical rationale.

I have written two general books (*Wonderful Life*, 1989, and *Life's Grandeur*, 1996) dedicated to viewing the history of life as a sensible and interpretable unfolding of one actual pattern among the countless alternative (and equally sensible) scenarios that just didn't happen to attain the privilege of empirical realization. We live in a basically unpredictable world, featuring histories dominated by contingency – that is, actual patterns that make good sense and become subject to interesting and sensible explanation once they unfold as they did, but that could have proceeded along innumerable alternative routes that would have yielded just as sensible a history, but that did not gain the good fortune of actual occurrence.

Thus, if it be true that intellectually inclined American sports fans tend to enjoy and follow baseball at a higher frequency than other popular national sports, I don't for a moment attribute such favoritism to any inherent property of the game itself. Baseball became America's pastime for a complex set of reasons, but the game is not intrinsically more difficult or inherently harder to fathom than any other major sport – and I therefore reject the common assumption that strong rootership among intellectuals can be related in any important way to the nature of the activity itself.

Rather, I would argue that sport plays an important role in the lives of many people (either by direct participation or by following it as a fan), and that intellectuals roughly match the norms of any other group in their predisposition to such avocational interests. Thus, if baseball has captured the serious attention of many scholarly fans, I would seek no special cause beyond the general appeal of the game among all aficionados of sport. By this argument, baseball holds its favored place as a general phenomenon in the history of American sport, and not because the game holds any special or intrinsic appeal to the intellectually minded fans.

Modern baseball coagulated from a variety of stick-and-ball games, played in England and perhaps in other European nations and imported to this country by early settlers. In essentials, the modern form of the game coalesced by the mid-nineteenth century, having evolved as a truly popular sport, played by farm boys and city slickers alike. By contrast, other currently popular team sports, football and basketball in particular, arose primarily as university activities at a time when only a small percentage of Americans achieved any tertiary education. During my childhood in the late 1940s and early 1950s, for example, professional football and basketball played short seasons and commanded only quite restricted popular attention. Hockey was, and to a large extent remains, an import from a great land just to our north. (I remember going to a hockey game at Madison Square Garden in the late 1940s and reading in the program that almost every starting player for the New York Rangers was Canadian by nationality.)

And so, I would argue (at least for myself and, I suspect, for most baseball fans of scholarly bent as well) that a serious personal affection for the sport does not follow, either logically or intrinsically, from any particular inherent property of the game's uniqueness, but rather needs to be explained in the same basic mode as most autobiographical phenomena – that is, as a contingent circumstance that did not have to unfold as it did, but that makes perfectly good sense as a reasonable outcome among a set of possibilities. In this general sense, and for a large array of excellent reasons, baseball became America's 'signature sport.' I can think of no reason why its appeal should be any less or greater among intellectuals than among any other segment of our population. So I would suspect that the appeal of baseball should, at least as an initial hypothesis, be equally strong among intellectually minded fans as among any other group of Americans.

I am not, however, trying to advance general explanations of the appeal or success of our national pastime. Thus, I can only speak for myself and from my own life. If any of my personal reasons apply more generally, then we will need the confirming testimony of others. I view the major features of my own odyssey as a set of mostly fortunate contingencies. I was not destined by inherited mentality or family tradition to become a paleontologist. I can locate no tradition for scientific or intellectual careers anywhere on either side of my Eastern European Jewish background. I myself am the oldest member of the third cohort, the offspring of immigrant grandparents who passed through Ellis Island – that is, the generation destined for university education and professional careers outside the garment district and the world of small shopkeepers.

I accepted this circumstance gladly (not that we have much choice in such matters). And I view my serious and lifelong commitment to baseball in entirely the same manner: purely as a contingent circumstance of numerous, albeit not entirely capricious, accidents. In other words, my affection for baseball does not predictably follow from any generality of my being (in a 'laws of nature' type of explanation preferred by scientists like myself), but rather from a set of 'accidents' arising from the particulars of my personal life.

Among these particulars, I would single out two for special emphasis. In fact, I rather suspect that versions of these two factors tend to rank high on the list of contingencies for explaining the inclinations and commitments of many serious fans.

1. Issues of how, or whether, to assimilate to the language and customs of an adopted land stood in the forefront of consciousness for the millions of immigrants (including all members of both sides of my family) who arrived in America during the great wave of the late nineteenth and early twentieth centuries. Some chose to retain native

languages and customs so far as they could, and to assimilate only to the minimal degree required for basic success and solvency. Others consciously abjured their natal tongues and traditions and struggled to speak only English and to learn and practice the history and customs of their adopted land. This second assimilationist group tended to dismiss the traditionalists and newcomers who had not yet made up their minds as 'green horns' or 'greenies.' My maternal Hungarian grandparents (the relatives I know best and who served as my surrogate parents during World War II, when my father fought in Europe and Northern Africa) were devoted assimilationists who spoke Hungarian only when they didn't want me to understand, and who took great pride in their accommodation to America. I doubt that they ever understood the limits to their success, particularly as expressed in strong accents that they actively denied, but never lost nonetheless.

Immigrants who opted for assimilation tended to choose particular American institutions or customs as public foci for their commitments. Some veered toward politics of democratic systems that they and their families could enjoy for the first time – as in, for example, the domination of local governments in several major cities by new Irish and Italian citizens. (The WASPs of old Brahmin Boston feared the death of their beloved city when poor Irish immigrants took local political power away from traditional sources, but the Hub persevered and prospered.)

Particularly for men, and especially commonly for Jewish men, a dedication to a distinctively American sport provides the major tactic for assimilation. The three Bs in particular (boxing, basketball, and especially baseball) assumed great importance in the lives of many Jewish and other immigrants. Few Jews grow very tall, so we were probably not destined for basketball triumph as players (although Abe Saperstein put together and coached the Harlem Globetrotters, and many important teams in the early history of basketball – notably the SPHAs, for South Philadelphia Hebrew Association – were formed and staffed by Jewish players).

Boxing and baseball offered stronger possibilities, where champions like Benny Leonard, Max Baer, Moe Berg (a mediocre player, but absolutely outstanding character) and especially Hank Greenberg and, later, Sandy Koufax could become heroes and role models for entire generations, and where even your average city street kid (like me) could play with reasonable success. As a happenstance of personal contingencies, the men of my strongly assimilationist maternal side took up baseball as their major sign and symbol of 'Americanization,' and became serious and knowledgeable fans, eager to pass this new tradition to subsequent generations.

Thus, baseball became a centerpiece of family life in my household – and I take pride in being enmeshed within an unbroken string of four generations of serious baseball rooting. The sequence began with my maternal grandfather, Papa Joe, a dedicated Yankee fan from, by his testimony, 1904, when he thrilled to Jack Chesbro's forty-one victories in a single season – a pitching record that will stand unless the game undergoes radical changes in scheduling and counting – until his death in 1953. All three of his sons followed in serious fandom, although his only daughter, my mother, never really caught the bug despite a personal crush on Mel Ott. ('She thinks a foul ball is a chicken dance,' if I may quote a misogynist line from my father.)

Moving to the other side, my father passed his boyhood rooting for the great Yankee team of Ruth and Gehrig – and his dramatic and detailed memories provided me, a lifelong skeptic in religion, with my closest insight into the potential nature of a deity. In this familial context, my own adoption of serious interest can scarcely be deemed surprising! I must also confess to a feeling of personal pride that my youngest son, Ethan, has now continued a family tradition into a fourth generation – although he grew up in Boston, and three generations of Yankee fandom have now been eclipsed, understandably of course, by his exquisite pain of rooting for the Red Sox, and hoping to live long enough to win another World Series – last achieved in 1918!

2. If we honor the entertainment and real estate industries' cliché that the three most important factors for success are 'location, location, location,' then we must aver, I think, that my being a baseball fan requires no special explanation and should evoke no surprise, whereas we might become puzzled and feel the need for some resolution if I were indifferent to the game. As a pure contingency of my own life, I happened to come of baseball fandom's age in the greatest conjunction of time and place that the game has ever known: in New York City during the late 1940s and early 1950s. (I was born on September 10, 1941.)

The situation was entirely unfair to the rest of the country, but – hey – you can't possibly cast any blame on me, so I owe no one any apology. From 1947 to 1957, New York City had the three greatest teams in major league baseball. (Many fans do not understand why a county or borough, rather than a full city, had its own team as the Brooklyn Dodgers. But New York City did not incorporate its outer boroughs into a single city until 1897, so the Dodgers represented an independent city when the team first formed and played major league ball.)

During these eleven years, one of the three New York teams (the Yankees of the American League and the New York Giants and Brooklyn Dodgers of the National League) won the World Series in all but two years, when the Cleveland Indians prevailed in 1948 and Milwaukee in 1957. Only in 1948 – Cleveland vs. the Boston Braves – did a New York team not play in the Series at all. In seven of these years ('47, '49, '51, '52, '53, '55, and '56) two New York teams played each other in the World Series – all won by my beloved Yankees except for the ultimate tragedy of '55, when the Dodgers won their single victory over the Yanks as a Brooklyn team. We got them back the next year, though, in '56!

My earliest vague memories of baseball date to the 1946 or 1947 season. I remember the great 1948 season in substantial detail – the year that should have been the Boston subway series, but the Indians tied the Red Sox and then won the single game playoff for the right to play the Boston Braves in the World Series. Starting in 1949, I suspect that I could narrate at least the major events of all World Series games through the Yankees' revenge on Milwaukee in the 1958 contest.

But my point is simply this – and plausible though the claim may be as an abstraction, one really had to 'live it' to know the full extent of the pull and the virtual inevitability – during these years, nearly all boys in New York City and quite a few girls, as well, became passionate baseball fans, spending a good bit of each day, from April to early October, tracing the developing fate of one's favorites.

Patterns of rooting were neither entirely capricious nor entirely predictable. Nearly all of Brooklyn's two million citizens rooted passionately for the Dodgers. I'm still mad at my cousin Steve Sosland for failing to protect me, as he promised he would, when I admitted to being a Yankee fan while playing stickball with his Brooklyn neighborhood friends – the worst street beating I ever received, but a rite of passage in the coming of age for any New York street kid.

The still solid Jewish and Italian ethnic communities of the Bronx lived and died with the Yankees (a.k.a. The Bronx Bombers), of course.

The Polo Grounds, home of the New York Giants, located in northeastern Manhattan and literally within sight of Yankee Stadium across the Harlem River, did not command so clear a geographic region of nearly exclusive rooting – and Giants fans tended to be scattered throughout the city. Many kids, myself included, rooted for two New York teams, one from each league. Affection for the Yankees and Dodgers proved difficult, for they played each other too often in the World Series ('41, '47, '49, '52, '53, '55, and '56, all won by the Yanks except 1955). By contrast, the Yankees and Giants only met in 1951 – on my watch at least, for several Yankee–Giant Series had been played before my birth in the 1920s and 1930s.

I grew up in Queens, the most 'neutral' borough, with no team of its own (the expansion Mets did not begin until the early 1960s, and I have never been able truly to view them as a 'home team,' despite substantial affection based on pure accidents of birth and upbringing – and some wonderful memories of going to Shea Stadium whenever Sandy Koufax of the Los Angeles Dodgers pitched against the Mets, and invariably won).

Memory, of course, is the ultimate trickster, but I do have a very clear impression that at least 50 percent of boy-talk between April and October in Queens focused on the fates of our three teams, with constant bets, threats, and bickerings about pennant races and World Series outcomes. I rooted for the Yanks and Giants. But I even managed to tolerate the Dodger fanaticism of some of my best friends.

The final point is simply this: All New York City boys of the late 1940s and early 1950s were baseball nuts, barring mental deficiency or incomprehensible idiosyncrasy. How could one not be? This decade was the greatest conjunction of quality and place that the game has ever known. Grossly unfair to the rest of the country, of course, but a fabulous piece of luck that made the 'coming of age' for me and a million other New York kids ever so much easier – and what purely contingent blessing of ontogeny could be more precious?

TROUBLE IN OUR OWN HOUSE: A BRIEF LEGAL SURVEY FROM SCOPES TO SCALIA

THE fundamentalist movement may be as old as America, and its opposition to teaching evolution must be as old as Darwin. But this marginal, politically disenfranchised, and largely regional movement could muster no clout to press a legislative agenda until one of the great figures of American history, William Jennings Bryan, decided to make his last hurrah on this issue. Bryan gave the creationist movement both influence and contacts. In the early 1920s, several Southern states passed flat-out anti-evolution statutes. The Tennessee law, for example, declared it a crime to teach that 'man had descended from a lower order of animals.'

American liberals, including many clergymen, were embarrassed and caught off guard by the quick (if local) successes of this movement. In a challenge to the constitutionality of these statutes, the American Civil Liberties Union (ACLU) instigated the famous Scopes trial in Dayton, Tennessee, in 1925. John Scopes, a young free-thinker, but quite popular among his largely fundamentalist students, worked as the physics teacher and track coach of the local high school. He had substituted for the fundamentalist biology teacher during an illness, and had assigned the chapters on evolution from the class textbook, *A Civic Biology*, by George William Hunter. Scopes consented to be the guinea pig or stalking horse (choose your zoological metaphor) for a legal challenge to the constitutionality of Tennessee's recently enacted anti-evolution law – and the rest is history, largely filtered and distorted, for most Americans, through the fictionalized account of a wonderful play, *Inherit the Wind*, written in 1955 by Jerome Lawrence and Robert Edwin Lee, and performed by some of America's best actors in several versions. (I had the great privilege, as a teenager, to see Paul Muni at the end of his career playing Clarence Darrow in the original Broadway production, with an equally impressive Ed Begley as William Jennings Bryan. Two film versions featured similar talent, with Spencer Tracy as Darrow and Fredric March as Bryan in the first, and Kirk Douglas as Darrow with Jason Robards as Bryan in the later remake for television.)

Contrary to the play, Scopes was not persecuted by Bible-thumpers, and never spent a second in jail. The trial did have its epic moments – particularly when Bryan, in his major speech, virtually denied that humans were mammals; and, in the most famous episode, when Judge Raulston convened his court on the lawn (for temperatures had risen into triple digits and cracks had developed in the ceiling on the floor below the crowded courtroom), and allowed Darrow to put Bryan on the stand as a witness for the defense. But the usual reading of the trial as an epic struggle between benighted Yahooism and resplendent virtue simply cannot suffice – however strongly this impression has been fostered both by *Inherit the Wind* and by the famous reporting of H. L. Mencken, who attended the trial and, to say the least, professed little respect for Bryan, whom he called 'a tinpot pope in the Coca-Cola belt.'

Scopes was recruited for a particular job – both by the ACLU and by Dayton fundamentalists, who saw the trial as an otherwise unobtainable opportunity to put their little town 'on the map' – and not proactively persecuted in any way. The ACLU wanted a quick process and a sure conviction, not a media circus. (The Scopes trial initiated live broadcast by radio, and might therefore be designated as the inception of a trajectory leading to O. J. Simpson and other extravaganzas of arguable merit.) The local judge held no power to determine the constitutionality of the statute, and the ACLU therefore sought an unproblematical conviction, designated for appeal to a higher court. They may have loved Clarence Darrow as a personality, but they sure as hell didn't want him in Dayton. However, when Bryan announced that he would appear for the state of Tennessee to rout Satan from Dayton, the die was cast, and Darrow's counteroffer could hardly be refused.

The basic facts have been well reported, but the outcome has almost always been misunderstood. Darrow did bring several eminent scientists to testify, and the judge did refuse to let them take the stand. In so deciding, he was not playing the country bumpkin, but making a proper ruling that his court could judge Scopes's guilt or innocence only under the given statute – and Scopes was guilty as charged – not the legitimacy or constitutionality of the law itself. Testimony by experts about the validity or importance of evolution therefore became irrelevant. In this context, historians have never understood why Judge Raulston then allowed Bryan to testify as an expert for the other side. But this most famous episode has also been misread. First of all, the judge later struck the entire testimony from the record. Second, Darrow may have come out slightly ahead, but Bryan parried fairly well, and certainly didn't embarrass himself. The most celebrated moment – when Darrow supposedly forced Bryan to admit that the days of creation might have spanned more than twenty-four hours – represented Bryan's free-will statement about his own and well-known personal beliefs (he had never been a strict biblical literalist), not a fatal inconsistency, exposed by Darrow's relentless questioning.

To correct the other most famous incident of the trial, Bryan did indeed drop dead of heart failure in Dayton – not dramatically on the courtroom floor (as fiction requires for maximal effect), but rather a week later, after stuffing himself at a church dinner. However, the most serious misunderstanding lies with the verdict itself, and the subsequent history of creationism. *Inherit the Wind* presents a tale of free inquiry triumphant over dogmatism. As an exercise in public relations, the Scopes trial may be read as a victory for our side. But the legal consequences could hardly have been more disastrous. Scopes was, of course, convicted – no surprises there. But the case was subsequently declared moot – and therefore unappealable – by the judge's error of fining Scopes one hundred dollars (as the creationism statute specified), whereas Tennessee law required that all fines over fifty dollars be set by the jury. (Perhaps sleepy little towns like Dayton never fined anyone more than fifty bucks for anything, and the judge had simply forgotten this detail of unapplied law.) In any case, this error provides a good argument against using 'outside agitators' like Darrow as sole representatives in local trials. The fancy plaintiff's team, led by Darrow and New York lawyer Dudley Field Malone, included no one with enough local knowledge to challenge the judge and assure proper procedure.

Thus, Scopes's conviction was overturned on a technicality – an outcome that has usually been depicted as a victory, but was actually a bitter procedural defeat that stalled the real purpose of the entire enterprise: to test the law's constitutionality. In order to reach the appropriate higher court, the entire process would have to start again, with a retrial of Scopes. But history could not be rolled back, for Bryan was dead, and Scopes, now enrolled as a graduate student in geology at the University of Chicago, had no desire to revisit this part of his life. (Scopes, a splendidly modest and honorable man, became a successful oil geologist in Shreveport, Louisiana. He never sought any profit from what he recognized as his accidental and transitory fame, and he never wavered in defending freedom of inquiry and the rights of teachers.)

So the Tennessee law (and similar statutes in other states) remained on the books – not actively enforced, to be sure, but ever-present as a weapon against the proper teaching of biology. Textbook publishers, the most cowardly arm of the printing industry, generally succumbed and either left evolution out or relegated the subject to a small chapter at the back of the book. I own a copy of the text that I used in 1956 at a public high school in New York City, a liberal constituency with no compunction about teaching evolution. This text, *Modern Biology*, by Moon, Mann, and Otto, dominated the market and taught more than half of America's high school students. Evolution occupies only 18 of the book's 662 pages – as chapter 58 out of 60. (Many readers, remembering the realities of high school, will immediately know that most classes never got to this chapter at all.) Moreover, the text never mentions the dreaded 'E' word, and refers to Darwin's theory as 'the hypothesis of racial development.' But the first edition of this textbook, published in 1921, before the Scopes trial, featured Darwin on the frontispiece (my 1956 version substitutes a crowd of industrious beavers for the most celebrated of all naturalists), and includes several chapters treating evolution as both a proven fact and the primary organizing theme for all biological sciences.

This sorry situation persisted until 1968, when Susan Epperson, a courageous teacher from Arkansas, challenged a similar statute in the Supreme Court – and won the long-sought verdict of unconstitutionality on obvious First Amendment grounds. (A lovely woman approached me after a talk in Denver last year. She thanked me for my work in fighting creationism and then introduced herself as Susan Epperson. She had attended my lecture with her daughter, who, as a graduate student in evolutionary biology, had reaped the fruits of her mother's courage. I could only reply that the major thrust of thanks must flow in the other direction.)

But nothing can stop a true believer. The creationists regrouped, and came back fighting with a new strategy designed to circumvent constitutional problems. They had always honorably identified their alternative system as explicitly theological, and doctrinally based in a literal reading of the Bible. But now they expurgated their texts, inventing the oxymoronic concept of 'creation science.' Religion, it seems, and contrary to all previous pronouncements, has no bearing upon the subject at all. The latest discoveries of pure science now reveal a factual world that just happens to correlate perfectly with the literal pronouncements of the Book of Genesis. If virtually all professionally trained scientists regard such a view as nonsensical, and based on either pure ignorance or outright prevarication, then we can only conclude that credentialed members of this discipline cannot recognize the cutting edge of their own subject. In such a circumstance, legislative intervention becomes necessary. And besides, the creationists continued, we're not asking schools to ban evolution anymore (that argument went down the tubes with the Epperson decision). Now we are only demanding 'equal time' for 'creation science' in any classroom that also teaches evolution. (Of course, if they decide not to teach evolution at all . . . well, then . . .)

However ludicrous such an argument might be, and however obviously self-serving as a strategy to cloak a real aim (the imposition of fundamentalist theological doctrine) in new language that might pass constitutional muster, two states actually did pass nearly identical 'equal time' laws in the late 1970s – Arkansas and Louisiana. A consortium of the ACLU and many professional organizations, both scientific and religious, challenged the Arkansas statute in a trial labeled by the press (not inappropriately) as 'Scopes II,' before Federal Judge William R. Overton in Little Rock during December 1981. Judge Overton, in a beautifully crafted decision (explaining the essence of science, and the proper role of religion, so well that *Science*, our leading professional journal, published the text verbatim), found the Arkansas equal-time law unconstitutional in January 1982.

The state of Arkansas, now back under the liberal leadership of Bill Clinton, decided not to appeal. Another federal judge then voided the nearly identical Louisiana law by summary judgment, stating that the case had been conclusively made in Arkansas. Louisiana, however, did appeal to the U. S. Supreme Court in *Edwards v. Aguillard*, where, in 1987, we won a strong and final victory by a seven-to-two majority, with (predictably) Rehnquist and Scalia in opposition (Thomas, a probable third vote today, had not yet joined the court).

I testified at the Arkansas trial as one of six 'expert witnesses' in biology, philosophy of science, and theology – with my direct examination centered upon creationist distortion of scientific work on the length of geological time and the proof of evolutionary transformation in the fossil record, and my cross-examination fairly perfunctory. (The attorney general of Arkansas, compelled by the ethics of his profession to defend a law that he evidently deemed both silly and embarrassing to his state, did a competent job, but just didn't have his heart in the enterprise.)

As a group, by the way, we did not try to prove evolution in our testimony. Courtrooms are scarcely the appropriate venue for adjudicating such issues under the magisterium of science. We confined our efforts to the only legal issue before us: to proving, by an analysis of their texts and other activities, that 'creation science' is nothing but a smoke screen, a meaningless and oxymoronic phrase invented as sheep's clothing for the old wolf of Genesis literalism, already identified in the Epperson case as a partisan theological doctrine, not a scientific concept at all – and clearly in violation of First Amendment guarantees for separation of church and state if imposed by legislative order upon the science curricula of public schools.

I can't claim that the trial represented any acme of tension in my life. The outcome seemed scarcely in doubt, and we held our victory party on the second day of a two-week trial. But cynicism does not run strongly in my temperament – and I expect that when I am ready to intone my *Nunc Dimittis*, or rather my *Sh'ma Yisroel*, I will list among my sources of pride the fact that I joined a group of scholars to present the only testimony ever provided by expert witnesses before a court of law during this interesting episode of American cultural history – the legal battle over creationism that raged from Scopes in 1925 to *Edwards v. Aguillard* in 1987. (Judge Raulston did not allow Darrow's experts to testify at the Scopes trial, and the Louisiana law was dismissed by summary judgment and never tried; live arguments before the Supreme Court last only for an hour, and include no witnesses.) It was, for me, a great joy and privilege to play a tiny role in a historical tale that featured such giant figures as Bryan and Darrow.

The Arkansas trial may have been a no-brainer, but many anecdotes, both comic and serious, still strike me as illuminating or instructive. In the former category, I may cite my two favorite moments of the trial.

First, I remember the testimony of a second-grade teacher who described an exercise he uses to convey the immense age of the earth to his students: he stretches a string across his classroom, and then places the children at appropriate points to mark the origin of life, the death of dinosaurs, and human beginnings right next to the wall at the string's end. In cross-examination, the assistant attorney general asked a question that he later regretted: What would you do under the equal-time law if you had to present the alternative view that the earth is only ten thousand years old? 'I guess I'd have to get a short string,' the teacher replied. The courtroom burst into laughter, evidently all motivated by the same image that had immediately popped into my mind: the thought of twenty earnest second-graders all scrunched up along one millimeter of string.

In a second key moment, the creationist side understood so little about the subject of evolution that they brought, all the way from Sri Lanka, a fine scientist named Chandra Wickramasinghe, who happens to disagree with Darwinian theory (but who is not an anti-evolutionist, and certainly not a young-earth creationist – a set of distinctions that seemed lost on intellectual leaders of this side). Their lawyer asked him, 'What do you think of Darwin's theory?' and Wickramasinghe replied, in the crisp English of his native land, 'Nonsense.' In cross-examination, our lawyer asked him: 'And what do you think of the idea that the earth is only ten thousand years old?' 'Worse nonsense,' he tersely replied.

On the plane back home, I got up to stretch my legs (all right, I was going to take a pee), and a familiar-looking man, sitting in an aisle seat of the coach section, stopped me and said in the local accent, 'Mr Gould, I wanna thank you for comin' on down here and heppin' us out with this little problem.' 'Glad to do it,' I replied, 'but what's your particular interest in the case? Are you a scientist?' He chuckled and denied the suggestion. 'Are you a businessman?' I continued. 'Oh no,' he finally replied, 'I used to be the governor. I'd have vetoed that bill.' I had been talking with Bill Clinton. In an odd contingency of history that allowed this drama to proceed to its end at the Supreme Court, Clinton had become a bit too complacent as boy-wonder governor, and had not campaigned hard enough to win reelection in 1980 – a mistake that he never made again, right up to the presidency. The creationism bill, which he would surely have vetoed, passed during his interregnum, and was signed by a more conservative governor.

But such humor served only as balance for the serious and poignant moments of the trial – none so moving as the dignity of committed teachers who testified that they could not practice their profession honorably if the law were upheld. One teacher pointed to a passage in his chemistry text that attributed great age to fossil fuels. Since the Arkansas act specifically included 'a relatively recent age of the earth' among the definitions of creation science requiring 'balanced treatment,' this passage would have to be changed. The teacher claimed that he did not know how to make such an alteration. Why not? retorted the assistant attorney general in his cross-examination. You only need to insert a simple sentence: 'Some scientists, however, believe that fossil fuels are relatively young.' Then, in the most impressive statement of the entire trial, the teacher responded: I could, he argued, insert such a sentence in mechanical compliance with the act. But I cannot, as a conscientious teacher, do so. For 'balanced treatment' must mean 'equal dignity,' and I would therefore have to justify the insertion. And this I cannot do, for I have heard no valid arguments that would support such a position.

Another teacher spoke of similar dilemmas in providing balanced treatment in a conscientious rather than a mechanical way. What then, he was asked, would he do if the law were upheld? He looked up and said, in a calm and dignified voice: It would be my tendency not to comply. I am not a revolutionary or a martyr, but I have responsibilities to my students, and I cannot forgo them.

And now, led back by this serious note, I realize that I have been a bit too sanguine during this little trip down memory lane. Yes, we won a narrow and specific victory after sixty years of contention: creationists can no longer hope to realize their aims by official legislation. But these well-funded and committed zealots will not therefore surrender. Instead they have changed their tactics, often to effective strategies that cannot be legally curtailed. They continue to pressure textbook publishers for deletion or weakening of chapters about evolution. (But we can also fight back – and have done so effectively in several parts of the country – by urging school boards to reject textbooks that lack adequate coverage of this most fundamental topic in the biological sciences.) They agitate before local school boards, or run their own candidates in elections that rarely inspire large turnouts, and can therefore be controlled by committed minorities who know their own voters and get them to the polls. (But scientists are also parents, and 'all politics is local,' as my own former congressman from Cambridge, MA, used to say.)

Above all – in an effective tactic far more difficult to combat because it works so insidiously and invisibly – they can simply agitate in vociferous and even mildly threatening ways. Most of us, including most teachers, are not particularly courageous, and do not choose to become martyrs. Who wants trouble? If little Billy tells his parents that I'm teaching evolution, and they then cause a predictable and enormous public fuss (particularly in parts of America where creationism is strong and indigenous) . . . well, then, what happens to me, my family, and my job? So maybe I just won't teach evolution this year. What the hell. Who needs such a mess?

Which leads me to reiterate an obvious and final point: We misidentify the protagonists of this battle in the worst possible way when we depict evolution versus creationism as a major skirmish in a general war between science and religion. Almost all scientists and almost all religious leaders have joined forces *on the same side* – against the creationists. And the chief theme of this book provides the common currency of agreement – NOMA*, and the call for respectful and supportive dialogue between two distinct magisteria, each inhabiting a major mansion of human life, and each operating best by shoring up its own home while admiring the other guy's domicile and enjoying a warm friendship filled with illuminating visits and discussions.

Creationists do not represent the magisterium of religion. They zealously promote a particular theological doctrine – an intellectually marginal and demographically minority view of religion that they long to impose upon the entire world. And the teachers of Arkansas represent far more than 'science.' They stand for toleration, professional competence, freedom of inquiry, and support for the Constitution of the United States – a worthy set of goals shared by the vast majority of professional scientists and theologians in modern America. The enemy is not religion but dogmatism and intolerance, a tradition as old as humankind, and impossible to extinguish without eternal vigilance, which is, as a famous epigram proclaims, the price of liberty. We may laugh at a marginal movement like young-earth creationism, but only at our peril – for history features the principle that risible stalking horses, if unchecked at the starting gate, often grow into powerful

* Non-overlapping magisteria; see pp. 584–597.

champions of darkness. Let us give the last word to Clarence Darrow, who stated in his summation at the Scopes trial in 1925:

> If today you can take a thing like evolution and make it a crime to teach it in the public schools, tomorrow you can make it a crime to teach it in the private schools and next year you can make it a crime to teach it to the hustings or in the church. At the next session you may ban books and the newspapers . . . Ignorance and fanaticism are ever busy and need feeding. Always feeding and gloating for more. Today it is the public school teachers; tomorrow the private. The next day the preachers and the lecturers, the magazines, the books, the newspapers. After a while, Your Honor, it is the setting of man against man and creed against creed until with flying banners and beating drums we are marching backward to the glorious ages of the sixteenth century when bigots lighted fagots to burn the men who dared to bring any intelligence and enlightenment and culture to the human mind.

OF TWO MINDS AND ONE NATURE

(*with Rhonda Shearer*)

OUR propensity for thinking in dichotomies may lie deeply within human nature itself. In his *Lives and Opinions of Eminent Philosophers* (written circa A.D. 200), Diogenes Laertius quotes a much older maxim of Protagoras: 'there are two sides to every question, exactly opposite to each other.' But we can also utilize another basic trait of our common humanity – our mental flexibility, and our consequent potential for overcoming such innate limitations by education.

Our tendency to parse complex nature into pairings of 'us versus them' should not only be judged as false in our universe of shadings and continua, but also (and often) harmful, given another human propensity for judgment – so that 'us versus them' easily becomes 'good versus bad,' or even, when zealotry fans our xenophobic flames, 'chosen for martyrdom versus ripe for burning.'

The contingent and largely arbitrary nature of disciplinary boundaries has unfortunately been reinforced, and even made to seem 'natural,' by our drive to construct dichotomies – with science versus art as perhaps the most widely accepted of all. Moreover, given our tendencies to clannishness and parochiality, this false division becomes magnified as the two, largely noncommunicating, sides then develop distinct cultural traditions that evoke mutual stereotyping and even ridicule. (Scientists, who nearly always speak extemporaneously in public presentations, note that humanists almost always read papers at professional meetings, and rarely show slides – except for art historians, who always use two screens simultaneously – even for the most visual subjects. Why, 'we' ask, do 'they' not realize that written and spoken English are different languages, and that very few people can read well in public

– a particular irony since humanists supposedly hold language as their primary tool of professional competence. But 'they,' on the other hand, rightly ridicule 'our' tendencies to darken a lecture room even before we reach the podium and to rely almost entirely upon a string of pictures thereafter. (A stale joke proclaims that if Galileo had first presented the revolutionary results of *Siderius Nuncius* as a modern scientific talk, his opening line could only have been: 'First slide please.')

The worst and deepest stereotypes drive a particularly strong wedge between art (viewed as an ineffably 'creative' activity, based on personal idiosyncrasy and subject only to hermeneutical interpretation) and science (viewed as a universal and rational enterprise, based on factual affirmation and analytical coherence). We do not, of course, deny the differences in subject matters and criteria (empirical versus aesthetic judgment) in these two realms of human achievement, but we do believe that the common ground of methods for mental creativity and innovation, and the pedagogic virtues of unified nurturing for all varieties of human creativity, should inspire collaboration for mutual reinforcement.

At least we should recognize, if only for practical reasons, that both fields meet resistance in educational lobbies of primary and secondary public schooling – with art classes viewed as superfluous icing on a cake already stripped to a bare minimum of supposedly essential nutrients, and science classes regarded as 'too hard' for most students, and too expensive for most constituencies. (How can we forget the infamous words that Teen Barbie once spoke – 'Math class is tough' – before a public outcry led her makers to eliminate this philistine aspersion upon half of America's students?) If art and science could join forces by stressing our common methods in critical thinking, our common search for innovation, and our common respect for historical achievement – rather than emphasizing our disparate substrates and trying to profit from the differences in playing a zero-sum game at the other's expense – then we might, in Benjamin Franklin's remarkably relevant pun, truly hang together rather than hang separately.

Rather than indulging in such general, and tendentious, preaching, we can best illustrate the potential junction of art and science in the work of creative people whose innovations cannot be neatly slotted into either camp but can only be understood as a reinforcing unification of goals usually parsed between the two realms under Kipling's motto 'never the twain shall meet.'

The standard examples of Leonardo and other Renaissance figures have been well and justly referenced. But our best cases should not be sought in an earlier age that did not recognize our modern disciplinary boundaries and did not even possess a word for the enterprise now called 'science.' If we look instead to twentieth-century figures who suffered the penalties of mistrust and misunderstanding for working in both domains simultaneously, we can make our major point in more immediate terms.

Marcel Duchamp (1887–1968) may even surpass Picasso in his influence upon the history of twentieth-century art – especially in his conventional image as the ultimate Dada jokester, the *enfant terrible* who festooned the Mona Lisa with a beard, a moustache and a salacious caption, and then called the product art under his own signature; the man who submitted an ordinary urinal as his own sculpture, entitled 'Fountain,' to a major art show. But Duchamp, as a disciple of Henri Poincaré, also understood the mathematics of non-Euclidean geometry and higher dimensionality in a far more serious and technical way than any other artist of his time. He maintained a passionate interest in science throughout his life, and he made several innovations, in optics, mathematics and perception, that we have not understood both because Duchamp himself chose to be maddeningly cryptic about his intentions and achievements, and because we have not been open to the possibility that an acknowledged genius, once categorized as an 'artist,' could also be innovative in science.

Among his many hybrid ventures – experiments in optics and perception, mixed with aesthetic achievements in what he called 'nonretinal' art or the beauty of the mind or 'gray matter' – Duchamp devoted considerable attention and expense (he even trademarked the name) to developing a series of twelve discs, called 'Rotoreliefs,' and designed for spinning in circular motion on a record turntable (preferably mounted on a wall, so that an observer can view the spinning discs face on).

Although Italian scientists (unaware of Duchamp's work) found and named this particular form of illusion as 'the stereo-kinetic effect' in 1924, Duchamp apparently discovered this perceptual phenomenon independently in the early 1920s, and completed his first set of discs in 1923. Duchamp recognized that by spinning designs composed as sets of eccentric but concentric circles, a viewer would see the resulting pattern as a three dimensional form even through one eye alone, without the supposedly necessary benefit of stereoscopy! By the 1930s, Duchamp had constructed from his experiments a wonderfully whimsical set of twelve spinning images – from a goldfish in a bowl, to the eclipsed sun seen through a tube, to a cocktail glass, to a light bulb – in order to emphasize his discovery of these three-dimensional effects. (Ironically, as another example of harmful separation between truly unified aspects of art and science, art museums almost invariably exhibit these discs as framed, static objects on a wall – whereas they have no meaning, either artistic or scientific, unless they spin.

Duchamp knew what he had done, and he explicitly regarded the Rotoreliefs as a contribution to science. He wrote to Katherine Dreier in 1935: 'I showed it to scientists (optical people) and they say it is a new form, unknown before, of producing the illusion of volume or relief . . . That serious side of the play toy is very interesting.' Moreover, Duchamp took great pleasure in the efforts of a professor who wished to use his Rotorelief discs to retrain the three-dimensional insights of soldiers who had lost one eye in the First World War. [At a recent talk, one of us (R. R. S.) demonstrated the rotating discs to a physics professor, blind in one eye for more than a decade, who almost wept for joy at his first sight of three dimensions in so many years]. Duchamp also understood the general basis of his illusion when he wrote in a letter: 'I only had to use two circumferences – eccentric – and make them turn on a third center.'

We could cite many other examples of innovators, labeled as 'artists,' who used the tools of their trade to make discoveries that had eluded official 'scientists' within their own parochial world. In the eighteenth century, the Dutch artist Petrus Camper established rules for depicting characteristic differences in the physiognomies of human groups (sexes, ages, and ethnicities) after he noticed that many Renaissance paintings of the Three Kings had depicted Balthazar, the black magus, as a European painted dark, rather than a native of sub-Saharan Africa. (European artists could find few African models at the time.) At the beginning of our century, the celebrated American artist (and amateur ornithologist) A. H. Thayer discovered the adaptive value of countershading [not for concealment by cryptic coloration, as evolutionary biologists had previously assumed, but rather for making a three-dimensional object fade into invisibility because countershaded organisms appear entirely flat (two dimensional) against their background] – a solution that had eluded scientists but seemed starkly clear to an artist who had spent his life promoting the opposite illusion of making flat paintings look three-dimensional. Abbott's work led to important advances in naval camouflage and saved countless lives in twentieth-century warfare.

What could be more precious, or more difficult, than conceptual innovation? We need to access all the tools at our command – even when linguistic and sociological convention parcels out these common mental devices among noncommunicating disciplinary camps – if we wish to triumph in this hardest, yet most rewarding, of all intellectual pursuits. In a key passage from one of the most influential books of our times (*The Structure of Scientific Revolutions*), T. S. Kuhn bridged the disciplinary gap between visual representation and conceptual innovation when he used the famous gestalt illusion of the duck-rabbit as a primary symbol for the meaning and nature of scientific revolution: 'It is as elementary prototypes for these transformations of the scientist's world that the familiar demonstrations of a switch in visual gestalt prove so suggestive. What were ducks in the scientist's world before the revolution are rabbits afterwards.'

PART II

BIOGRAPHIES

If Steve was sparing when it came to committing details of his own life to print, he was nonetheless fascinated by the lives of other scientists significant to the history of evolutionary theory, and many of the finest of his *Natural History* essays show evidence of this fascination. Mostly he approaches his subject crabwise, so that the reader is a few pages into the essay before its theme becomes apparent. Or – better – themes, because each of these essays serves a dual purpose. One is to 'rescue' from oblivion a person or episode otherwise lost in the Whiggish narrative adopted by most popular histories of science; the second is to use that episode to illuminate often quite profound questions of evolutionary theory.

The choice of material for this section has been particularly hard as so many of the essays are worthy of rereading. Those we have chosen are arranged chronologically, in the historical sense of the dates of the central figure in each. But in doing so it also becomes apparent how certain issues recur throughout the sequence – notably efforts to reconcile geology with the Bible, mistakes, disputes and fakery, the stuff of scientific and detective romance. Thus we begin in the seventeenth century with Thomas Burnet and his suitably illustrated attempts, as a rationalist, to derive a physically acceptable explanation for the biblical flood. Half a century later, Gould offers us Dr Johann Beringer and the ludicrously fake fossils that provide the title for the volume *The Lying Stones of Marrakech* and the jumping-off point for some ruminations on fakery to which we return in the final essay in this section, 'The Piltdown Conspiracy'.

Skip a century, and we are in the company of one of the greatest pre-Darwinian biologists, the founder of paleontology, Georges Cuvier, and his use of – this time, genuine – fossils to deduce the geological history of our planet. A contemporary of Cuvier, though far more obscure, Johann Fischer von Waldheim is the subject of 'The Razumovsky Duet'. The linkage of a little-known German fossil collector with the subject of one of Beethoven's dedications typifies Gould's essayist style in the making of unlikely but thought-provoking connections, and his musical enthusiasms. Another side to these enthusiasms, a love of Gilbert and Sullivan, is the crabwise opening to our next choice, ostensibly a book review but one that becomes a history lesson in itself over the great Devonian controversy: a mid-nineteenth century battle over the ages and sequencing of geological strata, and a battle royal between the respectable surveyor Henry de la Beche and the aristocratic Roderick Murchison. Typically Gould uses the story to make a series of points about the nature of scientific disputes and the intersections of scientific authority, class and conviction with deep modern resonances.

A better-known nineteenth-century dispute, this time between Charles Darwin and one of his first disciples, and subsequently one of his first critics, forms the next of our choices. The post-Darwinian controversies are often presented simply as a conflict between science and religion. But as is usual the truth is more complex. St George Mivart, a zoologist and Anglican turned Catholic, was one of those posing what has always been seen as a fundamental challenge to Darwinian gradualism. A bird's wing is well adapted for flight, but what use would a partial wing be? How can one get from no-wing to working-wing by gradual evolutionary change? Darwin was an emphatic gradualist, and the dilemma presented him with great problems (he trembled, he confessed, when he thought of how such a perfect structure as the eye might evolve). Mivart's solution, which Darwin rejected, was to argue that nature makes jumps – saltations. Gould's own alternative, a process for which he coined the term exaptation, is spelled out in later selections, but in this essay he shows, with the aid of some recent experiments, just how gradualism might work in producing flying creatures from early, nonflying ancestors.

The great Charles himself appears in both our next two selections. A theorist Darwin was, but also a consummate observer and incredibly patient experimentalist. He worked for years on a study of barnacles. And visitors to Down House, in Kent, where he lived for some forty years, settling down after his epic voyage in the *Beagle* and a brief period in London, can still see in the lawn a rock, now half buried in the ground, which he used to explore how the action of worms, churning the soil and throwing up their casts, could result in the slow burying of surface objects. In this essay, 'Worm for a Century, and All Seasons', published for the centenary of Darwin's death, Gould uses the worm study to pay tribute to Darwin the experimentalist. Darwin is also a hidden presence in the essay that follows. The relationship, or non-relationship, between Darwin and Karl Marx has been the subject of much speculation, including the oft-repeated but disputed story that Marx wished to dedicate *Das Kapital* to Darwin. Gould takes a sideways look at this story in his reflection on the handful of mourners at Marx's funeral in London's Highgate Cemetery in 1883. Perhaps the most unlikely amongst them, he discovers, was the biologist and Darwinian popularizer Ray Lankester, and Steve uses this conjunction both to rescue the now largely forgotten Lankester, and to try to put the Marx–Darwin record straight.

Our final choice in this section brings into focus one of the most famous of all frauds in the history of biology, the notorious Piltdown Conspiracy. 'Piltdown Man', a supposed 'missing link' in human evolution, was discovered by the amateur archaeologist Charles Dawson in 1908 in a gravel pit in Piltdown, Sussex. The finding achieved enormous publicity and the endorsement of a number of eminent paleontologists, notably Arthur Woodward of the British Museum and the Catholic priest Teilhard de Chardin. Nearly half a century elapsed until the crude nature of the forgery, which stuck together fragments of ape and human skulls, was uncovered. But who was responsible? The mystery resonates even now. Gould points the finger at Teilhard as being primarily responsible. Teilhard himself is revered as a mystic of genius by some, but amongst most biologists is now seen as little more than a charlatan. Needless to say, Steve's essay provoked indignant responses from amongst the Teilhardians, to which he responded in a lengthy coda to the essay, a reply to his critics, which we have omitted here for reasons of space but is reprinted in the original collection, *Hen's Teeth and Horse's Toes*.

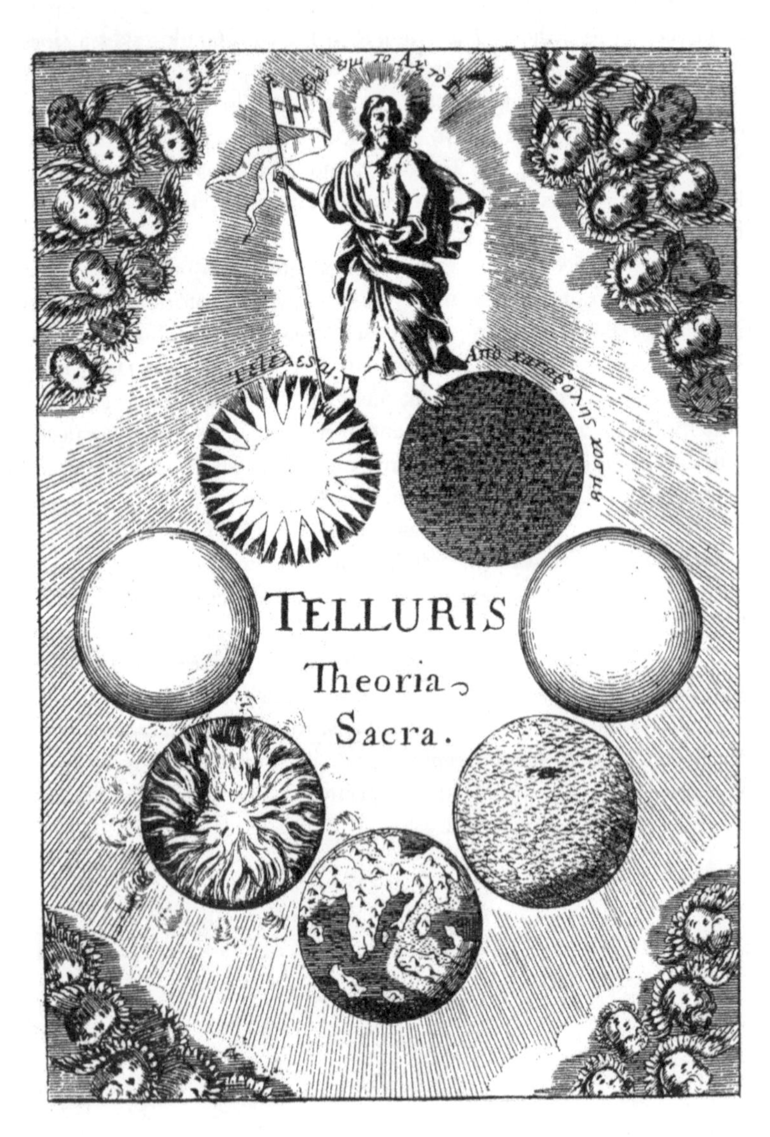

Figure 2. The frontispiece from the first edition of Thomas Burnet's *Telluris theoria sacra*, or *Sacred Theory of the Earth.*

THOMAS BURNET'S BATTLEGROUND OF TIME

Burnet's Frontispiece

THE frontispiece to Thomas Burnet's *Telluris theoria sacra* (*The Sacred Theory of the Earth*) may be the most comprehensive and accurate epitome ever presented in pictorial form – for it presents both the content of Burnet's narrative and his own internal debate about the nature of time and history (Fig. 2).

Below the requisite border of cherubim (for Burnet's baroque century), we see Jesus, standing atop a circle of globes, his left foot on the beginning, his right on the culmination of our planet's history. Above his head stands the famous statement from the Book of Revelation: I am alpha and omega (the beginning and the end, the first and the last). Following conventions of the watchmakers' guild, and of eschatology (with bad old days before salvation to the left, or sinister, side of divinity), history moves clockwise from midnight to high noon.

We see first (under Christ's left foot) the original chaotic earth 'without form and void,' a jumble of particles and darkness upon the face of the deep. Next, following the resolution of chaos into a series of smooth concentric layers, we note the perfect earth of Eden's original paradise, a smooth featureless globe. But the deluge arrives just in time to punish our sins, and the earth is next consumed by a great flood (yes, the little figure just above center is Noah's ark upon the waves). The waters retreat, leaving the cracked crust of our current earth, 'a broken and confused heap of bodies.' In times to come, as the prophets foretold, the earth shall be consumed by fire, then made smooth again as descending soot and ashes reestablish concentric perfection. Christ shall reign for a thousand years with his resurrected saints on this new globe. Finally, after a last triumphant battle against evil forces, the final judgment shall allocate all bodies to their proper places, the just shall ascend to heaven, and the earth (under Christ's right foot), no longer needed as a human abode, shall become a star.

This tale embodies time's arrow at its grandest – a comprehensive rip-roaring narrative, a distinctive sequence of stages with a definite beginning, a clear trajectory, and a particular end. Who could ask for a better story?

But Burnet's frontispiece records more than time's arrow. The globes are arranged as a circle, not a line or some other appropriate metaphor of exclusively sequential narrative – and Christ, the Word who was with God at the beginning, straddles the inception and culmination. Consider also, the careful positioning of globes, with our current earth in the center between two symmetrical flanks. Note the conscious correspondences between right and left flanks: the perfect earth following descent of the elements from chaos (at 3:00), and directly across at 9:00, the earth made perfect again after particles descend from the conflagration; or the earth *in extremis*, first by water then by fire, on either side of its current ruined state.

In other words, Burnet displays his narrative (time's arrow) in the context of time's cycle – an eternal divine presence at top, a circular arrangement of globes beginning and ending in immanence, a complex set of correspondences between our past and future.

This same picture also embodies, and equally well, the dubious reasons for Burnet's status as a primal villain of the history of geology, a symbol of the major impediment to its discovery of deep time. For we see the earth's history intimately entwined with, indeed dictated by, a strictly literal reading of the sacred text.

The Burnet of Textbooks

Burnet emerges from our textbooks as the archetype of a biblical idolatry that reined the progress of science. We may extend this tradition of commentary right back to the other two protagonists of this book – to James Hutton, who wrote of Burnet, 'This surely cannot be considered in any other light than as a dream, formed upon a poetic fiction of a golden age'; and to Charles Lyell, who remarked that 'even Milton had scarcely ventured in his poem to indulge his imagination so freely . . . as this writer, who set forth pretensions to profound philosophy'.

No one professed the empiricist faith in purer form than the leading Scottish geologist, Archibald Geikie. His *Founders of Geology* (1897) promoted the tradition of heroes as field workers, and villains as speculators. As a 'standard' history of geology for several generations, this book became the source for much continuing textbook dogma. Geikie included Burnet's book among the 'monstrous doctrines' that infested late-seventeenth-century science: 'Nowhere did speculation run so completely riot as in England with regard to theories of the origin and structure of our globe'. Geikie then presented his empiricist solution – that facts must precede theory – to this retrospective dilemma: 'It was a long time before men came to understand that any true theory of the earth must rest upon evidence furnished by the globe itself, and that no such theory could properly be framed until a large body of evidence had been gathered together'.

Horace B. Woodward, in his official history of the Geological Society of London (1911) placed Burnet's work among the 'romantic and unprofitable labors' of its time. From a peculiar source came the most interesting of all critiques. George McCready Price, grandfather and originator of the pseudoscience known to its adherents by the oxymoron 'scientific creationism,' considered Burnet a special threat to his system. Price wished to affirm biblical literalism by an inductive approach based strictly on fieldwork. On the old principle that the enemy within is more dangerous than the enemy without, Price wanted to distance himself as far as possible from men like Burnet, who told their scriptural history of the earth from their armchairs:

> Their wild fancies deserve to be called travesties alike on the Bible and on true science; and the word 'diluvial' has been a term to mock at ever since. Happy would it have been for the subsequent history of all the sciences, if the students of the rocks had all been willing patiently to investigate the records, and hold their fancies sternly in leash until they had gathered sufficient facts upon which to found a true induction or generalization.

This characterization persists into our generation. Fenton and Fenton, in their popular work *Giants of Geology* (1952) dismiss Burnet's theory as 'a series of queer ideas about earth's development,' and misread his mechanism as a series of divine interventions: 'Thomas Burnet thought an angry God had used the sun's rays as a chisel to split open the crust and let the central waters burst forth upon an unrepentant mankind.' Davies, in his excellent history of British geomorphology, states that the scriptural geologies of Burnet and others 'have always had a peculiar fascination for historians as bizarre freaks of pseudo-science.'

Science versus Religion?

The matrix that supports this canonical mischaracterization of Burnet is the supposed conflict, or war, between science and religion. Though scholars have argued *ad nauseam* that no such dichotomy existed – that the debate, if it expressed any primary division, separated traditionalists (mostly from the church) and modernists (including most scientists, but always many churchmen as well) – this appealing and simplistic notion persists.

The *locus classicus* of 'the warfare of science with theology' is the two-volume work (1896) of the same name by Andrew Dickson White, president of Cornell University. White, although personally devout, held an even stronger commitment to the first amendment and sought to establish a nondenominational university. Speaking of his work with Ezra Cornell, he wrote: 'Far from wishing to injure Christianity, we both hoped to promote it; but we did not confound religion with sectarianism'. White then presented his central thesis as a paragraph in bold italics:

> In all modern history, interference with science in the supposed interest of religion, no matter how conscientious such interference may have been, has resulted in the direst evils both to religion and to science, and invariably; and, on the other hand, all untrammelled scientific investigation, no matter how dangerous to religion some of its stages may have seemed for the time to be, has invariably resulted in the highest good both of religion and of science.

White began his book with a metaphor. As a member of the United States embassy in Russia, he watches from his room above the Neva in St Petersburg as a crowd of Russian peasants breaks the ice barrier that still dams the river as the April thaws approach. The peasants are constructing hundreds of small channels through the ice, so that the pent-up river may discharge gradually and not vent its fury in a great flood caused by sudden collapse of the entire barrier:

> The waters from thousands of swollen streamlets above are pressing behind [the ice dam]; wreckage and refuse are piling up against it; every one knows that it must yield. But there is a danger that it may . . . break suddenly, wrenching even the granite quays from their foundations, bringing desolation to a vast population . . . The patient mujiks are doing the right thing. The barrier, exposed more and more to the warmth of spring by the scores of channels they are making, will break away gradually, and the river will flow on beneficent and beautiful.

The rising waters, White tells us, represent 'the flood of increased knowledge and new thought'; the dam is dogmatic religion and unyielding convention (White then confesses the hope that his book might act as a mujik's channel to let light through gently). For if dogma stands fast, and the dam breaks (as truth cannot be forestalled forever), then the flood of goodness, by its volume alone, will overwhelm more than darkness: '. . . a sudden breaking away, distressing and calamitous, sweeping before it not only outworn creeds and noxious dogmas, but cherished principles and ideals, and even wrenching out most precious religious and moral foundations of the whole social and political fabric'.

Burnet, in White's view, was part of the dam – an example of religion's improper intrusion into scientific matters and, therefore, a danger to gentle enlightenment. This interpretation underlies the short-takes of our textbooks and classroom lectures. Modern scholars know better, but the world of textbooks is a closed club, passing its errors directly from generation to generation.

Burnet's Methodology

The Reverend Thomas Burnet was a prominent Anglican clergyman who became the private chaplain of King William III. Between 1680 and 1690, Burnet published, first in Latin then in English, the four books of *Telluris theoria sacra*, or *The Sacred Theory of the Earth: Containing an Account of the Original of the Earth, and of all the General Changes which it hath already undergone, or is to undergo Till the Consummation of all Things*. In Book I on the deluge, Book II on the preceding paradise, Book III on the forthcoming 'burning of the world,' and Book IV 'concerning the new heavens and new earth,' or paradise regained after the conflagration, Burnet told our planet's story as proclaimed by the unfailing concordance of God's words (the sacred texts) and his works (the objects of nature).

In previous hints of my affection for Burnet, I hope I did not convey the impression that I would defend him as a scientist under contemporary standards invoked by his textbook critics. In these terms, he clearly fails just as his detractors insist. *The Sacred Theory of the Earth* contains precious few appeals to empirical information. It speaks with as much confidence, and at comparable length, about an unobservable future as about a confirmable past. Its arguments cite scripture as comfortably and as often as nature. But how can we criticize Burnet for mixing science and religion when the taxonomy of his times recognized no such division and didn't even possess a word for what we now call science? Burnet, who won Newton's high praise for his treatise, was an exemplary representative of a scholarly style valued in his own day. To be sure, that style imposed stringent limits upon what we would now call empirical truth, but retrospective history, with its anachronistic standards, can only lead us to devalue (and thereby misunderstand) our predecessors – for time's arrow asserts its sway upon human history primarily through the bias of progress and leads us to view the past as ever more inadequate the further back we go.

I propose to treat Burnet with clementary respect, to take the logic of his argument seriously and at face value.* Burnet proceeded by a

* I know that motives are ever so much more complex than the logic of argument. I accept many of the arguments advanced by scholars to untangle the hidden agenda of Burnet's conclusions – that, for example, his insistence upon resurrection only *after* a future conflagration served as a weapon against religious radicals who preached an imminent end to the world. Yet I find personal merit in taking unfamiliar past arguments at face value and working through their logic and implications. These exercises have taught me more about thinking in general than any explicit treatise on principles of reasoning.

method used in our era only by Immanuel Velikovsky (among names well known). Velikovsky began his radical, and now disproven, reconstruction of cosmology and human history with a central premise that reversed our current tradition of argument: suppose, for the sake of investigation, that everything in the written documents of ancient civilizations is true. Can we then invent a physics that would yield such results?* (If Joshua said that the sun stood still upon Gibeon, then something stopped the earth's rotation – the close passage of wandering Mars or Venus in Velikovsky's reconstruction.)

Burnet began by assuming that only one document – the Bible – is unerringly true.† His treatise then becomes a search for a physics of natural causes to render these certain results of history. (Burnet, of course, differs from Velikovsky in a fundamental way. Velikovsky took the veracity of ancient texts only as a heuristic beginning. For Burnet, the necessary concordance of God's words and works established harmony between physics and scripture as necessary a priori.)

Within this constraint of concordance, Burnet followed a strategy that placed him among the rationalists ('good guys' for the future development of science, if we must follow Western-movie scenarios of retrospective history). As the centerpiece of his logic, Burnet insists again and again that the earth's scripturally specified history will be adequately explained only when we identify natural causes for the entire panoply of biblical events. Moreover, he urges, in apparent conflicts (they cannot be real) between reason and revelation, choose reason first and then untangle the true meaning of revelation:

> 'Tis a dangerous thing to engage the authority of scripture in disputes about the natural world, in opposition to reason; lest time, which brings all things to light, should discover that to be evidently false which we had made scripture to assert . . . We are not to suppose that any truth concerning the natural world can be an enemy to religion; for truth

* While all scientists now argue that the possibilities of physics set prior limits upon what claims of the ancient texts might be historically true.

† This commitment led Burnet into arguments that we, with different assumptions, might regard as the height of folly – that, for example, Noah's flood must have been truly global, not merely local, because Noah would not have built an ark, but simply fled to safety in a neighboring land, if the entire earth had not been drowned.

> cannot be an enemy to truth, God is not divided against himself.

Burnet strenuously attacked those who would take the easy road and call upon miraculous intervention whenever a difficult problem presented itself to physics – for such a strategy cancels reason as a guide and explains nothing by its effortless way of resolving everything. In rejecting a miraculous creation of extra water to solve the central problem that motivated his entire treatise – how could the earth drown in its limited supply of water? – Burnet invoked the same metaphor later used by Lyell against the catastrophists: easy and hard ways to untie the Gordian knot. 'They say in short, that God Almighty created waters on purpose to make the deluge, and then annihilated them again when the deluge was to cease; And this, in a few words, is the whole account of the business. This is to cut the knot when we cannot loose it'. Likewise, for the second greatest strain on physical credulity, a worldwide conflagration, Burnet again insists that ordinary properties of fire must do the job: 'Fire is the instrument, or the executive power, and hath no more force given it, than what it hath naturally'.

Burnet's basic position has been advanced by nearly every theistic scientist since the Newtonian revolution: God made it right the first time. He ordained the laws of nature to yield an appropriate history; he needn't intervene later to patch and fix an imperfect cosmos by miraculous alteration of his own laws. In a striking passage, Burnet invokes the standard metaphor of clockwork to illustrate this primary principle of science – the invariance of natural laws in space and time.

> We think him a better artist that makes a clock that strikes regularly at every hour from the springs and wheels which he puts in the work, than he that hath so made his clock that he must put his finger to it every hour to make it strike: and if one should contrive a piece of clock-work so that it should beat all the hours, and make all its motions regularly for such a time, and that time being come, upon a signal given, or a spring touched, it should of its own accord fall all to pieces; would not this be looked upon as a piece of greater art, than if the workman came at

> that time prefixed, and with a great hammer beat it into pieces?

Only late in the book, when he must specify the earth's future following the conflagration, does Burnet admit that reason must fail – for how can one reconstruct the details of an unobservable future? Yet he abandons reason with much tenderness and evident regret:

> Farewell then, dear friend, I must take another guide: and leave you here, as Moses upon Mount Pisgah, only to look into that land, which you cannot enter. I acknowledge the good service you have done, and what a faithful companion you have been, in a long journey: from the beginning of the world to this hour . . . We have travelled together through the dark regions of a first and second chaos: seen the world twice shipwrecked. Neither water nor fire could separate us. But now you must give place to other guides. Welcome, holy scriptures, the oracles of God, a light shining in darkness.

The Physics of History

I have already presented the content of Burnet's scenario in outline by discussing his frontispiece; but what physics did he invoke to produce such an astonishing sequence of events?

Burnet viewed the flood as central to his methodological program. The *Sacred Theory*, therefore, does not proceed chronologically, but moves from deluge to preceding paradise, for Burnet held that if he could find a rational explanation for this most cataclysmic and difficult event, his method would surely encompass all history. He tried to calculate the amount of oceanic water, grossly underestimating both the average depth (100 fathoms) and extent (half the earth's surface) of the seas.*

* Burnet, who was not the armchair speculator of legend, lamented the absence of adequate maps to make assessments and calculations for these key elements of his theory: 'To this purpose I do not doubt but that it would be of very good use to have natural maps of the earth . . . Methinks every prince should have such a draught of his own country and dominions, to see how the ground lies . . . which highest, which lowest . . . how the rivers flow, and why; how the mountains stand . . .'

Concluding that the seas could not nearly bury the continents, calculating that forty days and nights of rain would add little (and only recycle seawater in any case), and rejecting, as methodologically destructive to his rational program, the divine creation of new water, Burnet had to seek another source. He fixed upon a worldwide layer of water, underlying and concentric with the original crust of the earth's surface. The flood, he declared, occurred when this original crust cracked, permitting the thick, underlying layer of water to rise from the abyss.

This interpretation of the flood allowed Burnet to specify conditions both before and after. Nothing much has happened since the deluge, only some inconsequent erosion of postdiluvian topography. (Burnet's geology lacked a concept of repair; processes of ordinary times could only follow the dictates of Isaiah 40 and erode the mountain to fill the valleys, thus smoothing and leveling the surface.)

The earth's current surface was fashioned by the deluge. It is, in short, a gigantic ruin made of cracked fragments from the original crust. Ocean basins are holes, mountains the edges of crustal fragments broken and turned upon their side. 'Say but that they are a ruin, and you have in one word explained them all'. Burnet's descriptions and metaphors all record his view of our current earth as a remnant of destruction – a 'hideous ruin,' 'a broken and confused heap of bodies,' 'a dirty little planet.'

Burnet then proceeded backward (in Book II) to reconstruct the perfect earth before the deluge. Scripture specifies an original chaos of particles, and physics dictates their sorting as a series of concentric layers, denser at the center. (Since Burnet regarded the solid crust as a thin and light froth, denser water formed a layer beneath – and a source for the deluge.)

This perfect earth housed the original paradise of Eden. Its surface was featureless and smooth. Rivers ran from high latitudes and dissipated in the dry tropics. (They flowed, in Burnet's reversed concept of the earth's shape, because the poles stood slightly higher above the center than the equator.) A planet with such perfect radial symmetry bore no irregularity to tilt its axis. Hence the earth rotated bolt upright and Eden, located at a mid-latitude, enjoyed perpetual spring. The salubrious conditions of this earthly paradise nurtured the early patriarchal life-spans of more than nine hundred years. But the deluge was truly paradise lost. The earth, made asymmetric, tilted to its present angle of some twenty degrees. An unhealthy change of seasons commenced, and life-spans declined to the currently prescribed three score years and ten.

If this reconstruction strikes modern readers as fanciful and scripture-burdened (I do not deny it, but only urge different criteria of judgment), recall Burnet's commitment to a rational explanation based on natural laws. We might contrast Burnet's account of the change in axial tilt with the words of a celebrated near-contemporary who did not shrink from attributing the work directly to angels:

> Some say he bid his angels turn askance
> The poles of Earth twice ten degrees and more
> From the sun's axle; they with labor pushed
> Oblique the centric globe: some say the sun
> Was bid turn reins from the equinoctial road
> Like distant breadth to Taurus with the seven
> Atlantic Sisters, and the Spartan Twins,
> Up to the Tropic Crab; thence down amain
> By Leo and the Virgin and the Scales,
> As deep as Capricorn, to bring in change
> Of seasons to each clime; else had the spring
> Perpetual smiled on Earth with vernant flowers,
> Equal in days and nights, . . .
> The sun, as from Thyestean banquet, turned
> His course intended; else how had the World
> Inhabited, though sinless, more than now
> Avoided pinching cold and scorching heat?
>
> Milton, *Paradise Lost*

In Book III Burnet presents a set of arguments, guided more by scripture since physics treats pasts more securely than futures, for a coming worldwide conflagration that shall completely consume the upper layers of the earth and remobilize all resulting particles into a new chaos. Burnet continues to demand a rational, physical explanation. In successive chapters, he discusses how such a wet and rocky mass can burn (the waters will first evaporate in a major drought), how the jumbled surface of our ruined earth will abet the flames by including so much nurturing air from internal vacuities), and where the fire will start. The torches of Vesuvius and Aetna specify an Italian origin, and God also knows the home of antichrist, the Bishop of Rome (Burnet was nothing if not a committed Anglican). Still, in an ecumenical spirit, Burnet tells us that Britain, with its deposits of coal, will burn brightly, if a bit later.

If the conflagration replays the flood in a different manner, the succeeding earth also reproduces the original paradise, and for the same physical reason: the descent of burned particles into concentric layers sorted by density. On this earth made perfect again, Christ shall reign for a thousand years with Satan bound in chains. Following this millennium, Gog and Magog shall herald the final battle of good against evil; trumpets shall announce the last judgment; the saints shall ascend (the sinners go elsewhere); and the superseded earth shall become a star.

Burnet's commitment to explanation by natural law, and his allegiance to historical narrative, can be best appreciated in explicit contrast with alternatives proposed by his friend Isaac Newton in a fascinating exchange of long letters (thank goodness they didn't have a telephone, or even a train) between London and Cambridge, in January 1681.*

Newton had made two suggestions that troubled Burnet: first, that the earth's current topography arose during its initial formation from primeval chaos, and was not sculpted by Noah's flood; second, that the paradox of creation in but six days might be resolved by arguing that the earth rotated much more slowly then, producing a 'day' of enormous extent. Burnet excused his long and impassioned response, stating (in modern orthography):

> Your kindness hath brought upon you the trouble of this long letter, which I could not avoid seeing you have insisted upon two such material points, the possibility (as you suppose) of forming the earth as it now is, immediately from the Chaos or without a dissolution [as the flood had produced in Burnet's scheme]; and the necessity of adhering to Moses' Hexameron as a physical description [Burnet actually wrote 'Moses his Hexameron' in that delightful construction used before English codified the current form of the genitive]; to show the contrary to these two hath swelled my letter too much.

* I thank Rhoda Rappoport for sending me these documents.

Burnet objected to Newton's first proposal because it removed the role of extended history by forming all the earth's essential features at the outset – see quote above with its plea for the flood as an agent of topography. Burnet then rejected Newton's long initial days because he suspected that a later acceleration in rotation would require supernatural intervention. (Burnet favored an allegorical interpretation of Genesis I, arguing that the notion of a 'day' could not be defined before the sun's creation on the fourth day.) 'But if the revolutions of the earth were thus slow at first, how came they to be swifter? From natural causes or supernatural?' Burnet also objected that long early days would stretch the lives of patriarchs even beyond the already problematical 969 of Methuselah and his compatriots – and that while organisms might enjoy sunny days of such extended length, the long nights might become unbearable: 'If the day was thus long what a doleful night would there be'.

Newton's response confirms Burnet's reading of their differences. Newton argues that a separation of parts from original chaos might produce irregular topography, not the smooth concentric layers of Burnet's system – therefore requiring no subsequent narrative to explain the current face of our earth: 'Moses teaches a subdivision . . . of the miry waters under the firmament into clear water and dry land on the surface of the whole globous mass, for which separation nothing more was requisite than that the water should be drained from the higher parts of the limus to leave them dry and gather together into the lower to compose seas. And some parts might be made higher than others'. As for an early speeding of rotation, Newton confirmed Burnet's fears by allowing a direct supernatural boost: 'Where natural causes are at hand God uses them as instruments in his works, but I do not think them alone sufficient for the creation and therefore may be allowed to suppose that amongst other things God gave the earth its motion by such degrees and at such times as was most suitable to the creatures'. Newton also disregarded the problem of 'doleful nights' for the earth's first inhabitants, arguing: 'And why might not birds and fishes endure one long night as well as those and other animals endure many in Greenland'.

Burnet therefore emerges from this correspondence with the greatest of all scientific heroes as more committed to the reign of natural law, and more willing to embrace historical explanations. He ends his letter to Newton by describing a singular event of time's arrow, the great comet of 1680 then hanging over the skies of London. 'Sir we are all so busy in gazing upon the comet, and what do you say at Cambridge can be the cause of such a prodigious coma as it had'. Mr Halley, mutual friend of Newton and Burnet, also gazed in awe at this comet. Two years later, still inspired by this spectacular sight, he observed a smaller comet, and eventually predicted its return on a seventy-six-year cycle. This smaller object, Halley's Comet, now resides in my sky as I write this chapter – a primary signal of time's cycle.

THE LYING STONES OF MARRAKECH

We tend to think of fakery as an activity dedicated to minor moments of forgivable fun (from the whoopie cushion to the squirting lapel flower), or harmless embellishment (from my grandfather's vivid eyewitness tales of the Dempsey–Firpo fight he never attended, to the 250,000 people who swear they were there when Bobby Thomson hit his home run in a stadium with a maximal capacity of some fifty thousand).

But fakery can also become a serious and truly tragic business, warping (or even destroying) the lives of thousands, and misdirecting entire professions into sterility for generations. Scoundrels may find the matrix of temptation irresistible, for immediate gains in money and power can be so great, while human gullibility grants the skillful forger an apparently limitless field of operation. The Van Gogh *Sunflowers*, bought in 1987 by a Japanese insurance company for nearly 25 million pounds sterling – then a record price for a painting – may well be a forged copy made around 1900 by the stockbroker and artist manqué Emile Schuffenecker. The phony Piltdown Man, artlessly confected from the jaw of an orangutan and a modern human cranium, derailed the profession of paleoanthropology for forty years, until exposed as a fake in the early 1950s.

Earlier examples cast an even longer and broader net of disappointment. A large body of medieval and Renaissance scholarship depended upon the documents of Hermes Trismegistus (Thrice-Great Hermes), a body of work attributed to Thoth, the Egyptian god of wisdom, and once viewed as equal in insight (not to mention antiquity) to biblical and classical sources – until exposed as a set of forgeries compiled largely in the third century A.D. And how can we possibly measure the pain of so many thousands of pious Jews, who abandoned their possessions and towns to follow the false messiah Shabbetai Tzevi to Jerusalem in the apocalyptic year of 1666 – only to learn that their leader, imprisoned by the sultan and threatened with torture, had converted to Islam, been renamed Mehmed Efendi, and made the sultan's personal doorkeeper.

The most famous story of fraud in my own field of paleontology may not qualify for this first rank in the genre, but has surely won both general fame and staying power by persistence for more than 250 years. Like all great legends, this story has a canonical form, replete with conventional moral messages, and told without any variation in content across the centuries. Moreover, this standard form bears little relationship to the actual course of events as best reconstructed from available evidence. Finally, to cite the third common property of such legends, a correction of the conventional tale wins added and general value in teaching us important lessons about how we use and abuse our own history. Thus, the old story merits yet another retelling – which I first provide in the canonical (and false) version known to so many generations of students (and no doubt remembered by many readers from their college courses in natural science).

In 1726, Dr Johann Bartholomew Adam Beringer, an insufferably pompous and dilettantish professor and physician from the town of Würzburg, published a volume, the *Lithographiae Wirceburgensis* (Würzburg lithography), documenting in copious words and twenty-one plates a remarkable series of fossils that he had found on a mountain adjacent to the city. These fossils portrayed a large array of objects, all neatly exposed in three-dimensional relief on the surface of flattened stones. The great majority depicted organisms, nearly all complete, including remarkable features of behaviour and soft anatomy that had never been noted in conventional fossils – lizards in their skin, birds complete with beaks and eyes, spiders with their webs, bees feeding on flowers, snails next to their eggs, and frogs copulating. But others showed heavenly objects – comets with tails, the crescent moon with rays, and the sun all effulgent with a glowing central face of human form. Still others depicted Hebrew letters, nearly all spelling out the tetragrammaton, the ineffable name of God – YHVH, usually transliterated by Christian Europe as 'Jehovah.'

Beringer did recognize the difference between his stones and conventional fossils, and he didn't state a dogmatic opinion about their nature. But he didn't doubt their authenticity either, and he did dismiss claims that they had been carved by human hands, either recently in an attempt to defraud, or long ago for pagan purposes. Alas, after publishing his book and trumpeting the contents, Beringer realized that he had indeed been duped, presumably by his students playing a prank. (Some sources say that he finally acknowledged the trickery when he found his own name written in Hebrew letters on one stone.) According to legend, the brokenhearted Beringer then impoverished himself by attempting to buy back all copies of his book – and died dispirited just a few years later. Beringer's false fossils have been known ever since as *Lügensteine*, or 'lying stones.'

To illustrate the pedigree of the canonical tale, I cite the version given in the most famous paleontological treatise of the early nineteenth century, Dr James Parkinson's *Organic Remains of a Former World* (volume 1, 1804). Parkinson, a physician by training and a fine paleontologist by avocation, identified and gave his name to the degenerative disease that continues to puzzle and trouble us today. He wrote of his colleague Beringer:

> One work, published in 1726, deserves to be particularly noticed; since it plainly demonstrates, that learning may not be sufficient to prevent an unsuspecting man, from becoming the dupe of excessive credulity. It is worthy of being mentioned on another account: the quantity of censure and ridicule, to which its author was exposed, served, not only to render his cotemporaries [*sic*] less liable to imposition; but also more cautious in indulging in unsupported hypotheses. . . . We are here presented with the representation of stones said to bear petrifactions of birds; some with spread, others with closed, wings: bees and wasps, both resting in their curiously constructed cells, and in the act of sipping honey from expanded flowers . . . and, to complete the absurdity, petrifactions representing the sun, moon, stars, and comets: with many others too monstrous and ridiculous to deserve even mention. These stones, artfully prepared, had been intentionally deposited in a mountain, which he was in the habit of exploring, purposely to dupe the enthusiastic collector. Unfortunately, the silly and cruel trick, succeeded in so far,

> as to occasion to him, who was the subject of it, so great a degree of mortification, as, it is said, shortened his days.

All components of the standard story line, complete with moral messages, have already fallen into place – the absurdity of the fossils, the gullibility of the professor, the personal tragedy of his undoing, and the two attendant lessons for aspiring young scientists: do not engage in speculation beyond available evidence, and do not stray from the empirical method of direct observation.

In this century's earlier and standard work on the history of geology (*The Birth and Development of the Geological Sciences*, published in 1934), Frank Dawson Adams provides some embellishments that had accumulated over the years, including the unforgettable story, for which not a shred of evidence has ever existed, that Beringer capitulated when he found his own name in Hebrew letters on one of his stones. Adam's verbatim 'borrowing' of Parkinson's last line also illustrates another reason for invariance of the canonical tale: later retellings copy their material from earlier sources:

> Some sons of Belial among his students prepared a number of artificial fossils by moulding forms of various living or imaginary things in clay which was then baked hard and scattered in fragments about on the hillsides where Beringer was wont to search for fossils. . . . The distressing climax was reached, however, when later he one day found a fragment bearing his own name upon it. So great was his chagrin and mortification in discovering that he had been made the subject of a cruel and silly hoax, that he endeavored to buy up the whole edition of his work. In doing so he impoverished himself and it is said shortened his days.

Modern textbooks tend to present a caricatured 'triumphalist' account in their 'obligatory' introductory pages on the history of their discipline – the view that science marches inexorably forward from dark superstition toward the refining light of truth. Thus, Beringer's story tends to acquire the additional moral that his undoing at least had the good effect of destroying old nonsense about the inorganic or mysterious origin of fossils – as in this text for first-year students, published in 1961:

> The idea that fossils were merely sports of nature was finally killed by ridicule in the early part of the eighteenth century. Johann Beringer, a professor at the University of Würzburg, enthusiastically argued against the organic nature of fossils. In 1726, he published a paleontological work . . . which included drawings of many true fossils but also of objects that represented the sun, the moon, stars, and Hebraic letters. It was not till later, when Beringer found a 'fossil' with his own name on it, that he realized that his students, tired of his teachings, had planted these 'fossils' and carefully led him to discover them for himself.

A recent trip to Morocco turned my thoughts to Beringer. For several years, I have watched, with increasing fascination and puzzlement, the virtual 'takeover' of rock shops throughout the world by striking fossils from Morocco – primarily straight-shelled nautiloids (much older relatives of the coiled and modern chambered nautilus) preserved in black marbles and limestones, and usually sold as large, beautifully polished slabs intended for table or dresser tops. I wondered where these rocks occurred in such fantastic abundance; had the High Atlas Mountains been quarried away to sea level? I wanted to make sure that Morocco itself still existed as a discrete entity and not only as disaggregated fragments, fashioning the world's coffee tables.

I discovered that most of these fossils come from quarries in the rocky deserts, well and due east of Marrakech, and not from the intervening mountains. I also learned something else that alleviated my fears about imminent dispersal of an entire patrimony. Moroccan rock salesmen dot the landscape in limitless variety – from young boys hawking a specimen or two at every hairpin turn on the mountain roads, to impromptu stands at every lookout point, to large and formal shops in the cities and towns. The aggregate volume of rock must be immense, but the majority of items offered for sale are either entirely phony or at least strongly 'enhanced.' My focus of interest shifted dramatically: from worrying about sources and limits to studying the ranges and differential expertises of a major industry dedicated to the manufacture of fake fossils.

I must judge some 'enhancements' as quite clever – as when the strong ribs on the shell of a genuine ammonite are extended by carving into the smallest and innermost whorls and then 'improved' in regular expression on the outer coil. But other 'ammonites' have simply been carved from scratch on a smoothed rock surface, or even cast in clay and then glued into a prepared hole in the rock. Other fakes can only be deemed absurd – as in my favorite example of a wormlike 'thing' with circles on its back, grooves on both sides, eyes on a head shield, and a double projection, like a snake's forked tongue, extending out in front. (In this case the forger, too clever by half, at least recognized the correct principle of parts and counterparts – for the 'complete' specimen includes two pieces that fit together, the projecting 'fossil' on one slab, and the negative impression on the other, where the animal supposedly cast its form into the surrounding sediment. The forger even carved negative circles and grooves into the counterpart image, although these impressions do not match the projecting, and supposedly corresponding, embellishments on the 'fossil' itself.)

But one style of fakery emerges as a kind of 'industry standard,' as defined by constant repetition and presence in all shops. (Whatever the unique and personal items offered for sale in any shop, this *vin ordinaire* of the genre always appears in abundance.) These 'standards' feature small (up to four or six inches in length) flattened stones with a prominent creature spread out in three dimensions on the surface. The fossils span a full range from plausible 'trilobites,' to arthropods (crabs, lobsters, and scorpions, for example) with external hard parts that might conceivably fossilize (though never in such complete exactitude), to small vertebrates (mostly frogs and lizards) with a soft exterior, including such delicate features as fingers and eyes that cannot be preserved in the geological record.

After much scrutiny, I finally worked out the usual mode of manufacture. The fossil fakes are plaster casts, often remarkably well done. The forger cuts a flat surface on a real rock and then cements the plaster cast to this substrate. (If you look carefully from the side, you can always make out the junction of rock and plaster.) Some fakes have been crudely confected, but the best examples match the color and form of rock to overlying plaster so cleverly that distinctions become nearly invisible.

When I first set eyes on these fakes. I experienced the weirdest sense of déjà vu, an odd juxtaposition of old and new that sent shivers of fascination and discomfort up and down my spine – a feeling greatly enhanced by a day just spent in the medina of Fez, the ancient walled town that has scarcely been altered by a millennium of surrounding change, where only mules and donkeys carry the goods of commerce, and where high walls, labyrinthine streets, tiny open shops, and calls to prayer, enhanced during the fast of Ramadan, mark a world seemingly untouched by time, and conjuring up every stereotype held by an uninformed West about a 'mysterious East.' I looked at these standard fakes, and I saw Beringer's *Lügensteine* of 1726. The two styles are so uncannily similar that I wondered, at first, if the modern forgers had explicity copied the plates of the *Lithographiae Wirceburgensis* – a silly idea that I dropped as soon as I returned and consulted my copy of Beringer's original. But the similarities remain overwhelming. I purchased two examples – a scorpion of sorts and a lizard – as virtual dead ringers for Beringer's *Lügensteine*, two sets of fakes separated by 250 years and a different process of manufacture (carved in Germany, cast in Morocco). I only wonder if the proprietor believed my assurances, rendered in my best commercial French, that I was a professional paleontologist, and that his wares were *faux, absolument et sans doute* – or if he thought that I had just devised a bargaining tactic more clever than most.

But an odd similarity across disparate cultures and centuries doesn't provide a rich enough theme for an essay. I extracted sufficient generality only when I realized that this maximal likeness in appearance correlates with a difference in meaning that couldn't be more profound. A primary strategy of the experimental method in science works by a principle known since Roman times as *ceteris paribus* ('all other things being equal') – that is, if you wish to understand a controlling difference between two systems, keep all other features constant, for the difference may then be attributed to the only factor that you have allowed to vary. If, for example, you wish to test the effect of a new diet pill, try to establish two matched groups – folks of the same age, sex, weight, nutrition, health, habits, ethnicity, and so on. Then give the pill to one group and a placebo to the other (without telling the subjects what they have received, for such knowledge would, in itself, establish inequality based on differing psychological expectations). The technique, needless to say, does not work perfectly (for true *ceteris paribus* can never be obtained), but if the pill group loses a lot of weight, and the placebo group remains as obese as before, you may conclude that the pill probably works as hoped.

Ceteris paribus represents a far more distant pipe dream in trying to understand two different contexts in the developing history of a profession – for we cannot now manipulate a situation of our own design, but must study past circumstances in complex cultures not subject to regulation by our experimental ideals at all. But any constancy between the two contexts increases our hope of illustrating and understanding their variations in the following special way: if we examine the different treatment of the same object in two cultures, worlds apart, then at least we can attribute the observed variation to cultural distinctions, for the objects treated do not vary.

The effectively identical *Lügensteine* of early-eighteenth-century Würzburg and modern Marrakech embody such an interesting difference in proposed meaning and effective treatment by two cultures – and I am not sure that we should be happy about the contrast of then and now. But we must first correct the legend of Beringer and the original *Lügensteine* if we wish to grasp the essential difference.

As so often happens when canonical legends arise to impart moral lessons to later generations, the standard tale distorts nearly every important detail of Beringer's sad story. (I obtained my information primarily from an excellent book published in 1963 by Melvin E. Jahn and Daniel J. Woolf, *The Lying Stones of Dr Beringer*, University of California Press. Jahn and Woolf provide a complete translation of Beringer's volume, along with extensive commentary about the paleontology of Beringer's time. I used original sources from my own library for all quotations not from Beringer in this essay.)

First of all, on personal issues not directly relevant to the theme of this essay, Beringer wasn't tricked by a harmless student prank but rather purposely defrauded by two colleagues who hated his dismissive pomposity and wished to bring him down. These colleagues – J. Ignatz Roderick, professor of geography and algebra at the University of Würzburg, and Georg von Eckhart, librarian to the court and the university – 'commissioned' the fake fossils (or, in Roderick's case, probably did much of the carving himself), and then hired a seventeen-year-old boy, Christian Zänger (who may also have helped with the carving), to plant them on the mountain. Zänger, a double agent of sorts, was then hired by Beringer (along with two other boys, both apparently innocent of the fraud) to excavate and collect the stones.

This information for revising the canonical tale lay hidden for two hundred years in the incomplete and somewhat contradictory records of hearings held in April 1726 before the Würzburg cathedral chapter and the city hall of Eivelstadt (the site of Beringer's mountain just outside Würzburg). The German scholar Heinrich Kirchner discovered these documents in 1934 in the town archives of Würzburg. These hearings focus on testimony of the three boys. Zänger, the 'double agent,' states that Roderick had devised the scheme because he 'wished to accuse Dr Beringer . . . because Beringer was so arrogant and despised them all.' I was also impressed by the testimony of the two brothers hired by Beringer. Their innocence seems clear in the wonderfully ingenuous statement of Nicklaus Hahn that if he and his brother 'could make such stones, they wouldn't be mere diggers.'

The canonical tale may require Beringer's ruin to convey a desired moral, but the facts argue differently. I do not doubt that the doctor was painfully embarrassed, even mortified, by his exposed gullibility; but he evidently recovered, kept his job and titles, lived for another fourteen years, and published several more books (including, though probably not by his design or will, a posthumous second edition of his *Würzburg Lithography*!). Eckhart and Roderick, on the other hand, fell into well-earned disgrace. Eckhart died soon thereafter, and Roderick, having left Würzburg (voluntarily or not, we do not know), then wrote a humbling letter to the prince-bishop begging permission to return – which his grace allowed after due rebuke for Roderick's past deeds – and to regain access to the library and archives so that he could write a proper obituary for his deceased friend Eckhart.

But on the far more important intellectual theme of Beringer's significance in the history of paleontology, a different kind of correction inverts the conventional story in a particularly meaningful way. The usual carboard tale of progressive science triumphant over past ignorance requires that benighted 'bad guys,' who upheld the old ways of theological superstition against objective evidence of observational science, be branded as both foolish and stubbornly unwilling to face nature's factuality. Since Beringer falls into this category of old and bad, we want to see him as hopelessly duped by preposterous fakes that any good observer should have recognized – hence the emphasis, in the canonical story, on Beringer's mortification and on the ridiculous character of the *Lügensteine* themselves.

The Würzburg carvings are, of course, absurd by modern definitions and understanding of fossils. We know that spiders' webs and lizards' eyes – not to mention solar rays and the Hebrew name of God – cannot fossilize, so the *Lügensteine* can only be human carvings. We laugh at Beringer for not making an identification that seems so obvious to us. But in so doing, we commit the greatest of all historical errors: arrogantly judging our forebears in the light of modern knowledge perforce unavailable to them. Of course the *Lügensteine* are preposterous, once we recognize fossils as preserved remains of ancient organisms. By this criterion, letters and solar emanations cannot be real fossils, and anyone who unites such objects with plausible images of organisms can only be a fool.

But when we enter Beringer's early-eighteenth-century world of geological understanding, his interpretations no longer seem so absurd. First of all, Beringer was puzzled by the unique character of his *Lügensteine*, and he adopted no dogmatic position about their meaning. He did regard them as natural and not carved (a portentous error, of course), but he demurred on further judgment and repeatedly stated that he had chosen to publish in order to provide information so that others might better debate the nature of fossils – a tactic that scientists supposedly value. We may regard the closing words of his penultimate chapter as a tad grandiose and self-serving, but shall we not praise the sentiment of openness?

> I have willingly submitted my plates to the scrutiny of wise men, desiring to learn their verdict, rather than to proclaim my own in this totally new and much mooted question. I address myself to scholars, hoping to be instructed by their most learned responses. . . . It is my fervent expectation that illustrious lithographers will shed light upon this dispute which is as obscure as it is unusual. I shall add thereto my own humble torch, nor shall I spare any effort to reveal and declare whatever future yields may rise from the Würzburg field under the continuous labors of my workers, and whatever opinion my mind may embrace.

More importantly, Beringer's hoaxers had not crafted preposterous objects but had cleverly contrived – for their purposes, remember, were venomous, not humorous – a fraud that might fool a man of decent will and reasonable intelligence by standards of interpretation then current. Beringer wrote his treatise at the tail end of a debate that had engulfed seventeenth-century science and had not yet been fully resolved:

what did fossils represent, and what did they teach us about the age of the earth, the nature of our planet's history, and the meaning and definition of life?

Beringer regarded the *Lügensteine* as 'natural' but not necessarily as organic in origin. In the great debate that he knew and documented so well, many scientists viewed fossils as inorganic products of the mineral realm that somehow mimicked the forms of organisms but might also take the shapes of other objects, including planets and letters. Therefore, in Beringer's world, the *Lügensteine* could not be dismissed as preposterous prima facie. This debate could not have engaged broader or more crucial issues for the developing sciences of geology and biology – for if fossils represent the remains of organisms, then the earth must be ancient, life must enjoy a long history of consistent change, and rocks must form from the deposition and hardening of sediments. But if fossils can originate as inorganic results of a 'plastic power' in the mineral kingdom (that can fashion other interesting shapes like crystals, stalactites, and banded agates in different circumstances), then the earth may be young and virtually unchanged (except for the ravages of Noah's flood), while rocks, with their enclosed fossils, may be products of the original creation, not historical results of altered sediments.

If pictures of planets and Hebrew letters could be 'fossils' made in the same way as apparent organisms, then the inorganic theory gains strong support – for a fossilized aleph or moonbeam could not be construed as a natural object deposited in a streambed and then fossilized when the surrounding sediment became buried and petrified. The inorganic theory had been fading rapidly in Beringer's time, while the organic alternative gained continually in support. But the inorganic view remained plausible, and the *Lügensteine* therefore become clever and diabolical, not preposterous and comical.

In Beringer's day, many scientists believed that simple organisms arose continually by spontaneous generation. If a polyp can originate by the influence of sunshine upon waters, or a maggot by heat upon decaying flesh, why not conjecture that simple images of objects might form upon rocks by natural interactions of light or heat upon the inherent 'lapidifying forces' of the mineral kingdom? Consider, moreover, how puzzling the image of a fish *inside* a rock must have appeared to people who viewed these rocks as products of an original creation, not as historical outcomes of sedimentation. How could an organism get inside; and how could fossils be organisms if they frequently occur petrified, or made of the same stone as their surroundings? We now have simple and 'obvious' answers to these questions, but Beringer and his colleagues still struggled – and any sympathetic understanding of early-eighteenth-century contexts should help us to grasp the centrality and excitement of these debates and to understand the *Lügensteine* as legitimately puzzling.

I do not, however, wish to absolve Beringer of all blame under an indefensibly pluralistic doctrine that all plausible explanations of past times may claim the same weight of judicious argument. The *Lügensteine* may not have been absurd, but Beringer had also encountered enough clues to uncover the hoax and avoid embarrassment. However, for several reasons involving flaws in character and passable intelligence short of true brilliance, Beringer forged on, finally trumping his judgment by his desire to be recognized and honored for a great discovery that had consumed so much of his time and expense. How could he relinquish the fame he could almost taste in writing:

> Behold these tablets, which I was inspired to edit, not only by my tireless zeal for public service, and by your wishes and those of my many friends, and by my strong filial love for Franconia, to which, from these figured fruits of this previously obscure mountain, no less glory will accrue than from the delicious wines of its vine-covered hills.

I am no fan of Dr Beringer. He strikes me, first of all, as an insufferable pedant – so I can understand his colleagues' frustration, while not condoning their solutions. (I pride myself on always quoting from original sources, and I do own a copy of Beringer's treatise. I am no Latin scholar, but I can read and translate most works in this universal scientific language of Beringer's time. But I cannot make head or tail of the convoluted phrasings, the invented words, the absurdly twisted sentences of Beringer's prose, and I have had to rely on Jahn and Woolf's translation.)

Moreover, Beringer saw and reported more than enough evidence to uncover the hoax, had he been inclined to greater judiciousness. He noted that his *Lügensteine* bore no relationship to any other objects known to the burgeoning science of paleontology, not even to the numerous 'real' fossils also found on his mountain. But instead of alerting him to possible fraud, these differences only fueled Beringer's hopes for fame. He made many observations that should have clued him in (even by standards of his own time) to the artificial carving of his fossils: why were they nearly always complete, and not usually fragmentary like most other finds; why did each object seem to fit so snugly and firmly on the enclosing rock; why did only the top sides protrude, while the lower parts merged with the underlying rock; why had letters and sunbeams not been found before; why did nearly all fossils appear in the same orientation, splayed out and viewed from the top, never from the side or bottom? Beringer's own words almost shout out the obvious and correct conclusion that he couldn't abide or even discern: 'The figures expressed on these stones, especially those of insects, are so exactly fitted to the dimensions of the stones, that one would swear that they are the work of a very meticulous sculptor.'

Beringer's arrogance brought him down in a much more direct manner as well. When Eckhart and Roderick learned that Beringer planned to publish his work, they realized that they had gone too far and became frightened. They tried to warn Beringer, by hints at first but later quite directly as their anxiety increased. Roderick even delivered some stones to Beringer and later showed his rival how they had been carved – hoping that Beringer would then draw an obvious inference about the rest of his identical collection.

Beringer, however, was now committed and would not be derailed. He replied with the argument of all true believers, the unshakable faith that resists all reason and evidence: yes, you have proven that *these* psychics are frauds, but *my* psychics are the real McCoy, and I must defend them even more strongly now that you have heaped unfair calumnies upon the entire enterprise. Beringer never mentions Eckhart and Roderick by name (so their unveiling awaited the 1934 discovery in the Würzburg town archives), but he had been forewarned of their activities. Beringer wrote in chapter 12 of his book:

> Then, when I had all but completed my work, I caught the rumor circulating throughout the city . . . that every one of these stones . . . was recently sculpted by hand, made to look as though at different periods they had been resurrected from a very old burial, and sold to me as to one indifferent to fraud and caught up in the blind greed of curiosity.

Beringer then tells the tale of Roderick's warning but excoriates his rival as an oafish modern caricature of Praxiteles (the preeminent Greek sculptor), out to discredit a great discovery by artificial mimicry:

> Our Praxiteles has issued, in an arrogant letter, a declaration of war. He has threatened to write a small treatise exposing my stones as supposititious [*sic*] – I should say, his stones, fashioned and fraudulently made by his hand. Thus does this man, virtually unknown among men of letters, still but a novice in the sciences, make a bid for the dawn of his fame in a shameful calumny and imposture.

If only Beringer had realized how truly and comprehensively he had spoken about 'a shameful calumny and imposture.' But Roderick succeeded because he had made his carvings sufficiently plausible to inspire belief by early-eighteenth-century standards. The undoing of all protagonists then followed because Beringer, in his overweening and stubborn arrogance, simply could not quench his ambition once a clever and plausible hoax had unleashed his ardor and vanity.

In summary, the *Lügensteine* of Würzburg played a notable role in the most important debate ever pursued in paleontology – a struggle that lasted for centuries and that placed the nature of reality itself up for grabs. By Beringer's time, this debate had largely been settled in favor of the organic nature of fossils, and this resolution would have occurred even if Beringer had never been born and the *Lügensteine* never carved. Beringer may have been a vain and arrogant man of limited talent, working in an academic backwater of his day, but at least he struggled with grand issues – and he fell because his hoaxers understood the great stakes and fashioned frauds that could be viewed as cleverly relevant to this intellectual battle, however preposterous they appear to us today with our additional knowledge and radically altered theories about the nature of reality and causation.

(One often needs a proper theory to set a context for the exposure of fraud. Piltdown Man fooled some of the world's best scientists for generations. I will never forget what W. E. Le Gros Clark, one of the three scientists who exposed the fraud in the early 1950s, said to me when I asked him why the hoax had stood for forty years. Even an amateur in vertebrate anatomy – as this snail man can attest from personal experience – now has no trouble seeing the Piltdown bones for what they are. The staining is so crude, and the recent file marks on the orangutan teeth in the lower jaw so obvious – yet so necessary to make them look human in the forgers' plan, for the cusps of ape and human teeth differ so greatly. Le Gros Clark said to me: 'One needed to approach the bones with the hypothesis of fraud already in mind. In such a context, the fakery immediately became obvious.')

The *Lügensteine* of Marrakech are, by contrast – and I don't know how else to say this – merely ludicrous and preposterous. No excuse save ignorance – and I do, of course, recognize the continued prevalence of this all-too-human trait – could possibly inspire a belief that the plaster blobs atop the Moroccan stones might be true fossils, the remains of ancient organisms. Beringer was grandly tricked in the pursuit of great truths, however inadequate his own skills. We are merely hoodwinked for a few dollars that mean little to most tourists but may make or break the lives of local carvers. *Caveat emptor.*

In contrasting the conflicting meanings of these identical fakes in such radically different historical contexts, I can only recall Karl Marx's famous opening line to *The Eighteenth Brumaire of Louis Napoleon*, his incisive essay on the rise to power of the vain and cynical Napoleon III after the revolution of 1848, in contrast with the elevated hopes and disappointments inspired by the original Napoleon. (The French revolutionary calendar had renamed the months and started time again at the establishment of the Republic. In this system, Napoleon's coup d'état occurred on the eighteenth of Brumaire, a foggy month in a renamed autumn, of year VIII – or November 9, 1799. Marx, now justly out of fashion for horrors later committed in his name, remains a brilliant analyst of historical patterns.) Marx opened his polemical treatise by noting that all great events in history occur twice – the first time as tragedy, and the second as farce.

Beringer was a pompous ass, and his florid and convoluted phrases represent a caricature of true scholarship. Still, he fell in the course of a great debate, using his limited talents to defend an inquiry that he loved and that even more pompous fools of his time despised – those who argued that refined people wouldn't dirty their hands in the muck of mountains but would solve the world's pressing issues under their wigs in their drawing rooms. Beringer characterized this opposition from the pseudo-elegant glitterati of his day:

> They pursue [paleontology] with an especially censorious rod, and condemn it to rejection from the world of erudition as one of the wanton futilities of intellectual idlers. To what purpose, they ask, do we stare fixedly with eye and mind at small stones and figured rocks, at little images of animals or plants, the rubbish of mountain and stream, found by chance amid the muck and sand of land and sea?

He then defended his profession with the greatest of geological metaphors:

> any [paleontologist], like David of old, would be able with one flawless stone picked from the bosom of Nature, to prostrate, by one blow on the forehead, the gigantic mass of objections and satires and to vindicate the honor of this sublime science from all its calumniators.

Beringer, to his misfortune and largely as a result of his own limitations, did not pick a 'flawless stone,' but he properly defended the importance of paleontology and of empirical science in general. As a final irony, Beringer could not have been more wrong about the *Lügensteine*, but he couldn't have been more right about the power of paleontology. Science has so revolutionized our view of reality since 1726 that we, in our current style of arrogance, can only regard the Würzburg *Lügensteine* as preposterous, because we unfairly impose our modern context and fail to understand Beringer's world, including the deep issues that made his hoaxing a tragedy rather than a farce.

Our current reality features an unslayable Goliath of commercialism, and modern scientific Davids must make an honorable peace, for a slingshot cannot win this battle. I may be terribly old-fashioned (shades, I hope not, of poor Beringer) – but I continue to believe that such honor can only be sought in separation and mutual respect. Opportunities for increasing fusion with the world of commerce surround us with almost overwhelming temptation, for the immediate and palpable 'rewards' are so great. So scientists go to work for competing pharmaceutical or computer companies, make monumental salaries, but cannot choose their topics of research or publish their work. And museums expand their gift shops to the size of their neglected exhibit halls, and purvey their dinosaurs largely for dollars in the form of images on coffee mugs and T-shirts, or by special exhibits, at fancy prices, of robotic models, built by commercial companies, hired for the show, and featuring, as their come-on, the very properties – mostly hideous growls and lurid colors – that leave no evidence in the fossil record and therefore remain a matter of pure conjecture to science.

I am relieved that Sue the *Tyrannosaurus*, sold at auction by Sotheby's for more than 8 million dollars, will go to Chicago's Field Museum and not to the anonymity of some corporate boardroom, to stand (perhaps) next to a phony Van Gogh. But I am not happy that no natural history museum in the world can pony up the funds for such a purpose – and that McDonald's had to provide the cash. McDonald's is not, after all, an eleemosynary institution, and they will legitimately want their piece for their price. Will it be the Happy Meal Hall of Paleontology at the Field Museum? (Will we ever again be able to view a public object with civic dignity, unencumbered by commercial messages? Must city buses be fully painted as movable ads, lampposts smothered, taxis festooned, even seats in concert halls sold one by one to donors and embellished in perpetuity with their names on silver plaques?) Or will we soon see Sue the Robotic Tyrannosaur – the purchase of the name rather than the thing, for Sue's actual skeleton cannot improve the colors or sounds of the robots, and her value, in this context, lies only in the recognition of her name (and the memory of the dollars she attracted), not in her truly immense scientific worth.

I am neither an idealist nor a Luddite in this matter. I just feel that the world of commerce and the world of intellect, by their intrinsic natures, must pursue different values and priorities – while the commercial world looms so much larger than our domain that we can only be engulfed and destroyed if we make a devil's bargain of fusion for short-term gain. The worth of fossils simply cannot be measured in dollars. But the *Lügensteine* of Marrakech can only be assessed in this purely symbolic way – for the Moroccan fakes have no intellectual value and can bring only what the traffic (and human gullibility) will bear. We cannot possibly improve upon Shakespeare's famous words for this sorry situation – and this ray of hope for the honor and differences of intellect over cash:

> Who steals my purse steals trash . . .
> But he that filches from me my good name

Robs me of that which not enriches him,
And makes me poor indeed.

But we must also remember that these words are spoken by the villainous Iago, who will soon make Othello a victim, by exploiting the Moor's own intemperance, of the most poignant and tragic deception in all our literature. Any modern intellectual, to avoid Beringer's sad fate, must hold on to the dream – while keeping a cold eye on immediate realities. Follow your bliss, but remember that handkerchiefs can be planted for evil ends and fossils carved for ready cash.

THE STINKSTONES OF OENINGEN

In his manifesto for a science of paleontology, Georges Cuvier compared our ignorance of geological time with our mastery of astronomical space. He wrote, in 1812, in the preliminary discourse to his great four-volume work on the bones of fossil vertebrates:

> Genius and science have burst the limits of space, and . . . have unveiled the mechanism of the universe. Would it not also be glorious for man to burst the limits of time. . . . Astronomers, no doubt, have advanced more rapidly than naturalists; and the present period, with respect to the theory of the earth, bears some resemblance to that in which some philosophers thought that the heavens were formed of polished stone, and that the moon was no larger than the Peloponnesus; but, after Anaxagoras, we have had our Copernicuses, and our Keplers, who pointed out the way to Newton; and why should not natural history also have one day its Newton? [I have followed the famous Jameson translation of 1817, which is as canonical for Cuvier's *Discours préliminaire* as its namesake King James's is for Moses – hence some pleasant archaisms throughout, although I have checked the original in all cases for accuracy.]

Cuvier, an ambitious man, may have held personal hopes, though Darwin (whose earthly remains do lie next to Newton's in Westminster Abbey) has generally commandeered the proffered title. Still, Cuvier didn't do badly. His immediate successors, at least in France, usually referred to him as the Aristotle of biology.

The centenary of Darwin's death (April 1882) has prompted a round of celebrations throughout the world. But 1982 is also the sesquicentenary of Cuvier's demise (1769–1832), and our erstwhile Aristotle has attracted scant notice. Why has Cuvier, surely the greater giant in his own day, been eclipsed (at least in the public eye) during our own? In power of intellect, and range and breadth of output, Cuvier easily matched Darwin. He virtually founded the modern sciences of paleontology and comparative anatomy and produced some of the first (and most beautiful) geological maps. Moreover, and so unlike Darwin, he was a major public and political figure, a brilliant orator, and a high official in governments ranging from revolution to restoration. Charles Lyell, the great English geologist, visited Cuvier at the height of his influence and described the order and system that yielded such a prodigious output from a single man:

> I got into Cuvier's sanctum sanctorum yesterday, and it is truly characteristic of the man. In every part it displays that extraordinary power of methodising which is the grand secret of the prodigious feats which he performs annually without appearing to give himself the least trouble. . . . There is first the museum of natural history opposite his house, and admirably arranged by himself, then the anatomy museum connected with his dwelling. In the latter is a library disposed in a suite of rooms, each containing works on one subject. There is one where there are all the works on ornithology, in another room all on ichthyology, in another osteology, in another *law* books! etc., etc. . . . The ordinary studio contains no bookshelves. It is a longish room, comfortably furnished, lighted from above, with eleven desks to stand to, and two low tables, like a public office for so many clerks. But all is for the one man, who multiplies himself as author, and admitting no one into this room, moves as he finds necessary, or as fancy inclines him, from one occupation to another. Each desk is furnished with a complete establishment of inkstand, pens, etc. . . . There is a separate bell to several desks. The low tables are to sit to when he is tired. The collaborators are not numerous, but always chosen well. They save him

> every mechanical labour, find references, etc., are rarely admitted to the study, receive orders and speak not.

Cuvier has suffered primarily because posterity has deemed incorrect the two cardinal conclusions that motivated his work in biology and geology – his belief in the fixity of species and his catastrophism. Since being wrong is a primary intellectual sin when we judge the past by its approach to current wisdom, dubious motives must be ascribed to Cuvier. How else can one explain why such a brilliant man went so far astray? Cuvier then becomes an object lesson for aspiring scientists. Cuvier must have failed because he allowed prejudice to cloud objective truth. Conventional theology must have dictated both his creationism and the geological catastrophism that supposedly squeezed our earth into the Mosaic chronology. Consider this assessment of Cuvier presented by a leading modern textbook in geology:

> Cuvier believed that Noah's flood was universal and had prepared the earth for its present inhabitants. The Church was happy to have the support of such an eminent scientist, and there is no doubt that Cuvier's great reputation delayed the acceptance of the more reasonable views that ultimately prevailed.

I devote this essay to defending Cuvier (who ranks, in my judgment, with Darwin and Karl Ernst von Baer as the greatest of nineteenth-century natural historians). But I do not choose to do so in the usual manner of historians – by showing that Cuvier's beliefs were not rooted in irrational prejudice, but both arose from and advanced beyond the social and scientific context of his own time. Nor (obviously) will I defend Cuvier's creationism or more than a sliver of his catastrophism. Instead, I want to argue that Cuvier used the very doctrines for which he stands condemned – creationism and catastrophism – as specific and highly fruitful research strategies for establishing the basis of modern geology – the stratigraphic record of fossils and its attendant long chronology for earth history. Some types of truth may require pursuit on the straight and narrow, but the pathways to scientific insight are as winding and complex as the human mind.

Cuvier is often portrayed as an armchair speculator because his conclusions are now regarded as incorrect and error supposedly arises from aversion to hard data. In fact, he was a committed empiricist. He railed against the prevalent tradition in geology for constructing comprehensive 'theories of the earth' with minimal attention to actual rocks and fossils. 'Naturalists,' he wrote, 'seem to have scarcely any idea of the propriety of investigating facts before they construct their systems.' (Cuvier correctly includes Hutton among the system builders, although he confesses more sympathy for his Scottish colleague than for most of his ilk.)

Instead, Cuvier argues, we must seek some empirical criterion for unraveling the earth's history. But what shall it be? What has changed with sufficient regularity and magnitude to serve as a marker of time? Cuvier recognized that the lithology of rocks would not do, since limestones and shales look pretty much alike whether they occur at the tops or bottoms of stratigraphic sequences. What about the fossils entombed in rocks?

The idea that fossils reflect history is now so commonplace, we tend to regard it as an ancient truth. It was, however, a contentious issue in Cuvier's day, when debate centered on whether or not species could become extinct – for without extinction, all creatures are coeval and fossils cannot measure time (unless new forms keep accumulating and we can date rocks by first appearances. But a finite earth would seem to preclude continual addition with no subtraction).

Many of Cuvier's illustrious contemporaries (including Thomas Jefferson who, when not preoccupied with other matters, devoted a paper to the subject) argued strongly that extinction could not occur. Cuvier decided that the a priori (and often explicitly biblical) defenses of nonextinction were worthless and that the issue would have to be decided empirically. But previous studies of fossil vertebrates (his specialty) had been undertaken in the mindless manner of mere collection. Fossils had been gathered primarily as curiosities – but scientists must *ask questions* and collect systematically in their light.

> Other naturalists, it is true, have studied the fossil remains of organized bodies; they have collected and represented them by thousands, and their works will certainly serve as a valuable storehouse of materials. But, considering these fossil plants and animals merely in themselves, instead of viewing them in their connection with the theory of the earth; or regarding their petrifactions . . . as mere curiosities, rather

> than historical documents . . . they have almost always neglected to investigate the general laws affecting their position, or the relation of the extraneous fossils with the strata in which they are found.

Cuvier then provides a two-page compendium of questions, an empiricist's *vade mecum* to combat the older speculative tradition.

> Are there certain animals and plants peculiar to certain strata and not found in others? What are the species that appear first in order, and those which succeed? Do these two kinds of species ever accompany one another? Are there alternations in their appearance; or, in other words, does the first species appear a second time, and does the second species then disappear?

But this research program for establishing a geological record cannot work unless extinction is a common fact of nature – and ancient creatures are therefore confined to rocks of definite and restricted ages. Cuvier's great four-volume work (*Recherches sur les ossemens fossiles*, 'studies on fossil bones') is a long demonstration that fossil bones belong to lost worlds of extinct species.

Cuvier used the comparative anatomy of living vertebrates to assign his fossils to extinct species. Since fossils come in bits and pieces, a tooth here or a femur there, some method must be devised to reconstruct a whole from scrappy parts and to ascertain whether that whole still walks among the living. But what principles shall govern the reconstruction of wholes from parts? Can it be done at all? Cuvier recognized that he must study the anatomy of modern organisms – where we have unambiguous wholes – to learn how to interpret fragments of the past. The second paragraph of his essay presents this program for research:

> As an antiquary of a new order, I have been obliged to learn the art of deciphering and restoring these remains, of discovering and bringing together, in their primitive arrangement, the scattered and mutilated fragments of which they are composed. . . . I had . . . to prepare myself for these enquiries by others of a far more extensive kind, respecting the animals

> which still exist. Nothing, except an almost complete review of creation in its present state, could give a character of demonstration to the results of my investigations into its ancient state; but that review has afforded me, at the same time, a great body of rules and affinities which are no less satisfactorily demonstrated; and the whole animal kingdom has been subjected to new laws in consequence of this Essay on a small part of the theory of the earth.

As his cardinal rule for reconstruction, Cuvier devised a principle that he called 'correlation of parts.' Animals are exquisitely designed and integrated structures – perfect Newtonian machines of a sort. Each part implies the next, and a whole lies embodied in the implications of any fragment – a grand version of that immortal commentary on Ezekiel's vision, 'the foot bone's connected to the ankle bone. . . .'

Cuvier presents the law of correlation as if it could be applied by reason alone, using the principles of animal mechanics:

> Every organized individual forms an entire system of its own, all the parts of which mutually correspond and concur. . . . Hence none of these separate parts can change their forms without a corresponding change in the other parts of the same animal, and consequently each of these parts, taken separately, indicates all the other parts to which it had belonged. . . . If the viscera of an animal are so organized as only to be fitted for the digestion of recent flesh, it is also requisite that the jaws should be constructed as to fit them for devouring prey; the claws must be constructed for seizing and tearing it to pieces; the teeth for cutting and dividing its flesh; the entire system of the limbs, or organs of motion, for pursuing and overtaking it; and the organs of sense, for discovering it at a distance. . . . Thus, commencing our investigation by a careful survey of any one bone by itself, a person who is sufficiently master of the laws of organic structure, may, as it were, reconstruct the whole animal to which that bone had belonged.

Cuvier's principle of correlation lies behind the popular myth that paleontologists can see an entire dinosaur in a single neck bone. (I believed this legend as a child and once despaired of entering my chosen profession because I could not imagine how I could ever obtain such arcane and wondrous knowledge.) Cuvier's principle may well apply in the most general sense: if I find a jaw with weak peglike teeth, I do not expect to find the sharp claws of a carnivore on the accompanying legs. But a single tooth will not tell me how long the legs were, how sharp the claws, or even how many other teeth the jaw held. Animals are bundles of historical accidents, not perfect and predictable machines.

When a paleontologist does look at a single tooth and says, 'Aha, a rhinoceros,' he is not calculating through laws of physics, but simply making an empirical association: teeth of this peculiar form (and rhino teeth are distinctive) have never been found in any animal but a rhino. The single tooth implies a horn and a thick hide only because all rhinos share these characters, not because the deductive laws of organic structure declare their necessary connection. Cuvier, in fact, knew perfectly well that he operated by empirical association (and not by logical inference), although he regarded his observational method as an imperfect way station to a future rational morphology:

> As all these relative conformations are constant and regular, we may be assured that they depend upon some sufficient cause; and, since we are not acquainted with that cause, we must here supply the defect of theory by observation, and in this way lay down empirical rules on the subject, which are almost as certain as those deduced from rational principles, especially if established upon careful and repeated observation. Hence, any one who observes merely the print of a cloven hoof, may conclude that it has been left by a ruminant animal, and regard the conclusion as equally certain with any other in physics or in morals.

Since Cuvier didn't know the laws of rational morphology (we now suspect that they do not exist in the form he anticipated), he proceeded by his favorite method of empirical cataloging. He amassed an enormous collection of vertebrate skeletons, and noted an invariant association of parts by repeated observation. He could then use his catalog of recent skeletons to decide whether fossils belong to extinct species. The earth, he argued, has been explored with sufficient care (for large terrestrial mammals at least) that fossil bones outside the range of modern skeletons must represent vanished species.

The four volumes of the 1812 treatise form a single long argument for the fact of extinction, the resultant utility of fossil vertebrates for ascertaining the relative ages of rocks, and the consequent antiquity of the earth. The introductory *Discours préliminaire* sets out basic principles. In the first technical monograph, on mummified remains of the Egyptian ibis, Cuvier finds no difference between modern birds and fossils from the beginning of recorded history as then construed. The present creation therefore has considerable antiquity; if extinct species inhabited still earlier worlds, then the earth must be truly ancient. The next set of monographs discusses the detailed anatomy of large mammals found in the uppermost geological strata – Irish elks, woolly rhinos, and a variety of fossil elephants (mammoths and mastodons). They are similar to modern relatives, but the sizes and shapes of their fragmentary bones lie outside modern ranges and will not correlate with the normal skeletons of living forms (no modern deer could hold up the antlers of an Irish elk). Hence, extinction has occurred and life on earth has a history. The final monographs demonstrate that still older bones belonged to creatures even more unlike modern species. Life's history has a direction – and great antiquity if it has passed through so many cycles of creation and destruction.

Cuvier did not give an evolutionary interpretation to the direction that he discerned, for the very principle that he used to establish extinction – the correlation of parts – precluded evolution in his mind. If an animal's parts are so interdependent that each one implies the exact form of all others, then any change would require a total remodeling of an entire body, and what process can accomplish such a complete and harmonious change all at once? The direction of life's history must reflect a sequence of creations (and subsequent extinctions), each more modern in character. (We would not deny Cuvier's inference today, but only his initial premise of tight and ubiquitous correlation. Evolution is mosaic in character, proceeding at different rates in different structures. An animal's parts are largely dissociable, thus permitting historical change to proceed.)

Thus, ironically, the incorrect premise that has sealed Cuvier's poor reputation today – his belief in the fixity of species – was the basis for his greatest contribution to human thought and hard-nosed empirical science: a proof that extinction grants life a rich history and the earth a great antiquity. (I note the further irony that Cuvier's creationism – good science in his time – disproved, more than 150 years ago, the linchpin of modern fundamentalist creationism: an age of but a few thousand years for the earth.)

Cuvier's reputation took a second strike from his adherence to (and partial invention of) the geological theory of catastrophism, a complex doctrine of many parts, but focusing on the claim that geological change is concentrated in rare episodes of paroxysm on a global or nearly global scale: floods, fires, the rise of mountains, the cracking and foundering of continents – in short, all the components of traditional fire and brimstone. Cuvier, of course, linked his catastrophism to his theory of successive creations and extinctions by identifying geological paroxysms as the agent of faunal debacles.

A perverse reading of history had led to the usual claim – as in the textbook assessment of Cuvier cited earlier – that catastrophism was an antiscientific feint by a theological rear guard laboring to place Noah's flood under the aegis of science, and to justify a compression of earth's history into the Mosaic chronology. Of course, if the earth is but a few thousand years old, then we can only account for its vast panoply of observed changes by telescoping them into a few episodes of worldwide destruction. But the converse does not hold: a claim that paroxysms sometimes engulf the earth dictates no conclusion about its age. The earth might be billions of years old, and its changes might still be concentrated in rare episodes of destruction.

Cuvier's eclipse is awash in irony, but no element of his denigration is more curiously unfair than the charge that his catastrophism reflects a theological compromise with his scientific ideals. In the great debates of early-nineteenth-century geology, catastrophists followed the stereotypical method of objective science – empirical literalism. They believed what they saw, interpolated nothing, and read the record of the rocks directly. This record, read literally, is one of discontinuity and abrupt transition: faunas disappear; terrestrial rocks lie under marine rocks with no recorded transitional environments between; horizontal sediments overlie twisted and fractured strata of an earlier age. Uniformitarians, the traditional opponents of catastrophism, did not triumph because they read the record more objectively. Rather, uniformitarians, like Lyell and Darwin, advocated a more subtle and *less* empirical method: use reason and inference to supply the missing information that imperfect evidence cannot record. The literal record is discontinuous, but gradual change lies in the missing transitions. To cite Lyell's thought experiment: if Vesuvius erupted again and buried a modern Italian town directly atop Pompeii, would the abrupt transition from Latin to Italian, or clay tablets to television, record a true historical jump or two thousand years of missing data? I am no partisan of gradual change, but I do support the historical method of Lyell and Darwin. Raw empirical literalism will not adequately map a complex and imperfect world. Still, it seems unjust that catastrophists, who almost followed a caricature of objectivity and fidelity to nature, should be saddled with a charge that they abandoned the real world for their Bibles.

Cuvier's methodology may have been naïve, but one can only admire his trust in nature and his zeal for building a world by direct and patient observation, rather than by fiat or unconstrained feats of imagination. His rejection of received doctrine as a source of necessary truth is, perhaps, most apparent in the very section of the *Discours préliminaire* that might seem, superficially, to tout the Bible as infallible – his defense of Noah's flood. He does argue for a worldwide flood some five thousand years ago, and he does cite the Bible as support. But his thirty-page discussion is a literary and ethnographic compendium of all traditions, from Chaldean to Chinese. And we soon realize that Cuvier has subtly reversed the usual apologetic tradition. He does not invoke geology and non-Christian thought as window dressing for 'how do I know, the Bible tells me so.' Rather, he uses the Bible as a single source among many of equal merit as he searches for clues to unravel the earth's history. Noah's tale is but one local and highly imperfect rendering of the last major paroxysm.

As a rough rule of thumb, I always look to closing paragraphs as indications of a book's essential character. General treatises in the pontifical mode proclaim a union of all knowledge, or tell us, in no uncertain terms, what *it* all means for man's physical future and moral development. Cuvier's conclusion is revealing in its starkly contrasting style. No drum rolls, no statements about the implications of catastrophism for human history. Cuvier simply presents a ten-page list of outstanding problems in stratigraphic geology. 'It appears to me,' he writes, 'that a consecutive history of such singular deposits would be infinitely more valuable than so many contradictory conjectures respecting the first origin of the world and other planets.' He ranges across Europe, up and down the geological column, offering suggestions for empirical work: study recent alluvial deposits of the Po and the Arno, dig in the gypsum quarries of Aix and Paris, collect 'gryphites, the cornua ammonis and the entrochi' that may abound in the Black Forest. 'We are as yet uninformed of the real position of the stinkstone slate of Oeningen, which is also said to be full of the remains of fresh-water fish.'

A man who could end one of the greatest *theoretical* treatises in natural history with a plea for unraveling the stratigraphic position and faunal content of the Oeningen stinkstones knew, in the most profound way, what science is about. We may wallow forever in the thinkable; science traffics in the doable.

THE RAZUMOVSKY DUET

I live on a small street of some twenty houses, all closely packed together. I presume that most of my neighbors share my mental geography of this terrain – structural divisions into roadway, sidewalks, houses, and gardens, with primary taxonomic separations set by property lines of ownership. But the street also features more cats than people – and I know that these members of the mammalian majority divide the space differently. They clearly have some sense of ownership and territory – for charges, howlings, and spats occur on a daily basis – but their parsings do not match the human separations. Think what we might understand about mammalian mentality if we could ever obtain the cat map of Crescent Street. A potential informant does share my living quarters, but he has been persistently unresponsive and uncooperative (and he still gets fed!).

In a zoologically more restricted framework, this important theme of alternate mappings might give us great insight into differences among human cultures, times, and mentalities – as the French *Annales* school of historians has taught us, with their emphasis on changing patterns in ordinary life, featuring working men and women rather than kings and conquerors. In school, I learned the conventional history of dates, nations, and battles. My mental maps of time and geography all follow the usual lines: temporal divisions by kings and presidents, spatial boundaries by nations and languages. But other systems make as much sense, and surely have more relevance to people whose primary activities enjoin different divisions.

I assume that sailors prefer size or function to place of registration as a criterion for classifying vessels (especially when so many ships, for reasons of taxation and licensing, officially carry the Liberian or Panamanian flag). My *Cerion* snails do not recognize a political separation between the Bahamas and the Turks and Caicos Islands, for these places form a unified climatic and geographic province well suited to molluscan lifestyles (yet I have long been frustrated by the difficulty of obtaining maps with both political entities shown in the same way, and to the same scale). This essay emphasizes the artificiality of conventional linguistic and national boundaries in European history – by telling a story about scientists and artists who might have parsed the early eighteenth century world along different lines of patronage.

So many projects, both little essays like this and lifetimes of effort, begin by accident; you can't, after all, explicitly search for the unexpected. I bought an old book a few months ago, largely because it was so underpriced and the opportunity would therefore not arise again. Its author – Johann Gotthelf Fischer von Waldheim (1771–1853) – cannot claim a place in the generally recognized pantheon of great scientists. But he occupies a small spot in the heart of paleontologists as one member of the pioneering generation who established the basic ordering of life's history when scientists codified the geological time scale. We also owe a debt of thanks to Gotthelf Fischer (to cite the abbreviated moniker that he used in most of his own publications) for popularizing, and perhaps inventing, our modern name of *paleontology*. (Otherwise we might still be stuck with one of the earlier, unpronounceable alternatives, like *oryctology*.)

Fischer was German by birth (from the town of Waldheim in Saxony, which Fischer added to his name when the Russian Czar ennobled him in 1817). He studied with Cuvier, became friendly with Goethe, traveled with the Humboldt brothers, taught in various German universities, and finally, in 1804, moved permanently to Russia as Professor of Natural History and director of the natural history museum at the University of Moscow. Lest his last post seem anomalous, Fischer's emigration followed a common pattern fostered in Russia ever since Peter the Great (1672–1725) sought to leap over Russia's scientific backwardness by importing specialists from other nations, and by purchasing foreign collections. (Peter developed his own great museum of natural history – still partly on view in the building constructed to house it, the Kunstkamera of St Petersburg – by acquiring two important Dutch collections.)

The growth of Russian universities during the late eighteenth and early nineteenth centuries provided another pathway for importation of foreign scientists and professors – for Russia, lacking universities in its past, could not subsist on home-grown specialists. Fischer's residency in Moscow, beginning in 1804, lies sandwiched between arrivals of the two greatest German-speaking biological emigrés to Russian universities – Peter Simon Pallas in 1768 and Karl Ernst von Baer in 1834. (Both men taught in St Petersburg. Pallas was born in Berlin, while von Baer belonged to an old Prussian family, then living in Estonia. Von Baer, the greatest embryologist of the nineteenth century, was imperial Russia's greatest academic 'catch.' He discovered the human ovum in 1827, but abandoned embryological research for an astonishing variety of brilliant studies in ethnology, anthropology, and geomorphology during his Russian years.)

Fischer had a remarkably fruitful and successful career in Russia. He founded three journals (published in French), and wrote nearly two hundred books and articles (also predominantly in French) during his Russian years. His wide-ranging publications spanned all of zoology, but concentrated on pioneering work in describing living Russian insects and fossils of all groups. Fischer was widely recognized by the international scientific community, and became an active or honorary member of nearly ninety institutions and academies, including the American Association of Arts and Sciences in Boston, and the American Philosophical Society in Philadelphia. At his Jubilee in 1847 (to celebrate the fiftieth anniversary of his doctorate), Alexander von Humboldt, the world's most popular living scientist at the time, extolled Fischer as *mein edler, ältester Freund* ('my noble and oldest friend'). Fischer wrote home to his friends in Waldheim, describing the ceremony. He spoke of the six carriages that led the way, five of them drawn by four horses and the last by six. He described the gifts and the encomia, ending with his reaction: *Hier brachen mir die Thränen der Dankbarkeit aus meinen Augen* ('this brought tears of thanks from my eyes').

The book that I bought is not one of Fischer's paleontological works, but a superb example, published in 1813, of a genre that computer technology has driven to extinction – bibliographic compendia of alternative systems for classifying objects, in this case all genera of the animal kingdom. Fischer uses Linnaeus's account of animal genera as a framework, but, in a series of ingenious tables, charts, and lists, displays the correspondences between Linnaeus and all other major systems proposed by leading zoologists, mostly French and German.

(The visually attractive result looks like a chart of comparisons among the three synoptic gospels, or among various biblical translations.) The effort may be primarily bibliographic, but we learn much of historical and theoretical interest from Fischer's compilation. For example, historians have often claimed that Lamarck's *Philosophie zoologique*, his major exposition of evolution, written in 1809, was widely ignored as a fatuous and speculative treatise. But one of Fischer's longer charts presents a *Tabula clarissimi Lamarck* ('list of the most celebrated Lamarck'), reproducing Lamarck's chain of being in full evolutionary order, just four years after its publication.

The title of Fischer's book records its intended audience and utility: *Zoognosia: tabulis synopticis illustrata, in usum praelectionum Academiae Imperialis Medico-Chiurgicae Mosquensis* (Zoological knowledge: illustrated with synoptic tables for lectures at the Royal Academy of Medicine and Surgery in Moscow). Such single-volume compendia must have been especially useful for students with little access to extensive libraries and collections. (My own copy went to an even more peripheral location with even more limited primary material, for Fischer has inscribed it 'to the Society of Arts and Sciences of Courland' – a duchy on the Baltic Sea, now part of Latvia, that became a Polish fiefdom in the sixteenth century, but passed to the Russian empire in 1795 after the third partition of Poland. Courland enjoyed a seventeenth-century moment of glory, and even held enough power to build a small colonial empire in the West Indies [Tobago] and Africa [Gabon].)

I love to read the dedications of old books written in monarchies – for they invariably honor some (usually insignificant) knight or duke with fulsome words of sycophantic insincerity, praising him as the light of the universe (in hopes, no doubt, for a few ducats to support future work); this old practice makes me feel like such an honest and upright man, by comparison, when I put a positive spin, perhaps ever so slightly exaggerated, on a grant proposal. Fischer's dedication to *Comes Illustrissime, Domine Clementissime*! (Most illustrious count and most compassionate lord!) first struck me as an example of *vin ordinaire* within the genre. But two features of the text sparked my interest and led to this essay. First, Fischer's lament over impediments to his research seemed more fervent and more extreme than usual – for evident reasons most easily inferred. He writes that he is dedicating his book to the illustrious count, and would also like to so dedicate the natural history collections of the museum, but *eheu, omnes perierunt, restant paucissimae.* ('Alas, everything has been destroyed, so very little remains.') He then asks what will happen *post tot calamitates luctuosas casusque tristissimos, quos Musarum cultores Mosquensis experti sumus* ('after so many sorrowful calamities, and so many most dismal events, that we supporters of the muses of Moscow have endured').

I happen to be writing this essay on July 4, so local events in Boston are primed to dramatize the source of Fischer's complaints. Just before sunset, as they do every year, the Boston Pops Orchestra, on the esplanade adjoining the Charles River, will play Tchaikovsky's *1812 Overture*, with glorious fireworks commencing at the booming of cannons (marking the conclusion of this loudest piece in the classical repertory). When I was a child, wallowing in patriotism, I couldn't quite figure out why Tchaikovsky had written an overture to celebrate a little skirmish of an American war that we didn't win. I discovered later that a few other events of passing importance occurred in 1812, most notably, and right in Tchaikovsky's bailiwick, the capture of Moscow by Napoleon, followed by his subsequent retreat and decisive defeat, as the old saying goes, by those greatest of Russian generals, November and December. Napoleon had entered Moscow on September 14, hoping to win quick and favorable peace terms from Czar Alexander. But the Czar would not deal with him and, more important, a massive fire had broken out in Moscow on the day of Napoleon's entry, eventually destroying more than two thirds of the city, and preventing Napoleon from feeding and housing his troops through the winter. The fire may have aided Russia by helping to force Napoleon's withdrawal and subsequent defeat, but the flames also wrecked most of Moscow, including the great libraries and museum collections of the university. Fischer, in other words, had been one of so many victims, largely anonymous to later history, of a signal event in the construction of our modern world.

As a second intriguing feature of Fischer's dedication, I then considered the identity of the illustrious count from whom he sought patronage – and not so subtly – by ending his dedication: *in Te, Comes Illustrissime, spes omnis nostra collocata est.* ('All our hope is placed in you, most illustrious count.') Fischer dedicated his book to Alexis Kirillovich Razumovsky, the Minister of Public Instruction.

Now, speaking of music, and going back a generation or two from Tchaikovsky, we meet a Russian reference known to all lovers of the classics – for Beethoven dedicated three of his most famous string quartets (Opus 59) to a gentleman of the same name. (They are, in fact, called the Razumovsky Quartets, and the first two feature Russian folk melodies in their composition.) I couldn't help wondering about the relationship between Beethoven's actual and Fischer's anticipated patron. Beethoven wrote the quartets in 1806, and Fischer sought help seven years later, so the events are nearly contemporary. The story, as so often happens in our fascinating world of complex interactions, turned out to be well worth reporting.

Beethoven's patron, Andrei Kirillovich Razumovsky, was the brother of Fischer's *Comes Illustrissime* – but the resemblance extends little beyond genealogy. Their story begins two generations further back, with a Ukrainian cossack named Grigor Rozum, who had two remarkable sons. Again, music served as the prod to success. One of his sons, Aleksei Grigorevich Razumovsky (1709–1771), became a singer in the court choir of St Petersburg. There he attracted the attention, and later the love, of Princess Elizabeth, who became Czarina of Russia in 1741. Aleksei secretly married Elizabeth in 1742. They had no children, and Aleksei took little interest in affairs of state. But he remained the favorite of his secret wife, and became enormously wealthy thanks to her largesse. His brother Kirill Grigorevich (1718–1803, and father of the Razumovskys in our tale of Fischer and Beethoven – as their common patronymic, Kirillovich, indicates) was a more ambitious and accomplished man. He served, for nearly twenty years, as president of the St Petersburg Academy of Sciences, but held much greater political power (including dominion over more than 100,000 serfs) as the last hetman of Little Russia (ruler of the Ukraine). His two sons therefore inherited all conceivable advantages of wealth and position.

The younger son – Andrei Kirillovich, Beethoven's Razumovsky (1752–1836) – was a warm, generous, and liberal man, and one of the great art patrons of Europe. He spent his professional life as a diplomat in central Europe, wooing women (most notoriously the Queen of Naples) and making deals. He served as Russian ambassador to Vienna from 1790 to 1799 and again from 1801 to 1807. Razumovsky continued to live in Austria for the rest of his life, serving as a leader of the Russian delegation at the Congress of Vienna in 1815 (where Napoleon's spoils were split among the victors, thereby establishing yet another connection to Fischer and the opposite side of our story). For his service, Andrei Razumovsky was promoted from count to prince by the Czar (as His Majesty passed through Vienna).

Music had always been Razumovsky's first love (or perhaps his second, following women). In his initial post as clerk for the Russian Embassy in Vienna, he certainly knew Mozart, and probably met Haydn. He was a more than merely competent violinist and often played in the quartet that he established, largely for Beethoven's use. But he performed his finest service for music as a patron. Paul Nettl's *Beethoven Encyclopedia* states that

> Razumovsky was the most general Maecenas of his time, supporting artists, musicians, and painters. His picture gallery and musical parties were famous throughout Europe. A well-educated, liberal, and generous aristocrat, and brilliant *causeur*, he was one of the most popular and renowned aristocrats of the late eighteenth and early nineteenth centuries.

Razumovsky certainly knew and supported Beethoven as early as 1796, for his name appears on the subscription list to Beethoven's Trios, Opus 1. When Beethoven composed the Razumovsky Quartets ten years later, the Russian nobleman took title to controversial music, not to pretty oompahs – so we must respect his support for artistic license. Many musicians could not grasp or stomach the score's unconventionalities. When one Italian player belligerently asked if Beethoven truly regarded the quartets as music, the master replied, 'They aren't for you, but for a later age.' Beethoven also dedicated his fifth and sixth symphonies jointly to Razumovsky and to Prince Lobkowitz.

As his finest testimony and favor to Beethoven, Razumovsky established and funded a permanent string quartet, led by Schuppanzigh, in 1808 – and placed the players at Beethoven's disposal. A contemporary observer noted,

> Beethoven was . . . cock of the walk in the princely establishment; everything that he composed was rehearsed hot from the griddle and performed to the nicety of a hair, according to his ideas, just as he wanted it and not otherwise, with affectionate interest, obedience and devotion, such as could spring only from such ardent admirers of his lofty genius.

This happy situation persisted until 1816, when Razumovsky held a gigantic New Year's Eve party to celebrate his elevation to princehood after the Council of Vienna. He could not accommodate all seven hundred guests in his capacious palace, so he built an adjacent wooden structure for supplementary space. This addition caught fire, and flames spread to the main palace, consuming the great library and all the works of art, and finally destroying the prince's most famous space – a room full of Canova's statues, reduced to dust and fragments as the ceiling collapsed. Razumovsky, devastated in both spirit and pocketbook, disbanded and pensioned off his quartet.

The older son – Aleksei Kirillovich, Fischer's Razumovsky (1748–1822) – lacked nearly all of his brother's admirable characteristics. Standard sources (not only the *Soviet Encyclopedia*, which might be accused of bias, but others likely to be both more objective and sympathetic) describe him as indolent, dyspeptic, litigious, domineering, and generally miserable. He married the richest heiress in Russia, but sent her packing several years later, after draining most of her wealth. Of his two sons, one was wildly dissolute and the other floridly mad; his two daughters sound quite admirable – one devoted herself to establishing hospitals for the poor – but sexist sources and limitations provide little information about their lives and fates.

Razumovsky hated court life and largely sought to avoid public responsibilities. But he did love botany and natural history, and he established, at his estate in Gorenski, near Moscow, a wonderful botanical garden (specializing in alpine plants) and the most extensive collection of books on natural history in Russia (including the library of P. S. Pallas, which he had purchased). Nonetheless, Razumovsky rarely used those resources to benefit either science or the public. A contemporary observer commented (probably not impartially):

> Count Aleksei Kirillovich has enclosed himself in his property at Gorenski in order to vegetate there with his plants. He is doing with his knowledge the same thing that he does with his immense fortune – that is, it all stays with him and confers no profit upon others.

Razumovsky did, generally after much persuasion, accept some governmental responsibilities, most notably as Minister of Public Instruction, beginning in 1810 (where sources describe him as largely inactive, but by no means incompetent). In this post, he advocated some reforms, notably the suppression of corporal punishment in the schools.

He did not, however, act as an advocate of the field he supposedly loved, for he argued that the study of natural history required no place in the training of statesmen and politicians. He adopted a similar hands-off policy in dealing with Russia's universities. In his 1988 book, *The University Reform of Tsar Alexander I*, James T. Flynn writes:

> Razumovsky gave so little active leadership that each university was left to its own devices to work out its foundation and development. This circumstance magnified the importance of each local curator, as well as the particular local situation, and minimized the importance of the statutes and the ministry.

Not necessarily a bad thing.

Razumovsky is best known as the confidant and supporter of the deeply conservative social thinker Joseph de Maistre, perhaps best remembered for his quip that the public executioner is the guardian of social order. Maistre left France during the revolution for a life in Switzerland. He spent many years in St Petersburg as diplomatic envoy from the King of Sardinia. There he met Razumovsky, directed his letters on public instruction to the Russian minister, and persuaded him to support a variety of conservative doctrines – including increased press censorship and greater integration of religious instruction into the curriculum.

In short, Aleksei K. Razumovsky seems an awfully poor choice as Fischer's hope for active support in rebuilding Moscow's natural history library and collections. Too bad his more affable and public-spirited brother was off playing second fiddle (literally) in a Vienna quartet. Still, I suppose one has to begin with the Minister of Public Instruction in such a dire situation, and hope for the best.

The full extent of Fischer's plight may be appreciated from an extraordinary letter that he wrote on November 20, 1812, to his German-speaking colleague Nikolaus Fuss, secretary of the Royal Academy of Science in St Petersburg (quoted in the only available biography, by J. W. E. Büttner, titled *Leben und Wirken des Naturforschers Johann Gotthelf Fischer von Waldheim*):

> All scientific institutions are destroyed. Our university has lost so much. Of our library and museum, I was able to save so little – only the best things that I could quickly pack into

> twenty boxes. What is this against the beautiful totality [*Was ist das gegen das schöne Ganze!*] Your Fischer has lost everything.

Fischer then details the destruction. He only saved five books (they happened to be in his carriage at the time) of five thousand in his library. He lost his most precious items, including a completed manuscript on fossils, and his personal copy of Linnaeus's *Systema naturae* with twenty years' worth of annotations. Nearly all the natural history collections were destroyed, and Fischer particularly lamented his beautiful collection of skulls, insects, and dried plants. He also lost 'all my anatomical instruments, my mineralogical apparatus,' and even 130 copper plates for engravings, 'including the large plate of the mammoth skeleton.'

But as he continues the letter to Fuss, a strange alchemy takes hold, and Fischer's zeal and optimism break through:

> We seek to console ourselves with the thought that we are healthy, and that we do not lack bread. I do not know if I will be able to finish my *Zoognosia*, as only nine sheets have been printed [I am delighted that my book – the completed *Zoognosia* – demonstrates the happy negation of this fear]. I am now working on the latest edition of the *Onomasticon Oryctognosiae* [list of names of rocks and fossils]. The work lets me forget all my misfortunes, and makes me most happy.

What more honorable and telling mark of a true scholar – to seek solace in intellectual work, and to rebuild, by painstaking skill, all that misfortune had destroyed. To an old friend, the rector of his school in Waldheim, Fischer wrote, 'To be sure, I have lost everything. But I judge myself as so much more lucky than many of my fellow sufferers, for my knowledge remains with me, and with its help, I hope to get everything back again.'

And so Fischer appealed to Razumovsky for administrative and financial help. And the indolent minister did nothing. Flynn writes:

> Neither Kutuzov nor Razumovsky helped the university during the calamity of the French invasion in 1812. Razumovsky joined the Committee of Ministers, the army,

> and several other agencies in issuing orders to the university to close, or stay open; to stay in Moscow or evacuate . . . to return to Moscow, or not return, and so on.

Fischer and the other professors therefore turned to the surest of all supports: their own bootstraps and self-help. By importuning local friends for money, gifts of books, and use of buildings; working their butts off; improvising; and even holding the equivalent of yard and bake sales, the professors managed to resume operations in September 1813, with a much-reduced cadre of 129 students. By 1815 the library had reached twelve thousand volumes (compared with twenty thousand lost to the flames). Fischer himself spent the remainder of his career successfully rebuilding the great collections of his museum – with no help from Razumovsky. As late as 1830 he was still making trips to Germany with explicit plans for replacing parts of the collection destroyed in 1812.

How, then, shall we taxonomize Europe for an optimal understanding of this complicated story about two noble brothers, their temperamental differences, and their beneficiaries, both actual and unrealized. The usual divisions of nations and languages don't seem to help much. Fischer was a German who worked in Moscow and published mostly in French. He wrote a dedication in Latin, hoping that a Russian nobleman would help to rebuild what the fire of a French invasion had destroyed. Meanwhile, this nobleman's brother lived in Austria, where he knew Mozart and then acted for years as Beethoven's most significant patron.

I do not know how students and members of other disciplines – musicians, diplomats, patrons – would choose to slice this particular cake; I can only speak from my own perspective as a scientist. My profession often gets bad press for a variety of sins, both actual and imagined: arrogance, venality, insensitivity to moral issues about the use of knowledge, pandering to sources of funding with insufficient worry about attendant degradation of values. As an advocate for science, I plead 'mildly guilty now and then' to all these charges. Scientists are human beings subject to all the foibles and temptations of ordinary life. Some of us are moral rocks; others are reeds. I like to think (though I have no proof) that we are better, on average, than members of many other callings on a variety of issues central to the practice of good science: willingness to alter received opinion in the face of uncomfortable data, dedication to discovering and publicizing our best and most honest account of nature's factuality, judgment of colleagues on the might of their ideas rather than the power of their positions.

But on one issue I do have confidence, based on sufficient personal experience, of adherence to a moral code not always followed in other professions. Science does tend to be international. We share information, try hard to communicate with each other, and deplore the parochialisms that stymie contact. (How, for example, can my field of paleontology prosper if scientists are not free to collect and study the fossils of their expertise wherever they occur?) We all know numerous stories of warm and continued cooperation between scientists in nations dedicated to blasting each other off the face of the earth. We have to work this way, for knowledge is universal.

Jingoism may do no serious harm, when expressed as boycotts of silk products, unofficial bans on German opera at the Met, or campaigns to change the name of a baseball team from the Cincinnati Reds to the Redlegs (yes, all these happened) – but science cannot and dare not ban a fact because colleagues in hostile nations discovered the information. Fischer published his *Zoognosia* just a year after French aggression provoked the holocaust of his lifetime's work, but he did not exclude the systems of Cuvier and Lamarck from his volume, for these Frenchmen were the world's greatest living taxonomists, and their work transcended the happenstance of their national origin. Fischer lost nearly all his books and specimens in the conflagration of Moscow, and then spent the rest of his life restocking his collection for future generations of students. In so doing, he called upon the generosity of a large network of scientists – who donated, sold, or traded replacements – throughout Europe and the rest of the world. The boundaries of Fischer's world did not lie at national or linguistic borders. His greatest impediment may have been an indolent and arrogant count, a countryman in his adopted land, who did not care, but whose brother in Austria served as midwife to some of the world's most glorious music.

The sciences of natural history have always been strong in international solidarity, and weak in attracting official support. Fischer discovered this fact for himself when Razumovsky wouldn't help, but scientists throughout the world pitched in. Fischer's own name should have taught him the virtues of self-sufficiency. The Austrian Razumovsky first met a great musician who included among his many names the weighty designation of Gottlieb (Theophilus as he was christened in

Greek, Amadeus as he later chose to render the name in Latin). They all mean 'God's love' – and celestial powers surely adored Wolfgang Amadeus Mozart. The Russian Razumovsky was importuned by a fine scientist named Gotthelf, or 'God's help.' The count declined, but Fischer perservered and succeeded regardless – as he, no doubt, was meant to do, for Poor Richard told us that God only helps those who help themselves.

THE POWER OF NARRATIVE[1]

To begin the second act of Gilbert and Sullivan's *Patience*, Lady Jane enters a bare set, seats herself before her cello, and in two verses bemoans the changes of increasing age. In the first, or conventional, account, she laments what she has lost with the years, but in the second (speaking mostly of weight) she reports a steady increase: 'There will be too much of me in the coming by and by!' The humor of this song plays upon our onesided notion that anything old must become battered, worn, and increasingly bereft of information.

Scholars often make the same false assumption that contemporary cases must provide optimal data, while the records of scientific work steadily decrease in breadth and reliability as they grow older and older. We might therefore suppose that, to understand science, a historian or sociologist should study debates and discoveries now in the making. Yet a moment's thought about our technological age should expose the fallacy in such an idea. Our machines have generally rendered data more ephemeral, or simply unrecorded. The telephone is the greatest single enemy of scholarship; for what our intellectual forebears used to inscribe in ink now goes once over a wire into permanent oblivion (barring such good fortune mixed with ethical dubiety as a White House taping machine).

Moreover, in losing the art of writing letters, many scientists have abandoned the written word in a great many previous applications, from diaries (now passed from fashion) to lab notebooks (now punched directly in 'machine-readable' format). The present can be a verbal wasteland. Paradoxically, then, our most copious data should, like Lady Jane, occupy a comfortable middle age – old enough to avoid our modern technological debasement, and young enough to forestall the inevitable losses of time's destruction.

'The great Devonian controversy' occurred during the 1830s, an optimal decade probably unmatched for density of recorded detail. The controversy began with an apparently minor problem in dating the strata of Devonshire; it ended with a new view of the history of the earth. Martin Rudwick can usually trace the course of its enormously varied and complex changes on a daily basis; little more than conversations over ale and coffee, or bedtime thoughts before candle snuffing, are missing. But density of data makes no case for significance on the fallacious premise that 'more is better.' The subject must also be important and expansive. After a superficial first glance, most readers of good will and broad knowledge might dismiss *The Great Devonian Controversy* as being too much about too little. They would be making one of the biggest mistakes of their intellectual lives.

The geological time scale is a layer cake of odd names, learned by generations of grumbling students with mnemonics either too insipid or too salacious for publication: Cambrian, Ordovician, Silurian, Devonian. . . . Their ubiquity in all geological writing has led students to suspect that these names, like the rocks they represent, have been present from time immemorial (*et nunc, et semper, et in saecula saeculorum, amen*). In fact, the geological time scale was established in an amazingly fruitful burst of research during the first half of the nineteenth century. In 1800, scientists knew that the earth was ancient, but had devised no scheme for ordering events into an actual history. The primary criterion for unraveling that history – the sequence of unique events forming the complex history of life as recorded by fossils – had not been developed. Indeed, many fine scientists still denied that species could become extinct at all on an earth properly made by a benevolent deity. Geologists of 1800 confronted a situation not unlike the hypothetical, almost unthinkable dilemma that historians would face if they knew that modern cultures had antecedents recorded by artifacts, but did not know whether Cheops preceded Chartres or, indeed, whether any culture, however old and different, might not still survive in some uncharted region.

By 1850, history had been ordered in a consistent, worldwide sequence of recognizable, unrepeated events, defined by the ever-changing history of life, and recorded by a set of names accepted and used in the same way from New York to Moscow. This 'establishment of history' was a

GEOLOGIC ERAS			
Era	*Period*	*Epoch*	*Approximate number of years ago (millions of years)*
Cenozoic	Quaternary	Holocene Pleistocene	
	Tertiary	Pliocene Miocene Oligocene Eocene Paleocene	
			65
Mesozoic	Cretaceous Jurassic Triassic		
			225
Paleozoic	Permian Carboniferous Devonian Silurian Ordovician Cambrian		
			600
Precambrian			

Table 1.

great event in the annals of human thought, surely equal in importance to the more theoretical, and much lauded 'discovery of time' by geologists of generations just preceding. Yet while we celebrate Galileo, Darwin, and Einstein, who beyond a coterie of professionals has ever heard of William Smith, Adam Sedgwick, and Roderick Impey Murchison, the architects of our geological time scale and, therefore, the builders of history? Why has their achievement – surely as important an accomplishment as any ever made in science – been so invisible?

We must resolve this vexatious question in order to appreciate *The Great Devonian Controversy*, for Rudwick's seminal book tells the story of how one major period of the earth's history was recognized and unraveled. That period, the Devonian, occupies the crucial time between 410 and 360 million years ago, when life flourished in the seas, and plants and vertebrates became abundant and diverse on land. We will not grasp the importance of this achievement if we follow a common impression and consider stratigraphic geology as 'mere description,' to be dismissed as 'narrative' in the primitive mode of unquantified storytelling. *The Great Devonian Controversy* challenges us to understand natural history as a worthy style of science, equal in rigor and importance to the more visible activities of measurement and experiment that set the stereotype of science in the public image. The success of this book might therefore prompt a broadscale reassessment of science itself as a human activity. *The Great Devonian Controversy*, in its unassuming and highly technical format, could become one of our century's key documents in understanding science and its history.

Since the history of science is usually written by scholars who do not practice the art of doing science, they usually impose upon this greatest of human adventures a subtle emphasis on theories and ideas over practice. (I exempt Rudwick, who had a first career as a distinguished paleontologist before switching to the history of science.) The late-eighteenth-century Scottish geologist James Hutton, for example, is usually praised as the instigator of modern geology because his rigidly cyclical theory of the earth established a basis for an immense span of time. But Hutton had precious little impact on the practice of geology; his name became an icon, but his theory remained on a periphery of speculation; the doers of geology largely ignored his contribution and went about their work.

Theory, of course, necessarily permeates everything we do. But theory may also be pushed below consciousness by groups of scientists who choose to view themselves as ardent recorders of nature's facts. Early in the nineteenth century, the founders of the Geological Society of London explicitly banned all theoretical discussion from their meetings, and dedicated themselves to what they called the 'Baconian' (or purely factual) recording of history. They relegated Hutton and other theorists of past generations to the shelves of speculation, and pledged themselves to fieldwork – specifically to the inductive construction of a stratigraphic standard for history.

When I was younger, and understood science poorly, I bemoaned what I considered the paltry spirit of these men. How could they abandon the exciting ideas, the expansive vision, that motivated Buffon and Hutton, and dedicate themselves instead to finding out what lay on top of what in the rocks? I now appreciate the motives of these men who, after all, forged the geological time scale with their own eyes and hands. Of course they were too extreme (even disingenuous) in their impossible rejection of theory; of course they overreacted to what they interpreted as past excesses of speculation. But they understood the cardinal principle of all science – that the profession, as an art, dedicates itself above all to fruitful doing, not clever thinking; to claims that can be tested by actual research, not to exciting thoughts that inspire no activity.

Most younger rocks of Britain and the Continent had yielded with relative ease to consistent stratigraphic ordering – for they are arranged nearly as an ideal layer cake, younger above older, with relatively little distortion by folding, breakage by faulting, or inversion when older layers thrust over younger strata. But older rocks, from the Paleozoic Era of our modern time scale (600 million to 225 million years ago), posed greater problems for three major reasons well known to all field geologists: rocks of this age are usually far more contorted by intense folding and faulting; alteration by heat and pressure has obliterated fossils from many units, thus removing the primary criterion of history; large stretches of time lie unrecorded by strata, because repeated episodes of mountain building and continental collision destroy evidence.

The complexity of Paleozoic rocks seriously threatened the stratigraphical research program. If older strata could not be ordered by resolving their structure through mapping, and by sorting their sequence according to fossils, then the science itself was doomed. This ordering, moreover, had great economic importance. The period of great fossil forests or 'Coal Measures' (the Carboniferous Period of our modern time scale dating from about 360 to 280 million years ago) lay near the top of this older pile. If geologists couldn't figure out what strata came before and after, vast sums of money would be wasted drilling through rocks misinterpreted as young but actually older than Coal Measures, in the vain hope of finding coal beneath.

The rocks of Devonshire became a focus for resolving the stratigraphy of older times. Most strata were dark, dense, compacted, and highly contorted – composed of rock known as 'greywacke' to quarrymen. Under the lingering tendency to infer age from rock type (an old hope that had failed the test in detail, but had not been abandoned as a rough guide), greywacke smelled old and seemed destined for a position at the bottom of the stratigraphic pile.

*

The Devonian controversy began in 1834 with a paradoxical claim about fossils supposedly found within the greywacke sequence of Devonshire. Henry De la Beche, later director of the Geological Survey of Great Britain, claimed the discovery of fossil plants *within* the Devonshire greywackes. (Don't let the Francophonic name fool you; De la Beche, né Beach, was a proper English gentleman, an important theme in Rudwick's analysis.) Roderick Impey Murchison, the aristocratic and wealthy ex-soldier who had parlayed a taste for outdoor adventure into a serious and full-time professional commitment, responded in a manner that seemed shockingly inconsistent with the professed empiricism of the Geological Society.

Murchison had never seen the Devonshire rocks, but he proclaimed, with vigor and total assurance, that De la Beche had made a monumental mistake, a kind of fool's error in basic mapping and observation. The greywackes, Murchison argued, were old, belonging to the Silurian system (that he had named and first described himself); but the plant fossils must hail from Coal Measure (much later) times, since terrestrial life had not yet made an appearance during the Silurian. The plant-bearing strata, Murchison proclaimed, must lie on top of the stratigraphic pile, not within as De la Beche claimed. And the sequence must contain a prominent 'unconformity,' or temporal gap in deposition, if such young rocks (Coal Measures) actually lie directly atop the ancient greywackes.

(I present, with great regret and almost with a sense of shame, this simplified caricature of such a complex argument and its eventual resolution. As his primary theme, Rudwick emphasizes the multiplicity of forces and arguments leading to the resolution of such debates; and he urges us to view scientific change as a social process of complex interaction based on class, status, age, and place of residence. I have been forced, by limitations of space, to depict the debate largely as a struggle between two men and a pair of sharply contradictory interpretations later forged into unity. You simply must read the book to sample the true richness of the controversy.)

Murchison based his bold and cocky response on his confidence in fossils as an ultimate criterion for assessing the relative ages of rocks. No issue could be more important in developing a proper methodology for reconstructing history. Geologists had to find a reliable criterion – some feature of rocks or their contents that changed in an absolutely reliable way through time to produce a sequence of unique and unrepeated states for marking the measuring rod of history. Nature does not yield her secrets easily. Nowhere on earth can geologists find a complete pile of unaltered rocks – where relative age might be assessed by simply observing what lies on top of what (a procedure called 'superposition' in geological jargon). Rock outcrops are fragmentary and jumbled; only a tiny part of the sequence graces any particular road-cut or stream bed. We need a criterion for linking these isolated pieces one to the other ('correlation' in the jargon). Just as archaeologists might use tree rings of support beams, styles of pottery, or forms of axheads to order a group of widely scattered pueblos into a temporal sequence, geologists needed, above all else, a criterion of history.

Geologists had toyed with a variety of criteria and rejected them. Rock type itself had been a favorite hope, the basis of the so-called Wernerian theory so popular in the generation just preceding the Devonian debate. (In Werner's system, all rocks precipitated from a universal ocean in order of density. Granite crystallized out first and more conventional sediments followed. The demonstration, by Hutton and others, that granitic cores of mountains had, in liquid form, intruded the overlying sediments from below, and did not form the base of a stratigraphic pile, turned Wernerian theory on its head – for the granite became younger than overlying sediment, not the oldest rock of all. Moreover, and more importantly, granite could be intruded at any time, and the mineralogical composition of a rock could not therefore specify its age.)

We can now understand Murchison's vehemence. William Smith in England (the surveyor and engineer adopted by the more patrician leaders of the Geological Society as their intellectual father) and Georges Cuvier in France had established fossils as a proper criterion of history. We now know, as Murchison and his colleagues did not, that evolution is the reason for the uniqueness of each fossil assemblage. Extinction is truly forever; once a group dies out, the hundred thousand unpredictable stages that led to its origin will never be repeated in precisely the same way. All genealogical systems of such complexity must share this property of generating unique sequences, and can therefore serve as criteria of history. But a scientist need not understand (or may incorrectly interpret) the causal basis of uniqueness. Its empirical demonstration will suffice.

By claiming that Coal Measure plant fossils could occur in Silurian or older rocks, De la Beche had made, in Murchison's view, the most retrograde possible step. He had abandoned the hard-fought and newly won fossil criterion of history, and had reverted to discredited rock type (the 'old' appearance of the greywacke). The plants, Murchison claimed, were conclusive; the strata containing them must date from the Coal Measures.

The Great Devonian Controversy traces the resolution of De la Beche's and Murchison's diametrically opposite views into 'a significant new piece of reliable knowledge about the natural world.' Through four hundred pages of the most dense and complex documentation I have ever read, Rudwick shows that the eventual resolution was not one of those dull compromises that mix a bit of this with a little of that and end up squarely in a lifeless middle between extreme alternatives. The basic resolution introduced a new dimension to the argument, a type of solution that neither side could have foreseen, and that emerged from a swirling context of debate when both initial positions had reached impasses. As Rudwick concludes in an apt metaphor:

> The battle lines defended by [the two initial interpretations] having initially faced each other in opposition, filtered silently through each other, as it were, until they faced outward, leaving at their rear a domain defended by them both. . . . The development of a successful interpretation . . . resulted in claimed knowledge that was unforeseen, unexpected, and above all *novel*.

De la Beche soon admitted that he had mismapped the structure of Devonshire. The strata with the disputed plants did lie on top of the stratigraphic pile, not within it. But De la Beche stuck to his central claim that the plants came from old Silurian rock. He insisted that the sequence contained no unconformities (temporal breaks in sedimentation); the plants lay atop the pile, but the entire pile was Silurian. With greater reluctance, Murchison finally dropped his a priori insistence that an unconformity must separate old rocks from Coal Measure plants atop the stratigraphic pile. He admitted continuity of sedimentation, but continued to insist, by the fossil criterion, that the plants could not be Silurian. These modified positions were closer, but still irreconcilable, in a way even more so because agreement had been reached about basic data of structure and sequence.

The eventual resolution required a novel interpretation: the rocks containing the plants, indeed most of the strata of Devonshire, were neither old (Silurian) nor young (Coal Measures) but representatives of a previously unrecognized time in between (the Devonian Period). The fossil criterion had been vindicated – for Murchison could not know that plants of Coal Measure type had arisen in an earlier unknown age, while he had stated correctly and for certain that these plants could not have inhabited the well documented Silurian strata. The rock-type criterion had failed again – greywacke did not mean old. The Devonian concept proved enormously fruitful. It quickly achieved firm status as a distinct, worldwide fauna, recognizable in rocks from Russia to New York. By the early 1840s, Devonian had entered the geological lexicon.

As the subtitle of his book attests, Rudwick views scientific knowledge as a social construction, uninterpretable as nature speaking directly to us through bits of fact in a logic divorced from human context. It mattered intensely that the Devonian was codified in Britain by Anglican gentlemen who viewed the Geological Society of London as their intellectual home. Rudwick takes no sides in the silly and fruitless debate between realism and relativism. The study of social setting does not imply either the irrelevance or nonexistence of a factual world out there. A lot happened between 410 and 360 million years ago, and our geological record has entombed the events in copious strata. But a good deal of arbitrariness, or rather social play, must accompany the parsing of continuous time into a series of discrete, named periods. The rocks didn't have to be called Devonian. And the boundaries are not naturally fixed at 410 and 360 million years; the strata contain no golden spikes. These are largely social decisions. (In fact, the boundaries of the Devonian are poorly defined by the general standard that major divisions of the time scale should be set by episodes of mass extinction to form more or less natural units – for one of the five major extinctions of life's history occurs *within* the Devonian.)

These kinds of decisions are also vitally important. Scientists are apt to say – but don't think for a moment that they believe it – 'Who cares what we call it so long as we know what happened.' Anyone with the slightest understanding of reward in science knows in his bones that accepted names and terms define status and priority – for actual history is soon forgotten. (Charles Lyell surely pulled the greatest 'fast one' in the history of science when he subsumed acceptable parts of both catastrophism

– the doctrine of periodic destruction by cataclysms – and his own excessively gradualistic and ahistorical world view under his favored name 'uniformitarianism,' with victory as the father of geology as his reward.)[2] Murchison, a keen debater and a self-promoter if ever there was one, understood this reality acutely and campaigned assiduously for his cherished term, Devonian. 'The perpetuity of a name affixed to any group of rocks through his original research,' he wrote, 'is the highest distinction to which any working geologist can aspire.'

I know from my own experience as a participant in major scientific debates that the explicit record of publication is utterly hopeless as a source of insight about shifts, forays, and resolutions. As Peter Medawar and others have argued, scientific papers are polite or self-serving fictions in their statements about doing science; they are, at best, logical reconstructions after the fact, written under the conceit that fact and argument shape conclusions by their own inexorable demands of reason. Levels of interacting complexity, contradictory motives, thoughts that lie too deep for either tears or even self-recognition – all combine to shape this most complex style of human knowledge.

Just consider some – a pitifully small sample – of the levels that must be grasped in order to understand the Devonian controversy (note also that all but the first do not enter the public record). We might begin with the pressures of data, for nature does speak to us in muted tones. Second, consider the ideological setting, so often unmentioned in published papers that emphasize the fiction of empirical determination. Both Murchison's allegiance to the fossil criterion and De la Beche's reluctance to grant it pride of place set the outlines of the debate. Basic attitudes to history itself lie embedded in these differing commitments.

Third, the overall social setting of science: The chief participants in the Devonian controversy were not academic geologists (for the profession did not yet exist in this form), but gentlemen of private means (or genteel poverty that threatened their commitment and forced them to scramble in an age that had not devised the concept of federal grants). Their published papers were archival, for the arguments had long before been aired and modified in open, but semiprivate debate at the Geological Society and other venues. (Modern science, with its 'invisible colleges,' proceeds no differently.) Although nearly one hundred men (yes, as another social reality, all men) participated, only a dozen or so really mattered, and they effectively excluded (or milked for their assessed value) all others who were not of the right class, degree of commitment or expertise, or simply couldn't get to London for the meetings.

Fourth, consider practical utility. Murchison persuaded the Czar of Russia that he was wasting resources by drilling for coal in a basin of Silurian rocks. If De la Beche had been right about the Devonshire plants, his majesty might have excavated much warmth.

Fifth, the social and economic position of scientists: Murchison was well-off and free to travel (he was also busy campaigning for a baronetcy, which he eventually obtained). De la Beche's social status was high enough, but his family fortunes had fallen on hard times. He persuaded the government to set him up as head of a geological survey – with salary. Government commitments placed him at a great disadvantage to the peripatetic Murchison; often, he could not even attend meetings since survey work required his presence in the field. When Murchison attacked his interpretation of Devonshire stratigraphy, De la Beche reacted with a fierceness hard to understand until you realize that a charge of incompetence in mapping truly threatened his livelihood and his ability to continue any geological work at all.

Sixth, differences in personality: Murchison, an aristocratic old soldier, was self-assured and curt almost to the point of egomania. He described all his activities in military metaphor. In his own words, he never did fieldwork simply to find out what happened, or had a friendly discussion with an adversary over a pipe and glass of wine; he waged campaigns. De la Beche, on the other hand, tended to modesty and self-effacement in his letters and public statements. He swore up and down, over and over again, that he didn't care who got the credit, so long as the truth became known. Taking them at face value, one could develop a strong fondness for De la Beche and an equally firm antipathy toward Murchison.

But here we must probe below the surface of private documents. In fact, both men behaved pretty much the same in their secretiveness and quest for status. De la Beche once privately sent some crucial rock samples to the British Museum, hoping to hide them from Murchison, for he could scarcely deny access if Murchison knew about them. Murchison found out and forced De la Beche's hand; De la Beche relented and swore (quite dishonestly) that he had meant no such thing. One ends up with a dubious feeling about De la Beche's two-facedness, and with a grudging admiration for Murchison's consistency, despite his brutal characterizations, as in this note to Adam Sedgwick:

> De la Beche is a dirty dog. . . . I know him to be a thorough jobber & a great intriguer & *we* have proved him to be thoroughly incompetent to carry on the survey. . . . *He writes in one style to you and in another to me.*

Rudwick somehow manages to encompass in his narrative all these influences, and all the swirling, sallying, trenching and retrenching of opinion in the debate itself. The book is organized as a very thick sandwich – sixty pages on context and methods, 340 pages of documented narrative, and sixty pages of interpretation and conclusions. Bowing to the reality of harried lives, Rudwick recognizes that not everyone will read every word of the meaty second section; he even explicitly gives us permission to skip if we get 'bogged down in the narrative.' Readers absolutely must not do such a thing; it should be illegal. The publisher should lock up the last sixty pages, and deny access to anyone who doesn't pass a multiple-choice exam inserted into the book between parts two and three. The value of this book lies directly in its detailed narrative.

I read this book, particularly its dense narrative, with great joy, but not without criticism. In my view, *The Great Devonian Controversy* suffers from two kinds of problems linked to its greatest strength – the enormous complexity of its tale. This is not easy to say without sounding flip or even anti-intellectual, but the brain, like the eye, cannot focus on all depths simultaneously. One can lose important aspects of the general pattern by concentrating too strictly upon intricate details. The old cliché about trees and forests is hackneyed because it has merit. I cherish what Rudwick has done; it is a monument to scholarship in an age of mediocrity. But I do think that the years immersed in detail, indeed the love one develops for each tiny facet, have led Rudwick to weigh some of the nuances wrongly, and to miss a major message that the Devonian resolution offered to the history of geology.

My first criticism is structural. Rudwick faced a monumental problem in deciding how to convey his unprecedented detail, so brilliantly worked out, in comprehensible form. He chose, unfortunately I think, a procedure that virtually forecloses full understanding of his book to all but probably a few hundred people in the world. He decided, in short, to follow with uncompromising strictness the historian's proviso that past events should not be read in the light of later knowledge, perforce unavailable to participants at the time. I would never dispute this credo as a general statement; I have fulminated against 'Whiggish' history with as much gusto as any man. But one can make almost a fetish of a good principle and find that it has turned against you.

Rudwick has chosen to organize his story by strict chronology, and absolutely never to mention, never even to drop the slightest hint about, any development that occurred later in time. I do appreciate the rationale. In 1835, no one knew that, a few years hence, a new block of ancient time would resolve the controversy. The injection of foreshadowing can only distort our understanding of people's positions as they developed. Thus Rudwick never breathes a word about the Devonian solution until Buckland, Sedgwick, and Murchison think of it themselves.

The result is faithful to narrative, but deeply confusing. And why not? The participants themselves were incredibly confused; tell it from their point of view, and what else can you derive? Is this style of presentation best for modern readers? I could make head or tail of Rudwick's narrative only because, as a professional geologist, I knew the eventual resolution. I don't think that this knowledge (which I couldn't expunge in any case) distorted my perspective; rather, it gave me an anchor to mesh the disparate threads into a coherent tale.

To cite just one example where modern information would clarify rather than distort, Devonian rocks had not been unknown in Britain before the controversy. One of the most famous and prominent of British formations, the Old Red Sandstone (another of those lovely quarrymen's terms), contains a rich fauna of Devonian fishes. The Old Red had not been resolved for an interesting reason: fossils of marine organisms form the standard stratigraphic sequence, but rocks of the Old Red are freshwater in origin. Thus, the Old Red cannot easily be correlated with the standard sequence. A simple point, easily made in a sentence – and so enormously helpful in understanding the difficulties that the Old Red presented to geologists of the 1830s (who assumed that these fishes had lived in the sea). The Old Red and its problems circulate through nearly every page of Rudwick's narrative, but he never tells us about the freshwater solution because it wasn't devised until after the 1840s – and we are left confused about the Old Red even after Rudwick's story ends. Only in a short appendix, 'The Devonian Modernized,' do we finally learn the solution (but how many people will read the appendices?). Again, as a professional geologist, I knew the answer from the start – and could spare myself the frustration of repeating the contemporary confusion in my own mind. Mystery writers don't tell the end at the beginning, but even their most complex stories are orders of magnitude simpler than the Devonian controversy. You do need a scorecard, at least partially filled in, to tell the players of Rudwick's drama.

Yet I strongly defend Rudwick's narrative style, storytelling in the grandest mode. Narrative has fallen from fashion; even historians are supposed to ape the stereotype of physics and be quantitative, or cliometric. Fine in its place, but not as a fetish. Narrative remains an art and science of the highest order, but of different form. How fitting that a book defending the importance of those scientists who established geological history should also defend so ably the narrative style of historical writing itself.*

My second criticism is conceptual, for I disagree fundamentally with Rudwick's interpretation of the controversy's resolution. Rudwick's analysis is so fine-grained, his interest in every item so intense, that he does not provide criteria for ranking the theoretical importance of various issues that the controversy resolved. Rudwick sees, because he knows the details so well, that nobody 'won' in the narrow sense of getting all the marbles. But how else are issues of truly great scope resolved? No one can possibly be right about everything the first time. When controversies pit first-rate scientists against one another, any resolution will take bits and pieces of all views. Who could be so misguided as to get everything absolutely wrong?

From this knowledge of intimate detail, Rudwick depicts the resolution as a grand compromise (with a novel solution, not a melding of original views). He even implies that social construction of anything this complex could scarcely, in principle, be judged by victories. He writes, for example: 'The finally consensual Devonian interpretation could be regarded as the analogue of a successfully negotiated treaty, precisely because it incorporated the non-negotiable positions of both sides and found a reconciliation between them.'

*Rudwick's last chapter contains several wondrously complex diagrams, outlining the changing views and their resolution, and the roles of various actors in the drama. Some will read these charts as a cliometric excursion. They will misunderstand Rudwick's intent. The charts are not a quantification; they have no scale except the chronology of years. One cannot quantify the magnitude of a changed opinion. The charts are pictorial models of narrative arguments, brilliantly conceived as epitomes.

I disagree. I read the Devonian resolution as a clear victory for Roderick Impey Murchison – as sweet and unalloyed as any I know in science. Rudwick's own data reinforce this claim. De la Beche largely drops out of his story about half way through. Murchison moves on to honor and recognition. He is hailed as the 'King of Siluria'; he receives medals from the Czar of all the Russias; his Devonian (name and concept) becomes a significant period of the earth's history. He views his own 'campaign' as victorious. His triumph, moreover, is no cynical result of his assiduous self-promotion. He wins because his concepts have succeeded as the foundation of historical reconstruction.

Murchison had begun with an outrageously audacious claim; he identified De la Beche's errors before he had seen any of the rocks. Of course Murchison was wrong about the Coal Measure age of the plants. But how could he have known that such plants also lived just before, in a time previously unrecognized? He did know that such plants, by the fossil criterion of history, could not be Silurian – and he was right.

When we step back far enough, we can view the Devonian debate as part of a larger struggle about proper criteria for the reconstruction of history. From this perspective, we can rank the relative importance of issues. The defense of the fossil criterion was paramount. Murchison said so again and again and directed his choicest invective at De la Beche's willingness to abandon this criterion for the convenience of resolving some rocks in Devonshire. Murchison was right, and his rightness remains the cornerstone of stratigraphic geology. It also permits us to understand the meaning of history.

Of course, the fossil criterion did not survive in the simplistic form that Murchison had advocated at the beginning – as I said, you never get everything right the first time. The correlation of Old Red fish with marine Devonian proved that different environments could sport different faunas at the same time – a point that De la Beche had urged. But this represents a refinement of the criterion, not a retreat. What you cannot have – or the criterion fails – is a wholesale extension, as De la Beche sought, of the same fauna through time; that is, you can find different faunas at the same time, but not the same fauna through greatly different times. History is a series of irreversible changes yielding a series of unique states.

I am not happy in the role of Miniver Cheevy. I have lived through the revolution of plate tectonics and know how excitement in geological theory feels. But some victories of knowledge are so central, so sweet, that once attained, we can never experience their like again. Could any intellectual thrill be greater than the codification of a proper criterion for history itself, and the subsequent discovery of a great chunk of distinctive time, valid and recognizable on a worldwide basis?

We must recover this commitment to history from the conventional and prejudicial ordering of sciences by status that dismisses the sequencing of events in time as mere narrative or description. What better way than Rudwick's elegant double achievement – an explicit defense of the wisdom and modernity of historical inquiry (narrative as a term of pride) applied to the codification of geological history as one of the greatest triumphs in human understanding.

Notes

1. A review of *The Great Devonian Controversy: The Shaping of Scientific Knowledge among Gentlemanly Specialists* by Martin J. S. Rudwick.
2. See my book *Time's Arrow, Time's Cycle* (Cambridge, MA: Harvard University Press, 1987).

NOT NECESSARILY A WING

FROM *Flesh Gordon* to *Alex in Wonderland*, title parodies have been a stock-in-trade of low comedy. We may not anticipate a tactical similarity between the mayhem of *Mad* magazine's movie reviews and the titles of major scientific works, yet two important nineteenth-century critiques of Darwin parodied his most famous phrases in their headings.

In 1887, E. D. Cope, the American paleontologist known best for his fossil feud with O. C. Marsh but a celebrated evolutionary theorist in his own right, published *The Origin of the Fittest* – a takeoff on Herbert Spencer's phrase, borrowed by Darwin as the epigram for natural selection: survival of the fittest. (Natural selection, Cope argued, could only preserve favorable traits that must arise in some other manner, unknown to Darwin. The fundamental issue of evolution cannot be the differential survival of adaptive traits, but their unexplained origin – hence the title parody.)

St George Mivart (1817–1900), a fine British zoologist, tried to reconcile his unconventional views on religion and biology but ended his life in tragedy, rejected by both camps. At age seventeen, he abandoned his Anglican upbringing, became a Roman Catholic, and consequently (in a less tolerant age of state religion) lost his opportunity for training in natural history at Oxford or Cambridge. He became a lawyer but managed to carve out a distinguished career as an anatomist nonetheless. He embraced evolution and won firm support from the powerful T. H. Huxley, but his strongly expressed and idiosyncratic anti-Darwinian views led to his rejection by the biological establishment of Britain. He tried to unite his biology with his religion in a series of books and essays, and ended up excommunicated for his trouble six weeks before his death.

Cope and Mivart shared the same major criticism of Darwin – that natural selection could explain the preservation and increase of favored traits but not their origin. Mivart, however, went gunning for a higher target than Darwin's epigram. He shot for the title itself, naming his major book *On the Genesis of Species* (1871). (Darwin, of course, had called his classic *On the Origin of Species*.)

Mivart's life may have ended in sadness and rejection thirty years later, but his *Genesis of Species* had a major impact in its time. Darwin himself offered strong, if grudging, praise and took Mivart far more seriously than any other critic, even adding a chapter to later editions of *The Origin of Species* primarily to counter Mivart's attack.

Mivart gathered, and illustrated 'with admirable art and force' (Darwin's words), all objections to the theory of natural selection – 'a formidable array' (Darwin's words again). Yet one particular theme, urged with special attention by Mivart, stood out as the centerpiece of his criticism. This argument continues to rank as the primary stumbling block among thoughtful and friendly scrutinizers of Darwinism today. No other criticism seems so troubling, so obviously and evidently 'right' (against a Darwinian claim that seems intuitively paradoxical and improbable).

Mivart awarded this argument a separate chapter in his book, right after the introduction. He also gave it a name, remembered ever since. He called his objection 'The Incompetency of 'Natural Selection' to Account for the Incipient Stages of Useful Structures.' If this phrase sounds like a mouthful, consider the easy translation: We can readily understand how complex and fully developed structures work and how their maintenance and preservation may rely upon natural selection – a wing, an eye, the resemblance of a bittern to a branch or of an insect to a stick or dead leaf. But how do you get from nothing to such an elaborate something if evolution must proceed through a long sequence of intermediate stages, each favored by natural selection? You can't fly with 2 percent of a wing or gain much protection from an iota's similarity with a potentially concealing piece of vegetation. How, in other words, can natural selection explain the incipient stages of structures that can only be used in much more elaborated form?

I take up this old subject for two reasons. First, I believe that Darwinism has, and has long had, an adequate and interesting resolution to Mivart's challenge (although we have obviously been mightily unsuccessful in getting it across). Second, a paper recently published in the technical journal *Evolution* has provided compelling experimental evidence for this resolution applied to its most famous case – the origin of wings.

The dilemma of wings – *the* standard illustration of Mivart's telling point about incipient stages – is set forth particularly well in a perceptive letter that I recently received from a reader, a medical doctor in California. He writes:

> How does evolutionary theory as understood by Darwin explain the emergence of items such as wings, since a small move toward a wing could hardly promote survival? I seem to be stuck with the idea that a significant quality of wing would have to spring forth all at once to have any survival value.

Interestingly, my reader's proposal that much or most of the wing must arise all at once (because incipient stages could have no adaptive value) follows Mivart's own resolution. Mivart first enunciated the general dilemma:

> Natural selection utterly fails to account for the conservation and development of the minute and rudimentary beginnings, the slight and infinitesimal commencements of structures, however useful those structures may afterwards become.

After fifty pages of illustration, he concludes: 'Arguments may yet be advanced in favor of the view that new species have from time to time manifested themselves with suddenness, and by modifications appearing at once.' Advocating this general solution for wings in particular, he concludes: 'It is difficult, then, to believe that the Avian limb was developed in any other way than by a comparatively sudden modification of a marked and important kind.'

Darwin's theory is rooted in the proposition that natural selection acts as the primary creative force in evolutionary change. This creativity will be expressed only if the fortuitous variation forming the raw material of evolutionary change can be accumulated sequentially in tiny doses, with natural selection acting as the sieve of acceptance. If new species arise all at once in an occasional lucky gulp, then selection has no creative role. Selection, at best, becomes an executioner, eliminating the unfit following this burst of good fortune. Thus, Mivart's solution – bypassing incipient stages entirely in a grand evolutionary leap – has always been viewed, quite rightly, as an anti-Darwinian version of evolutionary theory.

Darwin well appreciated the force, and potentially devastating extent, of Mivart's critique about incipient stages. He counterattacked with gusto, invoking the standard example of wings and arguing that Mivart's solution of sudden change presented more problems than it solved – for how can we believe that so complex a structure as a wing, made of so many coordinated and coadapted parts, could arise all at once:

> He who believes that some ancient form was transformed suddenly through an internal force or tendency into, for instance, one furnished with wings, will be . . . compelled to believe that many structures beautifully adapted to all the other parts of the same creature and to the surrounding conditions, have been suddenly produced; and of such complex and wonderful co-adaptations, he will not be able to assign a shadow of an explanation. . . . To admit all this is, as it seems to me, to enter into the realms of miracle, and to leave those of Science.

(This essay must now go in other directions but not without a small, tangential word in Mivart's defense. Mivart did appreciate the problem of complexity and coordination in sudden origins. He did not think that any old complex set of changes could arise all at once when needed – *that* would be tantamount to miracle. Most of Mivart's book studies the regularities of embryology and comparative anatomy to learn which kinds of complex changes might be possible as expressions and elaborations of developmental programs already present in ancestors. He advocates these changes as possible and eliminates others as fanciful.)

Darwin then faced his dilemma and developed the interestingly paradoxical resolution that has been orthodox ever since (but more poorly understood and appreciated than any other principle in evolutionary theory). If complexity precludes sudden origin, and the dilemma of incipient stages forbids gradual development in functional continuity, then how can we ever get from here to there? Darwin replies that we must reject an unnecessary hidden assumption in this argument – the notion of functional continuity. We will all freely grant that no creature can fly with 2 percent of a wing, but why must the incipient stages be used for flight? If incipient stages originally performed a different function suited to their small size and minimal development, natural selection might superintend their increase as adaptations for this original role until they reached a stage suitable for their current use. In other words, the problem of incipient stages disappears because these early steps were not inadequate wings but well-adapted something-elses. This principle of *functional change in structural continuity* represents Darwin's elegant solution to the dilemma of incipient stages.

Darwin, in a *beau geste* of argument, even thanked Mivart for characterizing the dilemma so well – all the better to grant Darwin a chance to elaborate his solution. Darwin writes: 'A good opportunity has thus been afforded [by Mivart] for enlarging a little on gradations of structure, often associated with changed functions – an important subject, which was not treated at sufficient length in the former editions of this work.' Darwin, who rarely added intensifiers to his prose, felt so strongly about this principle of functional shift that he wrote: 'In considering transitions of organs, it is so important to bear in mind the probability of conversion from one function to another.'

Darwin presented numerous examples in Chapters 5 and 7 of the final edition of *The Origin of Species*. He discussed organs that perform two functions, one primary, the other subsidiary, then relinquish the main use and elaborate the formerly inconspicuous operation. He then examined the flip side of this phenomenon – functions performed by two separate organs (fishes breathing with both lungs and gills). He argues that one organ may assume the entire function, leaving the other free for evolution to some other role (lungs for conversion to air bladders, for example, with respiration maintained entirely by gills). He does not, of course, neglect the classic example of wings, arguing that insects evolved their organs of flight from tracheae (or breathing organs – a minority theory today, but not without supporters). He writes: 'It is therefore highly probable that in this great class organs which once served for respiration have been actually converted into organs of flight.'

Darwin's critical theory of functional shift, usually (and most unfortunately) called the principle of 'preadaptation', has been with us for a century.* I believe that this principle has made so little headway not only because the basic formulation seems paradoxical and difficult, but mainly because we have so little firm, direct evidence for such functional shifts. Our technical literature contains many facile verbal arguments – little more than plausible 'just-so' stories. The fossil record also presents some excellent examples of sequential development through intermediary stages that could not work as modern organs do – but we lack a rigorous mechanical analysis of function at the various stages.

Let us return, as we must, to the classic case of wings. *Archaeopteryx*, the first bird, is as pretty an intermediate as paleontology could ever hope to find – a complex melange of reptilian and avian features. Scientists are still debating whether or not it could fly. If so, *Archaeopteryx* worked like the Wrights' biplane to a modern eagle's Concorde. But what did the undiscovered ancestors of *Archaeopteryx* do with wing rudiments that surely could not produce flight? Evolutionists have been invoking Darwin's principle of functional shift for more than 100 years, and the list of proposals is long. Proto-wings have been reconstructed as stabilizers, sexual attractors, or insect catchers. But the most popular hypothesis identifies thermoregulation as the original function of incipient stages that later evolved into

* This dreadful name has made a difficult principle even harder to grasp and understand. Preadaptation seems to imply that the proto-wing, while doing something else in its incipient stages, knew where it was going – predestined for a later conversion to flight. Textbooks usually introduce the word and then quickly disclaim any odor of foreordination. (But a name is obviously ill-chosen if it cannot be used without denying its literal meaning.) Of course, by 'preadaptation' we only mean that some structures are fortuitously suited to other roles if elaborated, not that they arise with a different future use in view – now there I go with the standard disclaimer. As another important limitation, preadaptation does not cover the important class of features that arise without functions (as developmental consequences of other primary adaptations, for example) but remain available for later co-optation. I suspect, for example, that many important functions of the human brain are co-opted consequences of building such a large computer for a limited set of adaptive uses. For these reasons, Elizabeth Vrba and I have proposed that the restrictive and confusing word 'preadaptation' be dropped in favor of the more inclusive term 'exaptation' – for any organ not evolved under natural selection for its current use – either because it performed a different function in ancestors (classical preadaptation) or because it represented a nonfunctional part available for later co-optation. See our technical article, 'Exaptation: A Missing Term in the Science of Form,' *Paleobiology*, 1981.

feathered wings. Feathers are modified reptilian scales, and they work very well as insulating devices. Moreover, if birds evolved from dinosaurs (as most paleontologists now believe), they arose from a lineage particularly subject to problems with temperature control. *Archaeopteryx* is smaller than any dinosaur and probably arose from the tiniest of dinosaur lineages. Small animals, with high ratios of surface area to volume, lose heat rapidly and may require supplementary devices for thermoregulation. Most dinosaurs could probably keep warm enough just by being large. Surface area (length x length, or length squared) increases more slowly than volume (length x length x length, or length cubed) as objects grow. Since animals generate heat over their volumes and lose it through their surfaces, small animals (with their relatively large surface areas) have most trouble keeping warm.

There I go again – doing what I just criticized. I have presented a plausible story about thermoregulation as the original function of organs that later evolved into wings. But science is tested evidence, not tall tales. This lamentable mode of storytelling has been used to illustrate Darwin's principle of functional shift only *faute de mieux* – because we didn't have the goods so ardently desired. At least until recently, when my colleagues Joel G. Kingsolver and M. A. R. Koehl published the first hard evidence to support a shift from thermoregulation to flight as a scenario for the evolution of wings. They studied insects, not birds – but the same argument has long been favored for nature's smaller and far more abundant wings (see their article, 'Aerodynamics, Thermoregulation, and the Evolution of Insect Wings: Differential Scaling and Evolutionary Change,' in *Evolution*, 1985).

In preparing this essay, I spent several days reading the classical literature on the evolution of insect flight – and emerged with a deeper understanding of just how difficult Darwin's principle of functional shift can be, even for professionals. Most of the literature hasn't even made the first step of applying functional shift at all, not to mention the later reform of substituting direct evidence for verbal speculation. Most reconstructions are still trying to explain the incipient stages of insect wings as somehow involved in airborne performance from the start – not for flapping flight, of course, but still for some aspect of motion aloft rather than, as Darwin's principle would suggest, for some quite different function.

To appreciate the dilemma of such a position (so well grasped by

Mivart more than 100 years ago), consider just one recent study (probably the best and most widely cited) and the logical quandaries that a claim of functional continuity entails. In 1964, J. W. Flower presented aerodynamic arguments for wings evolved from tiniest rudiment to elaborate final form in the interest of airborne motion. Flower argues, supporting an orthodox view, that wings evolved from tiny outgrowths of the body used for gliding prior to elaboration for sustained flight. But Flower recognizes that these incipient structures must themselves evolve from antecedents too small to function as gliding planes. What could these very first, slight outgrowths of the body be for? Ignoring Darwin's principle of functional shift, Flower searches for an aerodynamic meaning even at this very outset. He tries to test two suggestions: E. H. Hinton's argument that initial outgrowths served for 'attitude control,' permitting a falling insect to land in a suitable position for quick escape from predators; and a proposal of the great British entomologist Sir Vincent Wigglesworth (wonderful name for an insect man, I always thought) that such first stages might act as stabilizing or controlling devices during takeoff in small, passively aerial insects.

Flower proceeded by performing aerodynamic calculations on consequences of incipient wings for simple body shapes when dropped – and he quickly argued himself into an inextricable logical corner. He found, first of all, that tiny outgrowths might help, as Wigglesworth, Hinton, and others had suggested. But the argument foundered on another observation: The same advantages could be gained far more easily and effectively by another, readily available alternative route – evolution to small size (where increased surface/volume ratios retard falling and enhance the probability of takeoff). Flower then realized that he would have to specify a reasonably large body size for incipient wings to have any aerodynamic effect. But he then encountered another problem: At such sizes, legs work just as well as, if not better than, proto-wings for any suggested aerodynamic function. Flower admitted:

> The first conclusion to be drawn from these calculations is that the selective pressure in small insects is towards smaller insects, which would have no reason to evolve wings.

I would have stopped and searched elsewhere (in Darwin's principle of functional shift) at this point, but Flower bravely continued along an improbable path:

> The main conclusions, however, are that attitude control of insects would be by the use of legs or by very small changes in body shape [*i.e.*, by evolving small outgrowths, or proto-wings].

Flower, in short, never considered an alternative to his assumption of functional continuity based upon some aspect of aerial locomotion. He concluded:

> At first they [proto-wings] would affect attitude; later they could increase to a larger size and act as a true wing, providing lift in their own right. Eventually they could move, giving the insect greater maneuverability during descent, and finally they could 'flap,' achieving sustained flight.

As an alternative to such speculative reconstructions that work, in their own terms, only by uncomfortable special pleading, may I suggest Darwin's old principle of functional shift (preadaptation – ugh – for something else).

The physiological literature contains voluminous testimony to the thermodynamic efficiency of modern insect wings: in presenting, for example, a large surface area to the sun for quick heating. If wings can perform this subsidiary function now, why not suspect thermoregulation as a primary role at the outset? M. M. Douglas, for example, showed that, in *Colias* butterflies, only the basal one-third of the wing operates in thermoregulation – an area approximately equal to the thoracic lobes (proto-wings) of fossil insects considered ancestral to modern forms.

Douglas then cut down some *Colias* wings to the actual size of these fossil ancestral lobes and found that insects so bedecked showed a 55 percent greater increase in body temperature than bodies deprived of wings entirely. These manufactured proto-wings measured 5 by 3 millimeters on a body 15 millimeters long. Finally, Douglas determined that no further thermoregulatory advantage could be gained by wings longer than 10 millimeters on a 15-millimeter body.

Kingsolver and Koehl performed a host of elaborate and elegant experiments to support a thermoregulatory origin of insect proto-wings. As with so many examples of excellent science producing clear and interesting outcomes, the results can be summarized briefly and cleanly.

Kingsolver and Koehl begin by tabulating all the aerodynamic hypotheses usually presented in the literature as purely verbal speculations. They arrange these proposals of functional continuity (the explanations that do not follow Darwin's solution of Mivart's dilemma) into three basic categories: proto-wings for gliding (aerofoils for steady-state motion), for parachuting (slowing the rate of descent in a falling insect), and attitude stability (helping an insect to land right side up). They then transcended the purely verbal tradition by developing aerodynamic equations for exactly how proto-wings should help an insect under these three hypotheses of continuity in adaptation (increasing the lift/drag ratio as the major boost to gliding, increasing drag to slow the descent rate in parachuting, measuring the moment about the body axis produced by wings for the hypothesis of attitude stability).

They then constructed insect models made of wire, epoxy, and other appropriate materials to match the sizes and body shapes of flying and nonflying forms among early insect fossils. To these models, they attached wings (made of copper wire enclosing thin, plastic membranes) of various lengths and measured the actual aerodynamic effects for properties predicted by various hypotheses of functional continuity. The results of many experiments in wind tunnels are consistent and consonant: Aerodynamic benefits begin for wings above a certain size, and they increase as wings get larger. But at the small sizes of insect proto-wings, aerodynamic advantages are absent or insignificant and do not increase with growing wing length. These results are independent of body shape, wind velocity, presence or placement of legs, and mounting position of wings. In other words, large wings work well and larger wings work better – but small wings (at the undoubted sizes of Mivart's troubling incipient stages) provide no aerodynamic edge.

Kingsolver and Koehl then tested their models for thermoregulatory effects, constructing wings from two materials with different thermal conductivities (construction paper and aluminum foil) and measuring the increased temperature of bodies supplied with wings of various lengths versus wingless models. They achieved results symmetrically opposite to the aerodynamic experiments. For thermoregulation, wings work well at the smallest sizes, with benefits increasing as the wing grows. However, beyond a measured length, further increase of the wing confers no additional effect. Kingsolver and Koehl conclude:

> At any body size, there is a relative wing length above which there is no additional thermal effect, and below which there is no significant aerodynamic effect.

We could not hope for a more elegant experimental confirmation of Darwin's solution to Mivart's challenge. Kingsolver and Koehl have actually measured the functional shift by showing that incipient wings aid thermoregulation but provide no aerodynamic benefit – while larger wings provide no further thermoregulatory oomph but initiate aerodynamic advantage and increase the benefits steadily thereafter. The crucial intermediate wing length, where thermoregulatory gain ceases and aerodynamic benefits begin, represents a domain of functional shift, as aerodynamic advantages pick up the relay from waning thermoregulation to continue the evolutionary race to increasing wing size.

But what might push an insect across the transition? Why reach this crucial domain at all? If wings originally worked primarily for thermoregulation, why not just stop as the length of maximum benefit approached? Here, Kingsolver and Koehl present an interesting speculation based on another aspect of their data. They found that the domain of transition between thermal and aerial effects varied systematically with body size: The larger the body, the sooner the transition (in terms of relative wing length). For a body 2 centimeters long, the transition occurred with wings 40 to 60 percent of body length; but a 10-centimeter body switches to aerodynamic advantage at only 10 percent of body length.

Now suppose that incipient ancestral wings worked primarily for thermoregulation and had reached a stable, optimum size for greatest benefit. Natural selection would not favor larger wings and a transition to the available domain of aerodynamic advantage. But if body size increased for other reasons, an insect might reach the realm of aerial effects simply by growing larger, without any accompanying change of body shape or relative wing length.

We often think, naïvely, that size itself should make no profound difference. Why should just more of the same have any major effect beyond simple accumulation? Surely, any major improvement or alteration must require an extensive and explicit redesign, a complex reordering of parts with invention of new items.

Nature does not always match our faulty intuitions. Complex objects often display the interesting and paradoxical property of major effect for apparently trifling input. Internal complexity can translate a simple quantitative change into a wondrous alteration of quality. Perhaps that greatest and most effective of all evolutionary inventions, the origin of human consciousness, required little more than an increase of brain power to a level where internal connections became rich and varied enough to force this seminal transition. The story may be much more complex, but we have no proof that it must be.

Voltaire quipped that 'God is always for the big battalions.' More is not always better, but more can be very different.

WORM FOR A CENTURY, AND ALL SEASONS

IN the preface to his last book, an elderly Charles Darwin wrote: 'The subject may appear an insignificant one, but we shall see that it possesses some interest; and the maxim "de minimis lex non curat" [the law is not concerned with trifles] does not apply to science.'

Trifles may matter in nature, but they are unconventional subjects for last books. Most eminent graybeards sum up their life's thought and offer a few pompous suggestions for reconstituting the future. Charles Darwin wrote about worms – *The Formation of Vegetable Mould, Through the Action of Worms, With Observations on Their Habits* (1881).

This month* marks the one-hundredth anniversary of Darwin's death – and celebrations are under way throughout the world. Most symposiums and books are taking the usual high road of broad implication – Darwin and modern life, or Darwin and evolutionary thought. For my personal tribute, I shall take an ostensibly minimalist stance and discuss Darwin's 'worm book.' But I do this to argue that Darwin justly reversed the venerable maxim of his legal colleagues.

Darwin was a crafty man. He liked worms well enough, but his last book, although superficially about nothing else, is (in many ways) a covert summation of the principles of reasoning that he had labored a lifetime to identify and use in the greatest transformation of nature ever wrought by a single man. In analyzing his concern with worms, we may grasp the sources of Darwin's general success.

* Darwin died on April 19, 1882 and this column first appeared in *Natural History* in April 1982.

The book has usually been interpreted as a curiosity, a harmless work of little importance by a great naturalist in his dotage. Some authors have even used it to support a common myth about Darwin that recent scholarship has extinguished. Darwin, his detractors argued, was a man of mediocre ability who became famous by the good fortune of his situation in place and time. His revolution was 'in the air' anyway, and Darwin simply had the patience and pertinacity to develop the evident implications. He was, Jacques Barzun once wrote (in perhaps the most inaccurate epitome I have ever read), 'a great assembler of facts and a poor joiner of ideas . . . a man who does not belong with the great thinkers.'

To argue that Darwin was merely a competent naturalist mired in trivial detail, these detractors pointed out that most of his books are about minutiae or funny little problems – the habits of climbing plants, why flowers of different form are sometimes found on the same plant, how orchids are fertilized by insects, four volumes on the taxonomy of barnacles, and finally, how worms churn the soil. Yet all these books have both a manifest and a deeper or implicit theme – and detractors missed the second (probably because they didn't read the books and drew conclusions from the titles alone). In each case, the deeper subject is evolution itself or a larger research program for analyzing history in a scientific way.

Why is it, we may ask at this centenary of his passing, that Darwin is still so central a figure in scientific thought? Why must we continue to read his books and grasp his vision if we are to be competent natural historians? Why do scientists, despite their notorious unconcern with history, continue to ponder and debate his works? Three arguments might be offered for Darwin's continuing relevance to scientists.

We might honor him first as the man who 'discovered' evolution. Although popular opinion may grant Darwin this status, such an accolade is surely misplaced, for several illustrious predecessors shared his conviction that organisms are linked by ties of physical descent. In nineteenth-century biology, evolution was a common enough heresy.

As a second attempt, we might locate Darwin's primary claim upon continued scientific attention in the extraordinarily broad and radical implications of his proffered evolutionary mechanism – natural selection. Indeed, I have pushed this theme relentlessly in my two previous books, focusing upon three arguments: natural selection as a theory of local adaptation, not inexorable progress; the claim that order in nature arises as a coincidental by-product of struggle among individuals; and the materialistic character of Darwin's theory, particularly his denial of any causal role to spiritual forces, energies, or powers. I do not now abjure this theme, but I have come to realize that it cannot represent the major reason for Darwin's continued *scientific* relevance, though it does account for his impact upon the world at large. For it is too grandiose, and working scientists rarely traffic in such abstract generality.

Everyone appreciates a nifty idea or an abstraction that makes a person sit up, blink hard several times to clear the intellectual cobwebs, and reverse a cherished opinion. But science deals in the workable and soluble, the idea that can be fruitfully embodied in concrete objects suitable for poking, squeezing, manipulating, and extracting. The idea that counts in science must lead to fruitful work, not only to speculation that does not engender empirical test, no matter how much it stretches the mind.

I therefore wish to emphasize a third argument for Darwin's continued importance, and to claim that his greatest achievement lay in establishing principles of *useful* reason for sciences (like evolution) that attempt to reconstruct history. The special problems of historical science (as contrasted, for example, with experimental physics) are many, but one stands out most prominently: Science must identify processes that yield observed results. The results of history lie strewn around us, but we cannot, in principle, directly observe the processes that produced them. How then can we be scientific about the past?

As a general answer, we must develop criteria for inferring the processes we cannot see from results that have been preserved. This is the quintessential problem of evolutionary theory: How do we use the anatomy, physiology, behavior, variation, and geographic distribution of modern organisms, and the fossil remains in our geological record, to infer the pathways of history?

Thus, we come to the covert theme of Darwin's worm book, for it is both a treatise on the habits of earthworms and an exploration of how we can approach history in a scientific way.

Darwin's mentor, the great geologist Charles Lyell, had been obsessed with the same problem. He argued, though not with full justice, that his predecessors had failed to construct a science of geology because they had not developed procedures for inferring an unobservable past from a surrounding present and had therefore indulged in unprovable reverie and speculation. 'We see,' he wrote in his incomparable prose, 'the ancient spirit of speculation revived and a desire manifestly shown to cut, rather than patiently to untie, the Gordian Knot.' His solution, an aspect of the complex world view later called uniformitarianism, was to observe the work of present processes and to extrapolate their rates and effects into the past. Here Lyell faced a problem. Many results of the past – the Grand Canyon for example – are extensive and spectacular, but most of what goes on about us every day doesn't amount to much – a bit of erosion here or deposition there. Even a Stromboli or a Vesuvius will cause only local devastation. If modern forces do too little, then we must invoke more cataclysmic processes, now expired or dormant, to explain the past. And we are in catch-22: if past processes were effective and different from present processes, we might explain the past in principle, but we could not be scientific about it because we have no modern analogue in what we can observe. If we rely only upon present processes, we lack sufficient oomph to render the past.

Lyell sought salvation in the great theme of geology: time. He argued that the vast age of our earth provides ample time to render all observed results, however spectacular, by the simple summing of small changes over immense periods. Our failure lay, not with the earth, but with our habits of mind: we had been previously unwilling to recognize how much work the most insignificant processes can accomplish with enough time.

Darwin approached evolution in the same way. The present becomes relevant, and the past therefore becomes scientific, only if we can sum the small effects of present processes to produce observed results. Creationists did not use this principle and therefore failed to understand the relevance of small-scale variation that pervades the biological world (from breeds of dogs to geographical variation in butterflies). Minor variations are the stuff of evolution (not merely a set of accidental excursions around a created ideal type), but we recognize this only when we are prepared to sum small effects through long periods of time.

Darwin recognized that this principle, as a basic mode of reasoning in historical science, must extend beyond evolution. Thus, late in his life, he decided to abstract and exemplify his historical method by applying it to a problem apparently quite different from evolution – a project broad enough to cap an illustrious career. He chose earthworms and the soil. Darwin's refutation of the legal maxim 'de minimis lex non curat' was a conscious double entendre. Worms are both humble and interesting, and a worm's work, when summed over all worms and long periods of time, can shape our landscape and form our soils.

Thus, Darwin wrote at the close of his preface, refuting the opinions of a certain Mr Fish who denied that worms could account for much 'considering their weakness and their size':

> Here we have an instance of that inability to sum up the effects of a continually recurrent cause, which has often retarded the progress of science, as formerly in the case of geology, and more recently in that of the principle of evolution.

Darwin had chosen well to illustrate his generality. What better than worms: the most ordinary, commonplace, and humble objects of our daily observation and dismissal. If they, working constantly beneath our notice, can form much of our soil and shape our landscape, then what event of magnitude cannot arise from the summation of small effects. Darwin had not abandoned evolution for earthworms; rather, he was using worms to illustrate the general method that had validated evolution as well. Nature's mills, like God's, grind both slowly and exceedingly small.

Darwin made two major claims for worms. First, in shaping the land, their effects are directional. They triturate particles of rock into ever smaller fragments (in passing them through their gut while churning the soil), and they denude the land by loosening and disaggregating the soil as they churn it; gravity and erosive agents then move the soil more easily from high to low ground, thus leveling the landscape. The low, rolling character of topography in areas inhabited by worms is, in large part, a testimony to their slow but persistent work.

Second, in forming and churning the soil, they maintain a steady state amidst constant change. As the primary theme of his book (and the source of its title), Darwin set out to prove that worms form the soil's upper layer, the so-called vegetable mold. He describes it in the opening paragraph:

> The share which worms have taken in the formation of the layer of vegetable mould, which covers the whole surface of the land in every moderately humid country, is the subject

> of the present volume. This mould is generally of a blackish color and a few inches in thickness. In different districts it differs but little in appearance, although it may rest on various subsoils. The uniform fineness of the particles of which it is composed is one of its chief characteristic features.

Darwin argues that earthworms form vegetable mold by bringing 'a large quantity of fine earth' to the surface and depositing it there in the form of castings. (Worms continually pass soil through their intestinal canals, extract anything they can use for food, and 'cast' the rest; the rejected material is not feces but primarily soil particles, reduced in average size by trituration and with some organic matter removed.) The castings, originally spiral in form and composed of fine particles, are then disaggregated by wind and water, and spread out to form vegetable mold. 'I was thus led to conclude,' Darwin writes, 'that all the vegetable mould over the whole country has passed many times through, and will again pass many times through, the intestinal canals of worms.'

The mold doesn't continually thicken after its formation, for it is compacted by pressure into more solid layers a few inches below the surface. Darwin's theme here is not directional alteration, but continuous change within apparent constancy. Vegetable mold is always the same, yet always changing. Each particle cycles through the system, beginning at the surface in a casting, spreading out, and then working its way down as worms deposit new castings above; but the mold itself is not altered. It may retain the same thickness and character while all its particles cycle. Thus, a system that seems to us stable, perhaps even immutable, is maintained by constant turmoil. We who lack an appreciation of history and have so little feel for the aggregated importance of small but continuous change scarcely realize that the very ground is being swept from beneath our feet; it is alive and constantly churning.

Darwin uses two major types of arguments to convince us that worms form the vegetable mold. He first proves that worms are sufficiently numerous and widely spread in space and depth to do the job. He demonstrates 'what a vast number of worms live unseen by us beneath our feet' – some 53,767 per acre (or 356 pounds of worms) in good British soil. He then gathers evidence from informants throughout the world to argue that worms are far more widely distributed, and in a greater range of apparently unfavorable environments, than we usually imagine. He digs to see how deeply they extend into the soil, and cuts one in two at fifty-five inches, although others report worms at eight feet down or more.

With plausibility established, he now seeks direct evidence for constant cycling of vegetable mold at the earth's surface. Considering both sides of the issue, he studies the foundering of objects into the soil as new castings pile up above them, and he collects and weighs the castings themselves to determine the rate of cycling.

Darwin was particularly impressed by the evenness and uniformity of foundering for objects that had once lain together at the surface. He sought fields that, twenty years or more before, had been strewn with objects of substantial size – burned coals, rubble from the demolition of a building, rocks collected from the plowing of a neighboring field. He trenched these fields and found, to his delight, that the objects still formed a clear layer, parallel to the surface but now several inches below it and covered with vegetable mold made entirely of fine particles. 'The straightness and regularity of the lines formed by the embedded objects, and their parallelism with the surface of the land, are the most striking features of the case,' he wrote. Nothing could beat worms for a slow and meticulous uniformity of action.

Darwin studied the sinking of 'Druidical stones' at Stonehenge and the foundering of Roman bathhouses, but he found his most persuasive example at home, in his own field, last plowed in 1841:

> For several years it was clothed with an extremely scant vegetation, and was so thickly covered with small and large flints (some of them half as large as a child's head) that the field was always called by my sons 'the stony field.' When they ran down the slope the stones clattered together. I remember doubting whether I should live to see these larger flints covered with vegetable mould and turf. But the smaller stones disappeared before many years had elapsed, as did every one of the larger ones after a time; so that after thirty years (1871) a horse could gallop over the compact turf from one end of the field to the other, and not strike a single stone with his shoes. To anyone who remembered the appearance of the field in 1842, the transformation was wonderful. This was certainly the work of the worms.

In 1871, he cut a trench in his field and found 2.5 inches of vegetable mold, entirely free from flints: 'Beneath this lay coarse clayey earth full of flints, like that in any of the neighboring ploughed fields. . . . The average rate of accumulation of the mould during the whole thirty years was only .083 inch per year (i.e., nearly one inch in twelve years).'

In various attempts to collect and weigh castings directly, Darwin estimated from 7.6 to 18.1 tons per acre per year. Spread out evenly upon the surface, he calculated that from 0.8 to 2.2 inches of mold would form anew every ten years. In gathering these figures, Darwin relied upon that great, unsung, and so characteristically British institution – the corps of zealous amateurs in natural history, ready to endure any privation for a precious fact. I was particularly impressed by one anonymous contributor: 'A lady,' Darwin tells us, 'on whose accuracy I can implicitly rely, offered to collect during a year all the castings thrown up on two separate square yards, near Leith Hill Place, in Surrey.' Was she the analogue of a modern Park Avenue woman of means, carefully scraping up after her dog: one bag for a cleaner New York, the other for Science with a capital S?

The pleasure of reading Darwin's worm book lies not only in recognizing its larger point but also in the charm of detail that Darwin provides about worms themselves. I would rather peruse 300 pages of Darwin on worms than slog through thirty pages of eternal verities explicitly preached by many writers. The worm book is a labor of love and intimate, meticulous detail. In the book's other major section, Darwin spends 100 pages describing experiments to determine which ends of leaves (and triangular paper cutouts, or abstract 'leaves') worms pull into their burrows first. Here we also find an overt and an underlying theme, in this case leaves and burrows versus the evolution of instinct and intelligence, Darwin's concern with establishing a usable definition of intelligence, and his discovery (under that definition) that intelligence pervades 'lower' animals as well. All great science is a fruitful marriage of detail and generality, exultation and explanation. Both Darwin and his beloved worms left no stone unturned.

I have argued that Darwin's last book is a work on two levels – an explicit treatise on worms and the soil and a covert discussion of how to learn about the past by studying the present. But was Darwin consciously concerned with establishing a methodology for historical science, as I have argued, or did he merely stumble into such generality in his last book? I believe that his worm book follows the pattern of all his other works, from first to last: every compendium on minutiae is also a treatise on historical reasoning – and each book elucidates a different principle.

Consider his first book on a specific subject, *The Structure and Distribution of Coral-Reefs* (1842). In it, he proposed a theory for the formation of atolls, 'those singular rings of coral-land which rise abruptly out of the unfathomable ocean,' that won universal acceptance after a century of subsequent debate. He argued that coral reefs should be classified into three categories – fringing reefs that abut an island or continent, barrier reefs separated from island or continent by a lagoon, and atolls, or rings of reefs, with no platform in sight. He linked all three categories with his 'subsidence theory,' rendering them as three stages of a single process: the subsidence of an island or continental platform beneath the waves as living coral continues to grow upward. Initially, reefs grow right next to the platform (fringing reefs). As the platform sinks, reefs grow up and outward, leaving a separation between sinking platform and living coral (a barrier reef). Finally the platform sinks entirely, and a ring of coral expresses its former shape (an atoll). Darwin found the forms of modern reefs 'inexplicable, excepting on the theory that their rocky bases slowly and successively sank beneath the level of the sea, whilst the corals continued to grow upwards.'

This book is about coral, but it is also about historical reasoning. Vegetable mold formed fast enough to measure its rate directly; we capture the past by summing effects of small and observable present causes. But what if rates are too slow, or scales too large, to render history by direct observation of present processes? For such cases, we must develop a different method. Since large-scale processes begin at different times and proceed at diverse rates, the varied stages of different examples should exist simultaneously in the present. To establish history in such cases, we must construct a theory that will explain a series of present phenomena as stages of a single historical process. The method is quite general. Darwin used it to explain the formation of coral reefs. We invoke it today to infer the history of stars. Darwin also employed it to establish organic evolution itself. Some species are just beginning to split from their ancestors, others are midway through the process, still others are on the verge of completing it.

But what if evidence is limited to the static object itself? What if we can neither watch part of its formation nor find several stages of the process that produced it? How can we infer history from a lion? Darwin treated this problem in his treatise on the fertilization of orchids by insects (1862); the book that directly followed *The Origin of Species*. I have discussed his solution in several essays and will not dwell on it here: we infer history from imperfections that record constraints of descent. The 'various contrivances' that orchids use to attract insects and attach pollen to them are the highly altered parts of ordinary flowers, evolved in ancestors for other purposes. Orchids work well enough, but they are jury-rigged to succeed because flowers are not optimally constructed for modification to these altered roles. If God wanted to make insect attractors and pollen stickers from scratch, he would certainly have built differently.

Thus, we have three principles for increasing adequacy of data: if you must work with a single object, look for imperfections that record historical descent; if several objects are available, try to render them as stages of a single historical process; if processes can be directly observed, sum up their effects through time. One may discuss these principles directly or recognize the 'little problems' that Darwin used to exemplify them: orchids, coral reefs, and worms – the middle book, the first, and the last.

Darwin was not a conscious philosopher. He did not, like Huxley and Lyell, write explicit treatises on methodology. Yet I do not think he was unaware of what he was doing, as he cleverly composed a series of books at two levels, thus expressing his love for nature in the small and his ardent desire to establish both evolution and the principles of historical science. I was musing on this issue as I completed the worm book two weeks ago. Was Darwin really conscious of what he had done as he wrote his last professional lines, or did he proceed intuitively, as men of his genius sometimes do? Then I came to the very last paragraph, and I shook with the joy of insight. Clever old man; he knew full well. In his last words, he looked back to his beginning, compared those worms with his first corals, and completed his life's work in both the large and the small:

> The plough is one of the most ancient and most valuable of man's inventions; but long before he existed the land was in fact regularly ploughed, and still continues to be thus ploughed by earthworms. It may be doubted whether there are many other animals which have played so important a part in the history of the world, as have these lowly organ-

> ized creatures. Some other animals, however, still more lowly organized, namely corals, have done more conspicuous work in having constructed innumerable reefs and islands in the great oceans: but these are almost confined to the tropical zones.

At the risk of unwarranted ghoulishness. I cannot suppress a final irony. A year after publishing his worm book, Darwin died on April 19, 1882. He wished to be buried in the soil of his adopted village, where he would have made a final and corporeal gift to his beloved worms. But the sentiments (and politicking) of fellow scientists and men of learning secured a guarded place for his body within the well-mortared floor of Westminster Abbey. Ultimately the worms will not be cheated, for there is no permanence in history, even for cathedrals. But ideas and methods have all the immortality of reason itself. Darwin has been gone for a century, yet he is with us whenever we choose to think about time.

THE DARWINIAN GENTLEMAN AT MARX'S FUNERAL: RESOLVING EVOLUTION'S ODDEST COUPLING

WHAT could possibly be deemed incongruous on a shelf of Victorian bric-a-brac, the ultimate anglophonic symbol for miscellany? What, to illustrate the same principle on a larger scale, could possibly seem out of place in London's Highgate Cemetery – the world's most fantastic funerary park of overgrown vegetation and overblown statuary, described as a 'Victorian Valhalla . . . a maze of rising terraces, winding paths, tombs and catacombs . . . a monument to the Victorian age and to the Victorian attitude to death . . . containing some of the most celebrated – and often most eccentric – funerary architecture to be found anywhere' (from *Highgate Cemetery* by F. Barker and J. Gay, published in 1984 by John Murray in London, the same firm that printed all Darwin's major books – score one for British continuity!).

Highgate holds a maximal variety of mortal remains from Victoria's era – from eminent scientists like Michael Faraday, to literary figures like George Eliot, to premier pundits like Herbert Spencer, to idols of popular culture like Tom Sayers (one of the last champions of bare-knuckle boxing), to the poignancy of early death for ordinary folks – the young Hampstead girl 'who was burned to death when her dress caught fire,' or 'Little Jack,' described as 'the boy missionary,' who died at age seven on the shores of Lake Tanganyika in 1899.

But one monument in Highgate Cemetery might seem conspicuously out of place, to people who have forgotten an odd fact from their high-school course in European history. The grave of Karl Marx stands almost adjacent to the tomb of his rival and arch opponent of all state intervention (even for street lighting and sewage systems), Herbert Spencer.

The apparent anomaly only becomes exacerbated by the maximal height of Marx's monument, capped by an outsized bust. (Marx had originally been buried in an inconspicuous spot adorned by a humble marker, but visitors complained that they could not find the site, so in 1954, with funds raised by the British Communist Party, Marx's gravesite reached higher and more conspicuous ground. To highlight the anomaly of his presence, this monument, until the past few years at least, attracted a constant stream of the most dour, identically suited groups of Russian or Chinese pilgrims, all snapping their cameras, or laying their 'fraternal' wreaths.)

Marx's monument may be out of scale, but his presence could not be more appropriate. Marx lived most of his life in London, following exile from Belgium, Germany, and France for his activity in the revolution of 1848 (and for general political troublemaking: he and Engels had just published the *Communist Manifesto*). Marx arrived in London in August 1848, at age thirty-one, and lived there until his death in 1883. He wrote all his mature works as an expatriate in England; and the great (and free) library of the British Museum served as his research base for *Das Kapital*.

Let me now introduce another anomaly, not so easily resolved this time, about the death of Karl Marx in London. This item, in fact, ranks as my all-time-favorite, niggling little incongruity from the history of my profession of evolutionary biology. I have been living with this bothersome fact for twenty-five years, and I made a pledge to myself long ago that I would try to discover some resolution before ending this series of essays. Let us, then, return to Highgate Cemetery, and to Karl Marx's burial on March 17, 1883.

Friedrich Engels, Marx's lifelong friend and collaborator (also his financial 'angel,' thanks to a family textile business in Manchester), reported the short, small, and modest proceedings. Engels himself gave a brief speech in English that included the following widely quoted comment: 'Just as Darwin discovered the law of evolution in organic nature, so Marx discovered the law of evolution in human history.' Contemporary reports vary somewhat, but the most generous count places only nine mourners at the gravesite – a disconnect between immediate notice and later influence exceeded, perhaps, only by Mozart's burial in a pauper's grave. (I exclude, of course, famous men like Bruno and Lavoisier, executed by state power and therefore officially denied any funerary rite.)

The list, not even a minyan in length, makes sense (with one exception): Marx's wife and daughter (another daughter had died recently, thus increasing Marx's depression and probably hastening his end): his two French socialist sons-in-law, Charles Longuet and Paul Lafargue: and four nonrelatives with long-standing ties to Marx, and impeccable socialist and activist credentials: Wilhelm Liebknecht, a founder and leader of the German Social-Democratic Party (who gave a rousing speech in German, which, together with Engels's English oration, a short statement in French by Longuet, and the reading of two telegrams from workers' parties in France and Spain, constituted the entire program of the burial); Friedrich Lessner, sentenced to three years in prison at the Cologne Communist trial of 1852; G. Lochner, described by Engels as 'an old member of the Communist League'; and Carl Schorlemmer, a professor of chemistry in Manchester, but also an old communist associate of Marx and Engels, and a fighter at Baden in the last uprising of the 1848 revolution.

But the ninth and last mourner seems to fit about as well as that proverbial snowball in hell or that square peg trying to squeeze into a round hole: E. Ray Lankester (1847–1929), already a prominent young British evolutionary biologist and leading disciple of Darwin, but later to become – as Professor Sir E. Ray Lankester K.C.B. (Knight, Order of the Bath), M.A. (the 'earned' degree of Oxford or Cambridge), D.Sc. (a later honorary degree as doctor of science), F.R.S. (Fellow of the Royal Society, the leading honorary academy of British science) – just about the most celebrated, and the stuffiest, of conventional and socially prominent British scientists. Lankester moved up the academic ladder from exemplary beginnings to a maximally prominent finale, serving as professor of zoology at University College London, then as Fullerian Professor of Physiology at the Royal Institution, and finally as Linacre Professor of Comparative Anatomy at Oxford University. Lankester then capped his career by serving as director (from 1898 to 1907) of the British Museum (Natural History), the most powerful and prestigious post in his field. Why, in heaven's name, was this exemplar of British respectability, this basically conservative scientist's scientist, hanging out with a group of old (and mostly German) communists at the funeral of a man described by Engels, in his graveside oration, as 'the best hated and most calumniated man of his times'?

Even Engels seemed to sense the anomaly, when he ended his official report of the funeral, published in *Der Sozialdemokrat* of Zürich on March 22, 1883, by writing: 'The natural sciences were represented by two celebrities of the first rank, the zoology Professor Ray Lankester and the chemistry Professor Schorlemmer, both members of the Royal Society of London.' Yes, but Schorlemmer was a countryman, a lifelong associate, and a political ally. Lankester did not meet Marx until 1880, and could not, by any stretch of imagination, be called a political supporter, or even a sympathizer (beyond a very general, shared belief in human improvement through education and social progress). As I shall discuss in detail later in this essay, Marx first sought Lankester's advice in recommending a doctor for his ailing wife and daughter, and later for himself. This professional connection evidently developed into a firm friendship. But what could have drawn these maximally disparate people together?

We certainly cannot seek the primary cause for warm sympathy in any radical cast to Lankester's biological work that might have matched the tenor of Marx's efforts in political science. Lankester may rank as the best evolutionary morphologist in the first generation that worked through the implications of Darwin's epochal discovery. T. H. Huxley became Lankester's guide and mentor, while Darwin certainly thought well of his research, writing to Lankester (then a young man of twenty-five) on April 15, 1872: 'What grand work you did at Naples! [at the marine research station]. I can clearly see that you will some day become our first star in Natural History.' But Lankester's studies now read as little more than an exemplification and application of Darwin's insights to several specific groups of organisms – a 'filling in' that often follows a great theoretical advance, and that seems, in retrospect, not overly blessed with originality.

As his most enduring contribution, Lankester proved that the ecologically diverse spiders, scorpions, and horseshoe crabs form a coherent evolutionary group, now called the Chelicerata, within the arthropod phylum. Lankester's research ranged widely from protozoans to mammals. He systematized the terminology and evolutionary understanding of embryology, and he wrote an important paper on 'degeneration,' showing that Darwin's mechanism of natural selection led only to local adaptation, not to general progress, and that such immediate improvement will often be gained (in many parasites, for example) by morphological simplification and loss of organs.

In a fair and generous spirit, one might say that Lankester experienced the misfortune of residing in an 'in between' generation that had imbibed Darwin's insights for reformulating biology, but did not yet possess the primary tool – an understanding of the mechanism of inheritance – so vitally needed for the next great theoretical step. But then, people make their own opportunities, and Lankester, already in his grumpily conservative maturity, professed little use for Mendel's insights upon their rediscovery at the outset of the twentieth century.

In the first biography ever published – the document that finally provided me with enough information to write this essay after a gestation period of twenty-five years! – Joseph Lester, with editing and additional material by Peter Bowler, assessed Lankester's career in a fair and judicious way (*E. Ray Lankester and the Making of British Biology*, 1995):

> Evolutionary morphology was one of the great scientific enterprises of the late nineteenth century. By transmuting the experiences gained by their predecessors in the light of the theory of evolution, morphologists such as Lankester threw new light on the nature of organic structures and created an overview of the evolutionary relationships that might exist between different forms. . . . Lankester gained an international reputation as a biologist, but his name is largely forgotten today. He came onto the scene just too late to be involved in the great Darwinian debate, and his creative period was over before the great revolutions of the early twentieth century associated with the advent of Mendelian genetics. He belonged to a generation whose work has been largely dismissed as derivative, a mere filling in of the basic details of how life evolved.

Lankester's conservative stance deepened with the passing years, thus increasing the anomaly of his early friendship with Karl Marx. His imposing figure only enhanced his aura of staid respectability (Lankester stood well over six feet tall, and he became quite stout, in the manner then favored by men of high station). He spent his years of retirement writing popular articles on natural history for newspapers, and collecting them into several successful volumes. But few of these pieces hold up well today, for his writing lacked both the spark and the depth of the great British essayists in natural history: T. H. Huxley, J. B. S. Haldane, J. S. Huxley, and P. B. Medawar.

As the years wore on, Lankester became ever more stuffy and isolated in his elitist attitudes and fealty to a romanticized vision of a more gracious past. He opposed the vote for women, and became increasingly wary of democracy and mass action. He wrote in 1900: 'Germany did not acquire its admirable educational system by popular demand . . . the crowd cannot guide itself, cannot help itself in its blind impotence.' He excoriated all 'modern' trends in the arts, especially cubism in painting and self-expression (rather than old-fashioned storytelling) in literature. He wrote to his friend H. G. Wells in 1919: 'The rubbish and self-satisfied bosh which pours out now in magazines and novels is astonishing. The authors are so set upon being "clever," "analytical," and "up-to-date," and are really mere prattling infants.'

As a senior statesman of science, Lankester kept his earlier relationship with Marx safely hidden. He confessed to his friend and near contemporary A. Conan Doyle (who had modeled the character of Professor Challenger in *The Lost World* upon Lankester), but he never told the young communist J. B. S. Haldane, whom he befriended late in life and admired greatly, that he had known Karl Marx. When, upon the fiftieth anniversary of the Highgate burial, the Marx-Engels Institute of Moscow tried to obtain reminiscences from all people who had known Karl Marx, Lankester, by then the only living witness of Marx's funeral, replied curtly that he had no letters and would offer no personal comments.

Needless to say, neither the fate of the world nor the continued progress of evolutionary biology depends in the slightest perceptible degree upon a resolution of this strange affinity between two such different people. But little puzzles gnaw at the soul of any scholar, and answers to small problems sometimes lead to larger insights rooted in the principles utilized for explanation. I believe that I have developed a solution, satisfactory (at least) for the dissolution of my own former puzzlement. But, surprisingly to me, I learned no decisive fact from the literature that finally gave me enough information to write this essay – the recent Lankester biography mentioned above, and two excellent articles on the relationship of Marx and Lankester: 'The Friendship of Edwin Ray Lankester and Karl Marx,' by Lewis S. Feuer, and 'Marx's Darwinism: A Historical Note,' by Diane B. Paul. Rather, my proposed solution invokes a principle that may seem disappointing and entirely uninteresting at first, but that may embody a generality worth discussing, particularly for the analysis of historical sequences – a common form of inquiry in both human biography and evolutionary biology. In short, I finally realized that I had been *asking the wrong question* all along.

A conventional solution would try to dissolve the anomaly by arguing that Marx and Lankester shared far more similarity in belief or personality than appearances would indicate, or at least that each man hoped to gain something direct and practical from the relationship. But I do not think that this ordinary form of argument can possibly prevail in this case.

To be sure, Lankester maintained a highly complex and, in some important ways, almost secretive personality beneath his aura of Establishment respectability. But he displayed no tendencies at all to radicalism in politics, and he surely included no Marxist phase in what he might later have regarded as the folly of youth. But Lankester did manifest a fierce independence of spirit, a kind of dumb courage in the great individualistic British tradition of 'I'll do as I see fit, and bugger you or the consequences' – an attitude that inevitably attracted all manner of personal trouble, but that also might have led Lankester to seek interesting friendships that more timid or opportunistic colleagues would have shunned.

Despite his basically conservative views in matters of biological theory, Lankester was a scrappy fighter by nature, an indomitable contrarian who relished professional debate, and never shunned acrimonious controversy. In a remarkable letter, his mentor T. H. Huxley, perhaps the most famous contrarian in the history of British biology, warned his protégé about the dangers of sapping time and strength in unnecessary conflict, particularly in the calmer times that had descended after the triumph of Darwin's revolution. Huxley wrote to Lankester on December 6, 1888:

> Seriously, I wish you would let an old man, who has had his share of fighting, remind you that battles, like hypotheses, are not to be multiplied beyond necessary. . . . You have a fair expectation of ripe vigor for twenty years; just think what may be done with that capital. No use to *tu quoque* me ['thou also' – that is, you did it yourself]. Under the circumstances of the time, warfare has been my business and duty.

To cite the two most public examples of his scrappy defense of science and skepticism, Lankester unmasked the American medium Henry Slade in September 1876. Slade specialized in seances (at high fees), featuring spirits that wrote messages on a slate. Lankester, recognizing Slade's modus operandi, grabbed the slate from the medium's hands just before the spirits should have begun their ghostly composition. The slate already contained the messages supposedly set for later transmission from a higher realm of being. Lankester then sued Slade for conspiracy, but a magistrate found the medium guilty of the lesser charge of vagrancy, and sentenced him to three months at hard labor. Slade appealed and won on a technicality. The dogged Lankester then filed a new summons, but Slade decided to pack up and return to a more gullible America. (As an interesting footnote in the history of evolutionary biology, the spiritualistically inclined Alfred Russel Wallace testified on Slade's behalf, while Darwin, on the opposite side of rational skepticism, quietly contributed funds for Lankester's efforts in prosecution.)

Three years later, in the summer of 1879, Lankester visited the laboratory of the great French physician and neurologist Jean-Martin Charcot. To test his theories on the role of electricity and magnetism in anesthesia, Charcot induced insensitivity by telling a patient to hold an electromagnet, energized by a bichromate battery, in her hand. Charcot then thrust large carpet needles into her affected arm and hand, apparently without causing any pain.

The skeptical Lankester, no doubt remembering the similar and fallacious procedures of Mesmer a century before, suspected psychological suggestion, rather than any physical effect of magnetism, as the cause of anesthesia. When Charcot left the room, Lankester surreptitiously emptied the chemicals out of the battery and replaced the fluid with ordinary water, thus disabling the device. He then urged Charcot to repeat the experiment – with the same result of full anesthesia! Lankester promptly confessed what he had done, and fully expected to be booted out of Charcot's lab *tout de suite*. But the great French scientist grabbed his hand and exclaimed, 'Well done, Monsieur,' and a close friendship then developed between the two men.

One additional, and more conjectural, matter must be aired as we try to grasp the extent of Lankester's personal unconventionalities (despite his conservative stance in questions of biological theory) for potential insight into his willingness to ignore the social norms of his time. The existing literature maintains a wall of total silence on this issue, but the pattern seems unmistakable. Lankester remained a bachelor, although he often wrote about his loneliness and his desires for family life. He was twice slated for marriage, but both fiancées broke their engagements for mysterious and unstated reasons. He took long European vacations nearly every year, and nearly always to Paris, where he maintained clear distance from his professional colleagues. Late in life, Lankester became an intimate platonic friend and admirer of the great ballerina Anna Pavlova. I can offer no proof, but if these behaviors don't point toward the love that may now be freely discussed, but then dared not speak its name (to paraphrase the one great line written by Oscar Wilde's paramour, Lord Alfred Douglas), well, then, Professor Lankester was far more mysterious and secretive than even I can imagine.

Still, none of these factors, while they may underscore Lankester's general willingness to engage in contentious and unconventional behavior, can explain any special propensity for friendship with a man like Karl Marx. (In particular, orthodox Marxists have always taken a dim view of personal, particularly sexual, idiosyncrasy as a self-centered diversion from the social goal of revolution.) Lankester did rail against the social conservatives of his day, particularly against hidebound preachers who opposed evolution, and university professors who demanded the standard curriculum of Latin and Greek in preference to any newfangled study of natural science.

But Lankester's reforming spirit centered only upon the advance of science – and his social attitudes, insofar as he discussed such issues at all, never transcended the vague argument that increasing scientific knowledge might liberate the human spirit, thus leading to political reform and equality of opportunity. Again, this common attitude of rational scientific skepticism only evoked the disdain of orthodox Marxists, who viewed this position as a bourgeois escape for decent-minded people who lacked the courage to grapple with the true depth of social problems, and the consequent need for political revolution. As Feuer states in his article on Marx and Lankester: 'Philosophically, moreover, Lankester stood firmly among the agnostics, the followers of Thomas Henry Huxley, whose standpoint Engels derided as a "shame-faced materialism."'

If Lankester showed so little affinity for Marx's worldview, perhaps we should try the opposite route and ask if Marx had any intellectual or philosophical reason to seek Lankester's company. Again, after debunking some persistent mythology, we can find no evident basis for their friendship.

The mythology centers upon a notorious, if understandable, scholarly error that once suggested far more affinity between Marx and Darwin (or at least a one-way hero-worshiping of Darwin by Marx) than corrected evidence can validate. Marx did admire Darwin, and he did send an autographed copy of *Das Kapital* to the great naturalist. Darwin, in the only recorded contact between the two men, sent a short, polite, and basically contentless letter of thanks. We do know that Darwin (who read German poorly and professed little interest in political science) never spent much time with Marx's *magnum opus*. All but the first 105 pages in Darwin's copy of Marx's 822-page book remain uncut (as does the table of contents), and Darwin, contrary to his custom when reading books carefully, made no marginal annotations. In fact, we have no evidence that Darwin ever read a word of *Das Kapital*.

The legend of greater contact began with one of the few errors ever made by one of the finest scholars of this, or any other, century – Isaiah Berlin in his 1939 biography of Marx. Based on a dubious inference from Darwin's short letter of thanks to Marx, Berlin inferred that Marx had offered to dedicate volume two of *Das Kapital* to Darwin, and that Darwin had politely refused. This tale of Marx's proffered dedication then gained credence when a second letter, ostensibly from Darwin to Marx, but addressed only to 'Dear Sir,' turned up among Marx's papers in the International Institute of Social History in Amsterdam. This letter, written on October 13, 1880, does politely decline a suggested dedication: 'I Shd. prefer the Part or Volume not be dedicated to me (though I thank you for the intended honor) as it implies to a certain extent my approval of the general publication, about which I know nothing.' This second document seemed to seal Isaiah Berlin's case, and the story achieved general currency.

To shorten a long story, two scholars, working independently and simultaneously in the mid-1970s, discovered the almost comical basis of the error – see Margaret A. Fay, 'Did Marx Offer to Dedicate *Capital* to Darwin' and Lewis S. Feuer, 'Is the "Darwin-Marx Correspondence" Authentic?'. Marx's daughter Eleanor became the common-law wife of the British socialist Edward Aveling. The couple safeguarded Marx's papers for several years, and the 1880 letter, evidently sent by Darwin to Aveling himself, must have strayed into the Marxian collection.

Aveling belonged to a group of radical atheists. He sought Darwin's official approval, and status as dedicatee, for a volume he had edited on Darwin's work and his (that is Aveling's, not necessarily Darwin's) view of its broader social meaning (published in 1881 as *The Student's Darwin*, volume two in the International Library of Science and Freethought). Darwin, who understood Aveling's opportunism and cared little for his antireligious militancy, refused with his customary politeness, but with no lack of firmness. Darwin ended his letter to Aveling (not to Marx, who did not treat religion as a primary subject in *Das Kapital*) by writing:

> It appears to me (whether rightly or wrongly) that direct arguments against christianity and theism produce hardly any effect on the public; and freedom of thought is best promoted by the gradual illumination of men's minds which follows from the advance of science. It has, therefore, been always my object to avoid writing on religion, and I have confined myself to science.

Nonetheless, despite this correction, Marx might still have regarded himself as a disciple of Darwin, and might have sought the company of a key Darwinian in the younger generation – a position rendered more plausible by Engels's famous comparison (quoted earlier) in his funerary oration. But this interpretation must also be rejected. Engels maintained far more interest in the natural sciences than did Marx (as best expressed in two books by Engels, *Anti-Dühring* and *The Dialectics of Nature*). Marx, as stated above, certainly admired Darwin as a liberator of knowledge from social prejudice, and as a useful ally, at least by analogy. In a famous letter of 1869, Marx wrote to Engels about Darwin's *Origin of Species*: 'Although it is developed in the crude English style, this is the book which contains the basis in natural history for our view.'

But Marx also criticized the social biases in Darwin's formulation, again writing to Engels, and with keen insight:

> It is remarkable how Darwin recognizes among beasts and plants his English society with its division of labor, competition, opening up of new markets, 'invention,' and the Malthusian 'struggle for existence.' It is Hobbes's *bellum omnium contra omnes* [the war of all against all].

Marx remained a committed evolutionist, of course, but his interest in Darwin clearly diminished through the years. An extensive scholarly literature treats this subject, and I think that Margaret Fay speaks for a consensus when she writes (in her article previously cited):

> Marx . . . though he was initially excited by the publication of Darwin's *Origin* . . . developed a much more critical stance towards Darwinism, and in his private correspondence of the 1860s poked gentle fun at Darwin's ideological biases. Marx's Ethnological Notebooks, compiled circa 1879–1881, in which Darwin is cited only once, provide no evidence that he reverted to his earlier enthusiasm.

To cite one final anecdote, the scholarly literature frequently cites Marx's great enthusiasm (until the more scientifically savvy Engels set him straight) for a curious book published in 1865 by the now (and deservedly) unknown French explorer and ethnologist P. Trémaux, *Origine et transformations de l'homme et des autres êtres* (Origin and Transformation of Man and Other Beings). Marx professed ardent admiration for this work, proclaiming it *einen Fortschritt über Darwin* (an advance over Darwin). The more sober Engels bought the book at Marx's urging, but then dampened his friend's ardor by writing: 'I have arrived at the conclusion that there is nothing to his theory if for no other reason than because he neither understands geology nor is capable of the most ordinary literary historical criticism.'

I had long been curious about Trémaux and sought a copy of his book for many years. I finally purchased one a while ago – and I must say that I have never read a more absurd or more poorly documented thesis. Basically, Trémaux argues that the nature of the soil determines national characteristics, and that higher civilizations tend to arise on more complex soils formed in later geological periods. If Marx really believed that such unsupported nonsense could exceed *The Origin of Species* in importance, then he could not have properly understood or appreciated the power of Darwin's facts and ideas.

We must therefore conclude that Lankester harbored no secret sympathy for Marxism, and that Marx sought no Darwinian inspiration in courting Lankester's friendship. Our puzzle only deepens: What brought these disparate men together; what kind of bond could have nurtured their friendship? The first question, at least, can be answered, and may even suggest a route toward resolving the second, and central, conundrum of this essay.

Four short letters from Lankester remain among Marx's papers. (Marx probably wrote to Lankester as well, but no evidence of such reciprocity has surfaced.) These letters clearly indicate that Marx first approached Lankester for medical advice in the treatment of his wife, who was dying, slowly and painfully, of breast cancer. Lankester suggested that Marx consult his dear friend (and co-conspirator in both the Slade and Charcot incidents), the physician H. B. Donkin. Marx took Lankester's advice, and proclaimed himself well satisfied with the result, as Donkin, whom Marx described as 'a bright and intelligent man,' cared, with great sensitivity, both for Marx's wife and then for Marx himself in their final illnesses.

We do not know for sure how Marx and Lankester first met, but Feuer develops an eminently plausible hypothesis in his article cited previously – one, moreover, that may finally lead us to understand the basis of this maximally incongruous pairing. The intermediary may well have been Charles Waldstein, born in New York in 1856, the son of a German Jewish immigrant. Waldstein, who later served as professor of classical archaeology at Cambridge, knew Lankester well when they both lived in London during the late 1870s. Waldstein became an intimate friend of Karl Marx, an experience that he remembered warmly in an autobiographical work written in 1917 (when he had attained eminence and respectability under the slightly, but portentously, altered name of Sir Charles Walston):

> In my young days, when I was little more than a boy, about 1877, the eminent Russian legal and political writer . . . Professor Kovalevsky, whom I had met at one of G. H. Lewes and George Eliot's Sunday afternoon parties in London, had introduced me to Karl Marx, then living in Hampstead. I had seen very much of this founder of modern theoretic socialism, as well as of his most refined wife; and, though he had never succeeded in persuading me to adopt socialist views, we often discussed the most varied topics of politics, science, literature, and art. Besides learning much from this great man, who was a mine of deep and accurate knowledge in every sphere, I learnt to hold him in high respect and to love the purity, gentleness, and refinement of his big heart.

> He seemed to find so much pleasure in the mere freshness of my youthful enthusiasm and took so great an interest in my own life and welfare, that one day he proposed that we should become *Dutz-freunde*.

The last comment is particularly revealing. Modern English has lost its previous distinction (*thou* versus *you*) between intimate and formal address, a difference that remains crucially important – a matter not to be taken lightly – in most European languages. In German, *Dutz-freunde* address each other with the intimate *Du*, rather than the formal *Sie* (just as the verb *tutoyer*, in French, means to use the intimate *tu* rather than the formal *vous*). In both nations, especially in the far more conservative social modes of nineteenth-century life, permission to switch from formal to intimate address marked a rare and precious privilege reserved only for one's family, one's God, one's pets, and one's absolutely dearest friends. If an older and established intellectual like Marx suggested such a change of address to a young man in his early twenties, he must have felt especially close to Charles Waldstein.

Lankester's first letter to Marx, written on September 19, 1880, mentions Waldstein, thus supporting Feuer's conjecture: 'I shall be very glad to see you at Wellington Mansions. I had been intending to return to you the book you kindly lent to me – but had mislaid your address and could not hear from Waldstein who is away from England.' Lankester and Waldstein remained close friends throughout their lives. Waldstein's son responded to Feuer's inquiry about his father's relationship with Lankester by writing, in 1978, that he retained a clear childhood memory of 'Ray Lankester . . . coming to dinner from time to time at my home – a very fat man with a face like a frog.'

Waldstein's memories of Marx as a kind man and a brilliant intellectual mentor suggest an evident solution to the enigma of Marx and Lankester – once we recognize that we had been asking the wrong question all along. No error of historical inquiry can match the anachronistic fallacy of using a known present to misread a past circumstance that could not possibly have been defined or influenced by events yet to happen. When we ask why a basically conservative biologist like Lankester could have respected and valued the company of an aging agitator like Karl Marx, we can hardly help viewing Marx through the glasses of later human catastrophes perpetrated in his name – from Stalin to Pol Pot. Even if we choose to blame Marx, in part, for not foreseeing these possible consequences of his own doctrines, we must still allow that when he died in 1883, these tragedies only resided in an unknowable future. Karl Marx, the man who met Lankester in 1880, must not be confused with Karl Marx, the posthumous standard-bearer for some of the worst crimes in human history. We err when we pose E. Ray Lankester, the stout and imposing relic of Victorian and Edwardian biology, with Karl Marx, cited as the rationale for Stalin's murderous career – and then wonder how two such different men could inhabit the same room, much less feel warm ties of friendship.

In 1880, Lankester was a young biologist with a broad view of life and intellect, and an independent mind that cared not a fig for conventional notions of political respectability, whatever his own basically conservative convictions. Showing a rare range of interest among professional scientists, he also loved art and literature, and had developed fluency in both German and French. Moreover, he particularly admired the German system of university education, then a proud model of innovation, especially in contrast with the hidebound classicism of Oxford and Cambridge, so often the object of his greatest scorn and frustration.

Why should Lankester not have enjoyed, even cherished, the attention of such a remarkable intellect as Karl Marx – for that he was, whatever you may think of his doctrines and their consequences? What could possibly have delighted Lankester more than the friendship of such a brilliant older man, who knew art, philosophy, and the classics so well, and who represented the epitome of German intellectual excellence, the object of Lankester's highest admiration? As for the ill, aging, and severely depressed Karl Marx, what could bring more solace in the shadow of death than the company of bright, enthusiastic, optimistic young men in the flower of their intellectual development?

Waldstein's memories clearly capture, in an evocative and moving way, this aspect of Marx's persona and final days. Many scholars have emphasized this feature of Marx's later life. Diane Paul, for example, states that 'Marx had a number of much younger friends. . . . The aging Marx became increasingly difficult in his personal relationships, easily offended and irritated by the behavior of old friends, but he was a gracious mentor to younger colleagues who sought his advice and support.' Seen in this appropriate light of their own time, and not with anachronistic distortion of later events that we can't escape but they couldn't know, Marx and Lankester seem ideally suited, indeed almost destined, for the warm friendship that actually developed.

All historical studies – whether of human biography or of evolutionary lineages in biology – potentially suffer from this 'presentist' fallacy. Modern chroniclers know the outcomes that actually unfolded as unpredictable consequences of past events – and they often, and inappropriately, judge the motives and actions of their subjects in terms of futures unknowable at the time. Thus, and far too frequently, evolutionists view a small and marginal lineage of pond-dwelling Devonian fishes as higher in the scale of being and destined for success because we know, but only in retrospect, that these organisms spawned all modern terrestrial vertebrates, including our exalted selves. And we overly honor a peculiar species of African primates as central to the forward thrust of evolution because our unique brand of consciousness arose, by contingent good fortune, from such a precarious stock. And as we northerners once reviled Robert E. Lee as a traitor, we now tend to view him, in a more distant and benevolent light, as a man of principle and a great military leader – though neither extreme position can match or explain this fascinating man in the more appropriate context of his own time.

A little humility before the luck of our present circumstances might serve us well. A little more fascination for past realities, freed from judgment by later outcomes that only we can know, might help us to understand our history, the primary source for our present condition. Perhaps we might borrow a famous line from a broken man, who died in sorrow, still a stranger in a strange land, in 1883 – but who at least enjoyed the solace of young companions like E. Ray Lankester, a loyal friend who did not shun the funeral of such an unpopular and rejected expatriate.

History reveals patterns and regularities that enhance our potential for understanding. But history also expresses the unpredictable foibles of human passion, ignorance, and dreams of transcendence. We can only understand the meaning of past events in their own terms and circumstances, however legitimately we may choose to judge the motives and intentions of our forebears. Karl Marx began his most famous historical treatise, his study of Napoleon III's rise to power, by writing, 'Men make their own history, but they do not make it just as they please.'

THE PILTDOWN CONSPIRACY*

Introduction and Background

Of Conspiracies

In his great aria 'La calunnia,' Don Basilio, the music master of Rossini's *Barber of Seville*, graphically describes how evil whispers grow, with appropriate watering, into truly grand and injurious calumnies. For the less conniving among us, the same lesson may be read with opposite intent: in adversity, try to contain. The desire to pin evil deeds upon a single soul acting alone reflects this strategy: conspiracy theories have a terrible tendency to ramify like Basilio's whispers until the runaway solution to 'whodunit' becomes 'everybodydunit.' But conspiracies do occur. Even the pros and pols now doubt that Lee Harvey Oswald acted alone: and everybody did do it on the Orient Express.†

* This essay has been the subject of many commentaries, most positive, but some (to put it mildly) excoriatingly negative (the most brutal, if you'll pardon the aroma of suspicion by association, from devotees of the 'Teilhard cult'). In this light, I have decided to reprint the essay (which first appeared in *Natural History Magazine* for August, 1980) without any changes – for it would be unfair to improve it by correcting errors and ambiguities and then to turn on my detractors with a product better than the original that they first criticized. Since it is simply immoral to publish known error, I shall correct a few mistakes by footnote, so everyone can see where I goofed the first time round. I shall let all interpretations stand without comment.

† I guess this should now read 'Many pros and pols,' the wheel of fortune spinning quickly, as it does. Yet fiction has, ironically, the rock-hard permanence that fact must lack – and everyone did do it, now and forever more, on the Orient Express.

The Piltdown case, surely the most famous and spectacular fraud of twentieth-century science, has experienced this tension ever since its exposé in 1953. The semiofficial, contained version holds that Charles Dawson, the lawyer and amateur archeologist who 'found' the first specimens, devised and executed the entire plot himself. Since J. S. Weiner's elegant case virtually precludes Dawson's innocence (*The Piltdown Forgery*, Oxford University Press, 1955), conspiracies become the only reasonable refuge for challengers. And proposals for co-conspirators abound, ranging from the great anatomist Grafton Elliot Smith to W. J. Sollas, professor of geology at Oxford. I regard these claims as far-fetched and devoid of reasonable evidence. But I do believe that a conspiracy existed at Piltdown and that, for once, the most interesting hypothesis is actually true. I believe that a man who later became one of the world's most famous theologians, a cult figure for many years after his death in 1955, knew what Dawson was doing and probably helped in no small way – the French Jesuit priest and paleontologist Pierre Teilhard de Chardin.

Teilhard and Piltdown

Teilhard, born in Auvergne (central France) in 1881, belonged to an old, conservative, and prosperous family. Entering the Society of Jesus in 1902, he studied on the English island of Jersey from 1902 to 1905 and then spent three years as a teacher of physics and chemistry at a Jesuit school in Cairo. In 1908, he returned to finish his theological training at the Jesuit seminary of Ore Place in Hastings, providentially located right next to Piltdown on England's southeast coast. Here he stayed for four years, and here he was ordained a priest in 1912.* As a theological student. Teilhard was talented enough, but lackadaisical. His passion at Hastings was, as it always had been, natural history. He scoured the countryside for butterflies, birds, and fossils. And, in 1909, he met Charles Dawson at the focus of their common interests – in a stone quarry, hunting for fossils. The

* I thank Rev. Thomas M. King, S. J. of Georgetown University for pointing out two inconsequential but highly embarrassing errors in this paragraph. Teilhard entered the Society of Jesus in 1899, not 1902, and he was ordained in 1911, not 1912.

two men became good friends and colleagues in pursuit of their interest. Teilhard described Dawson to his parents as 'my correspondent in geology.'

Dawson claimed that he had recovered the first fragment of Piltdown's skull in 1908, after workmen at a gravel pit told him of a 'coconut' (the entire skull) they had unearthed and smashed at the site. Dawson kept poking about, collecting a few more skull pieces and some fragments of other fossil mammals. He did not bring his specimens to Arthur Smith Woodward, keeper of paleontology at the British Museum, until the middle of 1912. Thus, for three years before any professional ever heard of the Piltdown material, Dawson and Teilhard were companions in natural history in the environs of Piltdown.

Smith Woodward was not a secretive man, but he knew the value of what Dawson had brought and the envy it might inspire. He clamped a tight lid upon Dawson's information prior to its publication. He wanted none of Dawson's lay friends at the site, and only one naturalist accompanied Dawson and Smith Woodward in their first joint excavations at Piltdown – Teilhard de Chardin, whom Dawson had described as 'quite safe.' More specimens came to light during 1912, including the famous jaw with its two molar teeth, artificially filed to simulate human patterns of wear. In December, Smith Woodward published and the controversy began.

The skull fragments, although remarkably thick, could not be distinguished from those of modern humans. The jaw, on the other hand, except for the wear of its teeth, loudly said 'chimpanzee' to many experts (in fact, it once belonged to an orangutan). No one smelled fraud, but many professionals felt that parts of two creatures had been mixed together at the Piltdown site. Smith Woodward stoutly defended the integrity of his creature, arguing, with flawed logic, that the crucial role of brain power in our mastery of the earth today implies a precocious role for large brains in evolutionary history as well. A fully vaulted skull still attached to an apish jaw vindicated such a brain-centered view of human evolution.

Teilhard left England late in 1912* to begin his graduate studies

* Again I thank Rev. King for calling my attention to an error, this one of some potential consequence and therefore more embarrassing. The longer Teilhard remained in England, the more opportunity he had to work with Dawson. He left, in fact, not 'late in 1912,' as I stated, but on July 16.

with Marcellin Boule, the greatest physical anthropologist of France. But in August 1913, he was back in England for a retreat at Ore Place. He also spent several days prospecting with Dawson and on August 30 made a major discovery himself – a canine tooth of the lower jaw, apish in appearance but worn in a human fashion. Smith Woodward continued his series of publications on the new material, but critics persisted in their belief that Piltdown man represented two animals improperly united.

The impasse broke in Smith Woodward's favor in 1915. Dawson had been prospecting at another site, two miles from Piltdown, for several years. He probably took Teilhard there in 1913; we know that he searched the area several times with Smith Woodward in 1914. Then, in January 1915, he wrote to Smith Woodward. The second site, later called Piltdown 2, had yielded its reward: 'I believe we are in luck again. I have got a fragment of the left side [it was actually the right] of a frontal bone with a portion of the orbit and root of nose.' In July of the same year, he announced the discovery of a lower molar, again, apish in appearance but worn in a human fashion. The bones of a human and an ape might wash into the same gravel pit once, but the second, identical association of vaulted skull and apish jaw surely proved the integrity of a single bearer, despite the apparent anatomical incongruity. H. F. Osborn, America's leading paleontologist and critic of the first Piltdown find, announced a conversion in his usual grandiloquent fashion. Even Teilhard's teacher Marcellin Boule, leader of the doubters, grumbled that the new finds had tipped the balance, albeit slightly, in Smith Woodward's favor. Dawson did not live to enjoy his triumph, for he died in 1916. Smith Woodward stoutly supported Piltdown for the rest of his long life, devoting his last book (*The Earliest Englishman*, 1948) to its defense. He died, mercifully, before his bubble burst.

Meanwhile, Teilhard pursued his calling with mounting fame, frustration, and exhilaration. He served with distinction as a stretcher bearer in World War I and then became professor of geology at the Institut Catholique of Paris. But his unorthodox (although always pious) thinking soon led him into irrevocable conflict with ecclesiastical authority. Ordered to abandon his teaching post and to leave France, Teilhard departed for China in 1926. There he remained for most of his life, pursuing distinguished research in geology and paleontology and writing the philosophical treatises on cosmic history and the reconciliation of science with religion that later made him so famous. (They all remained unpublished, by ecclesiastical fiat, until his death.) Teilhard died in 1955, but his passing only marked the beginning of his meteoric rise to fame. His treatises, long suppressed, were published and quickly translated into all major languages. *The Phenomenon of Man* became a bestseller throughout the world. Harvard's Widener Library now houses an entire tier of books devoted to Teilhard's writing and thinking. Two journals that were established to discuss his ideas still flourish.

Of the original trio – Dawson, Teilhard, and Smith Woodward – only Teilhard was still living when Kenneth Oakley, J. S. Weiner, and W. E. Le Gros Clark proved that the Piltdown bones had been chemically stained to mimic great age, the teeth artificially filed to simulate human wear, the associated mammal remains all brought in from elsewhere, and the flint 'implements' recently carved. The critics had been right all along, more right than they had dared to imagine. The skull bones did belong to a modern human, the jaw to an orangutan. As the shock of revelation gave way to the fascination of whodunit, suspicion quickly passed from two members of the trio. Smith Woodward had been too dedicated and too gullible; moreover, he knew nothing of the site before Dawson brought him the original bones in 1912. (I have no doubt whatsoever of Smith Woodward's total innocence.) Teilhard was too famous and too present for any but the most discreet probing. He was dismissed as a naïve young student who, forty years before, had been duped and used by the crafty Dawson. Dawson acting alone became the official theory; professional science was embarrassed, but absolved.

Doubts

I was just the right age for primal fascination – twelve years old – and a budding paleontologist when news of the fraud appeared on page one of the *New York Times* one morning at breakfast. My interest has never abated, and I have, over the years, asked many senior paleontologists about Piltdown. I have also remarked, both with amusement and wonder, that very few believed the official tale of Dawson acting alone. I noted, in particular, that several of the men I most admire suspected Teilhard, not so much on the basis of hard evidence (for their suspicions rested on what I regard as a weak point among the arguments), but from an intuitive feeling about this man whom they knew well, loved, and respected, but who seemed to hide passion, mystery, and good humor behind a garb of piety. A. S. Romer and Bryan Patterson, two of America's leading vertebrate paleontologists and my former colleagues at Harvard, often voiced their suspicions to me. Louis Leakey voiced them in print, without naming the name, but with no ambiguity for anyone in the know (see his autobiography, *By the Evidence*).*

I finally decided to get off my butt and probe a bit after I wrote a column on Piltdown for other reasons (*Natural History*, March 1979). I read all the official documents and concluded that nothing excluded Teilhard, although nothing beyond his presence at Piltdown from the start particularly implicated him either. I intended to drop the subject or to pass it along to someone with a greater zeal for investigative reporting. But at a conference in France last September, I happened to meet two of Teilhard's closest colleagues, the leading paleontologist J. Piveteau and the great zoologist P. P. Grassé. They greeted my suspicions with a blustering '*incroyable*.' Then Père François Russo, Teilhard's friend and fellow Jesuit, heard of my inquiries and promised to send me a document that would prove Teilhard's innocence – a copy of the letter that Teilhard had written to Kenneth Oakley on November 28, 1953. I received this letter in printed French translation (Teilhard wrote it in English) in October 1979 and realized immediately that it contained an inconsistency (a slip on Teilhard's part) most easily resolved by the hypothesis of Teilhard's complicity. When I visited Oakley at Oxford in April 1980, he showed me the original letter along with several others that Teilhard had written to him. We studied the documents and discussed Piltdown for the better part of a day, and I left convinced that Romer, Patterson, and Leakey had been right. Oakley, who had noted the inconsistency but interpreted it differently, agreed with me and stated as we parted: 'I think it's right that Teilhard was in it.' (Let me here express my deep appreciation for Dr Oakley's hospitality, his openness, and his simple, seemingly inexhaustible kindness and helpfulness. I always

* I have since learned that Louis Leakey was far more serious in his probings than I had realized. He was convinced of Teilhard's guilt and was writing a book on the subject when he died.

feel so exhilarated when I discover – and it is not so rare as many people imagine – that a great thinker is also an exemplary human being.) Since then, I have sharpened the basic arguments and read through Teilhard's published work, finding a pattern that seems hard to reconcile with his innocence. My case is, to be sure, circumstantial (as is the case against Dawson or anyone else), but I believe that the burden of proof must now rest with those who would hold Father Teilhard blameless.

The Case against Teilhard

The Letters to Kenneth Oakley

The main virtue of truth, quite apart from its ethical value (which I hold to be considerable), is that it represents an infallible guide for keeping your story straight. The problem with prevarication is that, when the going gets complex or the recollection misty, it becomes very difficult to remember all the details of your invented scheme. Richard Nixon finally succumbed on a minor matter, and Sir Walter Scott spoke truly when he wrote the famous couplet: 'Oh, what a tangled web we weave, / When first we practise to deceive!'

Teilhard made just such a significant slip on a minor point in his letter to Oakley. Teilhard offered no spontaneous recollections about Piltdown and responded only to Oakley's direct inquiries for help in establishing the forger's identity. He begins by congratulating Oakley 'most sincerely on your solution of the Piltdown problem. Anatomically speaking, "*Eoanthropus*" [Smith Woodward's name for the Piltdown animal] was a kind of monster. . . . Therefore I am fundamentally pleased by your conclusions, in spite of the fact that, sentimentally speaking, it spoils one of my brightest and earliest paleontological memories.'

Teilhard then stonewalls on the question of fraud. He refuses to believe it at all, declaring that Smith Woodward and Dawson (and, by implication, himself) were not the kind of men who could conceivably do such a thing. Is it not possible, he asks, that some collector discarded the ape bones in a gravel pit that legitimately contained a human skull, the product of a recent interment? Could not the iron staining have been natural, since the local water 'can stain (with iron) at a remarkable speed'? But Teilhard's notion can explain neither the artificial filing of the teeth to simulate human wear nor the crucial discovery of a second combination of ape and human at the Piltdown 2 site. In fact, Teilhard admits: 'The idea sounds fantastic. But, in my opinion, no more fantastic than to make Dawson the perpetrator of a hoax.'

Teilhard then goes on to discuss Piltdown 2 and, in trying to exonerate Dawson, makes his fatal error. He writes:

> He [Dawson] just brought me to the site of Locality 2 and explained me [*sic*] that he had found the isolated molar and the small pieces of skull in the heaps of rubble and pebbles raked at the surface of the field.

But this cannot be. Teilhard did visit the second site with Dawson in 1913, but they did not find anything. Dawson 'discovered' the skull bones at Piltdown 2 in January 1915, and the tooth not until July 1915. And now, the key point: Teilhard was mustered into the French army in December 1914 and was shipped immediately to the front, where he remained until the war ended. He could not have seen the remains of Piltdown 2 with Dawson, unless they had manufactured them together before he left (Dawson died in 1916).

Oakley caught the inconsistency immediately when he received Teilhard's letter in 1953, but he read it differently and for good reason. At that time, Oakley and his colleagues were just beginning their explorations into whodunit. They rightly suspected Dawson and had written to Teilhard to gather evidence. Oakley read Teilhard's statement when he was simply trying to establish the basic fact of Dawson's guilt. In that context, he assumed that Dawson had shown the specimens to Teilhard in 1913, but had withheld them from Smith Woodward until 1915 – more evidence for Dawson's complicity.

Oakley wrote back immediately, and Teilhard, realizing that he had tripped, began to temporize. In his second letter of January 29, 1954, he tried to recoup:

> Concerning the point of 'history' you ask me, my 'souvenirs' are a little vague. Yet, by elimination (and since Dawson died *during* the First World War, if I am correct) my visit with Dawson to the *second* site (where the two small fragments

> of skull and the isolated molar were found in the rubble) must have been in late July 1913 [it was probably in early August].

Obviously troubled, he then penned the following postscript.

> When I visited the site no. 2 (in 1913?) the two small fragments of skull and tooth had already been found, I believe. But your very question makes me doubtful! Yes, I think definitely they *had* been already found: and that is the reason why Dawson pointed to me the little heaps of raked pebbles as the place of the 'discovery.'

In a final letter to Mable Kenward, daughter of the owner of Barkham Manor, site of the first Piltdown find, Teilhard drew back even further: 'Dawson showed me the field where the second skull (fragments) were found. But, as I wrote to Oakley, I cannot remember whether it was after or before the find' (March 2, 1954).

I can devise only four interpretations for Teilhard's slip.

1. I thought initially, when I had only read the first letter, that one might interpret Teilhard's statement thus: Dawson took me to the site in 1913 and later stated in wartime correspondence that he had found the fragments in the rubble. But Teilhard's second letter states explicitly that Dawson, in the flesh, had pointed to the spot at Piltdown 2 where he had found the specimens.

2. Oakley's original hypothesis: Dawson showed the specimens to an innocent Teilhard in 1913, but withheld them from Smith Woodward until 1915. But Dawson would not blow his cover in such a crude way. For Dawson took Smith Woodward to the second site on several prospecting trips in 1914, always finding nothing. Now Teilhard and Smith Woodward were also fairly close. Dawson had introduced them in 1909 by sending to London some important mammal specimens (having nothing to do with Piltdown) that Teilhard had collected. Smith Woodward was delighted with Teilhard's work and praised him lavishly in a publication. He accepted Teilhard as the only other member of their initial collecting trips at Piltdown. Moreover, Teilhard was a house guest of the Smith Woodwards when he visited London in September 1913, following his discovery of the canine. If Dawson had shown Teilhard the Piltdown 2 finds in 1913, then led Smith Woodward

extensively astray during several field trips in 1914, and if an innocent Teilhard had told Smith Woodward about the specimens (and I can't imagine why he would have held back), then Dawson would have been exposed.

3. Teilhard never did hear about the Piltdown 2 specimens from Dawson, but simply forgot forty years later that he had never actually viewed the fossils he had read about later. This is the only alternative (to Teilhard's complicity) that I view as at all plausible. Were the letters not filled with other damaging points, and the case against Teilhard not supported on other grounds, I would take this possibility more seriously.

4. Teilhard and Dawson planned the Piltdown 2 discovery before Teilhard left England. Forty years later, Teilhard misconstructed the exact chronology, forgot that he could not have seen the specimens when they were officially 'found,' and slipped in writing to Oakley.

Teilhard's letters to Oakley contain other curious statements, each insignificant (or subject to other interpretations) by itself, but forming in their ensemble a subtle attempt to direct suspicion away from himself.

1. In his letter of November 28, 1953, Teilhard states that he first met Dawson in 1911. In fact, they met in May 1909, for Teilhard describes the encounter in a vivid letter to his parents. Moreover, this meeting was an important event in Teilhard's career, for Dawson befriended the young priest and personally forged his path to professional notice and respect by sending some important specimens he had collected to Smith Woodward. When Smith Woodward described this material before the Geological Society of London in 1911, Dawson, in the discussion following Smith Woodward's talk, paid tribute to the 'patient and skilled assistance' given to him by Teilhard since 1909. I don't regard this, in itself, as a particularly damning point. A first meeting in 1911 would still be early enough for complicity (Dawson 'found' his first piece of the Piltdown skull in 1911, although he states that a workman had given him a fragment 'several years before'), and I would never hold a mistake of two years against a man who tried to remember the event forty years later. Still, the later (and incorrect) date, right upon the heels of Dawson's first 'find,' certainly averts suspicion from Teilhard.

2. Oakley wrote again in February 1954, probing further into Dawson's first contact with the Piltdown material, wondering in particular what

had happened in 1908. Teilhard simply replied (March 1, 1954): 'In 1908 I did not know Dawson.' True enough, but they met just a few months later, and Teilhard might have mentioned it. A small point, to be sure.

3. In the same letter, Teilhard tries further to avert suspicion by writing of his years at Hastings: 'You know, at that time, I was a young student in theology – not allowed to leave much his cell of Ore Place (Hastings).' But this description of a young, pious, and restricted man stands in stark contrast with the picture that Teilhard painted of himself at the time in a remarkable series of letters to his parents. These letters speak little of theology, but they are filled with charming and detailed accounts of Teilhard's frequent wanderings all over southern England. Eleven letters refer to excursions with Dawson,* and no other naturalist is mentioned so frequently. If he spent much time at Ore Place, he didn't choose to write about it. On August 13, 1910, for example, he exclaims: 'I have travelled up and down the coast, to the left and right of Hastings; thanks to the cheap trains [*les cheaptrains* as he writes in French] so common at this time of year, it is easy to go far with minimal expense.'

Perhaps I am now too blinded by my own attraction to the hypothesis of Teilhard's complicity. Perhaps all these points are minor and unrelated, testifying only to the faulty memory of an aging man. But they do form an undeniable pattern. Still, I would not now come forward with my case were it not for a second argument, more circumstantial to be sure, but somehow more compelling in its persistent pattern of forty years – the record of Teilhard's letters and publications.

Piltdown in Teilhard's Writing

I remember a jokebook I had as a kid. The index listed 'mule, sex life,' but the indicated page was blank (ridiculous, in any case, for mules do not abstain just because the odd arrangement of their hybrid chromosomes debars them from bearing offspring). Teilhard's published record on Piltdown is almost equally blank. In 1920, he

* Several critics have pointed out that some of the letters refer to visits that Dawson made to Teilhard's seminary at Ore Place rather than to field trips, or 'excursions' proper. I goofed to be sure, but I don't see how my point is weakened.

wrote one short article in French for a popular journal on *Le cas de l'homme de Piltdown*. After this, virtually all is silence. Piltdown never again received as much as a full sentence in all his published work (except once in a footnote). Teilhard mentioned Piltdown only when he could scarcely avoid it – in comprehensive review articles that discuss all outstanding human fossils. I can find fewer than half a dozen references in the twenty-three volumes of his complete works. In each case, Piltdown appears either as an item listed without comment in a footnote or as a point (also without comment) on a drawing of the human evolutionary tree or as a partial phrase within a sentence about Neanderthal man.*

* (This footnote, and only this one, was part of the original essay.) Teilhard's complete works are spread over two editions – a thirteen-volume edited compendium of his general articles (Paris: Éditions de Seuil), and a more extensive ten-volume facsimile reprint of his professional publications (Olten: Walter-Verlag). To my annoyance, I discovered that the Paris edition has expurgated all Piltdown references without so stating and without even inserting ellipses. In trying to spiff up Teilhard's record, they have made him appear even more culpable by accentuating the impression of guilty silence. I therefore consulted likely originals whenever I could find them. One expurgation is particularly infuriating in its downright misrepresentation. A posthumous volume of essays, *Le Coeur de la matière* ('The heart of matter,' Seuil, 1976), reprints Teilhard's application for the chair of paleontology at the Collège de France in 1948 (ecclesiastical authority did not permit him to accept). In this autobiographical essay, Teilhard discusses his role in human paleontology: 'My first stroke of good luck in this area of ancient human paleontology came in 1923 when I was able to establish, with Emile Licent, the existence, hitherto contested, of paleolithic man in Northern China.' If Teilhard had thus truly suppressed Piltdown in presenting himself for review, what but complicity could we infer? By sheer good fortune, I found a copy of this unpublished, mimeographed document in the reprint collection of my late colleague A. S. Romer. It reads: 'My first stroke of good luck in this area of ancient human paleontology was to be included, when still young, in the excavation of *Eoanthropus dawsoni* in England. The second was, in 1923, to be able to establish, with Emile Licent. . . .' I doubt that half a dozen copies of the unpublished original exist in America, and I might easily have been led from the published and doctored version to an even stronger indication of Teilhard's complicity based on his silence. (It has, of course, occurred to me that the pattern of silence I detect in Teilhard's writing might represent more his editor's posthumous expurgation than Teilhard's own preference. Yet the completely honest ten-volume facsimile edition contains more than enough material to establish the pattern, and I have checked enough original versions of potentially expurgated texts to be confident that references to Piltdown are fleeting and exceedingly sparse throughout Teilhard's writing.)

Consider just how exceedingly curious this is. In his first letter to Oakley, Teilhard described his work at Piltdown as 'one of my brightest and earliest paleontological memories.' Why, then, such silence? Was Teilhard simply too diffident or saintly to toot his own trumpet? Scarcely, since no theme receives more voluminous attention, in scores of later articles, than his role in unearthing the legitimate Peking man in China.

As I began my investigation into this extraordinary silence, and trying to be as charitable as I could, I constructed two possible exonerating reasons for Teilhard's failure to discuss the major event of his paleontological youth. Kenneth Oakley then told me of the 1920 article, the only analysis of Piltdown that Teilhard ever published. I found a copy in the ten-volume edition of Teilhard's *oeuvre scientifique*, and realized that its content invalidated the only exculpatory arguments I could construct.

The first argument: Marcellin Boule, Teilhard's revered teacher, was a leading critic of Piltdown. He regarded it as a mixture of two creatures (not as a fraud), although he softened his opposition after he learned of the subsequent discovery at Piltdown 2. Perhaps Boule upbraided his young student for gullibility, and Teilhard, embarrassed to the quick, never spoke of the infernal creature, or of his role in discovering it, again.

The 1920 article invalidates such a conjecture, for in this work, Teilhard comes down squarely on the right side. He mentions that his English companions, convinced by finding the jaw so close to the skull fragments, never doubted the integrity of their fossil. Teilhard then notes, with keen insight, that experts who had not seen the specimens *in situ* would be swayed primarily by the formal anatomy of the bones themselves, and that these bones loudly proclaimed: human skull, ape's jaw. Which emphasis, then, shall prevail, geology or anatomy? Although he had witnessed the geology, Teilhard opted for anatomy:

> In order to admit such a combination of forms [a human skull and an ape's jaw in the same creature], it is necessary that we be forced to such a conclusion. Now this is not the case here. . . . The reasonable attitude is to grant primacy to the intrinsic morphological probability over the extrinsic probability of geological conditions. . . . We must suppose

> that the Piltdown skull and jaw belong to two different subjects.

Teilhard called it once, and he called it right. He had no reason to be embarrassed.

The second argument: Perhaps Teilhard had reveled in his role at Piltdown, cherished the memory, but simply found that the man he had helped to unearth could offer no support for, or even contact with, the concerns of his later career. On a broad level, this argument is implausible, if only because Teilhard wrote several general reviews about human fossils; however controversial or dubious, Piltdown should have been discussed. Even the leading doubters never failed to air their suspicions. Boule wrote chapters about Piltdown. Teilhard listed it without comment a few times and only when he had no choice.

In a more specific area, Teilhard's silence about Piltdown becomes inexplicable to the point of perversity (unless guilt and knowledge of fraud engendered it) – for Piltdown provided the best available support that fossils could provide for the most important argument of Teilhard's cosmic and mystical views about evolution, the dominant theme of his career and the source of his later fame. Teilhard never availed himself of his own best weapon, partly provided by his own hand.

The conclusion that skull and jaw belonged to different creatures did not destroy the scientific value of Piltdown, provided that both animals legitimately lay in the strata that supposedly entombed them. For these strata were older than any housing Neanderthal man, Europe's major claim to anthropological fame. Neanderthal, although now generally considered as a race of our species, was a low-vaulted, beetle-browed fellow of decidedly 'primitive' cast. Piltdown, despite the thickness of its skull bones, looked more modern in its globular vault. The assignment of the jaw to a fossil ape further enhanced the skull's advanced status. Humans of modern aspect must have lived in England even before Neanderthal man evolved on the continent. Neanderthal, therefore, cannot be an ancestral form; it must represent a side branch of the human tree. Human evolution is not a ladder but a series of lineages evolving along separate paths.

In the 1920 article, Teilhard presented the Piltdown skull, divorced from its jaw, in just that light – as proof that hominids evolved as a bundle of lineages moving in similar directions. He wrote:

> Above all, it is henceforth proved that even at this time [of Piltdown] a race of men existed, already included in our present human line, and very different from those that would become Neanderthal. . . . Thanks to the discovery of Mr Dawson, the human race appears to us even more distinctly, in these ancient times, as formed of strongly differentiated bundles, already quite far from their point of divergence. For anyone who has an idea of paleontological realities, this light, tenuous as it appears, illuminates great depths.

To what profundity, then, did Teilhard refer Piltdown as evidence? Teilhard believed that evolution moved in an intrinsic direction representing the increasing domination of spirit over matter. Under the thrall of matter, lineages would diverge to become more unlike, but all would move upward in the same general direction. With man, evolution reached its crux. Spirit had begun its domination over matter, adding a new layer of thought – the noosphere – above the older biosphere. Divergence would be stemmed; indeed, convergence had already begun in the process of human socialization. Convergence will continue as spirit prevails. When the last vestiges of matter have been discarded, spirit will involute upon itself at a single point called Omega and identified with God – the mystical evolutionary apocalypse that secured Teilhard's fame.

But convergence is a thing of the future. Scientists seeking evidence for such a scheme must look to the past for twin signs of divergence accompanied by similar upward direction – in other words, for *multiple, parallel lineages* within larger groups.

I have read all of Teilhard's papers from the early 1920s. No theme receives more emphasis than the search for multiple, parallel lineages. In an article on fossil tarsiers, written in 1921, he argues that three separate primate lineages extend back to the dawn of the age of mammals, each evolving in the same direction of larger brains and smaller faces. In a review published in 1922 of Marcellin Boule's *Les hommes fossiles*, Teilhard writes: 'Evolution is no more to be represented in a few simple strokes for us than for other living things; but it resolves itself into innumerable lines which diverge at such length that they appear parallel.' In a general essay on evolution, printed in 1921, he speaks continually of oriented evolution in multiple, parallel lines within mammals.

But where was Piltdown in this extended paean of praise for multiple, parallel lineages? Piltdown provided proof, the only available proof, of multiple, parallel lineages within human evolution itself – for its skull belonged to an advanced human older than primitive Neanderthal. Piltdown was the most sublime argument that Teilhard possessed, and he never breathed it again after the 1920 article.

These two arguments have been abstract. A third feature of the 1920 article is stunning in its directness. For I believe that Teilhard fleetingly tried to tell his colleagues, too subtly perhaps, that Piltdown was a phony. In discussing whether the Piltdown remains represent one or two animals. Teilhard laments that the direct and infallible test cannot be applied. One skull fragment contained a perfect glenoid fossa, the point of articulation for the upper jaw upon the lower. Yet the corresponding point of the lower jaw, the condyle, was missing on a specimen otherwise beautifully preserved at its posterior end. Teilhard writes: 'Since the glenoid fossa exists in perfect state on the temporal bone, we could simply have tried to articulate the pieces, *if the mandible had preserved its condyle*: we could have learned, without possible doubt, if the two fit together.' I read this statement in a drowsy state one morning at two o'clock, but the next line – set off by Teilhard as a paragraph in itself terminated by an exclamation point – destroyed any immediate thought of sleep: 'As if on purpose [*comme par exprès*], the condyle is missing!'

'*Comme par exprès*.' I couldn't get those words out of my mind for two days. Yes, it could be a literary line, a permissible metaphor for emphasis. But I think that Teilhard was trying to tell us something he didn't dare reveal directly.

Other arguments

1. *Teilhard's embarrassment at Oakley's disclosure*. Kenneth Oakley told me that, although he had not implicated Teilhard in his thoughts, one aspect of Teilhard's reaction had always puzzled him. All other scientists, including those who had cause for the most profound embarrassment (like the aged Sir Arthur Keith, who had used Piltdown for forty years as the bedrock of his thought), expressed keen interest amidst their chagrin. They all congratulated Oakley spontaneously and thanked him for resolving an issue that had always been puzzling, even though the solution hurt so deeply. Teilhard said nothing. His congratulations arrived

only when they could not be avoided – in the preface to a letter responding to Oakley's direct inquiries. When Teilhard visited London, Oakley tried to discuss Piltdown, but Teilhard always changed the subject. He took Teilhard to a special exhibit at the British Museum illustrating how the hoax had been uncovered. Teilhard glumly walked through as fast as he could, eyes averted, saying nothing. (A. S. Romer told me several years ago that he also tried to conduct Teilhard through the same exhibit, and with the same strange reaction.) Finally, Teilhard's secretary took Oakley aside and explained that Piltdown was a sensitive subject with Father Teilhard.

But why? If he had been gulled by Dawson at the site, he had certainly recouped his pride. Smith Woodward had devoted his life to Dawson's concoction. Teilhard had written about it but once, called it as correctly as he could, and then shut up. Why be so embarrassed? Unless, of course, the embarrassment arose from guilt about another aspect of his silence – his inability to come clean while he watched men he loved and respected make fools of themselves, partly on his account. Marcellin Boule, his beloved master, for example, correctly called Smith Woodward's *Eoanthropus* 'an artificial and composite being' in the first edition of *Les hommes fossiles* (1921). The skull, he said, could belong to *'un bourgeois de Londres'*: the jaw belonged to an ape. But he pondered the significance of Piltdown 2 and changed his mind in the second edition of 1923: 'In the light of these new facts, I cannot be as sure as I was before. I recognize that the balance has now tipped a bit in the direction of Smith Woodward's hypothesis – and I am happy for this scientist whose knowledge and character I esteem equally.' How did Teilhard feel as he watched his beloved master, Boule, falling into the abyss – when he contained tools for extraction that he could not use.

2. *The elephant and the hippo*. Bits and pieces of other fossil mammals were salted into the Piltdown gravels in order to set a geologic matrix for the human finds. All but two of these items could have been collected in England. But the hippo teeth, belonging to a distinctive dwarfed species, probably came from the Mediterranean island of Malta. The elephant tooth almost surely came from a distinctive spot at Ichkeul, Tunisia, for it is highly radioactive as a result of seepage from surrounding sediments rich in uranium oxide. This elephant species has been found in several other areas, but nowhere else in such highly radioactive sediments. Moreover, the Ichkeul site was only discovered

by professionals in 1947; the doctored specimen at Piltdown could not have come from a cataloged museum collection.

Teilhard taught physics and chemistry at a Jesuit school in Cairo from 1905 to 1908, just before coming to Piltdown. His volume of *Letters from Egypt* again records little about theology and teaching, but much about travel, natural history, and collecting. He did not call at Tunisia or Malta on his passage down, but I can find no record of his passage back, and the two areas are right on his route from Cairo to France. In any case. Teilhard's letters from Cairo abound in tales of swapping and exchange with other natural historians of several North African nations. He was plugged into an amateur network of information and barter and might have received the teeth from a colleague.

This argument formed the base of evidence among my senior colleagues who suspected Teilhard – A. S. Romer, Bryan Patterson, and Louis Leakey (Leakey also mentioned Teilhard's knowledge of chemistry and the clever staining of the Piltdown bones). According to hearsay, Le Gros Clark himself, a member of the trio that exposed the hoax, also suspected Teilhard on this basis. I regard this argument as suggestive, but not compelling. Dawson too was plugged into a network of amateur exchange.

3. *Teilhard's good luck at Piltdown.* Although records are frustratingly vague, I believe that all the Piltdown pieces were found by the original trio – Dawson, Smith Woodward, and Teilhard. (In the official version, a workman may have given Dawson the first piece in 1908.) Dawson, of course, unearthed most of the material himself. Smith Woodward, so far as I can tell, found only one cranial fragment. Teilhard, who spent less time at Piltdown than his two colleagues, was blessed. He found a fragment of the elephant tooth, a worked flint, and the famous canine.

People who have never collected in the field probably do not realize how difficult and chancy the operation is when fossils are sparse. There is no magic to it, just hard work. A tooth in a gravel pit is about as conspicuous as the proverbial needle in a haystack. The hoaxer worked hard on his Piltdown material. He filed the canine and painted it to simulate age. Apes' teeth are not easy to come by. If I had but one precious item, I would not stick it into a large gravel heap and then hope that some innocent companion would find it. It would probably be lost forever, not triumphantly recovered. I doubt that I would ever find it again myself, after someone else had mucked about extensively in the pile.

Teilhard described his discovery in the first letter to Oakley: 'When I found the canine, it was so inconspicuous amidst the gravels which had been spread on the ground for sifting that it seems to me quite unlikely that the tooth could have been planted. I can even remember Sir Arthur congratulating me on the sharpness of my eyesight.' Smith Woodward's recollection (from his last book of 1948) is more graphic:

> We were excavating a rather deep and hot trench in which Father Teilhard, in black clothing, was especially energetic; and, as we thought he seemed a little exhausted, we suggested that he should leave us to do the hard labor for a time while he had comparative rest in searching the rain-washed spread gravel. Very soon he exclaimed that he had picked up the missing canine tooth, but we were incredulous, and told him we had already seen several bits of ironstone, which looked like teeth, on the spot where he stood. He insisted, however, that he was not deceived, so we both left our digging to go and verify his discovery. There could be no doubt about it, and we all spent the rest of that day until dusk crawling over the gravel in the vain quest for more.

I also have some doubt about Teilhard's flint, for it is the only Piltdown item indubitably found in situ. All the other specimens either came from gravel heaps that had been dug up and spread upon the ground or cannot be surely traced. Now in situ can signify one of two things (and the records do not permit a distinction). It may mean that the gravel bed lay exposed in a ditch, cliff, or road cut – in which case, anyone might have stuck the flint in. But it may mean that Teilhard dug into the layer from undisturbed, overlying ground – in which case, he could only have planted the flint himself.

Again, I regard this argument only as suggestive, not as definitive. It is the weakest point of all, hence its place at the bottom of my list. Perhaps Teilhard was simply a particularly keen observer.

Conclusions

What shall we make of all this? I can only imagine three conclusions. First, perhaps Piltdown has simply deluded another gullible victim, this time myself. Maybe I have just encountered an incredible string of coincidences. Could all the slips in the letters have been innocent errors of an aging man; the *comme par exprès* merely a literary device; the failure to use his best argument a simple oversight; his conspicuous silence beyond a few fleeting and unavoidable mentions only an aspect of a complex personality that no one has fathomed; his profound embarrassment just another facet of the same personality; the elephant and hippo Dawson's property . . . ? I just can't believe it. Coincidences recede into improbability as more and more independent items coagulate to form a pattern. The mark of any good theory is that it makes coordinated sense of a string of observations otherwise independent and inexplicable. Let us then assume that Teilhard knew Piltdown was a hoax, at least from 1920.

We are left with two possibilities. Was Teilhard innocent in the field at Piltdown? Did he tumble to the hoax later (perhaps when he deciphered the inconsistencies in Piltdown 2)? Did he then maintain silence out of loyalty to Dawson who had befriended him or because he didn't wish to stir a hornet's nest when he was not completely sure? But why, then, did he try so hard to exonerate Dawson in the letters to Oakley? For Dawson had used him and played on his youthful innocence as cruelly as he had deceived Smith Woodward. And why did he write a series of slips and half-truths to Oakley that embody, as their only pattern, an attempt to extract himself alone? And why such intense embarrassment and such conspicuous silence if he had guessed right but had been too unsure to say so?

Alan Ternes, editor of *Natural History*, made the interesting suggestion that Teilhard, as a priest, might have heard of the hoax through a confession by Dawson that he could not subsequently reveal. I have not been able to ascertain whether Dawson was Catholic; I do not think that he was. But I am told that priests may regard statements of contrition by other baptized Christians as privileged information. This is the most sensible version I have heard of the hypothesis that Teilhard knew about the fraud but did not participate in it. It would explain his silence, his embarrassment, even the '*comme par exprès*.' But, in this case, why would Teilhard have tried to construct such an elaborate and far-fetched theory of Dawson's innocence in his first letter to Oakley? Confession may have required silence, but surely not sheltering by falsehood. And why the slips and half-truths for his own exoneration in the subsequent letters?

This leaves a third explanation – that Teilhard was an active co-conspirator with Dawson at Piltdown. Only in this way can I make sense of the pattern in Teilhard's letters to Oakley, the 1920 article, the subsequent silence, the intense embarrassment.

This conclusion raises two final issues. First, to cycle back to my introduction, conspiracies have a tendency to spread. Once we admit Teilhard into the plot, should we not wonder about others as well? In fact, several knowledgeable people have strong suspicions about some young subordinates in the British Museum. I have confined my work to Teilhard's role; I think that others may have participated.

Second, what about motive? However overwhelming, the evidence cannot satisfy us without a reasonable explanation for why Teilhard might have done such a thing. Here I see no great problem, although we must recast Piltdown (at least from Teilhard's standpoint) as a joke that went too far, not as a malicious attempt to defraud.

Teilhard was not the dour ascetic or transported mystic that his publications sometimes suggest. He was a passionate man – a genuine hero in war, a true adventurer in the field, a man who loved life and people, who strove to experience the world in all its pleasures and pains. I assume that Piltdown was merely a delicious joke for him – at first. At Hastings, he was an amateur natural historian, with no expectation of a professional career in paleontology. He probably shared the attitude toward professionals so common among his colleagues – there but for the vagaries of life go I. Why do they have the fame, the reputation, and the cash? Why do they sit at their desks and reap rewards while we, with deeper knowledge born of raw experience, amuse ourselves? Why not play a joke to see how far a gullible professional could be taken? And what a wonderful joke for a Frenchman, for England at the time boasted no human fossils at all, while France, with Neanderthal and Cro-Magnon, stood proudly as the queen of anthropology. What an irresistible idea – to salt English soil with this preposterous combination of a human skull and an ape's jaw and see what the pros could make of it.

But the joke quickly went sour. Smith Woodward tumbled too fast and too far. Teilhard was posted to Paris to become, after all, a *professional* paleontologist. The war broke out, and Teilhard had to leave just as the last act to quell skepticism, Piltdown 2, approached the stage. Then Dawson died in 1916, the war dragged on to 1918, and professional English paleontology fell further and further into the quagmire of acceptance. What could Teilhard say by the war's end? Dawson could not corroborate his story. The jobs and careers of other conspirators may have been on the line. Any admission on Teilhard's part would surely have wrecked irrevocably the professional career he had desired so greatly, dared so little to hope for, and at whose threshold he now stood with so much promise. What could he say beyond *comme par exprès*.

Shall we then blame Teilhard or shall we forgive him? We cannot simply laugh and forget. Piltdown absorbed the professional attention of many fine scientists. It led millions of people astray for forty years. It cast a false light upon the basic processes of human evolution. Careers are too short and time too precious to view so much waste with equanimity.

But who among us would or could have come clean in Teilhard's position? Unfortunately, intent does not always correlate with effect in our complicated world – yet I believe that we must judge a man primarily by intent. If Teilhard had acted for malice or in hope of reward, I would have no sympathy. But I cannot view his participation as more than an intended joke that unexpectedly turned to a galling bitterness almost beyond belief. I think that Teilhard suffered for Piltdown throughout his life. I believe that he must have cried inwardly as he watched Smith Woodward and even Boule himself make fools of themselves – the very men who had befriended and taught him. Could the anguish of Piltdown have been on his mind, when he made the following pledge from the trenches during World War I?

> I have come, these days, to realize one very elementary fact: that the best way to win some sort of recognition for my ideas would be for me myself to attain, in the truest possible sense of the word, to a 'sanctity' that will be manifest to others – not only because of the particular force God would then give to whatever good is in my aspirations and influence – but also because nothing can give me more authority over men than for them to see me as someone who speaks to them from close to God. With God's help, I must live my

> 'vision' fully, logically, and without deviation. There is nothing more infectious than the example of a life governed by conviction and principle. And now I feel sufficiently drawn to and sufficiently equipped for, such a life.

Teilhard paid his debt and lived a full life; may we all do so well.

PART III

EVOLUTIONARY THEORY

WITH this section we move to the central focus of Gould's intellectual life, the exegesis, defence and expansion of Darwin's theory of evolution by natural selection. A committed Darwinian, he was nonetheless critical of the rigid orthodoxies of a neo-Darwinism which reduces organisms to genes (Richard Dawkins's 'replicators') and empty organisms (Dawkins's 'vehicles'). In this way of thinking, evolution becomes simply a change in gene frequencies in a population; phenotypes disappear. Gould argued instead that selection operates at many levels, on genes and genomes, phenotypes, populations and species. He emphasized Darwin's pluralism and replaced his gradualism with the theory of punctuated equilibrium that he and Niles Eldredge developed in the early seventies. What had once been discussed as 'preadaptation' (see 'Not Necessarily a Wing' in Part II) Gould developed into a full-blown theory of exaptation. And finally, to the anger of many among the orthodox, he emphasized, again and again, the role of chance, contingency, in evolutionary history. Many of these ideas were developed in technical papers beyond the scope of this collection, but they were reflected in his popular writings, and above all in his magisterial book, *The Structure of Evolutionary Theory*, published in 2002, not long before his death.

Here, we include extracts from that book, together with others from the essays which reflect on evolutionary processes. We begin with an overview of the evolution of life on earth, an exemplary popular account from a special issue of the *Scientific American,* and his most succinct statement of evolutionary principles, given as one of a series of Tanner Lectures on Human Values in 1985. Then to one of our hardest tasks, finding appropriate extracts from the 1,400 pages of *The Structure of Evolutionary Theory*. That book begins with a ninety-one page summary of its later arguments, followed by two major sections, the first historical, the second expounding Gould's own revisionist programme. When Darwin published *The Origin* in some haste, spurred on by the simultaneous discovery of selection theory by Alfred Russel Wallace, he described it as merely an abstract of the longer book he intended to write (but never got round to). Gould wrote the big book, and never the shorter version that would have been so helpful. So we have chosen two short extracts from the summary chapter of *Structure*, a brief scene-setting, followed by a longer account of punctuated equilibrium theory and macroevolution.

There's no doubt that in his more academic writing Gould took no prisoners, and made few concessions to a nonspecialist audience, so we follow this text with a more accessible account of punctuated equilibrium, the idea that over evolutionary history there have been long periods of relative stasis punctuated by briefer intervals of rapid change. Orthodox neo-Darwinism argues that every change in the genes must be reflected in some phenotypic change that might be either deleterious or adaptive. This view was challenged in the 1960s by the Japanese evolutionary geneticist Motoo Kimura, who argued that much genetic change was selectively neutral. His ideas were taken up enthusiastically by Gould and others like Richard Lewontin as one aspect of the revisionist programme. In 'Betting on Chance', Gould summarises the neutralist position. And in 'The Power of the Modal Bacter', he disposes of one of the most pervasive of evolutionary misconceptions, that it is in some way progressive – a theme that recurs in later essays too.

Evolution is not just about the appearance of new species, but also the extinction of older ones. The three final essays in this section deal with this question: 'The Great Dying', and the mass extinction of the dinosaurs, has been a subject of much scientific speculation over the years. Smuggled in beside it by a trick of contiguity Steve would have appreciated we include his reflections on the struggle for the acceptance of the theory of continental drift. Finally we were unable to resist a spectacular example of the Gouldian ability to find accessible analogies for even the most esoteric of phenomena, the grave topic of the putative extinction of that well-loved American candy, the Hershey bar.

THE EVOLUTION OF LIFE ON THE EARTH

SOME creators announce their inventions with grand éclat. God proclaimed, 'Fiat lux,' and then flooded his new universe with brightness. Others bring forth great discoveries in a modest guise, as did Charles Darwin in defining his new mechanism of evolutionary causality in 1859: 'I have called this principle, by which each slight variation, if useful, is preserved, by the term Natural Selection.'

Natural selection is an immensely powerful yet beautifully simple theory that has held up remarkably well, under intense and unrelenting scrutiny and testing, for 135 years. In essence, natural selection locates the mechanism of evolutionary change in a 'struggle' among organisms for reproductive success, leading to improved fit of populations to changing environments. (Struggle is often a metaphorical description and need not be viewed as overt combat, guns blazing. Tactics for reproductive success include a variety of nonmartial activities such as earlier and more frequent mating or better cooperation with partners in raising offspring.) Natural selection is therefore a principle of local adaptation, not of general advance or progress.

Yet powerful though the principle may be, natural selection is not the only cause of evolutionary change (and may, in many cases, be overshadowed by other forces). This point needs emphasis because the standard misapplication of evolutionary theory assumes that biological explanation may be equated with devising accounts, often speculative and conjectural in practice, about the adaptive value of any given feature in its original environment (human aggression as good for hunting, music and religion as good for tribal cohesion, for example). Darwin himself strongly emphasized the multifactorial nature of evolutionary change and warned against too exclusive a reliance on natural selection, by placing the following statement in a maximally conspicuous place at the very end of his introduction: 'I am convinced that Natural Selection has been the most important, but not the exclusive, means of modification.'

Natural selection is not fully sufficient to explain evolutionary change for two major reasons. First, many other causes are powerful, particularly at levels of biological organization both above and below the traditional Darwinian focus on organisms and their struggles for reproductive success. At the lowest level of substitution in individual base pairs of DNA, change is often effectively neutral and therefore random. At higher levels, involving entire species or faunas, punctuated equilibrium can produce evolutionary trends by selection of species based on their rates of origin and extirpation, whereas mass extinctions wipe out substantial parts of biotas for reasons unrelated to adaptive struggles of constituent species in 'normal' times between such events.

Second, and the focus of this article, no matter how adequate our general theory of evolutionary change, we also yearn to document and understand the actual pathway of life's history. Theory, of course, is relevant to explaining the pathway (nothing about the pathway can be inconsistent with good theory, and theory can predict certain general aspects of life's geologic pattern). But the actual pathway is strongly underdetermined by our general theory of life's evolution. This point needs some belaboring as a central yet widely misunderstood aspect of the world's complexity. Webs and chains of historical events are so intricate, so imbued with random and chaotic elements, so unrepeatable in encompassing such a multitude of unique (and uniquely interacting) objects, that standard models of simple prediction and replication do not apply. History can be explained, with satisfying rigor if evidence be adequate, after a sequence of events unfolds, but it cannot be predicted with any precision beforehand. Pierre-Simon Laplace, echoing the growing and confident determinism of the late eighteenth century, once said that he could specify all future states if he could know the position and motion of all particles in the cosmos at any moment, but the nature of universal complexity shatters this chimerical dream. History includes too much chaos, or extremely sensitive dependence on minute and unmeasurable differences in initial conditions, leading to massively divergent outcomes based on tiny and unknowable disparities in starting points.

And history includes too much contingency, or shaping of present results by long chains of unpredictable antecedent states, rather than immediate determination by timeless laws of nature.

Homo sapiens did not appear on the earth, just a geologic second ago, because evolutionary theory predicts such an outcome based on themes of progress and increasing neural complexity. Humans arose, rather, as a fortuitous and contingent outcome of thousands of linked events, any one of which could have occurred differently and sent history on an alternative pathway that would not have led to consciousness. To cite just four among a multitude: (1) If our inconspicuous and fragile lineage had not been among the few survivors of the initial radiation of multicellular animal life in the Cambrian explosion 530 million years ago, then no vertebrates would have inhabited the earth at all. (Only one member of our chordate phylum, the genus Pikaia, has been found among these earliest fossils. This small and simple swimming creature, showing its allegiance to us by possessing a notochord, or dorsal stiffening rod, is among the rarest fossils of the Burgess Shale, our best preserved Cambrian fauna.) (2) If a small and unpromising group of lobe-finned fishes had not evolved fin bones with a strong central axis capable of bearing weight on land, then vertebrates might never have become terrestrial. (3) If a large extraterrestrial body had not struck the earth 65 million years ago, then dinosaurs would still be dominant and mammals insignificant (the situation that had prevailed for 100 million years previously). (4) If a small lineage of primates had not evolved upright posture on the drying African savannas just two to four million years ago, then our ancestry might have ended in a line of apes that, like the chimpanzee and gorilla today, would have become ecologically marginal and probably doomed to extinction despite their remarkable behavioral complexity.

Therefore, to understand the events and generalities of life's pathway, we must go beyond principles of evolutionary theory to a paleontological examination of the contingent pattern of life's history on our planet – the single actualized version among millions of plausible alternatives that happened not to occur. Such a view of life's history is highly contrary both to conventional deterministic models of Western science and to the deepest social traditions and psychological hopes of Western culture for a history culminating in humans as life's highest expression and intended planetary steward.

Science can, and does, strive to grasp nature's factuality, but all science is socially embedded, and all scientists record prevailing 'certainties,' however hard they may be aiming for pure objectivity. Darwin himself, in the closing lines of *The Origin of Species*, expressed Victorian social preference more than nature's record in writing: 'As natural selection works solely by and for the good of each being, all corporeal and mental endowments will tend to progress towards perfection.'

Life's pathway certainly includes many features predictable from laws of nature, but these aspects are too broad and general to provide the 'rightness' that we seek for validating evolution's particular results – roses, mushrooms, people and so forth. Organisms adapt to, and are constrained by, physical principles. It is, for example, scarcely surprising, given laws of gravity, that the largest vertebrates in the sea (whales) exceed the heaviest animals on land (elephants today, dinosaurs in the past), which, in turn, are far bulkier than the largest vertebrate that ever flew (extinct pterosaurs of the Mesozoic era).

Predictable ecological rules govern the structuring of communities by principles of energy flow and thermodynamics (more biomass in prey than in predators, for example). Evolutionary trends, once started, may have local predictability ('arms races,' in which both predators and prey hone their defenses and weapons, for example – a pattern that Geerat J. Vermeii of the University of California at Davis has called 'escalation' and documented in increasing strength of both crab claws and shells of their gastropod prey through time). But laws of nature do not tell us why we have crabs and snails at all, why insects rule the multicellular world and why vertebrates rather than persistent algal mats exist as the most complex forms of life on the earth.

Relative to the conventional view of life's history as an at least broadly predictable process of gradually advancing complexity through time, three features of the paleontological record stand out in opposition and shall therefore serve as organizing themes for the rest of this article: the constancy of modal complexity throughout life's history; the concentration of major events in short bursts interspersed with long periods of relative stability; and the role of external impositions, primarily mass extinctions, in disrupting patterns of 'normal' times. These three features, combined with more general themes of chaos and contingency, require a new framework for conceptualizing and drawing life's history, and this article therefore closes with suggestions for a different iconography of evolution.

The primary paleontological fact about life's beginnings points to predictability for the onset and very little for the particular pathways thereafter. The earth is 4.6 billion years old, but the oldest rocks date to about 3.9 billion years because the earth's surface became molten early in its history, a result of bombardment by large amounts of cosmic debris during the solar system's coalescence, and of heat generated by radioactive decay of short-lived isotopes. These oldest rocks are too metamorphosed by subsequent heat and pressure to preserve fossils (though some scientists interpret the proportions of carbon isotopes in these rocks as signs of organic production). The oldest rocks sufficiently unaltered to retain cellular fossils – African and Australian sediments dated to 3.5 billion years old – do preserve prokaryotic cells (bacteria and cyanophytes) and stromatolites (mats of sediment trapped and bound by these cells in shallow marine waters). Thus, life on the earth evolved quickly and is as old as it could be. This fact alone seems to indicate an inevitability, or at least a predictability, for life's origin from the original chemical constituents of atmosphere and ocean.

No one can doubt that more complex creatures arose sequentially after this prokaryotic beginning – first eukaryotic cells, perhaps about two billion years ago, then multicellular animals about 600 million years ago, with a relay of highest complexity among animals passing from invertebrates, to marine vertebrates and, finally (if we wish, albeit parochially, to honor neural architecture as a primary criterion), to reptiles, mammals and humans. This is the conventional sequence represented in the old charts and texts as an 'age of invertebrates,' followed by an 'age of fishes,' 'age of reptiles,' 'age of mammals,' and 'age of man' (to add the old gender bias to all the other prejudices implied by this sequence).

I do not deny the facts of the preceding paragraph but wish to argue that our conventional desire to view history as progressive, and to see humans as predictably dominant, has grossly distorted our interpretation of life's pathway by falsely placing in the center of things a relatively minor phenomenon that arises only as a side consequence of a physically constrained starting point. The most salient feature of life has been the stability of its bacterial mode from the beginning of the fossil record until today and, with little doubt, into all future time so long as the earth endures. This is truly the 'age of bacteria' – as it was in the beginning, is now and ever shall be.

For reasons related to the chemistry of life's origin and the physics of self-organization, the first living things arose at the lower limit of life's conceivable, preservable complexity. Call this lower limit the 'left wall' for an architecture of complexity (see Fig. 3 on p. 281). Since so little space exists between the left wall and life's initial bacterial mode in the fossil record, only one direction for future increment exists – toward greater complexity at the right. Thus, every once in a while, a more complex creature evolves and extends the range of life's diversity in the only available direction. In technical terms, the distribution of complexity becomes more strongly right-skewed through these occasional additions.

But the additions are rare and episodic. They do not even constitute an evolutionary series but form a motley sequence of distantly related taxa, usually depicted as eukaryotic cell, jellyfish, trilobite, nautiloid, eurypterid (a large relative of horseshoe crabs), fish, an amphibian such as *Eryops*, a dinosaur, a mammal and a human being. This sequence cannot be construed as the major thrust or trend of life's history. Think rather of an occasional creature tumbling into the empty right region of complexity's space. Throughout this entire time, the bacterial mode has grown in height and remained constant in position. Bacteria represent the great success story of life's pathway. They occupy a wider domain of environments and span a broader range of biochemistries than any other group. They are adaptable, indestructible and astoundingly diverse. We cannot even imagine how anthropogenic intervention might threaten their extinction, although we worry about our impact on nearly every other form of life. The number of *Escherichia coli* cells in the gut of each human being exceeds the number of humans that has ever lived on this planet.

One might grant that complexification for life as a whole represents a pseudo-trend based on constraint at the left wall but still hold that evolution within particular groups differentially favors complexity when the founding lineage begins far enough from the left wall to permit movement in both directions. Empirical tests of this interesting hypothesis are just beginning (as concern for the subject mounts among paleontologists), and we do not yet have enough cases to advance a generality. But the first two studies – by Daniel W. McShea of the University of Michigan on mammalian vertebrae and by George F. Boyajian of the University of Pennsylvania on ammonite suture lines – show no evolutionary tendencies to favor increased complexity.

Moreover, when we consider that for each mode of life involving greater complexity, there probably exists an equally advantageous style based on greater simplicity of form (as often found in parasites, for example), then preferential evolution toward complexity seems unlikely a priori. Our impression that life evolves toward greater complexity is probably only a bias inspired by parochial focus on ourselves, and consequent overattention to complexifying creatures, while we ignore just as many lineages adapting equally well by becoming simpler in form. The morphologically degenerate parasite, safe within its host, has just as much prospect for evolutionary success as its gorgeously elaborate relative coping with the slings and arrows of outrageous fortune in a tough external world.

Even if complexity is only a drift away from a constraining left wall, we might view trends in this direction as more predictable and characteristic of life's pathway as a whole if increments of complexity accrued in a persistent and gradually accumulating manner through time. But nothing about life's history is more peculiar with respect to this common (and false) expectation than the actual pattern of extended stability and rapid episodic movement, as revealed by the fossil record.

Life remained almost exclusively unicellular for the first five sixths of its history – from the first recorded fossils at 3.5 billion years to the first well-documented multicellular animals less than 600 million years ago. (Some simple multicellular algae evolved more than a billion years ago, but these organisms belong to the plant kingdom and have no genealogical connection with animals.) This long period of unicellular life does include, to be sure, the vitally important transition from simple prokaryotic cells without organelles to eukaryotic cells with nuclei, mitochondria and other complexities of intracellular architecture – but no recorded attainment of multicellular animal organization for a full three billion years. If complexity is such a good thing, and multicellularity represents its initial phase in our usual view, then life certainly took its time in making this crucial step. Such delays speak strongly against general progress as the major theme of life's history, even if they can be plausibly explained by lack of sufficient atmospheric oxygen for most of Precambrian time or by failure of unicellular life to achieve some structural threshold acting as a prerequisite to multicellularity.

More curiously, all major stages in organizing animal life's multicellular architecture then occurred in a short period beginning less than 600 million years ago and ending by about 530 million years ago – and the steps within this sequence are also discontinuous and episodic, not gradually accumulative. The first fauna, called Ediacaran to honor the Australian locality of its initial discovery but now known from rocks on all continents, consists of highly flattened fronds, sheets and circlets composed of numerous slender segments quilted together. The nature of the Ediacaran fauna is now a subject of intense discussion. These creatures do not seem to be simple precursors of later forms. They may constitute a separate and failed experiment in animal life, or they may represent a full range of diploblastic (two-layered) organization, of which the modern phylum Cnidaria (corals, jellyfishes and their allies) remains as a small and much altered remnant.

In any case, they apparently died out well before the Cambrian biota evolved. The Cambrian then began with an assemblage of bits and pieces, frustratingly difficult to interpret, called the 'small shelly fauna.' The subsequent main pulse, starting about 530 million years ago, constitutes the famous Cambrian explosion, during which all but one modern phylum of animal life made a first appearance in the fossil record. (Geologists had previously allowed up to 40 million years for this event, but an elegant study, published in 1993, clearly restricts this period of phyletic flowering to a mere five million years.) The Bryozoa, a group of sessile and colonial marine organisms, do not arise until the beginning of the subsequent, Ordovician period, but this apparent delay may be an artifact of failure to discover Cambrian representatives.

Although interesting and portentous events have occurred since, from the flowering of dinosaurs to the origin of human consciousness, we do not exaggerate greatly in stating that the subsequent history of animal life amounts to little more than variations on anatomical themes established during the Cambrian explosion within five million years. Three billion years of unicellularity, followed by five million years of intense creativity and then capped by more than 500 million years of variation on set anatomical themes can scarcely be read as a predictable, inexorable or continuous trend toward progress or increasing complexity.

We do not know why the Cambrian explosion could establish all major anatomical designs so quickly. An 'external' explanation based on ecology seems attractive: the Cambrian explosion represents an initial filling of the 'ecological barrel' of niches for multicellular organisms, and any experiment found a space. The barrel has never emptied since; even the great mass extinctions left a few species in each principal role, and their occupation of ecological space forecloses opportunity for fundamental novelties. But an 'internal' explanation based on genetics and development also seems necessary as a complement: the earliest multicellular animals may have maintained a flexibility for genetic change and embryological transformation that became greatly reduced as organisms 'locked in' to a set of stable and successful designs.

In any case, this initial period of both internal and external flexibility yielded a range of invertebrate anatomies that may have exceeded (in just a few million years of production) the full scope of animal form in all the earth's environments today (after more than 500 million years of additional time for further expansion). Scientists are divided on this question. Some claim that the anatomical range of this initial explosion exceeded that of modern life, as many early experiments died out and no new phyla have ever arisen. But scientists most strongly opposed to this view allow that Cambrian diversity at least equaled the modern range – so even the most cautious opinion holds that 500 million subsequent years of opportunity have not expanded the Cambrian range, achieved in just five million years. The Cambrian explosion was the most remarkable and puzzling event in the history of life.

Moreover, we do not know why most of the early experiments died, while a few survived to become our modern phyla. It is tempting to say that the victors won by virtue of greater anatomical complexity, better ecological fit or some other predictable feature of conventional Darwinian struggle. But no recognized traits unite the victors, and the radical alternative must be entertained that each early experiment received little more than the equivalent of a ticket in the largest lottery ever played out on our planet – and that each surviving lineage, including our own phylum of vertebrates, inhabits the earth today more by the luck of the draw than by any predictable struggle for existence. The history of multicellular animal life may be more a story of great reduction in initial possibilities, with stabilization of lucky survivors, than a conventional tale of steady ecological expansion and morphological progress in complexity.

Finally, this pattern of long stasis, with change concentrated in rapid episodes that establish new equilibria, may be quite general at several scales of time and magnitude, forming a kind of fractal pattern in self-similarity. According to the punctuated equilibrium model of speciation, trends within lineages occur by accumulated episodes of geologically instantaneous speciation, rather than by gradual change within continuous populations (like climbing a staircase rather than rolling a ball up an inclined plane).

Even if evolutionary theory implied a potential internal direction for life's pathway (although previous facts and arguments in this article cast doubt on such a claim), the occasional imposition of a rapid and substantial, perhaps even truly catastrophic, change in environment would have intervened to stymie the pattern. These environmental changes trigger mass extinction of a high percentage of the earth's species and may so derail any internal direction and so reset the pathway that the net pattern of life's history looks more capricious and concentrated in episodes than steady and directional. Mass extinctions have been recognized since the dawn of paleontology; the major divisions of the geologic time scale were established at boundaries marked by such events. But until the revival of interest that began in the late 1970s, most paleontologists treated mass extinctions only as intensifications of ordinary events, leading (at most) to a speeding up of tendencies that pervaded normal times. In this gradualistic theory of mass extinction, these events really took a few million years to unfold (with the appearance of suddenness interpreted as an artifact of an imperfect fossil record), and they only made the ordinary occur faster (more intense Darwinian competition in tough times, for example, leading to even more efficient replacement of less adapted by superior forms).

The reinterpretation of mass extinctions as central to life's pathway and radically different in effect began with the presentation of data by Luis and Walter Alvarez in 1979, indicating that the impact of a large extraterrestrial object (they suggested an asteroid seven to ten kilometers in diameter) set off the last great extinction at the Cretaceous-Tertiary boundary 65 million years ago. Although the Alvarez hypothesis initially received very skeptical treatment from scientists (a proper approach to highly unconventional explanations), the case now seems virtually proved by discovery of the 'smoking gun,' a crater of appropriate size and age located off the Yucatan peninsula in Mexico.

This reawakening of interest also inspired paleontologists to tabulate the data of mass extinction more rigorously. Work by David M. Raup, J. J. Sepkoski, Jr, and David Jablonski of the University of Chicago has established that multicellular animal life experienced five major (end of Ordovician, late Devonian, end of Permian, end of Triassic and end of Cretaceous) and many minor mass extinctions during its 530 million-year history. We have no clear evidence that any but the last of these events was triggered by catastrophic impact, but such careful study leads to the general conclusion that mass extinctions were more frequent, more rapid, more extensive in magnitude and more different in effect than paleontologists had previously realized. These four properties encompass the radical implications of mass extinction for understanding life's pathway as more contingent and chancy than predictable and directional.

Mass extinctions are not random in their impact on life. Some lineages succumb and others survive as sensible outcomes based on presence or absence of evolved features. But especially if the triggering cause of extinction be sudden and catastrophic, the reasons for life or death may be random with respect to the original value of key features when first evolved in Darwinian struggles of normal times. This 'different rules' model of mass extinction imparts a quirky and unpredictable character to life's pathway based on the evident claim that lineages cannot anticipate future contingencies of such magnitude and different operation.

To cite two examples from the impact-triggered Cretaceous–Tertiary extinction 65 million years ago: First, an important study published in 1986 noted that diatoms survived the extinction far better than other single-celled plankton (primarily coccoliths and radiolaria). This study found that many diatoms had evolved a strategy of dormancy by encystment, perhaps to survive through seasonal periods of unfavorable conditions (months of darkness in polar species as otherwise fatal to these photosynthesizing cells; sporadic availability of silica needed to construct their skeletons). Other planktonic cells had not evolved any mechanisms for dormancy. If the terminal Cretaceous impact produced a dust cloud that blocked light for several months or longer (one popular idea for a 'killing scenario' in the extinction), then diatoms may have survived as a fortuitous result of dormancy mechanisms evolved for the entirely different function of weathering seasonal droughts in ordinary times. Diatoms are not superior to radiolaria or other plankton that succumbed in far greater numbers; they were simply fortunate to possess a favorable feature, evolved for other reasons, that fostered passage through the impact and its sequelae.

Second, we all know that dinosaurs perished in the end Cretaceous event and that mammals therefore rule the vertebrate world today. Most people assume that mammals prevailed in these tough times for some reason of general superiority over dinosaurs. But such a conclusion seems most unlikely. Mammals and dinosaurs had coexisted for 100 million years, and mammals had remained rat-sized or smaller, making no evolutionary 'move' to oust dinosaurs. No good argument for mammalian prevalence by general superiority has ever been advanced, and fortuity seems far more likely. As one plausible argument, mammals may have survived partly as a result of their small size (with much larger, and therefore extinction-resistant, populations as a consequence, and less ecological specialization with more places to hide, so to speak). Small size may not have been a positive mammalian adaptation at all, but more a sign of inability ever to penetrate the dominant domain of dinosaurs. Yet this 'negative' feature of normal times may be the key reason for mammalian survival and a prerequisite to my writing and your reading this article today.

Sigmund Freud often remarked that great revolutions in the history of science have but one common, and ironic, feature: they knock human arrogance off one pedestal after another of our previous conviction about our own self-importance. In Freud's three examples, Copernicus moved our home from center to periphery, Darwin then relegated us to 'descent from an animal world'; and, finally (in one of the least modest statements of intellectual history), Freud himself discovered the unconscious and exploded the myth of a fully rational mind. In this wise and crucial sense, the Darwinian revolution remains woefully incomplete because, even though thinking humanity accepts the fact of evolution, most of us are still unwilling to abandon the comforting view that evolution means (or at least embodies a central principle of) progress defined to render the appearance of something like human consciousness either virtually inevitable or at least predictable. The pedestal is not smashed until we abandon progress or complexification as a central principle and come to entertain the strong possibility that *H. sapiens* is but a tiny, late-arising twig on life's enormously arborescent bush – a small bud that would almost surely not appear a second time if we could replant the bush from seed and let it grow again.

Primates are visual animals, and the pictures we draw betray our deepest convictions and display our current conceptual limitations. Artists have always painted the history of fossil life as a sequence from invertebrates, to fishes, to early terrestrial amphibians and reptiles, to dinosaurs, to mammals and, finally, to humans. There are no exceptions; all sequences painted since the inception of this genre in the 1850s follow the convention.

Yet we never stop to recognize the almost absurd biases coded into this universal mode. No scene ever shows another invertebrate after fishes evolved but invertebrates did not go away or stop evolving! After terrestrial reptiles emerge, no subsequent scene ever shows a fish (later oceanic tableaux depict only such returning reptiles as ichthyosaurs and plesiosaurs). But fishes did not stop evolving after one small lineage managed to invade the land. In fact, the major event in the evolution of fishes, the origin and rise to dominance of the teleosts, or modern bony fishes, occurred during the time of the dinosaurs and is therefore never shown at all in any of these sequences – even though teleosts include more than half of all species of vertebrates. Why should humans appear at the end of all sequences? Our order of primates is ancient among mammals, and many other successful lineages arose later than we did.

We will not smash Freud's pedestal and complete Darwin's revolution until we find, grasp and accept another way of drawing life's history. J. B. S. Haldane proclaimed nature 'queerer than we can suppose,' but these limits may only be socially imposed conceptual locks rather then inherent restrictions of our neurology. New icons might break the locks. Trees – or rather copiously and luxuriantly branching bushes – rather than ladders and sequences hold the key to this conceptual transition.

We must learn to depict the full range of variation, not just our parochial perception of the tiny right tail of most complex creatures. We must recognize that this tree may have contained a maximal number of branches near the beginning of multicellular life and that subsequent history is for the most part a process of elimination and lucky survivorship of a few, rather than continuous flowering, progress and expansion of a growing multitude. We must understand that little twigs are contingent nubbins, not predictable goals of the massive bush beneath. We must remember the greatest of all Biblical statements about wisdom: 'She is a tree of life to them that lay hold upon her; and happy is every one that retaineth her.'

CHALLENGES TO NEO-DARWINISM AND THEIR MEANING FOR A REVISED VIEW OF HUMAN CONSCIOUSNESS

I. Basic Premises of the Synthetic Theory

THE odyssey of evolution in the history of ideas has been, in microcosm, much like the history of species, the macrocosm that it seeks to explain – peculiar, tortuous, unpredictable, complex, weighted down by past inheritances, and not moving in unilinear fashion toward any clear goal.

Darwin divided his life's work, explicitly and often, into two major goals: to demonstrate the fact that evolution had occurred, and to promote the theory of natural selection as its primary mechanism. In the first quest, his success was abundant, and he now lies in Westminster Abbey, at the feet of Isaac Newton, for this triumph. In the second, he made much less headway during his lifetime. By the close of the nineteenth century, natural selection was a strong contender in a crowded field of evolutionary theories, but it held no predominant position.

Darwinian concepts are now so canonical in evolutionary theory that students without historical perspective often assume it has been so since 1859. In fact, the triumph of natural selection as a centerpiece of evolutionary theory dates only to a major intellectual movement of the 1930s and 1940s, called by Julian Huxley the 'modern synthesis.'[1] The synthesis validated natural selection as a powerful causal agent and raised it from a former status as one contender among many to a central position among mechanisms of change (the role assigned to natural selection later hardened to near exclusivity).[2] The modern synthesis is, essentially, the central logic of Darwin's argument updated by the genetic theory of variation and inheritance that he, perforce, lacked.

Ernst Mayr, leading architect and historian of the modern synthesis, offered this definition of its primary claims at a conference that assembled all the leading originators:

> The term 'evolutionary synthesis' was introduced by Julian Huxley . . . to designate the general acceptance of two conclusions: gradual evolution can be explained in terms of small genetic changes ('mutations') and recombination, and the ordering of this genetic variation by natural selection; and the observed evolutionary phenomena, particularly macroevolutionary processes and speciation, can be explained in a manner that is consistent with the known genetic mechanisms.[3]

Several major tenets may be distilled from this paragraph. I shall select three as inspirations for major critiques of the hegemony of Neo-Darwinism. Each has significance for a revised view of human consciousness and its evolutionary meaning.

1. *Chance and necessity*. Randomness and determinism occupy separate and definite spheres in the central logic of Darwin's theory. As Mayr states above, genetic variation arises by mutation and recombination; it is then ordered by natural selection – chance for the origin of the raw material of change, determination for the selective incorporation of some of this variation into altered organisms.

The central logic of Darwinism requires that natural selection not merely operate, but that it be the creative force of evolutionary change. Selection wins its role as a creative force because the other component of evolutionary mechanics – the forces that produce the raw material of genetic variation – are random, in the special sense of 'not inherently directed toward adaptation.' That is, if local environments change and smaller organisms are now at an advantage, genetic variation does not produce more small individuals, thus imparting a direction to evolutionary change from the level of variation itself. Variation continues to occur 'at random,' in a broad spectrum about the average size. Selection must impart direction – and be the creative force of evolution – by differentially preserving those random variants yielding smaller than average phenotypes.

Randomness is a part of Darwinian theory, but it has a very definite and restricted role (lest the central premise of creativity for natural selection be compromised). It operates only in the genesis of raw material – genetic variation. It plays no role at all in the production of evolutionary change – the selective preservation of a portion of this variation to build altered organisms.

Critics of Darwinism, Arthur Koestler for example, have often misunderstood this central tenet of Darwinism. They charge that Darwinism cannot be correct because a world so ordered as ours cannot be built by random processes. But they fail to understand that Darwinism invokes randomness only to generate raw material. It agrees with the critics in arguing that the world's order could only be produced by a conventional deterministic cause – natural selection in this case.

2. *The reductionistic tradition.* The central claim of the synthesis, and the basis for its alleged unifying power, holds that the phenomena of macroevolution, at whatever scale, can be explained in terms of genetic processes that operate within populations. Organisms are the primary Darwinian actors and evolution at all levels is a result of natural selection, working by sorting out individuals within populations (differential reproductive success). This argument reflects a reductionistic tradition, not of course to atoms and molecules of the classic physical version, but rather of such macroevolutionary events as long-term trends to the extended struggle of individual organisms within local populations.

Reduction to struggles among organisms within populations is fundamental to Darwinism and underlies the logic of Darwin's own version of natural selection.[4] Darwin developed his theory as a conscious analog to the laissez-faire economics of Adam Smith,[5] which holds as its primary argument that order and harmony within economies does not arise from higher-order laws destined for such effect, but can be justly attained only by letting individuals struggle for personal benefits, thereby allowing order to arise as an unplanned consequence of sorting among competitors. The Darwinism of the modern synthesis is, therefore, a *one-level theory* that identifies struggle among organisms within populations as the causal source of evolutionary change, and views all other styles and descriptions of change as consequences of this primary activity.

3. *The hegemony of adaptation.* If evolutionary change proceeds via the struggle of individuals within populations, then its result must be adaptation. Natural selection operates by the differential reproductive success of individuals better suited to local environments (as a happy result of their combinations of genetic variation). The statistical accum-

ulation of these favored genes within populations must produce adaptation if evolutionary change is controlled by natural selection. Of course, all Darwinians admit that other processes – the random force of genetic drift in particular – can produce evolutionary change as well, but the synthetic theory carefully limits their range and efficacy, so that they play no statistically important role in the net amount of phenotypic change within lineages. Since (under the second argument for extrapolation) long-term trends are nothing but natural selection within populations extended, then the phenomena of macroevolution reduce to natural selection as well and must be similarly adaptive throughout.

II. Current Critiques of the Central Logic of the Synthesis

All three major premises of the synthetic theory have been criticized in recent years:

1. *Chance as an agent of evolutionary change.* In a major revision of Darwinian logic, chance has been elevated, from its traditional and restricted role as generator of raw material only, to a more active part as agent of evolutionary *change*.

Most debates in natural history center upon issues of relative frequency, not exclusive occurrence. Chance as an agent of evolutionary change in the phenomenon of 'genetic drift' has long been recognized as orthodox, but traditional theory so restricted its occurrence and importance that it could play no major role in life's history. The new arguments are distinctive in that they advocate a high relative frequency for chance and make it an important evolutionary agent of change in both the qualitative and quantitative sense. They also award an important role to chance at all levels of the hierarchy of evolutionary causation – at the molecular level of allelic substitution, the domain of speciation, and the largest scope of changing taxonomic composition in mass extinction.

The quasi-clocklike accumulation of DNA differences in phyletic lines, the empirical basis for the so-called 'neutral' theory, or 'non-Darwinian evolution',[6] only makes sense if selection does not 'see' the substitutions and if they, therefore, drift to fixation in a stochastic manner. Most models of sympatric speciation – though the relative frequency of this process remains unresolved and may be quite low – propose a genetic change quick enough to produce reproductive isolation (often by major alterations in number or form of chromosomes) prior to any selective revamping of the new form. The genetic trigger of speciation would therefore be random with respect to the demands of adaptation. I shall have more to say about mass extinction in the next section, but if these debacles really run on a 26-million year cycle[7] triggered by cometary showers,[8] then the reasons for differential survival cannot have much to do with – and must be random with respect to – the deterministic, adaptive struggles of organisms in the preceding normal geological times.

2. *The hierarchical perspective and the non-reducibility of macroevolution.* The material of biology is ordered into a genealogical hierarchy of ever more inclusive objects: genes, bodies, demes (local populations of a species), species, and monophyletic clades of species. Although our linguistic habits generally restrict the term 'individual' to bodies alone, each unit of this hierarchy maintains the two essential properties that qualify it as an 'individual' and therefore (under selectionist theories), as a potential causal agent in its own right – stability in time (with recognizable inception and extinction, and sufficient coherence of form between beginning and end) and ability to replicate with error (a prerequisite for units of selection in Darwin's world). Traditional Darwinian gradualists would deny individuality to species by arguing that they are mere abstractions, names we give to segments of gradually transforming lineages. But under the punctuated equilibrium model,[9] species are generally stable following their geologically rapid origin, and most evolutionary change occurs in conjunction with events of branching speciation, not by the transformation *in toto* of existing species. Under this model, therefore, species maintain the essential properties of individuals and may be so designated.[10]

Few evolutionists would deny this hierarchy in a descriptive sense, but traditions of the modern synthesis specify that causality be sought only at the level of organisms – for natural selection operates by sorting organisms within populations. Richard Dawkins has challenged this view, but in the interests of an even further and stricter reductionism.[11] He argues that genes are the only true causal agents and organisms merely their temporary receptacles. I strongly disagree with Dawkins,[12] since I feel that he has confused bookkeeping (which may be done efficiently in terms of genes) with causality. But I feel that he has inadvertently made an important contribution to the theory of causal hierarchy by establishing numerous cases of true gene-level selection –

that is, selection upon genes that occurs without a sorting of bodies and that has no effect upon the phenotypes of bodies. The hypothesis of 'selfish DNA' as an explanation for iteration of copies in middle-repetitive DNA (with no initial benefit or detriment to organisms at the next hierarchical level) represents the most interesting proposal for independent gene-level selection.[13]

If genes can be selected independently of organisms, then we may extend the causal hierarchy upward as well. Deme-level selection has long been advocated by Sewall Wright in his theory of 'shifting balance'.[14] Species selection may be a more potent force than traditional Darwinian sorting of organisms in both the spread of features within clades and the differential success of some clades over others. True species selection relies upon properties of species as entities – propensity to speciate in particular – that cannot be reduced to characteristics of organisms, and therefore cannot be explained by natural selection operating at its usual level. The expanded hierarchical theory remains Darwinian in spirit – since it advocates a process of selection at several levels of a hierarchy of individuals – but it confutes the central Darwinian logic that evolutionary events at all scales be reduced for causal explanation to the level of organisms within populations.[15]

3. *Critique of adaptation.* A potent critique against the hegemony of adaptation has arisen from the theory of neutralism – the claim that much genetic change accumulates in populations by genetic drift upon allelic variants that are irrelevant to adaptation, and that natural selection therefore cannot recognize. Although these critiques are valid and were historically important in breaking the hegemony of adaptation, I shall bypass them here because I wish to discuss the evolution of phenotypes, and neutral changes, by definition, do not affect phenotypes.

At the level of phenotypes, the critique of adaptation does not claim a discovery of new evolutionary processes that actively produce substantial phenotypic change without natural selection. The critique remains content with the conventional idea that natural selection is the only identified agent of substantial and persistent evolutionary *change*. In what sense, then, can we speak of a critique of adaptation?

Suppose that every adaptive change brings with it (since organisms are integrated entities) a set of non-adaptive sequelae far exceeding in number and extent the direct adaptation itself.[16] Suppose then that these sequelae serve as constraints and channels that powerfully determine the limits and directions of future evolutionary change. Natural selection may still be the force that pushes organisms down the channels, but if these channels are the only paths available, and if they themselves were not constructed as a direct result of adaptation, then phenotypes are as much determined by the limits and potentialities set by non-adaptation as by the direct change produced by natural selection itself.

Of course, traditional Darwinians do not deny that adaptation entails non-adaptive consequences. This theme is, for example, the classic material of allometry, a subject named and popularized by the great Darwinian Julian Huxley.[17] But these consequences are conventionally viewed as superficial, epiphenomenal and non-constraining; natural selection, after all, can break an allometric correlation when necessary. Moreover, although Darwinism does not deny the existence of powerful constraints upon pathways of evolutionary change, the constraints are attributed to past adaptations for different roles. Thus, features of the phenotype are either current adaptations or past adaptations to different circumstances that constrain current change. Adaptation reigns. Darwin himself, a careful student of constraints and correlations, wrestled long and hard with this problem and finally resolved it in favor of adaptive supremacy in a key but neglected passage in *The Origin of Species*:

> All organic beings have been formed on two great laws – Unity of Type and the Conditions of Existence. By unity of type is meant that fundamental agreement in structure, which we see in organic beings of the same class, and which is quite independent of their habits of life. . . . The expression of conditions of existence . . . is fully embraced by the principle of natural selection. For natural selection acts by either now adapting the varying parts of each being to its organic and inorganic conditions of life; or by having adapted them during long-past periods of time. . . . Hence, in fact, the law of Conditions of Existence is the higher law; as it includes, through the inheritance of former adaptations, that of Unity, of Type.[18]

I believe that our views on the causes of phenotypic change have become stalled in a strict Darwinism that has already offered its valid insights – and that each critique of Darwinism offers an important new perspective. I shall illustrate the potential of each critique to expand our view of evolution in specific cases by discussing their potential impact upon the event of most immediate concern and importance to us – the evolution of human consciousness.

III. Consciousness as Cosmic Accident

Four controlling biases of Western thought – progressivism, determinism, gradualism, and adaptationism – have combined to construct a view of human evolution congenial to our hopes and expectations. Since we evolved late and, by our consciousness, now seem well in control (for better or for worse), the four biases embody a view that we rule by right because evolution moves gradually and predictably toward progress, always working for the best. These four biases have long stood as the greatest impediments to a general understanding and appreciation of the Darwinian vision, with its explicit denial of inherent progress and optimality in the products of evolution.

Yet Darwinism does not confute all our hopes. It still smuggles the idea of progress back into empirical expectation, not by the explicit workings of its basic mechanism (which does, indeed, deny inherent advance), but by an accumulation of superior designs through successive local adaptations.[19] All the great modern Darwinians have come to terms with (and supported) the notion of evolutionary progress, even though they recognized that the basic mechanism of natural selection contains no explicit statement about it.[20] Moreover, in viewing selection as a deterministic process, Darwinism supports our hope that the directions of change have their good reasons. In this Darwinian climate, we may still view the evolution of human consciousness as the predictable end of a long history of increasing mentality. Yet our new ideas about the importance of randomness in evolutionary change – particularly at the highest level of mass extinction – seriously upset this comforting and traditional notion and strongly suggest that we must view the evolution of human consciousness as a lucky accident that occurred only by the fortunate (for us) concatenation of numerous improbabilities. The argument is not based on a waffling theoretical generality, but on a specific empirical claim about a single important event: the Cretaceous mass extinction.

We may summarize the exciting ferment now reorganizing our ideas on mass extinction[21] by stating that these major punctuations in life's history are more frequent, more sudden, more severe, and more qualitatively different than we had realized before. I believe that Alvarez et al.[22] have now proved their originally startling claim that a large extraterrestrial body struck the earth some 65 million years ago and must, therefore, be viewed as the major trigger of the Cretaceous extinction. Enhanced levels of presumably extraterrestrial iridium (the empirical basis of the Alvarez claim for the Cretaceous) have now been found at other extinction boundaries as well – so we may have the basis for a general theory of mass extinction, not merely a good story for the Cretaceous.

The meaning of the extraterrestrial theory for human consciousness as a cosmic accident begins with a basic fact that should be more widely known (but that will surprise most nonprofessionals, who assume something different): *dinosaurs and mammals evolved at the same time.* Mammals did not arise later, as superior forms that gradually replaced inferior dinosaurs by competition. Mammals existed throughout the 100 million years of dinosaurian domination – and they lived as small, mostly mouse-sized creatures in the ecological interstices of a world ruled by large reptiles. They did not get bigger; they did not get better (or at least their changes did nothing to drive dinosaurs toward extinction). They did nothing to dislodge the incumbents; they bided their time.

Structural or mental inferiority did not drive the dinosaurs to extinction. They were doing well, and showing no sign of ceding domination, right until the extraterrestrial debacle unleashed a set of sudden consequences (as yet to be adequately specified, although the 'nuclear winter' scenario of a cold, dark world has been proposed for the same reasons). Some mammals weathered the storm; no dinosaurs did. We have no reason to believe that mammals prevailed as a result of any feature traditionally asserted to prove their superiority – warm-bloodedness, live bearing, large brains, for example. Their 'success' might well be attributed to nothing more than their size – for nothing large and terrestrial got through the Cretaceous debacle, while many small creatures survived.

In any case, had the cometary shower (or whatever) not hit, we have no reason to think that dinosaurs, having dominated the earth for 100 million years, would not have held on for another sixty-five to continue their hegemony today. In such a case, mammals would probably still be mouse-sized creatures living on the fringes – after all, they had done nothing else for 100 million years before. Moreover, dinosaurs were not evolving toward any form of consciousness. In other words, those comets or asteroids were the *sine quibus non* of our current existence. Without the removal of dinosaurs that they engendered, consciousness would not have evolved on our earth.

IV. Exaptation and the Flexibility of Mind

Strict adaptation entails a paradox for students of evolutionary change. If all structures are well designed for immediate use, where is the flexibility for substantial change in response to severely altered environments? The conventional answer calls upon a phenomenon of 'preadaptation' – the idea that structures actively evolved for one use may be fortuitously fitted for easy modification to strikingly different functions (feathers, evolved for thermo-regulation, then available for flight, for example). But preadaptation speaks only of one-for-one substitutions based on previous adaptation. Can we identify a *pool* of flexibility in uncommitted structures?

Vrba and I have argued that strict adaptationism has blinded us to the absence of an important concept in our science of form.[23] Some evolutionists use 'adaptation' for any structure that performs a beneficial function, regardless of its origin. But a long tradition, dating from Darwin himself, restricts 'adaptation' to those structures evolved directly by natural selection for their current use. If we accept this stricter definition, what shall we call structures that contribute to fitness but evolved for other reasons and were later coopted for their current role? They have no name at present, and Vrba and I suggest that they be called 'exaptations.' Preadaptation is, of course, a related concept – a kind of exaptation before the fact (feathers on a running dinosaur are preadaptations for flight; unaltered on a bird, they are exaptations). But preadaptation does not cover the range of exaptation because it refers to structures *adapted* for one role that are fortuitously suited for another. Preadaptation does not cover the large class of structures that never were adaptations for anything, but arose as the numerous nonadaptive sequelae of primary adaptations. These are also available for later cooptation as exaptations.[24] Surely, since nonadaptive sequelae are more numerous than adaptations themselves, the range of exaptive possibility must be set primarily by nonadaptation. Thus, if flexibility is primarily a result of possibilities that remain labile either because they have no current function (potential exaptations) or because their currently adapted structure can do other things just as well (preadaptations), then the major basis of flexibility must lie in nonadaptation. The old adage that flexibility correlates positively with complexity is correct, but the reason is not primarily – as usually stated – that complexity is itself so highly adaptive, but rather that increased complexity implies a vastly greater range of nonadaptive sequelae for any change, and hence a greatly enlarged exaptive pool.

Flexibility and computing power are the interrelated keys to the power of human consciousness. Among the usual reasons cited for extreme flexibility of human consciousness are the biological neoteny that probably keeps our brain in a labile, juvenile state[25] and the unparalleled potential of the nonsomatic culture that our brains have made possible. These are indeed the two major reasons for human flexibility, but both are reflections of a single underlying theme: no biological structure has ever been so pregnant with exaptive possibilities as the human brain; no other biological structure has ever produced so many nonadaptive sequelae to its primary adaptation of increased size.

I do not doubt that the brain became large for an adaptive reason (probably a set of complex reasons) and that natural selection brought it to a size that made consciousness possible. But, surely, most of what our brain does today, most of what makes us so distinctively human (and flexible), arises as a consequence of the nonadaptive sequelae, not of the primary adaptation itself – for the sequelae must be so vastly greater in number and possibility. The brain is a complex computer constructed by natural selection to perform a tiny subset of its potential operations. An arm built for one thing can do others (I am now typing with fingers built for other purposes). But a brain built for some functions can do orders of magnitude more simply by virtue of its basic construction as a flexible computer. Never in biological history has evolution built a structure with such an enormous and ramifying set of exaptive possibilities. The basis of human flexibility lies in the unselected capacities of our large brain.

This perspective also suggests that we must radically revise our methodology for thinking about the biological basis of essential human institutions and behaviors. An enormous, and largely speculative, literature attempts to interpret anything important that our brains do today as direct adaptations to the environments that shaped our earlier evolution. Thus, for example, religion may be a modern reflection of behaviors that evolved to cement group coherence among savannah hunters. But religion might as well record our human response to that most terrifying fact that a large brain allowed us to learn (for no directly adaptive reason) – the inevitability of our personal mortality. I suspect that most of our current cognitive life uses the nonadaptive sequelae of a large brain as exaptations, and does not record the direct reasons why natural selection originally fashioned our large brain.

V. Hierarchy and the Simultaneous, Conscious Control of Levels

Hierarchies of inclusion, like the genealogical hierarchy under discussion here, maintain an important property of asymmetry. Sorting at any high level must produce effects at all lower levels by shuffling their units (individuals) as well.[26] This property of hierarchies is responsible for the causal confusion of reductionists who assume that because lowest-level units – genes in this case – are always sorted, then this sorting (which they confuse with, and call, selection) must record the causal locus of change. But, again, bookkeeping is not causality, and this argument is invalid. Thus, when species selection operates and certain kinds of species are removed from or differentially added to a clade, proportions of organisms and frequencies of genes must also change within the clade – although the cause of sorting resides at the species level.

The converse, however, is not true. Sorting at low levels does not necessarily produce any effect at all upon the character or relative numbers of higher-level individuals. Lower-level sorting may be effectively insulated from any effect upon higher levels. Thus, at least initially, mobile genes may increase their number of copies within genomes without producing any effect upon bodies, demes, or species. This invisibility is the basis of the selfish DNA hypothesis.

Organisms – the quintessential Darwinian actors – normally can only operate directly for themselves. This produces the paradox of overspecialization when benefits to individuals entail eventual extinction of species because bizarre specializations so limit flexibility in the face of environmental change. Imagine what evolutionary possibilities would be opened if this asymmetry could be broken, and if lower-level units could work simultaneously both for their own fitness and for the fitness of those higher units in which they reside. Yet this cannot happen in a world of unconscious objects, for how could a gene work actively for its body, or a body for its species, when individuals only 'see' selection at their own level and cannot know (because they are unaffected by) the forces and directions of higher-level selection?

But human consciousness has ruptured this system. We can use conscious thought to break through the bounds of our own level and to understand what we might do as individuals to enhance or injure the groups in which we reside. In short, we can work directly on our own higher-level fitness. We also have the genetic flexibility – since we are not programmed automata – to choose actions injurious to ourselves but beneficial to our groups, even though natural selection has been working only on our individual-level fitness for so long. Thus, we can behave altruistically not only because certain organism-level processes – kin selection and reciprocal altruism – select for self-abnegation as a good Darwinian strategy, but primarily because we can understand the importance of group-level fitness and have the genetic flexibility (probably for non-adaptive reasons, and not necessarily as the result of millennia of kin selection) to act accordingly. In this sense, the strict Darwinian explanations for altruism offered by sociobiology are inadequate.

For the first time in biological history, organisms can actively pursue fitness not only for themselves but at several levels of their own hierarchy. The gain in potential power and flexibility is staggering. We can now speed and alter the evolution of our species at unprecedented rates and effectiveness. We have broken the ordering principles of the evolutionary hierarchy.

This unique mode of evolution also presents new challenges. If we lived in a world of intrinsic harmony, where fitness at one level inevitably enhanced fitness at others, our new abilities would simply allow us to tap a positive feedback loop between individual and species-level fitness *ad majorem hominis generisque gloriam*. But our world is not so pleasant. The components of fitness at one level are just as likely to depress (as in overspecialization) as to enhance fitness at higher levels. Consciousness puts us in the uncomfortable position of being the only species that can directly affect the components of both its individual and species-level fitness – and of finding that they often conflict. What then are we to do? Shall our great athletes press for even higher salaries and imperil the health and finances of their game's organization?

I have argued that three criticisms of strict Darwinism – randomness, nonadaptation, and hierarchy – each has important implications for a revised view of the evolutionary meaning of human consciousness. Some readers might draw a pessimistic message from the coordinated theme that less predictability, less order, less design attended the evolution of our unique mentality. They may be justly reminded of the *Rubaiyat*'s famous couplet,

> Into this Universe, and Why not knowing
> Nor Whence, like Water willy-nilly flowing.

I draw no somber conclusions from these arguments. I do not believe, first of all, that the answer to moral dilemmas about meaning lies with the facts of nature, whatever they may be. Moreover, I see only hope in the flexibility offered to human consciousness by its evolutionary construction. If our mentality evolved for no particular predictable reasons, then we may make of it what we will. If the major activity of our brain records the nonadaptive sequelae of its construction as a powerful computer, then evolutionary adaptation does not specify how we must behave and what we must do. *Vita brevis* to be sure, but what possibilities.

> Ah, make the most of what we yet may spend,
> Before we too into the Dust descend . . .
> Here with a little Bread beneath the Bough . . .
> Oh, Wilderness were Paradise enow!

Notes

1. J. Huxley, *Evolution: The Modern Synthesis* (London: Allen & Unwin, 1942).
2. See S. J. Gould, 'Irrelevance, Submission and Partnership: The Changing Role of Palaeontology in Darwin's Three Centennials, and a Modest Proposal for Macroevolution', in J. Bendell (ed.) *Evolution from Molecules to Man* (Cambridge: Cambridge University Press, 1983), pp. 347–66.
3. In E. Mayr and W. Provine, *The Evolutionary Synthesis* (Cambridge, MA: Harvard University Press, 1980), p. 1.
4. S. J. Gould, 'Darwinism and the Expansion of Evolutionary Theory', *Science*, 216 (1982): 380–7.
5. S. S. Schweber, 'The Origin of the *Origin* Revisited', *Journal of the History of Biology*, 10 (1977): 229–316.

6. J. L. King and T. H. Jukes, 'Non-Darwinian Evolution', *Science*, 164 (1969): 788.
7. D. M. Raup and J. J. Sepkoski Jr., 'Mass Extinctions in the Marine Fossil Record', *Science*, 215 (1982): 1501–3.
8. W. Alvarez and R. A. Muller, 'Evidence from Crater Ages for Periodic Impacts on the Earth', *Nature*, 308 (1984): 718–20.
9. N. Eldredge and S. J. Gould, 'Punctuated Equilibria: An Alternative to Phyletic Gradualism', in T. J. M. Schopf (ed.), *Models in Paleobiology* (San Francisco: Freeman, Cooper and Co., 1972), pp. 82–115; S. J. Gould and N. Eldredge, 'Punctuated Equilibria: The Tempo and Mode of Evolution Reconsidered', *Paleobiology*, 3/2 (1977): 115–51.
10. N. Eldredge and J. Cracraft, *Phylogenetic Pattern and the Evolutionary Process* (New York: Columbia University Press, 1980); E. S. Vrba and N. Eldredge, 'Individuals, Hierarchies and Processes: Towards a More Complete Evolutionary Theory', *Paleobiology* (1984).
11. R. Dawkins, *The Selfish Gene* (New York: Oxford University Press, 1976); R. Dawkins, *The Extended Phenotype* (San Francisco: W. H. Freeman, 1982).
12. Gould, 'Irrelevance, Submission and Partnership'.
13. W. F. Doolittle and C. Sapienza, 'Selfish Genes, the Phenotype Paradigm, and Genome Evolution', *Nature*, 284 (1980): 601–3; L. E. Orgel and F. H. C. Crick, 'Selfish DNA: The Ultimate Parasite', *Nature*, 284 (1980): 604–7.
14. S. Wright, 'Evolution in Mendelian Populations', *Genetics*, 16 (1931): 97–159.
15. S. J. Gould, 'Darwinism and the Expansion of Evolutionary Theory', *Science*, 216 (1982): 380–7; S. J. Gould, 'The Paradox of the First Tier: An Agenda for Paleobiology', *Paleobiology* (1985).
16. For specific examples, see S. J. Gould, 'Covariance Sets and Ordered Geographic Variation in *Cerion* from Aruba, Bonaire and Curacao: A Way of Studying Non-Adaptation', *Systematic Zoology*, 33/2 (1984): 217–37; S. J. Gould, 'Morphological Channeling by Structural Constraint: Convergence in Styles of Dwarfing and Gigantism in *Cerion*, with a Description of Two New Fossil Species and a Report on the Discovery of the Largest *Cerion*', *Paleobiology* (1984).
17. J. Huxley, *Problems of Relative Growth* (London: MacVeagh, 1932); S. J. Gould, 'Allometry and Size in Ontogeny and Phylogeny', *Biology Reviews*, 41 (1966): 587–640.
18. C. Darwin, *On the Origin of Species* (London: John Murray, 1859), p. 206.
19. For a resolution of this apparent paradox, see Gould, 'The Paradox of the First Tier'.
20. J. Huxley, *Evolution in Action* (London: Chatto & Windus, 1953); B. Rensch, *Biophilosophy* (New York: Columbia University Press, 1971); G. L. Stebbins, *The Basis of Progressive Evolution* (Chapel Hill: University of North Carolina Press, 1969); Th. Dobzhansky, 'The Ascent of Man', *Social Biology*, 19 (1972): 367–78.
21. See summary in Gould, 'The Paradox of the First Tier'.

22. L. Alvarez, W. Alvarez, F. Asaro and H. V. Michel, 'Extraterrestrial Cause for the Cretaceous-Tertiary Extinction', *Science*, 208 (1980): 1095–1108.
23. S. J. Gould and Elisabeth S. Vrba, 'Exaptation: A Missing Term in the Science of Form', *Paleobiology*, 8/1 (1982): 4–15.
24. See examples, mostly from architecture and anthropology, where the concept does not threaten conventional thought and is therefore easier to grasp and accept, in S. J. Gould and R. C. Lewontin, 'The Spandrels of San Marco and the Panglossian Paradigm: A Critique of the Adaptionist Programme', *Proceedings of the Royal Society of London. Series B: Biological Sciences*, 205 (1979): 581–98.
25. S. J. Gould, *Ontogeny and Phylogeny* (Cambridge, MA: Harvard University Press, 1977).
26. On downward causation, see D. T. Campbell, '"Downward Causation" in Hierarchically Organised Biological Systems', in F. J. Ayala and Th. Dobzhansky (eds), *Studies in the Philosophy of Biology* (London: Macmillan, 1974), pp. 179–83.

THE STRUCTURE OF EVOLUTIONARY THEORY: REVISING THE THREE CENTRAL FEATURES OF DARWINIAN LOGIC

IN the opening sentence of *The Origin*'s final chapter, Darwin famously wrote that 'this whole volume is one long argument.'[1] The present book, on 'the structure of evolutionary theory,' despite its extravagant length, is also a brief for an explicit interpretation that may be portrayed as a single extended argument. Although I feel that our best current formulation of evolutionary theory includes modes of reasoning and a set of mechanisms substantially at variance with strict Darwinian natural selection, the logical structure of the Darwinian foundation remains remarkably intact – a fascinating historical observation in itself, and a stunning tribute to the intellectual power of our profession's founder. Thus, and not only to indulge my personal propensities for historical analysis, I believe that the best way to exemplify our modern understanding lies in an extensive analysis of Darwin's basic logical commitments, the reasons for his choices, and the subsequent manner in which these aspects of 'the structure of evolutionary theory' have established and motivated all our major debates and substantial changes since Darwin's original publication in 1859. I regard such analysis not as an antiquarian indulgence, but as an optimal path to proper understanding of our *current* commitments, and the underlying reasons for our decisions about them.

As a primary theme for this one long argument, I claim that an 'essence' of Darwinian logic can be defined by the practical strategy defended in the first section of this chapter: by specifying a set of minimal commitments, or broad statements so essential to the central logic of the enterprise that disproof of any item will effectively destroy the theory, whereas a substantial change to any item will convert the theory into something still recognizable as within the *Bauplan* of descent from its forebear, but as something sufficiently different to identify, if I may use the obvious taxonomic metaphor, as a new subclade within the monophyletic group. Using this premise, the long argument of this book then proceeds according to three sequential claims that set the structure and order of my subsequent chapters:

1. Darwin himself formulated his central argument under these three basic premises. He understood their necessity within his system, and the difficulty that he would experience in convincing his contemporaries about such unfamiliar and radical notions. He therefore presented careful and explicit defenses of all three propositions in *The Origin*. I devote the first substantive chapter (number 2) to an exegesis of *The Origin of Species* as an embodiment of Darwin's defense for this central logic.

2. As evolutionary theory experienced its growing pains and pursued its founding arguments in the late nineteenth and early twentieth centuries (and also in its pre-Darwinian struggles with more inchoate formulations before 1859), these three principles of central logic defined the themes of deepest and most persistent debate – as, in a sense, they must because they constitute the most interesting intellectual questions that any theory for causes of descent with modification must address. The historical chapters of this book's first half then treat the history of evolutionary theory as responses to the three central issues of Darwinian logic (Chapters 3–7).

3. As the strict Darwinism of the Modern Synthesis prevailed and 'hardened,' culminating in the overconfidences of the centennial celebrations of 1959, a new wave of discoveries and theoretical reformulations began to challenge aspects of the three central principles anew – thus leading to another fascinating round of development in basic evolutionary theory, extending throughout the last three decades of the twentieth century and continuing today. But this second round has been pursued in an entirely different and more fruitful manner than the nineteenth century debates. The earlier questioning of Darwin's three central principles tried to disprove natural selection by offering alternative theories based on confutations of the three items of central logic. The modern versions accept the validity of the central logic as a foundation, and introduce their critiques as helpful auxiliaries or additions that enrich, or substantially alter, the original Darwinian formulation, but that leave the kernel of natural selection intact. Thus, the modern

reformulations are helpful rather than destructive. For this reason, I regard our modern understanding of evolutionary theory as closer to Falconer's metaphor than to Darwin's, for the Duomo of Milan – a structure with a firm foundation and a fascinatingly different superstructure. (Chapters 8–12, the second half of this book on modern developments in evolutionary theory, treat this third theme.)

Thus, one might say, this book cycles through the three central themes of Darwinian logic at three scales – by brief mention of a framework in this chapter, by full exegesis of Darwin's presentation in Chapter 2, and by lengthy analysis of the major differences and effects in historical (Part 1) and modern critiques (Part 2) of these three themes in the rest of the volume.

The basic formulation, or bare-bones mechanics, of natural selection is a disarmingly simple argument, based on three undeniable facts (overproduction of offspring, variation, and heritability)* and one syllogistic inference (natural selection, or the claim that organisms enjoying differential reproductive success will, on average, be those variants that are fortuitously better adapted to changing local environments, and that these variants will then pass their favored traits to offspring by inheritance). As Huxley famously, and ruefully, remarked (in self-reproach for failing to devise the theory himself), this argument must be deemed elementary (and had often been formulated before, but in negative contexts, and with no appreciation of its power, and can only specify the guts of the operating machine, not the three principles that established the range and power of Darwin's revolution in human thought. Rather, these three larger principles, in defining the Darwinian essence, take the guts of the machine, and declare its simple operation sufficient to generate the entire history of life in a philosophical manner that could not have been more contrary to all previous, and cherished, assumptions of Western life and science.

The three principles that elevated natural selection from the guts of a working machine to a radical explanation of the mechanism of life's history can best be exemplified under the general categories of agency,

* Two of these three ranked as 'folk wisdom' in Darwin's day and needed no further justification – variation and inheritance (the mechanism of inheritance remained unknown, but its factuality could scarcely be doubted). Only the principle that all organisms produce more offspring than can possibly survive – superfecundity, in Darwin's lovely term – ran counter to popular assumptions about nature's benevolence, and required Darwin's specific defense in *The Origin*.

efficacy, and scope. I treat them in this specific order because the logic of Darwin's own development so proceeds (as I shall illustrate in Chapter 2), for the most radical claim comes first, with assertions of complete power and full range of applicability then following.

Agency. The abstract mechanism requires a locus of action in a hierarchical world, and Darwin insisted that the apparently intentional 'benevolence' of nature (as embodied in the good design of organisms and the harmony of ecosystems) flowed entirely as side-consequences of this single causal locus, the most 'reductionistic' account available to the biology of Darwin's time. Darwin insisted upon a virtually exceptionless, single-level theory, with organisms acting as the locus of selection, and all 'higher' order emerging, by the analog of Adam Smith's invisible hand, from the (unconscious) 'struggles' of organisms for their own personal advantages as expressed in differential reproductive success. One can hardly imagine a more radical reformulation of a domain that had unhesitatingly been viewed as the primary manifestation for action of higher power in nature – and Darwin's brave and single-minded insistence on the exclusivity of the organismic level, although rarely appreciated by his contemporaries, ranks as the most radical and most distinctive feature of his theory.

Efficacy. Any reasonably honest and intelligent biologist could easily understand that Darwin had identified a *vera causa* (or true cause) in natural selection. Thus, the debate in his time (and, to some extent, in ours as well) never centered upon the existence of natural selection as a genuine causal force in nature. Virtually all anti-Darwinian biologists accepted the reality and action of natural selection, but branded Darwin's force as a minor and negative mechanism, capable only of the headsman's or executioner's role of removing the unfit, once the fit had arisen by some other route, as yet unidentified. This other route, they believed, would provide the centerpiece of a 'real' evolutionary theory, capable of explaining the origin of novelties. Darwin insisted that his admittedly weak and negative force of natural selection could, nonetheless, under certain assumptions (later proved valid) about the nature of variation, act as the positive mechanism of evolutionary novelty – that is, could 'create the fit' as well as eliminate the unfit – by slowly accumulating the positive effects of favorable variations through innumerable generations.

Scope. Even the most favorably minded of contemporaries often admitted that Darwin had developed a theory capable of building up small changes (of an admittedly and locally 'positive' nature as adaptations to changing environments) within a 'basic type' – the equivalent, for example, of making dogs from wolves or developing edible corn from teosinte. But these critics could not grasp how such a genuine microevolutionary process could be extended to produce the full panoply of taxonomic diversity and apparent 'progress' in complexification of morphology through geological time. Darwin insisted on full sufficiency in extrapolation, arguing that his micro-evolutionary mechanism, extended through the immensity of geological time, would be fully capable of generating the entire pageant of life's history, both in anatomical complexity and taxonomic diversity – and that no further causal principles would be required.

Punctuated Equilibrium and the Validation of Macroevolutionary Theory

The clear predominance of an empirical pattern of stasis and abrupt geological appearance as the history of most fossil species has always been acknowledged by paleontologists, and remains the standard testimony (as documented herein) of the best specialists in nearly every taxonomic group. In Darwinian traditions, this pattern has been attributed to imperfections of the geological record that impose this false signal upon the norm of a truly gradualistic history. Darwin's argument may work in principle for punctuational origin, but stasis is data and cannot be so encompassed.

This traditional argument from imperfection has stymied the study of evolution by paleontologists because the record's primary (and operational) signal has been dismissed as misleading, or as 'no data.' Punctuated equilibrium, while not denying imperfection, regards this signal as a basically accurate record of evolution's standard mode at the level of the origin of species. In particular, before the formulation of punctuated equilibrium, stasis had been read as an embarrassing indication of absence of evidence for the desired subject of study – that is, of data for evolution itself, falsely defined as gradual change – and this eminently testable, fully operational, and intellectually fascinating (and positive) subject of stasis had never been subjected to quantitative empirical study, a situation that has changed dramatically during the last twenty-five years.

The key empirical ingredients of punctuated equilibrium – punctuation, stasis, and their relative frequencies – can be made testable and defined operationally. The theory only refers to the origin and development of species in geological time, and must not be misconstrued (as so often done) as a claim for true saltation at a lower organismal level, or for catastrophic mass extinction at a higher faunal level. Punctuation must be scaled relative to the later duration of species in stasis, and we suggest 1–2 percent (analogous to human gestation *vs.* the length of human life) as an upper bound. Punctuated equilibrium can be distinguished from other causes of rapid change (including anagenetic passage through bottlenecks and the traditional claim of imperfect preservation for a truly gradualistic event) by the criterion of ancestral survival following the branching of a descendant. Punctuations can be revealed by positive evidence (rather than inferred from compression on a single bedding plane) in admittedly rare situations, but not so infrequent in absolute number, of unusual fineness of stratigraphic resolution or ability to date the individual specimens of a single bedding plane. Stasis is not defined as absolute phenotypic immobility, but as fluctuation of means through time at a magnitude not statistically broader than the range of geographic variation among modern populations of similar species, and not directional in any preferred way, especially not towards the phenotype of descendants. Punctuated equilibrium will be validated, as all such theories in natural history must be (including natural selection itself), by predominant relative frequency, not by exclusivity. Gradualism certainly can and does occur, but at very low relative frequencies when all species of a fauna are tabulated, and when we overcome our conventional bias for studying only the small percentage of species qualitatively recognized beforehand as having changed through time.

Punctuated equilibrium emerges as the expected scaling of ordinary allopatric speciation into geological time, and does not suggest or imply radically different evolutionary mechanisms at the level of the origin of species. (Other proposed mechanisms of speciation, including most sympatric modes, envision rates of speciation even faster than conventional allopatry, and are therefore even more consistent with punctuated equilibrium.) The theoretically radical features of punctuated equilibrium flow from its proposals for macroevolution, with species treated as higher-level Darwinian individuals analogous to organisms in microevolution.

The difficulty of defining species in the fossil record does not threaten the validity of punctuated equilibrium for several reasons. First, in the few studies with adequate data for genetic and experimental resolution, paleospecies (even for such difficult and morphologically labile species as colonial cheilostome bryozoans) have been documented as excellent surrogates, comparable as units to conventional biospecies. Second, the potential underestimation of biospecies by paleospecies only imposes a bias that makes punctuated equilibrium harder to recognize. The fossil record's strongly positive signal for punctuated equilibrium, in the light of this bias, only increases the probability of the pattern's importance and high relative frequency. Third, the potential overestimation of biospecies by paleospecies is probably false in any case, and also of little practical concern because no paleontologist would assert punctuated equilibrium from the evidence of oversplit taxa in faunal lists, but only from direct biometric study of stasis and punctuation in actual data.

We originally, and probably wrongly, tried to validate punctuated equilibrium by asserting that, in principle, most evolutionary change should be concentrated at events of speciation themselves. Subsequent work in evolutionary biology has not confirmed any a priori preference for concentration in such episodes D. J. Futuyama's incisive macroevolutionary argument – that realized change will not become geologically stabilized and conserved unless such change can be 'tied up' in the unalienable individuality of a new species – offers a far richer, far more interesting, and theoretically justified rationale for correlating episodes of evolutionary change with speciation.

Section III presents a wide-ranging discussion of why proposed empirical refutations of punctuated equilibrium either do not hold in fact, or do not bear the logical weight claimed in their presentation. Refutations for single cases are often valid, but do not challenge the general hypothesis because we anticipate a low relative frequency for gradualism, and these cases may reside in this minor category. Claims for predominant gradualism in the entire clade of planktonic forams may hold as exceptional (although, even here, the majority of lineages remain unstudied, in large part because they seem, at least subjectively, to remain in stasis, and have therefore not attracted the attention of traditional researchers, who wish to study evolution, but then equate evolution with gradualism). However, in these asexual forms with vast populations, gradualism at this level may just represent the expected higher-level expression of punctuational clone selection, as R. E. Lenski has affirmed in the most thorough study of evolution in a modern bacterial species – and just as gradual cladal trends in multicellular lineages emerge as the expected consequences of sequential punctuated equilibrium at the species level (trends as stairsteps rather than inclined planes, so to speak). Claims for genetic gradualism do not challenge punctuated equilibrium, and may well be anticipated as the proper expression at the genic level (especially given the high relative frequency of random nucleotide substitutions) of morphological stasis in the phenotypic history of species. Punctuated equilibrium has done well in tests of conformity with general models, particularly in the conclusion that extensive polytomy in cladistic models may arise not only (as usually interpreted) from insufficient data to resolve a sequence of close dichotomies, but also as the expectation of punctuated equilibrium for successive branching of daughter species from an unchanged parental form in stasis. In fact, the frequency of polytomy *vs.* dichotomy may be used as a test for the relative frequency of punctuated equilibrium in well resolved cladograms – a test well passed in data presented by P. J. Wagner and D. H. Erwin.

Section IV then summarizes the data on empirical affirmations of punctuated equilibrium, first on documented patterns of stasis in unbranched lineages; second on punctuational cladogenesis affirmed by the criterion of ancestral survival; third on predominant relative frequencies for punctuated equilibrium in entire biotas (with particularly impressive affirmations by A. Hallam, P. H. Kelley, and S. M. Stanley and X. Yang for molluscs; and by D. R. Prothero and T. H. Heaton for Oligocene Big Badlands mammals, where a study of all taxa yielded 177 species that followed the expectations of punctuated equilibrium and three cases of potential gradualism, only one significant); fourth on predominant relative frequencies for punctuated equilibrium in entire clades, with emphasis on E. S. Vrba's antelopes and, especially, A. H. Cheetham's rigorously quantitative and multivariate data of evolution in the bryozoan genus *Metrarabdotos*, perhaps the best documented and most impressive case of exclusive punctuated equilibrium ever developed. Finally, we can learn much from variation in relative frequencies among taxa, times, and environments – and interesting inferences have been drawn from recorded differences, particularly in P. R. Sheldon's counterintuitive linkage of stasis to rapidly changing, and gradualism to stable, environments.

Among many reasons proposed to explain the predominance of stasis, a phenomenon not even acknowledged as a 'real' and positive aspect of evolution before punctuated equilibrium gave it some appropriate theoretical space, habitat tracking (favored by N. Eldredge), constraints imposed by the nature of subdivided populations (favored by B. S. Lieberman), and normalizing clade selection (proposed by G. L. Williams) represent the most novel and interesting proposals.

Among the implications of a predominantly punctuational origin of stable species-individuals for macroevolutionary theory, we must rethink trends (the primary phenomenon of macroevolution, at least in terms of dedicated discussion in existing literature) as products of the differential success of certain kinds of species, rather than as the adaptive anagenesis of lineages – a radical reformulation with consequences extending to a new set of explanations no longer rooted (as in all traditional resolutions) in the adaptive advantages conferred upon organisms, but potentially vested in such structural principles as sequelae (by hitchhiking or as spandrels) of fortuitous phenotypic linkage to higher speciation rates of certain taxa. In further extensions, macroevolution itself must be reconfigured in speciational terms, with attendant implications for a wide range of phenomena, including E. D. Cope's rule (structurally ordained biases of speciation away from a lower size limit occupied by founding members of the clade, rather than adaptive anagenesis towards organismal benefits of large size), living fossils (members of clades with persistently minimal rates of speciation, and therefore no capacity for ever generating much change in a speciational scheme, rather than forms that are either depauperate of variation, or have occupied morphological optima for untold ages), and reinterpretation of cladal trends long misinterpreted as triumphs of progressive evolution (and now reevaluated in terms of variational range in species numbers, rather than vectors of mean morphology across all species at any time – leading, for example, to a recognition that modern horses represent the single surviving twig of a once luxurious, and now depleted, clade, and not the apex of a continually progressing trend). By the same argument, generalized to all of life, we understand the stability and continued domination of bacteria as the outstanding feature of life's history, with the much vaunted progress of complexity towards mammalian elegance reinterpreted as a limited drift of a minor component of diversity into the only open space of complexity's theoretical distribution. But, to encompass this reformulation, we need to focus upon the diversity and variation among life's species, not upon the supposed vectors of its central tendencies, or even its peripheral superiorities. Hominid evolution must also be rethought as reduction of diversity to a single species of admittedly spectacular (but perhaps quite transient) current success. In addition, the last 50,000 years or more of human phenotypic stability becomes a theoretical expectation under punctuated equilibrium, and not the anomaly so often envisaged (and attributed to the suppression of natural selection by cultural evolution) both by the lay public and by many professionals as well.

Further extensions of punctuated equilibrium include the controversial phenomenon of 'coordinated stasis,' or the proposition that entire faunas, and not merely their component species, tend to remain surprisingly stable in composition over durations far longer than any model based on independent behavior of species (even under punctuated equilibrium) would allow, although other researchers attribute the same results to extended consequences of sudden external pulses and resulting faunal turnovers, while still others deny the empirics of coordination and continue to view species as more independent, one from the other, even in the classical faunas (like the Devonian Hamilton Group) that serve as 'types' for coordinated stasis.

Punctuated equilibrium has inspired several attempts, of varying success in my limited judgment, to construct mathematical models (or to simulate its central phenomena in simple computer systems of evolving 'artificial life') that may help us to identify the degree of generality in modes of change that this particular biological system, at this particular level of speciation, exemplifies and records. Punctuated equilibrium has also proved its utility in extension by meaningful analogy (based on common underlying principles of change) to the generation of punctuational hypotheses at other levels, and for other kinds of phenomena, where similar gradualistic biases had prevailed and had stymied new approaches to research. These extensions range from phyletic and ecological examples below the species level to interesting analogs of both stasis and punctuation above the species level. Nontrending, the analog of stasis in large clades, for example, had been previously disregarded – following the same fate as stasis in species – as a boring manifestation of nonevolution, but has now been recognized and documented as a real and fascinating phenomenon in itself. Punctuational analogs have proven their utility for understanding the differential pace of morphological innovation within large clades, and for resolving a variety of punctuational phenomena in ecological systems, including such issues of the immediate moment as rates of change in benthic faunas (previously the province of hypotheses about glacially slow and steady change in constantly depauperate environments), and such questions of broadest geological scale as the newly recognized stepped and punctuational 'morphology' (correcting the hypothetical growth through substantial time of all previous gradualistic accounts) of mutual biomechanical improvement in competing clades involved in 'arms race,' and generating a pattern known as 'escalation.'

Punctuational models have also been useful, even innovative in breaking conceptual logjams, in nonbiological fields ranging from closely cognate studies of the history of human tools (including extended stasis in the *Homo erectus* toolkit), and nontrending, despite classical (and false) claims to the contrary by both experts, the Abbé Breuil and André Leroi-Gourhan, for the 25,000-year history of elegance in parietal cave art of France and Spain – and extending into more distant fields like learning theory (plateaus and innovative punctuations), studies of the dynamics of human organizations, patterns of human history, and the evolution of technologies, including a fascinating account of the history of books, through punctuations of the clay tablet, the scroll, the codex, and our current electronic reformation (wherever it may lead), and long periods of morphological stasis (graced with such vital innovations as printing, imposed upon the unaltered phenotype of the codex, or standard 'book').

In a long and final section, I indulge myself, and perhaps provide some useful primary source material for future historians of scientific conflicts, by recording the plethora of nonscientific citations, ranging from the absurd to the insightful, for punctuated equilibrium (including creationist misuses and their politically effective exposure by scientists in courtroom trials that defeated creationist legislative initiatives; and the treatment of punctuated equilibrium, often very good but sometimes very bad, by journalists and by authors of textbooks – the primary arenas of vernacular passage). I also trace and repudiate the 'dark side' of nonscientific reactions by professional colleagues who emoted at challenges to their comfort, rather than reacting critically and sharply (as most others did, and as discussed extensively in the main body of the chapter) to the interesting novelty, accompanied by some prominent errors of inevitable and initial groping on our part, spawned by the basic hypothesis and cascading implications of punctuated equilibrium.

The Integration of Constraint and Adaptation: Historical Constraint and the Evolution of Development

Although the directing of evolutionary change by forces other than natural selection has loosely been described as 'constraint,' the term, even while acknowledged as a domain for exceptions to standard Darwinian mechanisms, has almost always been conceived as a 'negative' force or phenomenon, a mode of preventing (through lack of variation, for example) a population's attainment of greater adaptation. But constraint, both in our science (and in vernacular English as well), also has strongly positive meanings in two quite different senses: first, or empirically, as channeled directionality for reasons of past history (conserved as homology) or physical principles; and second, or conceptually, as a nonstandard force (therefore interesting *ipso facto*) acting differently from what orthodoxy would predict.

The classical and most familiar category of internal channeling (the first, or empirical, citation of constraint as a positive theme) resides in preferred directions for evolutionary change supplied by inherited allometries and their phylogenetic potentiation by heterochrony. As 'place holders' for an extensive literature, I present two examples from my own work: first, the illustration of synergy with natural selection (to exemplify the positive, rather than oppositional, meaning), where an inherited internal channel builds two important adaptations by means of one heterochronic alteration, as neoteny in descendant *Gryphaea* species of the English Jurassic produces shells of both markedly increased size (by retention of juvenile growth rates over an unchanged lifetime) and stabilized shape to prevent foundering in muddy environments (achieved by 'bringing forward' the proportions of attached juveniles into the unattached stage of adult ontogeny); second, an illustration of pervasiveness and equal (or greater) power than selective forces (to exemplify the strength and high relative frequency of such positive influences), as geographic variation of the type species, *Cerion uva*, on Aruba, Bonaire, and Curaçao, a subject of intense quantitative study and disagreement in the past, becomes resolved in multivariate terms, with clear distinction between local adaptive differences and the pervasive general pattern of an extensive suite of automatic sequelae, generated by nonadaptive variation in the geometry of coiling a continuous tube, under definite allometric regularities for the genus, around an axis.

For the second, or conceptually positive, meaning of constraint as a term for nonstandard causes of evolutionary change, I present a model that compares the conventional outcomes of direct natural selection, leading to local adaptation, with two sources that can also yield adaptive results, but for reasons of channeling by internal constraints rather than by direct construction under external forces of natural selection. In this triangular model for aptive structures, the *functional vertex* represents features conventionally built by natural selection for current utilities. At the *historical vertex*, currently aptive features probably originated for conventionally adaptive reasons in distant ancestors; but these features are now developmentally channeled as homologies that constrain and positively direct both patterns of immediate change and the inhomogeneous occupation of morphospace (especially as indicated by 'deep homologies' of retained developmental patterns among phyla that diverged from common ancestry more than 500 million years ago). At the *structural vertex*, two very different reasons underlie the origin of potentially aptive features for initially nonadaptive reasons: physical principles that build 'good' form by the direct action of physical laws upon plastic material (as in D'Arcy Thompson's theory of form), and architectural sequelae (spandrels) that arise as nonadaptive consequences of other features, and then become available for later cooptation (as exaptations) to aptive ends in descendant taxa. These two structural reasons differ strongly in the ahistoricist implications of direct physical production independent of phyletic context *vs.* the explicit historical analysis needed to identify the particular foundation for the origin of spandrels in any individual lineage.

As a conceptual basis for understanding the importance of recent advances in evo-devo (the study of the evolution of development), the largely unknown history of debate about categories of homology, particularly the distinction between convergence and parallelism, provides our best ordering device – for we then learn to recognize the key contrast between parallelism as a positive deep constraint of homology in underlying generators (and therefore as a structuralist theme in evolution) and convergence as the opposite sign of domination for external natural selection upon a yielding internal substrate that imposes no constraint (and therefore as a functionalist theme in evolution). As a beginning paradox, we must grasp why E. Ray Lankester coined the term homoplasy as a category of homology, whereas today's terminology ranks the concepts as polar opposites. Lankester wanted to contrast homology of overt structure (homogeny in his terms, or homology *sensu stricto*) with homology of underlying generators (later called parallelism) building the same structure in two separate lineages (homoplasy, or homology *sensu lato*, in Lankester's terms). Because parallelism could not be cashed out in operational terms (as science had no way, until our current revolution in evo-devo, to characterize, or even to recognize, these underlying generators), proper conceptual distinctions between parallelism and convergence have generally not been made, and the two terms have even (and often) been united as subtypes of homoplasy (now defined in the current, and utterly non-Lankesterian sense, as opposite to homology). I trace the complex and confused history of this discussion, and show that structuralist thinkers, with doubts about panadaptationism, have always been most sensitive to this issue, and most insistent upon separating and distinguishing parallelism as the chief category of positive developmental constraint – a category that has now, for the first time, become scientifically operational.

I summarize the revolutionary empirics and conceptualizations of evo-devo in four themes, united by a common goal: to rebalance constraint and adaptation as causes and forces of evolution, and to acknowledge the pervasiveness and importance – also the synergy with natural selection, rather than opposition to Darwinian themes – of developmental constraint as a positive, structuralist, and internal force. The first theme explores the implications – for internally directed evolutionary pathways and consequent clumping of taxa in morphospace – of the remarkable and utterly unanticipated discovery of extensive 'deep homology' among phyla separated at least since the Cambrian explosion, as expressed by shared and highly conserved genes regulating fundamental processes of development. I first discuss the role and action of some of these developmental systems – the *ABC* genes of *Arabidopsis* in regulating circlets of structures in floral morphology, the *Hox* genes of *Drosophila* in regulating differentiation of organs along the AP axis, and the role of the *Pax-6* system in the development of eyes – in validating (only partially, of course) the archetypal theories of nineteenth-century transcendental morphology, long regarded as contrary to strictly selectionist views of life's history – particularly Goethe's theory of the leaf archetype, and Geoffroy's idea of the vertebral groundplan of AP differentiation. I then discuss the even more exciting subject of homologically conserved systems across distant phyla, as expressed in high sequence similarity of important regulators, common rules of development (particularly the 'Hoxology' followed in both arthropod and vertebrate ontogeny), and similar action of homeotic mutations that impact Hoxological rules by loss or gain of function. Geoffroy was partially right in asserting segmental homology between arthropods and vertebrates, particularly for the comparison of insect metameres with rhombomeric segments in the developing vertebrate brain (a small part, perhaps, of the AP axis of most modern vertebrates, but the major component of the earliest fossil vertebrates), where the segments themselves may form differently, but where rules of Hoxology then work in the same manner during later differentiation. I also defend the substantial validity of Geoffroy's other 'crazy' comparison – the dorso-ventral inversion of the same basic body plan between arthropods and vertebrates.

The second theme stresses the even more positive role of parallelism, based on common action of regulators shared by deep homology, in directing the evolutionary pathways of distantly related phyla into similar channels of adaptations thus more easily generated (thereby defining this phenomenon as synergistic and consistent with an expanded Darwinian theory, and not confrontational or dismissive of selection). I discuss such broadscale examples as the stunning discovery of substantial parallelism in the supposedly classical, 'poster boy' expression of the opposite phenomenon of convergence – the development of eyes in arthropods, vertebrates, and cephalopods. The overt adult phenotype, of course, remains largely convergent, but homology of the underlying regulators demonstrates the strong internal channeling of parallelism. The vertebrate and squid version of *Pax-6* can, in fact, both rescue the development of eyes in *Drosophila* and produce ectopic expression of eyes in such odd places as limbs. I also discuss smaller-scale examples of 'convergence,' reinterpreted as parallelism, for even more precise similarities among separate lineages within coherent clades – particularly the independent conversion of thoracic limbs to maxillipeds, by identical homeotic changes in the same *Hox* genes, in several groups of crustaceans. Finally, I caution against overextension and overenthusiasm by pointing out that genuine developmental homologies may be far too broad in design, and far too unspecific in morphology, to merit a designation as parallelism, as in the role of *distal-less* in regulating 'outpouchings' so generalized in basic structure, yet so different in form, as annelid parapodia, tunicate ampullae and echinoderm tube feet. I designate these overly broad similarities (that should not be designated as parallelism, or used as evidence for constraint by internal channeling) as 'Pharaonic bricks' – that is, building blocks of such generality and multipurpose utility that they cannot be labeled as constraints (with the obvious *reductio ad absurdum* of DNA as the homological basis of all life). By contrast, the 'Corinthian columns' of more specific conservations define the proper category of important positive constraint by internal channelings of parallelism based on homology of underlying regulators (just as the specific form of a Corinthian column, with its acanthus-leafed capital, represents a tightly constrained historical lineage that strongly influences the particular shape and utility of the entire resulting building).

My third and shorter theme – for this subject, though 'classical' throughout the history of evolutionary thought, holds, I believe, less validity and scope than the others – treats the role of homologous regulators in producing rapid, even truly saltational, changes channeled into limited possibilities of developmental pathways (as in R. Goldschmidt's defense of discontinuous evolution based upon mutations in rate genes that control ontogenetic trajectories). I discuss the false arguments often invoked to infer such saltational changes, but then document some limited, but occasionally important, cases of such discontinuous, but strongly channeled, change in macroevolution.

The fourth theme of top-down channeling from full ancestral complements, rather than bottom-up accretion along effectively unconstrained pathways of local adaptation, explores the role of positive constraint in establishing the markedly nonrandom and inhomogeneous population of potential morphospace by actual organisms throughout the history of life. Ed Lewis, in brilliantly elucidating the action of *Hox* genes in the development of *Drosophila*, quite understandably assumed (albeit falsely, as we later discovered to our surprise) that evolution from initial homonomy to increasing complexity of AP differentiation had been achieved by addition of *Hox* genes, particularly to suppress abdominal legs and convert the second pair of wings to halteres. In fact, the opposite process of tinkering with established rules, primarily by increased localization of action and differentiation in timing (and also by duplication of sets, at least for vertebrate *Hox* genes), has largely established the increasing diversity and complexity of differentiation in bilaterian phyla. The (presumably quite homonomous) common ancestor of arthropods and vertebrates already possessed a full complement of *Hox* genes, and even the bilaterian common ancestor already possessed at least seven elements of the set. Moreover, the genomes of the most homonomous modern groups of onycophorans and myriapods also include a full set of *Hox* genes – so differentiation of phenotypic complexity must originate as a derived feature of *Hox* action, exapted from a different initial role. The Cambrian explosion remains a crucial and genuine phenomenon of phenotypic diversification, a conclusion unthreatened by a putatively earlier common ancestry of animal phyla in a strictly genealogical (not phenotypic) sense. The further evolution of admittedly luxuriant, even awesome, variety in major phyla of complex animals has followed definite pathways of internal channeling, positively abetted (as much as negatively constrained) by homologous developmental rules acting as potentiators for more rapid and effective selection (as in the loss of snake limbs and iteration of pre-pelvic segments), and not as brakes or limitations upon Darwinian efficacy.

The Integration of Constraint and Adaptation: Structural Constraints, Spandrels, and Exaptation

D'Arcy Thompson's idiosyncratic, but brilliantly crafted and expressed, theory of form[2] presents a twentieth-century prototype for the generalist, or ahistorical, form of structural constraint: adaptation produced not by a functionalist mechanism like natural selection (or Lamarckism), but directly and automatically impressed by physical forces operating under invariant laws of nature. This theory enjoyed some success in explaining the correlation of form and function in very simple and labile forms (particularly as influenced by scale-bound changes in surface/volume ratios). But similarly nongenetic (and nonphyletic) explanations do not apply to complex creatures, and even D'Arcy Thompson admitted that his mechanism could not encompass, say, 'hipponess,' but, at most, only the smooth transformations of these basic designs among closely related forms of similar *Bauplan* (the true theoretical significance of his much misunderstood theory of transformed coordinates). In summary, D'Arcy Thompson, the great student of Aristotle, erred in mixing the master's modes of causality – by assuming that the adaptive value (or final cause) of well designed morphology could specify the physical forces (or efficient causes) that actually built the structures.

Stuart Kauffman and Brian Goodwin have presented the most cogent modern arguments in this tradition of direct physical causation. These arguments hold substantial power for explaining some features of relatively simple biological systems, say from life's beginnings to the origin of prokaryotic cells, where basic organic chemistry and the physics of self-organizing systems can play out their timeless and general rules. Such models also have substantial utility in describing very broad features of the ecology and energy dynamics of living systems in general terms that transcend any particular taxonomic composition. But this approach founders, as did D'Arcy Thompson's as well, when the contingent and phyletically bound histories of particular complex lineages fall under scrutiny – and such systems do constitute the 'bread and butter' of macroevolution. Nonetheless, Kauffman's powerful notion of 'order for free,' or adaptive configurations that emerge from the ahistoric (even abiological) nature of systems, and need not be explained by particular invocations of some functional force like natural selection, should give us pause before we speculate about Darwinian causes only from evidence of functionality. This 'order for free' aids, and does not confute, such functional forces as selection by providing easier (even automatic) pathways towards a common desideratum of adaptive biological systems.

I then turn to the second, and (in my judgment) far more important, theme of structural constraint in the fully historicist and phyletic context of aptive evolution by cooptation of structures already present for other reasons (often nonadaptive in their origin), rather than by direct adaptation for current function via natural selection. The central principle of a fundamental logical difference between reasons for historical origin and current functional utility – a vital component in all historical analysis, as clearly recognized but insufficiently emphasized by Darwin, and then unfortunately underplayed or forgotten by later acolytes – was brilliantly identified and dissected by Friedrich Nietzsche in his *Genealogy of Morals*, where he contrasted the origin of punishment in a primal will to power, with the (often very different) utility of punishment in our current social and political systems.

Darwin himself invoked this principle of disconnection between historical origin and current utility both in *The Origin*'s first edition, and particularly in later responses to St George Mivart's critique (the basis for the only chapter that Darwin added to later editions of *The Origin*) on the supposed inability of natural selection to explain the incipient (and apparently useless) stages of adaptive structures. Darwin asserted the principle of functional shift to argue that, although incipient stages could not have functioned in the manner of their final form, they might still have arisen by natural selection for a different initial utility (feathers first evolved for thermoregulation and later coopted for flight, for example). Darwin used this principle of cooptation, or functional shift, in two important ways that enriched and expanded his theory away from a caricatured panselectionist version – as the primary ground of historical contingency in phyletic sequences (for one cannot predict the direction of subsequent cooptation from different primary utilities), and as a source of structural constraint upon evolutionary pathways. But these Darwinian invocations stopped short of a radical claim for frequent and important nonadaptive origins of structures coopted to later utility. That is, Darwin rarely proceeded beyond the principle of originally adaptive origin for different function, with later cooptation to altered utility.

This important principle of cooptation of preexisting structures originally built for different reasons has been so underemphasized in Darwinian traditions that the language of evolutionary theory does not even include a term for this central process – which Elisabeth Vrba and I called 'exaptation'.[3] (The available, but generally disfavored, term 'preadaptation' only speaks of potential before the fact, and has been widely rejected in any case for its unfortunate, but inevitable, linguistic implication of foreordination in evolution, the very opposite of the intended meaning!)

I present a list of criteria for recognizing exaptations and separating them from true adaptations. I also discuss some outstanding examples of exaptation from the recent literature, with particular emphasis on the multiple exaptation of lens crystallins (in part for their fortuitous transparency, but for many other cooptable characteristics as well) in so many vertebrates and from so many independent and different original functions.

The exaptation of structures that arose for different adaptive reasons remains within selectionist orthodoxy (while granting structural constraint a large influence over historical pathways, in contrast with crude panadaptationism) by confirming a Darwinian basis for the adaptive origin of structures, whatever their later history of exaptive shift. On the other hand, the theoretically radical version of this second, or historicist, style of structural constraint in evolution posits an important role for an additional phenomenon in macroevolution: the truly nonadaptive origin of structures that may later be exapted for subsequent utility. Many sources of such nonadaptive origin may be specified, but inevitable architectural consequences of other features – the spandrels of Gould and Lewontin's terminology[4] – probably rank as most frequent and most important in the history of lineages.

Spandrels (although unnamed and ungeneralized) have been acknowledged in Darwinian traditions, but relegated to insignificant relative frequencies by invalid arguments for their rarity, their structural inconsequentiality (the mold marks on an old bottle, for example), or their temporally subsequent status as sequelae – with the first two claims empirically false, and the last claim logically false as a further confusion between historical origin and current utility.

I affirm the importance and high relative frequency of spandrels, and therefore of nonadaptive origin, in evolutionary theory by two major arguments for ubiquity. First, for intrinsic structural reasons, the number of potential spandrels greatly increases as organisms and their traits become more complex. (The spandrels of the human brain must greatly outnumber the immediately adaptive reasons for increase in size; the spandrels of the cylindrical umbilical space of a gastropod shell, by contrast, may be far more limited, although exaptive use as a brooding chamber has been important in several lineages.) Second, under hierarchical models of selection, features evolved for any reason at one level generate automatic consequences at other levels – and these consequences can only be classified as cross-level spandrels (since they are 'injected into' the new level, rather than actively evolved there).

The full classification of spandrels and modes of exaptation offers a resolving taxonomy and solution – primarily through the key concept of the 'exaptive pool' – for the compelling and heretofore confusing (yet much discussed) problem of 'evolvability.' Former confusion has centered upon the apparent paradox that ordinary organismal selection, the supposed canonical mechanism of evolutionary change, would seem (at least as its primary overt effect) to restrict and limit future possibilities by specializing forms to complexities of immediate environments, and therefore to act against an 'evolvability' that largely defines the future macroevolutionary prospects of any lineage. The solution lies in recognizing that spandrels, although architecturally consequential, are not doomed to a secondary or unimportant status thereby. Spandrels, and all other forms of exaptive potential, define the ground of evolvability, and play as important a role in macroevolutionary potential as conventional adaptation does for the immediacy of microevolutionary success. I emphasize the centrality of the exaptive pool for solving the problem of evolvability by presenting a full taxonomy of categories for the pool's richness, focusing on a primary distinction between 'franklins' (or inherent potentials of structures evolved for other adaptive roles – that is, the classical Darwinian functional shifts that do not depart from adaptationism), and 'miltons' (or true nonadaptations, arising from several sources, with spandrels as a primary category, and then available for later cooptation from the exaptive pool – that is, the class of nonadaptive origins that does challenge the dominant role of panadaptationism in evolutionary theory).

I argue that the concept of cross-level spandrels vastly increases the range, power and importance of nonadaptation in evolution, and also unites the two central themes of this book by showing how the hierarchically expanded theory of selection also implies a greatly increased scope for nonadaptive structural constraint as an important factor in the potentiation of macroevolution.

Tiers of Time and Trials of Extrapolationism

Darwin clearly recognized the threat of catastrophic mass extinction to the extrapolationist and uniformitarian premises underlying his claim for full explanation of macroevolutionary results by microevolutionary causes (and not as a challenge to the efficacy of natural selection itself). Darwin therefore employed his usual argument about the imperfection of geological records to 'spread out' apparent mass extinction over sufficient time for resolution by ordinary processes working at maximal rates (and therefore only increasing the intensity of selection).

The transition of the impact scenario (as a catastrophic trigger for the K-T extinction) from apostasy at its proposal in 1980 to effective factuality (based on the consilience of disparate evidence from iridium layers, shocked quartz and, especially, the discovery of a crater of appropriate size and age at Chicxulub) has reinstated the global paroxysms of classical catastrophism (in its genuinely scientific form, not its dismissive Lyellian caricature) as a legitimate scientific mechanism outside the Darwinian paradigm, but operating in conjunction with Darwinian forces to generate the full pattern of life's history, and not, as previously (and unhelpfully) formulated, as an exclusive alternative to disprove or to trivialize Darwinian mechanisms.

If catastrophic causes and triggers for mass extinction prove to be general, or at least predominant in relative frequency (and not just peculiar to the K-T event), then this macroevolutionary phenomenon will challenge the crucial extrapolationist premise of Darwinism by being more frequent, more rapid, more intense and more different in effect than Darwinian biology (and Lyellian geology) can allow. Under truly catastrophic models, two sets of reasons, inconsistent with Darwinian extrapolationism by microevolutionary accumulation, become potentially important agents of macroevolutionary patterning: effectively random extinction (for clades of low N), and, more importantly, extinction under 'different rules' from reasons regulating the adaptive origin and success of autapomorphic cladal features in normal times.

Catastrophic mass extinction, while breaking the extrapolationist credo, may suggest an overly simplified and dichotomous macroevolutionary model based on alternating regimes of 'background' *vs.* 'mass' extinction. Rather, we should expand this insight about distinctive mechanisms at different scales into a more general model of several rising tiers of time – with conventional Darwinian microevolution dominating at the ecological tier of short times and intraspecific dynamics; punctuated equilibrium dominating at the geological tier of phyletic trends based on interspecific dynamics (with species arising in geological moments, and then treated as stable 'atoms,' or basic units of macroevolution, analogous to organisms in microevolution); and mass extinction (perhaps often catastrophic) acting as a major force of overall macroevolutionary pattern in the global history of relative waxing and waning of clades. (I also contrast this preferred model of time's tiering with the other possible style of explanation, which I reject but find interesting nonetheless, for denying full generality to smooth Darwinian upward extrapolation from the lowest level – namely, an equally smooth and monistic downward extrapolation from catastrophic mortality in mass extinction to diminishing, but equally random and sudden, effects at all scales, as proposed in D. M. Raup's 'field of bullets' model.)

In a paradoxical epilogue, I argue (despite my role as a longtime champion of the importance and scientific respectability of unpredictable contingency in the explanation of historical patterns) that the enlargement and reformulation of Darwinism, as proposed in this book, will recapture for general theory (by adding a distinctive and irreducible set of macroevolutionary causes to our armamentarium of evolutionary principles) a large part of macroevolutionary pattern that Darwin himself, as an equally firm supporter of contingency, willingly granted to the realm of historical unpredictability because he could not encompass these results within his own limited causal structure of strict reliance upon smooth extrapolation from microevolutionary processes by accumulation through the immensity of geological time.

A final thought. May I simply end by quoting the line that I wrote at the completion of a similar abstract (but vastly shorter, in a much less weighty book) for my first technical tome, *Ontogeny and Phylogeny*: 'This epitome is a pitiful abbreviation of a much longer and, I hope, more subtle development. Please read the book!'[5]

Notes

1. C. Darwin, *On the Origin of Species by Means of Natural Selection, or Preservation of Favored Races in the Struggle for Life* (London: John Murray, 1859), p. 459.
2. D'Arcy W. Thompson, *On Growth and Form* (Cambridge: Cambridge University Press, 1917; second edition, Cambridge: Cambridge University Press, 1942).
3. S. J. Gould and E. S. Vrba, 'Exaptation: A Missing Term in the Science of Form', *Paleobiology*, 8 (1982): 4–15.
4. S. J. Gould and R. C. Lewontin, 'The Spandrels of San Marco and the Panglossian Paradigm: A Critique of the Adaptationist Programme', *Proceedings of the Royal Society London. Series B: Biological Sciences*, 205 (1979): 581–98.
5. S. J. Gould, *Ontogeny and Phylogeny* (Cambridge, MA: Harvard University Press, 1977), p. 9.

THE EPISODIC NATURE OF EVOLUTIONARY CHANGE

ON November 23, 1859, the day before his revolutionary book hit the stands, Charles Darwin received an extraordinary letter from his friend Thomas Henry Huxley. It offered warm support in the coming conflict, even the supreme sacrifice: 'I am prepared to go to the stake, if requisite . . . I am sharpening up my claws and beak in readiness.' But it also contained a warning: 'You have loaded yourself with an unnecessary difficulty in adopting *Natura non facit saltum* so unreservedly.'

The Latin phrase, usually attributed to Linnaeus, states that 'nature does not make leaps.' Darwin was a strict adherent to this ancient motto. As a disciple of Charles Lyell, the apostle of gradualism in geology, Darwin portrayed evolution as a stately and orderly process, working at a speed so slow that no person could hope to observe it in a lifetime. Ancestors and descendants, Darwin argued, must be connected by 'infinitely numerous transitional links' forming 'the finest graduated steps.' Only an immense span of time had permitted such a sluggish process to achieve so much.

Huxley felt that Darwin was digging a ditch for his own theory. Natural selection required no postulate about rates; it could operate just as well if evolution proceeded at a rapid pace. The road ahead was rocky enough; why harness the theory of natural selection to an assumption both unnecessary and probably false? The fossil record offered no support for gradual change: whole faunas had been wiped out during disarmingly short intervals. New species almost always appeared suddenly in the fossil record with no intermediate links to ancestors in older rocks of the same region. Evolution, Huxley believed, could proceed so rapidly that the slow and fitful process of sedimentation rarely caught it in the act.

The conflict between adherents of rapid and gradual change had been particularly intense in geological circles during the years of Darwin's apprenticeship in science. I do not know why Darwin chose to follow Lyell and the gradualists so strictly, but I am certain of one thing: preference for one view or the other had nothing to do with superior perception of empirical information. On this question, nature spoke and continues to speak in multifarious and muffled voices. Cultural and methodological preferences had as much influence upon any decision as the constraints of data.

On issues so fundamental as a general philosophy of change, science and society usually work hand in hand. The static systems of European monarchies won support from legions of scholars as the embodiment of natural law. Alexander Pope wrote:

> Order is Heaven's first law; and this confessed.
> Some are, and must be, greater than the rest.

As monarchies fell and as the eighteenth century ended in an age of revolution, scientists began to see change as a normal part of universal order, not as aberrant and exceptional. Scholars then transferred to nature the liberal program of slow and orderly change that they advocated for social transformation in human society. To many scientists, natural cataclysm seemed as threatening as the reign of terror that had taken their great colleague Lavoisier.

Yet the geologic record seemed to provide as much evidence for cataclysmic as for gradual change. Therefore, in defending gradualism as a nearly universal tempo, Darwin had to use Lyell's most characteristic method of argument – he had to reject literal appearance and common sense for an underlying 'reality.' (Contrary to popular myths, Darwin and Lyell were not the heroes of true science, defending objectivity against the theological fantasies of such 'catastrophists' as Cuvier and Buckland. Catastrophists were as committed to science as any gradualist; in fact, they adopted the more 'objective' view that one should believe what one sees and not interpolate missing bits of a gradual record into a literal tale of rapid change.) In short, Darwin argued that the geologic record was exceedingly imperfect – a book with few remaining pages, few lines on each page, and few words on each line.

We do not see slow evolutionary change in the fossil record because we study only one step in thousands. Change seems to be abrupt because the intermediate steps are missing.

The extreme rarity of transitional forms in the fossil record persists as the trade secret of paleontology. The evolutionary trees that adorn our textbooks have data only at the tips and nodes of their branches; the rest is inference, however reasonable, not the evidence of fossils. Yet Darwin was so wedded to gradualism that he wagered his entire theory on a denial of this literal record:

> The geological record is extremely imperfect and this fact will to a large extent explain why we do not find interminable varieties, connecting together all the extinct and existing forms of life by the finest graduated steps. He who rejects these views on the nature of the geological record will rightly reject my whole theory.

Darwin's argument still persists as the favored escape of most paleontologists from the embarrassment of a record that seems to show so little of evolution directly. In exposing its cultural and methodological roots, I wish in no way to impugn the potential validity of gradualism (for all general views have similar roots). I wish only to point out that it was never 'seen' in the rocks.

Paleontologists have paid an exorbitant price for Darwin's argument. We fancy ourselves as the only true students of life's history, yet to preserve our favored account of evolution by natural selection we view our data as so bad that we almost never see the very process we profess to study.

For several years, Niles Eldredge of the American Museum of Natural History and I have been advocating a resolution of this uncomfortable paradox. We believe that Huxley was right in his warning. The modern theory of evolution does not require gradual change. In fact, the operation of Darwinian processes should yield exactly what we see in the fossil record. It is gradualism that we must reject, not Darwinism.

The history of most fossil species includes two features particularly inconsistent with gradualism:

1. *Stasis.* Most species exhibit no directional change during their tenure on earth. They appear in the fossil record looking much the same as when they disappear; morphological change is usually limited and directionless.

2. *Sudden appearance.* In any local area, a species does not arise gradually by the steady transformation of its ancestors; it appears all at once and 'fully formed.'

Evolution proceeds in two major modes. In the first, phyletic transformation, an entire population changes from one state to another. If all evolutionary change occurred in this mode, life would not persist for long. Phyletic evolution yields no increase in diversity, only a transformation of one thing into another. Since extinction (by extirpation, not by evolution into something else) is so common, a biota with no mechanism for increasing diversity would soon be wiped out. The second mode, speciation, replenishes the earth. New species branch off from a persisting parental stock.

Darwin, to be sure, acknowledged and discussed the process of speciation. But he cast his discussion of evolutionary change almost totally in the mold of phyletic transformation. In this context, the phenomena of stasis and sudden appearance could hardly be attributed to anything but imperfection of the record; for if new species arise by the transformation of entire ancestral populations, and if we almost never see the transformation (because species are essentially static through their range), then our record must be hopelessly incomplete.

Eldredge and I believe that speciation is responsible for almost all evolutionary change. Moreover, the way in which it occurs virtually guarantees that sudden appearance and stasis shall dominate the fossil record.

All major theories of speciation maintain that splitting takes place rapidly in very small populations. The theory of geographic, or allopatric, speciation is preferred by most evolutionists for most situations (allopatric means 'in another place').* A new species can arise when a

* I wrote this essay in 1977. Since then, a major shift of opinion has been sweeping through evolutionary biology. The allopatric orthodoxy has been breaking down and several mechanisms of sympatric speciation have been gaining both legitimacy and examples. (In sympatric speciation, new forms arise within the geographic range of their ancestors.) These sympatric mechanisms are united in their insistence upon the two conditions that Eldredge and I require for our model of the fossil record – *rapid* origin in a *small* population. In fact, they generally advocate smaller groups and more rapid change than conventional allopatry envisages (primarily because groups in potential contact with their forebears must move quickly towards reproductive isolation, lest their favorable variants be diluted by breeding with the more numerous parental forms).

small segment of the ancestral population is isolated at the periphery of the ancestral range. Large, stable central populations exert a strong homogenizing influence. New and favorable mutations are diluted by the sheer bulk of the population through which they must spread. They may build slowly in frequency, but changing environments usually cancel their selective value long before they reach fixation. Thus, phyletic transformation in large populations should be very rare – as the fossil record proclaims.

But small, peripherally isolated groups are cut off from their parental stock. They live as tiny populations in geographic corners of the ancestral range. Selective pressures are usually intense because peripheries mark the edge of ecological tolerance for ancestral forms. Favorable variations spread quickly. Small, peripheral isolates are a laboratory of evolutionary change.

What should the fossil record include if most evolution occurs by speciation in peripheral isolates? Species should be static through their range because our fossils are the remains of large central populations. In any local area inhabited by ancestors, a descendant species should appear suddenly by migration from the peripheral region in which it evolved. In the peripheral region itself, we might find direct evidence of speciation, but such good fortune would be rare indeed because the event occurs so rapidly in such a small population. Thus, the fossil record is a faithful rendering of what evolutionary theory predicts, not a pitiful vestige of a once bountiful tale.

Eldredge and I refer to this scheme as the model of *punctuated equilibria*. Lineages change little during most of their history, but events of rapid speciation occasionally punctuate this tranquility. Evolution is the differential survival and deployment of these punctuations. (In describing the speciation of peripheral isolates as very rapid, I speak as a geologist. The process may take hundreds, even thousands of years; you might see nothing if you stared at speciating bees on a tree for your entire lifetime. But a thousand years is a tiny fraction of one percent of the average duration for most fossil invertebrate species – 5 to 10 million years. Geologists can rarely resolve so short an interval at all; we tend to treat it as a moment.)

If gradualism is more a product of Western thought than a fact of nature, then we should consider alternate philosophies of change to enlarge our realm of constraining prejudices. In the Soviet Union, for example, scientists are trained with a very different philosophy of change – the so-called dialectical laws, reformulated by Engels from Hegel's philosophy. The dialectical laws are explicitly punctuational. They speak, for example, of the 'transformation of quantity into quality.' This may sound like mumbo jumbo, but it suggests that change occurs in large leaps following a slow accumulation of stresses that a system resists until it reaches the breaking point. Heat water and it eventually boils. Oppress the workers more and more and bring on the revolution. Eldredge and I were fascinated to learn that many Russian paleontologists support a model similar to our punctuated equilibria.

I emphatically do not assert the general 'truth' of this philosophy of punctuational change. Any attempt to support the exclusive validity of such a grandiose notion would border on the nonsensical. Gradualism sometimes works well. (I often fly over the folded Appalachians and marvel at the striking parallel ridges left standing by gradual erosion of the softer rocks surrounding them.) I make a simple plea for pluralism in guiding philosophies, and for the recognition that such philosophies, however hidden and unarticulated, constrain all our thought. The dialectical laws express an ideology quite openly: our Western preference for gradualism does the same thing more subtly.

Nonetheless, I will confess to a personal belief that a punctuational view may prove to map tempos of biological and geologic change more accurately and more often than any of its competitors – if only because complex systems in steady state are both common and highly resistant to change. As my colleague British geologist Derek V. Ager writes in supporting a punctuational view of geologic change: 'The history of any one part of the earth, like the life of a soldier, consists of long periods of boredom and short periods of terror.'

BETTING ON CHANCE – AND NO FAIR PEEKING

DOUBLE entendre can be delicious. Who does not delight in learning that Earnest, in Oscar Wilde's play, is a good chap, not a worthy attitude. And who has ever begrudged that tragic figure his little joke. But double meanings also have their dangers – particularly when two communities use the same term in different ways, and annoying confusion, rather than pleasant amusement or enlightenment, results.

Differences in scientific and vernacular definitions of the same word provide many examples of this frustrating phenomenon. 'Significance' in statistics, for example, bears little relation to the ordinary meaning of something that matters deeply. Mouse tails may be 'significantly' longer in Mississippi than in Michigan – meaning only that average lengths are not the same at some level of confidence – but the difference may be so small that no one would argue for significance in the ordinary sense. But the most serious of all misunderstandings between technical and vernacular haunts the concepts of probability and particularly the words *random* and *chance*.

In ordinary English, a random event is one without order predictability, or pattern. The word connotes disaggregation, falling apart, formless anarchy, and fear. Yet, ironically, the scientific sense of *random* conveys a precisely opposite set of associations. A phenomenon governed by chance yields maximal simplicity, order, and predictability – at least in the long run. Suppose that we are interested in resolving the forces behind a large-scale pattern of historical change. Randomness becomes our best hope for a maximally simple and tractable model. If we flip a penny or throw a pair of dice, once a second for days on end, we achieve a rigidly predictable distribution of outcomes. We can predict the margins of departure from 50–50 in the coins or the percentage of sevens for our dice, based on the total number of throws. When the number of tosses becomes quite large, we can give precise estimates and ranges of error for the frequencies and lengths of runs – heads or sevens in a row – all based on the simplest of mathematical formulas from the theory of probability. (Of course, we cannot predict the outcome of any particular trial or know when a run will occur, as any casual gambler should – but so few do – know.)

Thus, if you wish to understand patterns of long historical sequences, pray for randomness. Ironically, nothing works so powerfully against resolution as conventional forms of determinism. If each event in a sequence has a definite cause, then, in a world of such complexity, we are lost. If *A* happened because the estuary thawed on a particular day, leading to *B* because Billy the Seal swam by and gobbled up all those fishes, followed by *C* when Sue the Polar Bear sauntered through – not to mention ice age fluctuations, impacting asteroids, and drifting continents of the truly long run – then how can we hope to predict and resolve the outcome?

The beauty (and simplicity) of randomness lies in the absence of these maximally confusing properties. Coin flipping permits no distinctive personality to any time or moment; each toss can be treated in exactly the same way, whenever it occurs. We can date geological time with precision by radioactive decay because each atom has an equal probability of decaying in each instant. If causal individuality intervened – if 10:00 A.M. on Sunday differed from 5:00 P.M. on Wednesday, or if Joe the uranium atom, by dint of moral fiber, resisted decay better than his brother Tom, then randomness would fail and the method would not work.

One of the best illustrations for this vitally important, but counterintuitive, principle of maximal long-term order in randomness comes from my own field of evolutionary biology – and from a debate that has greatly contributed to making professional life more interesting during the past twenty years. Traditional Darwinism includes an important role for randomness – but only as a source of variation, or raw material, for evolutionary change, not as an agent for the direction of change itself. For Darwin, the predominant source of evolutionary change resides in the deterministic force of natural selection. Selection works for cause and adapts organisms to changing local environments. Random variation supplies the indispensable 'fuel' for natural selection but does not set the rate, timing, or pattern of change. Darwinism is a two-part theory of randomness for raw material and conventional causality for change – *Chance and Necessity*, as so well epitomized by Jacques Monod in the title of his famous book about the nature of Darwinism.

In the domain of organisms and their good designs, we have little reason to doubt the strong, probably dominant influence of deterministic forces like natural selection. The intricate, highly adapted forms of organisms – the wing of a bird or the mimicry of a dead twig by an insect – are too complex to arise as long sequences of sheer good fortune under the simplest random models. But this stricture of complexity need not apply to the nucleotide-by-nucleotide substitutions that build the smallest increments of evolutionary change at the molecular level. In this domain of basic changes in DNA, a 'neutralist' theory, based on simple random models, has been challenging conventional Darwinism with marked success during the past generation.

When the great Japanese geneticist Motoo Kimura formulated his first version of neutral theory in 1968, he was impressed by two discoveries that seemed difficult to interpret under the conventional view that natural selection overwhelms all other causes of evolutionary change. First, at the molecular level of substitutions in amino acids, measured rates indicated a constancy of change across molecules and organisms – the so-called molecular clock of evolution. Such a result makes no sense in Darwin's world, where molecules subject to strong selection should evolve faster than others, and where organisms exposed to different changes and challenges from the environment should vary their evolutionary rates accordingly. At most, one might claim that these deterministic differences in rate might tend to 'even out' over very long stretches of geological time, yielding roughly regular rates of change. But a molecular clock surely gains an easier interpretation from random models. If deterministic selection does not regulate most molecular changes – if, on the contrary, most molecular variations are neutral, and therefore rise and fall in frequency by the luck of the draw – then mutation rate and population size will govern the tempo of change. If most populations are large, and if mutation rates are roughly the same for most genes, then simple random models predict a molecular clock.

Second, Kimura noted the recent discovery of surprisingly high levels of variation maintained by many genes among members of populations. Too much variation poses a problem for conventional Darwinism because a cost must accompany the replacement of an ancestral gene by a new and more advantageous state of the same gene – namely, the differential death, by natural selection, of the now disfavored parental forms. This cost poses no problem if only a few old genes are being pushed out of a population at any time. But if hundreds of genes are being eliminated, then any organism must carry many of the disfavored states and should be ripe for death. Thus, selection should not be able to replace many genes at once. But the data on copious variability seemed to indicate a caldron of evolutionary activity at far too many genetic sites – too many, that is, if selection governs the changes in each varying gene. Kimura, however, recognized a simple and elegant way out of this paradox. If most of the varying forms of a gene are neutral with respect to selection, then they are drifting in frequency by the luck of the draw. Invisible to natural selection because they make no difference to the organism, these variations impose no cost in replacement.

In twenty years of copious writing, Kimura has always carefully emphasized that his neutral theory does not disprove Darwinism or deny the power of natural selection to shape the adaptations of organisms. He writes, for example, at the beginning of his epochal book *The Neutral Theory of Molecular Evolution* (1983):

> The neutral theory is not antagonistic to the cherished view that evolution of form and function is guided by Darwinian selection, but it brings out another facet of the evolutionary process by emphasizing the much greater role of mutation pressure and random drift at the molecular level.

The issue, as so often in natural history (and as I emphasize so frequently in these essays), centers upon the relative importance of the two processes. Kimura has never denied adaptation and natural selection, but he has tended to view these processes as quantitatively insignificant in the total picture – a superficial and minor ripple upon the ocean of neutral molecular change, imposed every now and again when selection casts a stone upon the waters of evolution. Darwinians, on the other hand, at least before Kimura and his colleagues advanced their potent challenge and reeled in the supporting evidence, tended to argue that neutral change occupied a tiny and insignificant corner of evolution – an odd process occasionally operating in small populations at the brink of extinction anyway.

This argument about relative frequency has raged for twenty years and has been, at least in the judgment of this bystander with no particular stake in the issue, basically a draw. More influence has been measured for selection than Kimura's original words had anticipated; Darwin's process is no mere pockmark on a sea of steady tranquility. But neutral change has been established at a comfortably high relative frequency. The molecular clock is neither as consistent nor as regular as Kimura once hoped, but even an imperfect molecular timepiece makes little sense in Darwin's world. The ticking seems best interpreted as a pervasive and underlying neutralism, the considerable perturbations as a substantial input from natural selection (and other causes).

Nonetheless, if forced to award the laurels in a struggle with no clear winners, I would give the nod to Kimura. After all, when innovation fights orthodoxy to a draw, then novelty has seized a good chunk of space from convention. But I bow to Kimura for another and more important reason than the empirical adequacy of neutralism at high relative frequency, for his theory so beautifully illustrates the theme that served as an introduction to this essay: the virtue of randomness in the technical as opposed to the vernacular sense.

Kimura's neutralist theory has the great advantage of simplicity in mathematical expression and specification of outcome. Deterministic natural selection yields no firm predictions for the histories of lineages – for you would have to know the exact and particular sequences of biotic and environmental changes, and the sizes and prior genetic states of populations, in order to forecast an outcome. This knowledge is not attainable in a world of imperfect historical information. Even if obtainable, such data would only provide a prediction for a particular lineage, not a general theory. But neutralism, as a random model treating all items and times in the same manner, yields a set of simple, general equations serving as precise predictions for the results of evolutionary change. These equations give us, for the first time, a base-level criterion for assessing any kind of genetic change. If neutralism holds, then actual outcomes will fit the equations. If selection predominates, then results will depart from predictions – and in a way that permits identification of Darwinian control. Thus, Kimura's equations have been as useful for selectionists as for neutralists themselves; the formulas provide a criterion for everyone, and debate can center upon whether or not the equations fit actual outcomes. Kimura has often emphasized this point about his equations, and about random models in general. He wrote, for example, in 1982:

> The neutral theory is accompanied by a well-developed mathematical theory based on the stochastic theory of population genetics. The latter enables us to treat evolution and variation quantitatively, and therefore to check the theory by observations and experiments.

The most important and useful of these predictions involves a paradox under older Darwinian views. If selection controls evolutionary rate, one might think that the fastest tempos of alteration would be associated with the strongest selective pressures for change. Speed of change should vary directly with intensity of selection. Neutral theory predicts precisely the opposite – for an obivious reason once you start thinking about it. The most rapid change should be associated with unconstrained randomness – following the old thermodynamic imperative that things will invariably go to hell unless you struggle actively to maintain them as they are. After all, stability is far more common than change at any moment in the history of life. In its ordinary everyday mode, natural selection must struggle to preserve working combinations against a constant input of deleterious mutations. In other words, natural selection, in our technical parlance, must usually be 'purifying' or 'stabilizing.' Positive selection for change must be a much rarer event than watchdog selection for tossing out harmful variants and preserving what works.

Now, if mutations are neutral, then the watchdog sees nothing and evolutionary change can proceed at its maximal tempo – the neutral rate of substitution. But if a molecule is being preserved by selection, then the watchdog inhibits evolutionary change. This originally counterintuitive proposal may be regarded as the key statement of neutral theory. Kimura emphasizes the point with italics in most of his general papers, writing for example (in 1982): '*Those molecular changes that are less likely to be subjected to natural selection occur more rapidly in evolution.*'

Both the greatest success, and the greatest modification, of Kimura's original theory have occurred by applying this principle that selection slows the maximal rate of neutral molecular change. For modification of the original theory, thousands of empirical studies have now shown that watchdog selection, measured by diminished tempo of change relative to predictions of randomness, operates at a far higher relative frequency than Kimura's initial version of neutralist theory had anticipated. For success, the firm establishment of the principle itself must rank as the greatest triumph of neutralism – for the tie of maximal rate to randomness (rather than to the opposite expectation of intense selection) does show that neutralism exerts a kind of base-level control over evolution as a whole.

The most impressive evidence for neutralism as a maximal rate has been provided by forms of DNA that make nothing of potential selective value (or detriment) to an organism. In all these cases, measured tempos of molecular change are maximal, thus affirming the major prediction of neutralism.

1. *Synonymous substitutions.* The genetic code is redundant in the third position. A sequence of three nucleotides in DNA codes for an amino acid. Change in either of the first two nucleotides alters the amino acid produced, but most changes in the third nucleotide – so-called synonymous substitutions – do not alter the resulting amino acid. Since natural selection works on features of organisms, in this case proteins built by DNA and not directly on the DNA itself, synonymous substitutions should be invisible to selection, and therefore neutral. Rates of change at the third position are usually five or more times as rapid as changes at the functional first and second positions – a striking confirmation of neutralism.

2. *Introns.* Genes come in pieces, with functional regions (called exons) interrupted by DNA sequences (called introns) that are snipped out and not translated into proteins. Introns change at a much higher evolutionary rate than exons.

3. *Pseudogenes.* Certain kinds of mutations can extinguish the function of a gene – for example, by preventing its eventual translation into protein. These so-called pseudogenes begin with nearly the same DNA sequence as the functional version of the gene in closely related species. Yet, being entirely free from function, these pseudogenes should exert no resistance against the maximal accumulation of changes by random drift. Pseudogenes become a kind of ultimate test for the proposition that absence of selection promotes maximal change at the neutral rate – and the test has, so far, been passed with distinction. In pseudogenes, rates of change are equal, and maximal, at all three positions of the triplet code, not only at the third site, as in functional genes.

I was inspired to write about neutral theory by a fascinating example of the value of this framework in assessing the causes of evolutionary rates. This example neither supports nor denies neutralism but forms a case in the middle, enlightened by the more important principle that random models provide simple and explicit criteria for judgment.

While supposedly more intelligent mammals are screwing up royally above ground, Near Eastern mole rats of the species *Spalax ehrenbergi* are prospering underneath. Subterranean mammals usually evolve reduced or weakened eyes, but *Spalax* has reached an extreme state of true blindness. Rudimentary eyes are still generated in embryology, but they are covered by thick skins and hair. When exposed to powerful flashes of light, *Spalax* shows no neurological response at all, as measured by electrodes implanted in the brain. The animal is completely blind.

What then shall we make of the invisible and rudimentary eye? Is this buried eye now completely without function, a true vestige on a path of further reduction to final disappearance? Or does the eye perform some other service not related to vision? Or perhaps the eye has no direct use, but must still be generated as a prerequisite in an embryological pathway leading to other functional features. How can we decide among these and other alternatives? The random models of neutral theory provide our most powerful method. If the rudimentary eye is a true vestige, then its proteins should be changing at the maximal neutral rate. If selection has not been relaxed, and the eye still functions in full force (though not for vision), then rates of change should be comparable to those for other rodents with conventional eyes. If selection has been relaxed due to blindness, but the eye still functions in some less constrained way, then an intermediate rate of change might be observed.

The eye of *S. ehrenbergi* still builds a lens (though the shape is irregular and cannot focus an image), and the lens includes a protein, called αA-crystallin. The gene for this protein has recently been sequenced and compared with the corresponding gene in nine other rodents with normal vision (by W. Hendriks, J. Leunissen, E. Nevo, H. Bloemendal, and W. W. de Jong).

Hendriks and colleagues obtained the most interesting of possible results from their study. The αA-crystallin gene is changing much faster in blind *Spalax* than in other rodents with vision as relaxation of selection due to loss of primary function would suggest. The protein coded by the *Spalax* gene, for example, has undergone nine amino acid replacements (of 173 possible changes), compared with the ancestral state for its group (the murine rodents, including rats, mice, and hamsters). All other murines in the study (rat, mouse, hamster, and gerbil) have identical sequences with no change at all from the ancestral state. The average tempo of change in αA-crystallin among vertebrates as a whole has been measured at about three amino acid replacements per 100 positions per 100 million years. *Spalax* is changing more than four times as fast, at about 13 percent per 100 million years. (Nine changes in 173 positions is 5.2 percent; but the *Spalax* lineage is only 40 million years old – and 5.2 percent in 40 million corresponds to 13 percent in 100 million years.) Moreover, *Spalax* has changed four amino acids at positions that are absolutely constant in all other vertebrates studied – seventy-two species ranging from dogfish sharks to humans.

'These findings,' Hendriks and colleagues conclude, 'all clearly indicate an increased tolerance for change in the primary structure of αA-crystallin in this blind animal.' So far so good. But the increased tempo of change in *Spalax*, though marked, still reaches only about 20 percent of the characteristic rate for pseudogenes, our best standard for the maximal, truly neutral pace of evolution. Thus, *Spalax* must still be doing enough with its eyes to damp the rate of change below the maximum for neutrality. Simple models of randomness have taught us something interesting and important by setting a testable standard, approached but not met in this case, and acting as a primary criterion for judgment.

What then is αA-crystallin doing for *Spalax*? What can a rudimentary and irregular lens, buried under skin and hair, accomplish? We do not know, but the established intermediate rate of change leads us to ask the right questions in our search for resolution.

Spalax is blind, but this rodent still responds to changes in photoperiod (differing lengths of daylight and darkness) – and apparently through direct influence of light regimes themselves, not by an indirect consequence that a blind animal might easily recognise (increase in temperature due to more daylight hours, for example). A. Haim, G. Heth, H. Pratt, and E. Nevo showed that *Spalax* would increase its tolerance for cold weather when exposed to a winterlike light regime of eight light followed by sixteen dark hours. These mole rats were kept at the relatively warm temperature of 22°C, and were therefore not adjusting to winter based on clues provided by temperature. Animals exposed to twelve light and twelve dark hours at the same temperature did not improve their thermoregulation as well. Interestingly, animals exposed to summerlike light regimes (sixteen light and eight dark), but at colder temperatures of 17°C, actually *decreased* their cold-weather tolerance. Thus, even though blind, *Spalax* is apparently using light, not temperature, as a guide for adjusting physiology to the cycle of seasons.

Hendriks and colleagues suggest a possible explanation, not yet tested. We know that many vertebrates respond to changes in photoperiod by secreting a hormone called melatonin in the pineal gland. The pineal responds to light on the basis of photic information transmitted via the retina. *Spalax* forms a retina in its rudimentary eye, yet how can the retina, which perceives no light in this blind mammal, act in concert with the pineal gland? But Hendriks and colleagues note that the retina can also secrete melatonin itself – and that the retina of *Spalax* includes the secreting layer. Perhaps the retina of *Spalax* is still functional as a source of melatonin or as a trigger of the pineal by some mechanism still unknown. (I leave aside the fascinating, and completely unresolved, issue of how a blind animal can respond, as *Spalax* clearly does, to seasonal changes in photoperiod.)

If we accept the possibility that *Spalax* may need and use its retina (in some nonvisual way) for adaptation to changing seasons, then a potential function for the lens, and for the αA-crystallin protein, may be sought in developmental pathways, not in direct utility. The lens cannot work in vision, and αA-crystallin focuses no image, but the retina does not form in isolation and can only be generated as part of a normal embryological pathway that includes the prior differentiation of other structures. The formation of a lens vesicle may be a prerequisite to the construction of a retina – and a functioning retina may therefore require a lens, even if the lens will be used for nothing on its own.

Evolution is strongly constrained by the conservative nature of embryological programs. Nothing in biology is more wondrously complex than the production of an adult vertebrate from a single fertilized ovum. Nothing much can be changed very radically without discombobulating the embryo. The intermediate rate of change in lens proteins of a blind rodent – a tempo so neatly between the maximal pace for neutral change and the much slower alteration of functioning parts – may point to a feature that has lost its own direct utility but must still form as a prerequisite to later, and functional, features in embryology.

Our world works on different levels, but we are conceptually chained to our own surroundings, however parochial the view. We are organisms and tend to see the world of selection and adaptation as expressed in the good design of wings, legs, and brains. But randomness may predominate in the world of genes – and we might interpret the universe very differently if our primary vantage point resided at this lower level. We might then see a world of largely independent items, drifting in and out by the luck of the draw – but with little islands dotted about here and there, where selection reins in tempo and embryology ties things together. What, then, is the different order of a world still larger than ourselves? If we missed the different world of genic neutrality because we are too big, then what are we not seeing because we are too small? We are like genes in some larger world of change among species in the vastness of geological time. What are we missing in trying to read this world by the inappropriate scale of our small bodies and minuscule lifetimes?

THE POWER OF THE MODAL BACTER, OR WHY THE TAIL CAN'T WAG THE DOG

An Epitome of the Argument

I believe that the most knowledgeable students of life's history have always sensed the failure of the fossil record to supply the most desired ingredient of Western comfort: a clear signal of progress measured as some form of steadily increasing complexity for life as a whole through time. The basic evidence cannot support such a view, for simple forms still predominate in most environments, as they always have. Faced with this undeniable fact, supporters of progress (that is, nearly all of us throughout the history of evolutionary thought) have shifted criteria and ended up grasping at straws. (The altered criterion may not have struck the graspers as such a thin reed, for one must first internalize the argument of this book – trends as changes in variation rather than things moving somewhere – to recognize the weakness.) In short, graspers for progress have looked exclusively at the history of the most complex organism through time – a myopic focus on extreme values only – and have used the increasing complexity of the most complex as a false surrogate for progress of the whole. But this argument is illogical and has always disturbed the most critical consumers.

Thus, James Dwight Dana, America's greatest naturalist in Darwin's era (at least after Agassiz's death), and a soul mate to Darwin in their remarkably parallel careers (both went on long sea voyages in their youth, and both became fascinated with coral reefs and the taxonomy of crustaceans), used this criterion when he finally converted to evolution in the mid-1870s. Dana's primary commitment to progress as the definition of life's organization held firm throughout his career, and in his personal transition from creationism to evolution. But Dana could validate progress only by looking at the history of extremes – 'the grand fact that the system of life began in the simple sea-plant and the lower forms of animals, and ended in man'. Julian Huxley, grandson of Thomas Henry, sensed the same unease, but could think of no other criterion in 1959. When Darwin's grandson challenged him to defend progress in the light of so many well-adapted but anatomically simplified parasites, Julian Huxley replied, 'I mean a higher degree of organization in general, as shown by the upper level attained.' But the 'upper level attained' (the extreme organism at the right tail) is not a measure of 'organization in general' – and Huxley's defense is illogical.

In debunking this conventional argument for progress in the history of life, I reach the crux of this book (although I will not disparage anyone who regards baseball as equal in importance to life's history, and therefore views the correct interpretation of 0.400 hitting as more vital to American life than understanding the central themes of 3.5 billion years in biological time)! Yet I can summarize my argument against progress in the history of life in just seven statements condensed into a few pages. I do not mean to be capricious or disrespectful in this brevity. If I have done my job in the rest of this book, I have already set the background and argument with sufficient thoroughness – so this focal application at grandest scale should follow quickly with just a few reminders and way stations for the new context.

I do not challenge the statement that the most complex creature has tended to increase in elaboration through time, but I fervently deny that this limited little fact can provide an argument for general progress as a defining thrust of life's history. Such a grandiose claim represents a ludicrous case of the tail wagging the dog, or the invalid elevation of a small and epiphenomenal consequence into a major and controlling cause.

I shall present, in seven arguments, my best sense of a proper case based on the history of expanding variation away from a beginning left wall. I shall then provide extended commentary for three of the statements that are particularly vital, and most generally misunderstood or unappreciated. Please note that the entire sequence of statements for life as a whole follows exactly the same logic, and postulates the same causes, as my previous story (at smallest scale) about the evolution of planktonic forams.

1. *Life's necessary beginning at the left wall.* The earth is about 4.5 billion years old. Life, as recorded in the fossil record, originated at least 3.5 billion years ago, and probably not much earlier because the earth passed through a molten period that ended about 3.8 billion years ago (the age of the oldest rocks). Life presumably began in primeval oceans as a result of sequential chemical reactions based on original constituents of atmospheres and oceans, and regulated by principles of physics for self-organizing systems. (The 'primeval soup' has long been a catchword for oceans teeming with appropriate organic compounds prior to the origin of life.) In any case, we may specify as a 'left wall' the minimal complexity of life under these conditions of spontaneous origin. (As a paleontologist, I like to think of this wall as the lower limit of 'conceivable, preservable complexity' in the fossil record.) For reasons of physics and chemistry, life had to begin right next to the left wall of minimal complexity – as a microscopic blob. You cannot begin by precipitating a lion out of the primeval soup.

2. *Stability throughout time of the initial bacterial mode.* If we are particularly parochial in our concern for multicellular creatures, we place the major division in life between plants and animals (as the Book of Genesis does in both creation myths of chapter 1 and 2). If we are more ecumenical, we generally place the division between unicellular and multicellular forms. But most professional biologists would argue that the break of maximal profundity occurs within the unicells, separating the prokaryotes (or cells without organelles – no nuclei, no chromosomes, no mitochondria, no chloroplasts) from the eukaryotes (organisms like amoebae and paramecia, with all the complex parts contained in the cells of multi-cellular organisms). Prokaryotes include the amazingly diverse groups collectively known as 'bacteria,' and also the so-called 'blue-green algae,' which are little more than photosynthesizing bacteria, and are now generally known as Cyanobacteria.

All the earliest forms of life in the fossil record are prokaryotes – or, loosely, 'bacteria.' In fact, more than half the history of life is a tale of bacteria only. In terms of preservable anatomy in the fossil

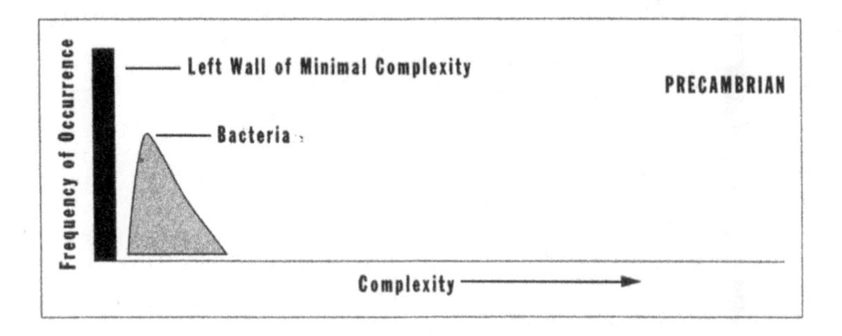

Figure 3. Life begins necessarily near the left wall of minimal complexity, and the bacterial mode soon develops.

record, bacteria lie right next to the left wall of minimal conceivable complexity. Life therefore began with a bacterial mode (see Fig. 3). Life still maintains a bacterial mode in the same position. So it was in the beginning, is now, and ever shall be – at least until the sun explodes and dooms the planet. How, then, using the proper criterion of variation in life's full house, can we possibly argue that progress provides a central defining thrust to evolution if complexity's mode has never changed? The modal bacter of this chapter's title has been life's constant paradigm of success.

3. *Life's successful expansion must form an increasingly right-skewed distribution*. Life had to begin next to the left wall of minimal complexity (see statement 1). As life diversified, only one direction stood open for expansion. Nothing much could move left and fit between the initial bacterial mode and the left wall. The bacterial mode itself has maintained its initial position and grown continually in height (see Fig. 4). Since space remains available away from the left wall and toward the direction of greater complexity, new species occasionally wander into this previously unoccupied domain, giving the bell curve of complexity for all species a right skew, with capacity for increased skewing through time.

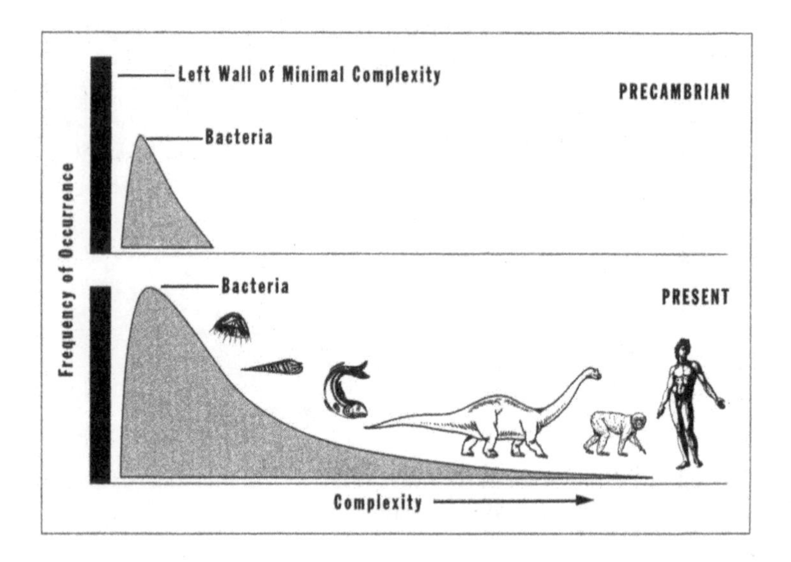

Figure 4. The frequency distribution for life's complexity becomes increasingly right skewed through time, but the bacterial mode never alters.

4. *The myopia of characterizing a full distribution by an extreme item at one tail.* Considering life's full house of Figure 4, the only conceivable argument for general progress must postulate that an expanding right tail demonstrates a predictable upward thrust of the whole. But such a claim only embodies the silly spectacle of a small tail wagging a large dog. (We have generally failed to grasp the evident absurdity because we have not visualized the dog properly; rather, in a move that recalls the Cheshire Cat of Wonderland, identified only by its smile, we have characterized the entire dog by its tail alone.)

A claim for general progress based on the right tail alone is absurd for two primary reasons: First, the tail is small and occupied by only a tiny percentage of species (more than 80 percent of multicellular animal species are arthropods, and we generally regard almost all members of this phylum as primitive and nonprogressive). Second, the occupants of the extreme right edge through time do not form an evolutionary sequence, but rather a motley series of disparate forms that have tumbled into this position, one after the other. Such a sequence through time might read: bacterium, eukaryotic cell, marine alga, jellyfish, trilobite, nautiloid, placoderm fish, dinosaur, saber-toothed cat, and *Homo sapiens*. Beyond the first two transitions, not a single form in this sequence can possibly be a direct ancestor of the next in line.

5. *Causality resides at the wall and in the spread of variation; the right tail is a consequence, not a cause.* The development of life's bell curve for complexity through time (Figs 3 and 4) does not represent a fully random phenomenon (though random elements play an important role). Two important causal influences shape the curve and its changes – but neither influence includes any statement about conventional progress. The two major causes are, first, necessary origin at the left wall of minimal complexity; and, second, increase of numbers and kinds, with predictable development of a right-skewed distribution. Given this point of origin at a wall and subsequent increase in variation, the right tail almost had to develop and extend. But this expansion of the right tail – the only (and myopic) source for any potential claim about progress – is an epiphenomenon and a side consequence of the two causes listed above, not a fundamental thrust produced by the superiority of complex forms under natural selection. In fact, as the paradigm of the drunkard's walk illustrates, such an extension of the right tail will occur in a regime of entirely random motion for each item, so long as the system begins at a wall. Thus, as the drunkard's walk shows in theory, and the evolution of planktonic forams confirms in fact, the expanding right tail of life's complexity may arise from random motion among all lineages. The vaunted progress of life is really *random motion away from simple beginnings*, not *directed impetus toward inherently advantageous complexity*.

6. *The only promising way to smuggle progress back into such a system is logically possible, but empirically false at high probability.* My argument for the whole system is sound: from a necessary beginning at the left wall, random motion of all items in a growing system will produce an increasingly right-skewed distribution. Thus, and with powerful irony, the most venerable evidence for general progress – the increasing complexity of the most complex – becomes a passive consequence of

growth in a system with no directional bias whatever in the motion of its components.

But one potential (though much vitiated) argument for general progress remains. The entire system is free to vary only in the direction of greater complexity from an initial position next to the left wall. But what about a smaller lineage that begins at some intermediary position with freedom to expand in *either* direction (the first living thing starts at the left wall, but the first mammal, or the first seed plant, or the first clam, begins in the middle and its descendants can move toward either tail). If we studied all the smaller lineages free to vary in any direction, perhaps we would then detect a clear bias for net movement to the right, or toward greater complexity. If we found such a bias, we could legitimately speak of a general trend to greater complexity in the evolutionary history of lineages. (This more subtle position would still not explain the general pattern of Figure 4, which would still arise as a consequence of random motion in a growing system constrained to begin at the left wall. But a rightward bias in individual lineages would function as a 'booster' or 'helpmate' in the general production of right skew. The entire system would then be built by two components: random motion from the left wall, and a rightward bias in individual lineages – and the second component would provide an argument for general progress.)

The logic of this argument is sound, but two strong reasons suggest (though not all the evidence is yet in) that the proposition is empirically false. I shall summarize the two reasons here. First, while I know of no proven bias for rightward motion under natural selection – a mechanism that yields only local adaptation to changing environments, not general progress – a good case can be made for leftward bias because parasitism is such a common evolutionary strategy, and parasites tend to be anatomically more simplified than their free-living ancestors. (Ironically, then, the full system of increasing right skew for the whole might actually be built with a slight bias toward *decreased* complexity in individual lineages!) Second, several paleontologists are now studying this issue directly by trying to quantify the elusive notion of progress and then tracing the changing spread of their measure in the history of individual lineages. Only a few studies have been completed so far, but current results show no rightward bias, and therefore no tendency to progress in individual lineages.

7. *Even a parochial decision to focus on the right tail alone will not yield the one, most truly desired conclusion, the psychological impetus to our yearning for general progress – that is, the predictable and sensible evolution to domination of a creature like us, endowed with consciousness*. We might adopt a position of substantial retreat from an original claim for general progress, but still a bastion of defense for what really matters to us. That is, we might say, 'Okay, you win. I understand your point that the evidence of supposed progress, the increasing right skew of life's bell curve, is only an epiphenomenal tail that cannot wag the entire dog – and that life's full house has never moved from its modal position. But I am allowed to be parochial. The right tail may be small and epiphenomenal, but I love the right tail because I dwell at its end – and I want to focus on the right tail alone because this little epiphenomenon is all that matters to me. Even you admit that the right tail had to arise, so long as life expanded. So the right tail had to develop and grow – and had to produce, at its apogee, something like me. I therefore remain the modern equivalent of the apple of God's eye: the predictably most complex creature that ever lived.'

Wrong again, even for this pitifully restricted claim (after advancing an initial argument for intrinsic directionality in the basic causal thrust of all evolution). The right tail had to exist, but the actual composition of creatures on the tail is utterly unpredictable, partly random, and entirely contingent – not at all foreordained by the mechanisms of evolution. If we could replay the game of life again and again, always starting at the left wall and expanding thereafter in diversity, we would get a right tail almost every time, but the inhabitants of this region of greatest complexity would be wildly and unpredictably different in each rendition – and the vast majority of replays would never produce (on the finite scale of a planet's lifetime) a creature with self-consciousness. Humans are here by the luck of the draw, not the inevitability of life's direction or evolution's mechanism.

In any case, little tails, no tails, or whoever occupies the tails, the outstanding feature of life's history has been the stability of its bacterial mode over billions of years!

THE GREAT DYING

ABOUT 225 million years ago, at the end of the Permian period, fully half the families of marine organisms died out during the short span of a few million years – a prodigious amount of time by most standards, but merely minutes to a geologist. The victims of this mass extinction included all surviving trilobites, all ancient corals, all but one lineage of ammonites, and most bryozoans, brachiopods, and crinoids.

This great dying was the most profound of several mass extinctions that have punctuated the evolution of life during the past 600 million years. The late Cretaceous extinction, some 70 million years ago, takes second place. It destroyed 25 percent of all families, and cleared the earth of its dominant terrestrial animals, the dinosaurs and their kin – thus setting a stage for the dominance of mammals and the eventual evolution of man.

No problem in paleontology has attracted more attention or led to more frustration than the search for causes of these extinctions. The catalog of proposals would fill a Manhattan telephone book and include almost all imaginable causes: mountain building of worldwide extent, shifts in sea level, subtraction of salt from the oceans, supernovae, vast influxes of cosmic radiation, pandemics, restriction of habitat, abrupt changes in climate, and so on. Nor has the problem escaped public notice. I remember well my first exposure to it at age five: the dinosaurs of Disney's *Fantasia* panting to their deaths across a desiccating landscape to the tune of Stravinsky's *Rite of Spring*.

Since the Permian extinction dwarfs all the others, it has long been the major focus of inquiry. If we could explain this greatest of all dyings, we might hold the key to understanding mass extinctions in general.

During the past decade, important advances in both geology and evolutionary biology have combined to provide a probable answer. This solution has developed so gradually that some paleontologists scarcely realize that their oldest and deepest dilemma has been resolved.

Ten years ago, geologists generally believed that the continents formed where they now stand. Large blocks of land might move up and down and continents might 'grow' by accretion of uplifted mountain chains at their borders, but continents did not wander about the earth's surface – their positions were fixed for all time. An alternative theory of continental drift had been proposed early in the century, but the absence of a mechanism for moving continents had assured its nearly universal rejection.

Now, studies of the ocean floor have yielded a mechanism in the theory of plate tectonics. The earth's surface is divided into a small number of plates bordered by ridges and subduction zones. New ocean floor is formed at the ridges as older portions of the plates are pushed away. To balance this addition, old parts of plates are drawn into the earth's interior at subduction zones.

Continents rest passively upon the plates and move with them; they do not 'plow' through solid ocean floor as previous theories proposed. Continental drift, therefore, is but one consequence of plate tectonics. Other important consequences include earthquakes at plate boundaries (like the San Andreas Fault running past San Francisco) and mountain chains where two plates bearing continents collide (the Himalayas formed when the Indian 'raft' hit Asia).

When we reconstruct the history of continental movements, we realize that a unique event occurred in the latest Permian: all the continents coalesced to form the single supercontinent of Pangaea. Quite simply, the consequences of this coalescence caused the great Permian extinction.

But which consequences and why? Such a fusion of fragments would produce a wide array of results, ranging from changes in weather and oceanic circulation to the interaction of previously isolated ecosystems. Here we must look to advances in evolutionary biology – to theoretical ecology and our new understanding of the diversity of living forms.

After several decades of highly descriptive and largely atheoretical work, the science of ecology has been enlivened by quantitative approaches that seek a general theory of organic diversity. We are gaining a better understanding of the influences of different environmental factors upon the abundance and distribution of life. Many studies now indicate that diversity – the numbers of different species present in a given area – is strongly influenced, if not largely controlled, by the amount of habitable area itself. If, for example, we count the number of ant species living on each of a group of islands differeing only in size (and otherwise similar in such properties as climate, vegetation, and distance from the mainland), we find that, in general, the larger the island, the greater the number of species.

It is a long way from ants on tropical islands to the entire marine biota of the Permian period. Yet we have good reason to suspect that area might have played a major role in the great extinction. If we can estimate organic diversity and area for various times during the Permian (as the continents coalesced), then we can test the hypothesis of control by area.

We must first understand two things about the Permian extinction and the fossil record in general. First, the Permian extinction primarily affected marine organisms. The relatively few terrestrial plants and vertebrates then living were not so strongly disturbed. Second, the fossil record is very strongly biased toward the preservation of marine life in shallow water. We have almost no fossils of organisms inhabiting the ocean depths. Thus, if we want to test the theory that reduced area played a major role in the Permian extinction, we must look to the area occupied by shallow seas.

We can identify, in a qualitative way, two reasons why a coalescence of continents would drastically reduce the area of shallow seas. The first is basic geometry: If each separate land mass of pre-Permian times were completely surrounded by shallow seas, then their union would eliminate all area at the sutures. Make a single square out of four graham crackers and the total periphery is reduced by half. The second reason involves the mechanics of plate tectonics. When oceanic ridges are actively producing new sea floor to spread outward, then the ridges themselves stand high above the deepest parts of the ocean. This displaces water from the ocean basins, world sea level rises, and continents are partly flooded. Conversely, if spreading diminishes or stops, ridges begin to collapse and sea level falls.

When continents collided in the late Permian, the plates that carried them 'locked' together. This set a brake upon new spreading. Ocean ridges sank and shallow seas withdrew from the continents. The drastic reduction in shallow seas was not caused by a drop in sea level per se, but rather by the configuration of sea floor over which the drop occurred. The ocean floor does not plunge uniformly from shoreline to ocean deep. Today's continents are generally bordered by a very wide continental shelf of persistently shallow water. Seaward of the shelf lies the continental slope of much greater steepness. If sea level fell far enough to expose the entire continental shelf, then most of the world's shallow seas would disappear. This may well have happened during the late Permian.

Thomas Schopf of the University of Chicago has recently tested this hypothesis of extinction by reduction in area. He studied the distribution of shallow water and terrestrial rocks to infer continental borders and extent of shallow seas for several times during the Permian as the continents coalesced. Then, by an exhaustive survey of the paleontological literature, he counted the numbers of different kinds of organisms living during each of these Permian times. Daniel Simberloff of Florida State University then showed that the standard mathematical equation relating numbers of species to area fits these data very well. Moreover, Schopf showed that the extinction did not affect certain groups differentially; its results were evenly spread over all shallow-water inhabitants. In other words, we do not need to seek a specific cause related to the peculiarities of a few animal groups. The effect was general. As shallow seas disappeared, the rich ecosystem of earlier Permian times simply lacked the space to support all its members. The bag became smaller and half the marbles had to be discarded.

Area alone is not the whole answer. Such a momentous event as the fusion of a single supercontinent must have entailed other consequences detrimental to the precariously balanced ecosystem of earlier Permian time. But Schopf and Simberloff have provided persuasive evidence for granting a major role to the basic factor of space.

It is gratifying that an answer to paleontology's outstanding dilemma has arisen as a by-product of exciting advances in two related disciplines – ecology and geology. When a problem has proved intractable for more than one hundred years, it is not likely to yield to more data collected in the old way and under the old rubric. Theoretical ecology allowed us to ask the right questions and plate tectonics provided the right earth upon which to pose them.

THE VALIDATION OF CONTINENTAL DRIFT

As the new Darwinian orthodoxy swept through Europe, its most brilliant opponent, the aging embryologist Karl Ernst von Baer, remarked with bitter irony that every triumphant theory passes through three stages: first it is dismissed as untrue; then it is rejected as contrary to religion; finally, it is accepted as dogma and each scientist claims that he had long appreciated its truth.

I first met the theory of continental drift when it labored under the inquisition of stage two. Kenneth Caster, the only major American paleontologist who dared to support it openly, came to lecture at my alma mater, Antioch College. We were scarcely known as a bastion of entrenched conservatism, but most of us dismissed his thoughts as just this side of sane. (Since I am now in von Baer's third stage, I have the distinct memory that Caster sowed substantial seeds of doubt in my own mind.) A few years later, as a graduate student at Columbia University, I remember the a priori derision of my distinguished stratigraphy professor toward a visiting Australian drifter. He nearly orchestrated the chorus of Bronx cheers from a sycophantic crowd of loyal students. (Again, from my vantage point in the third stage, I recall this episode as amusing, but distasteful.) As a tribute to my professor, I must record that he experienced a rapid conversion just two years later and spent his remaining years joyously redoing his life's work.

Today, just ten years later, my own students would dismiss with even more derision anyone who denied the evident truth of continental drift – a prophetic madman is at least amusing; a superannuated fuddy-duddy is merely pitiful. Why has such a profound change occurred in the short space of a decade?

Most scientists maintain – or at least argue for public consumption – that their profession marches toward truth by accumulating more and more data, under the guidance of an infallible procedure called 'the scientific method.' If this were true, my question would have an easy answer. The facts, as known ten years ago, spoke against continental drift; since then, we have learned more and revised our opinions accordingly. I will argue, however, that this scenario is both inapplicable in general and utterly inaccurate in this case.

During the period of nearly universal rejection, direct evidence for continental drift – that is, the data gathered from rocks exposed on our continents – was every bit as good as it is today. It was dismissed because no one had devised a physical mechanism that would permit continents to plow through an apparently solid oceanic floor. In the absence of a plausible mechanism, the idea of continental drift was rejected as absurd. The data that seemed to support it could always be explained away. If these explanations sounded contrived or forced, they were not half so improbable as the alternative – accepting continental drift. During the past ten years, we have collected a new set of data, this time from the ocean basins. With these data, a heavy dose of creative imagination, and a better understanding of the earth's interior, we have fashioned a new theory of planetary dynamics. Under this theory of plate tectonics, continental drift is an inescapable consequence. The old data from continental rocks, once soundly rejected, have been exhumed and exalted as conclusive proof of drift. In short, we now accept continental drift because it is the expectation of a new orthodoxy.

I regard this tale as typical of scientific progress. New facts, collected in old ways under the guidance of old theories, rarely lead to any substantial revision of thought. Facts do not 'speak for themselves'; they are read in the light of theory. Creative thought, in science as much as in the arts, is the motor of changing opinion. Science is a quintessentially human activity, not a mechanized, robot-like accumulation of objective information, leading by laws of logic to inescapable interpretation. I will try to illustrate this thesis with two examples drawn from the 'classical' data for continental drift. Both are old tales that had to be undermined while drift remained unpopular.

1. The late Paleozoic glaciation. About 240 million years ago, glaciers covered parts of what is now South America, Antarctica, India, Africa,

and Australia. If continents are stable, this distribution presents some apparently insuperable difficulties:

A. The orientation of striae in eastern South America indicates that glaciers moved onto the continent from what is now the Atlantic Ocean (striae are scratches on bedrock made by rocks frozen into glacier bottoms as they pass over a surface). The world's oceans form a single system, and transport of heat from tropical areas guarantees that no major part of the open ocean can freeze.
B. African glaciers covered what are now tropical areas.
C. Indian glaciers must have grown in semitropical regions of the Northern hemisphere; moreover, their striae indicate a source in tropical waters of the Indian Ocean.
D. There were no glaciers on any of the northern continents. If the earth got cold enough to freeze tropical Africa, why were there no glaciers in northern Canada or Siberia?

All these difficulties evaporate if the southern continents (including India) were joined together during this glacial period, and located farther south, covering the South Pole; the South American glaciers moved from Africa, not an open ocean; 'tropical' Africa and 'semi-tropical' India were near the South Pole; the North Pole lay in the middle of a major ocean, and glaciers could not develop in the Northern Hemisphere. Sounds good for drift; indeed, no one doubts it today.

2. The distribution of Cambrian trilobites (fossil arthropods living 500 to 600 million years ago). The Cambrian trilobites of Europe and North America divided themselves into two rather different faunas with the following peculiar distribution on modern maps. 'Atlantic' province trilobites lived all over Europe and in a few very local areas on the far eastern border of North America – eastern (but not western) Newfoundland and southeastern Massachusetts, for example. 'Pacific' province trilobites lived all over America and in a few local areas on the extreme western coast of Europe – northern Scotland and north-western Norway, for example. It is devilishly difficult to make any sense of this distribution if the two continents always stood 3,000 miles apart.

But continental drift suggests a striking resolution. In Cambrian times, Europe and North America were separated: Atlantic trilobites lived in waters around Europe; Pacific trilobites in waters around America. The continents (now including sediments with entombed trilobites) then drifted toward each other and finally joined together. Later, they split again, but not precisely along the line of their previous junction. Scattered bits of ancient Europe, carrying Atlantic trilobites, remained at the easternmost border of North America, while a few pieces of old North America stuck to the westernmost edge of Europe.

Both examples are widely cited as 'proofs' of drift today, but they were soundly rejected in previous years, not because their data were any less complete but only because no one had devised an adequate mechanism to move continents. All the original drifters imagined that continents plow their way through a static ocean floor. Alfred Wegener, the father of continental drift, argued early in our century that gravity alone could put continents in motion. Continents drift slowly westward, for example, because attractive forces of the sun and moon hold them up as the earth rotates underneath them. Physicists responded with derision and showed mathematically that gravitational forces are far too weak to power such a monumental peregrination. So Alexis du Toit, Wegener's South African champion, tried a different tack. He argued for a local, radioactive melting of oceanic floor at continental borders, permitting the continents to glide through. This ad hoc hypothesis added no increment of plausibility to Wegener's speculation.

Since drift seemed absurd in the absence of a mechanism, orthodox geologists set out to render the impressive evidence for it as a series of unconnected coincidences.

In 1932, the famous American geologist Bailey Willis strove to make the evidence of glaciation compatible with static continents. He invoked the deus ex machina of 'isthmian links' – narrow land bridges flung with daring abandon across 3,000 miles of ocean. He placed one between eastern Brazil and western Africa, another from Africa all the way to India via the Malagasy Republic, and a third from Vietnam through Borneo and New Guinea to Australia. His colleague, Yale professor Charles Schuchert, added one from Australia to Antarctica and another from Antarctica to South America, thus completing the isolation of a southern ocean from the rest of the world's waters. Such an isolated ocean might freeze along its southern margin, permitting glaciers to flow across into eastern South America. Its cold waters would also nourish the glaciers of southern Africa. The Indian glaciers, located above the equator 3,000 miles north of any southern ice, demanded a separate explanation. Willis wrote: 'No direct connection between the occurrences can reasonably be assumed. The case must be considered on the basis of a general cause and the local geographic and topographic conditions.' Willis's inventive mind was equal to the task: he simply postulated a topography so elevated that warm, wet southern waters precipitated their product as snow. For the absence of ice in temperate and Arctic zones of the Northern Hemisphere, Willis reconstructed a system of ocean currents that permitted him to postulate 'a warm, subsurface current flowing northward beneath cooler surface waters and rising in the Arctic as a warm-water heating system.' Schuchert was delighted with the resolution provided by isthmian links:

> Grant the biogeographer Holarctis, a land bridge from northern Africa to Brazil, another from South America to Antarctis (it almost exists today), still another from this polar land to Australia and from the latter across the Arafura Sea to Borneo and Sumatra and so on to Asia, plus the accepted means of dispersal along shelf seas and by wind and water currents and migratory birds, and he has all the possibilities needed to explain the life dispersion and the land and ocean realms throughout geological time on the basis of the present arrangement of the continents.

The only common property shared by all these land bridges was their utterly hypothetical status; not an iota of direct evidence supported any one of them. Yet, lest the saga of isthmian links be read as a warped fairy tale invented by dogmatists to support an untenable orthodoxy, I point out that to Willis, Schuchert, and any right-thinking geologist of the 1930s, one thing legitimately seemed ten times as absurd as imaginary land bridges thousands of miles long – continental drift itself.

In the light of such highly fertile imaginations, the Cambrian trilobites could present no insuperable problem. The Atlantic and Pacific provinces were interpreted as different environments, rather than different places – shallow water for the Pacific, deeper for the Atlantic. With a freedom to invent nearly any hypothetical geometry for Cambrian ocean basins, geologists drew their maps and hewed to their orthodoxy.

When continental drift came into fashion during the late 1960s, the classical data from continental rocks played no role at all: drift rode in on the coattails of a new theory, supported by new types of evidence.

The physical absurdities of Wegener's theory rested on his conviction that continents cut their way through the ocean floor. But how else could drift occur? The ocean floor, the crust of the earth, must be stable. After all, where could it go, if it moved in pieces, without leaving gaping holes in the earth? Nothing could be clearer. Or could it?

'Impossible' is usually defined by our theories, not given by nature. Revolutionary theories trade in the unexpected. If continents must plow through oceans, then drift will not occur; suppose, however, that continents are frozen into the oceanic crust and move passively as pieces of crust shift about. But we just stated that the crust cannot move without leaving holes. Here, we reach an impasse that must be bridged by creative imagination, not just by another field season in the folded Appalachians – we must model the earth in a fundamentally different way.

We can avoid the problem of holes with a daring postulate that seems to be valid. If two pieces of ocean floor move away from each other, they will leave no hole if material rises from the earth's interior to fill the gap. We can go further by reversing the causal implications of this statement: the rise of new material from the earth's interior may be the driving force that moves old sea floor away. But since the earth is not expanding, we must also have regions where old sea floor founders into the earth's interior, thus preserving a balance between creation and destruction.

Indeed, the earth's surface seems to be broken into fewer than ten major 'plates,' bounded on all sides by narrow zones of creation (oceanic ridges) and destruction (trenches). Continents are frozen into these plates, moving with them as the sea floor spreads away from zones of creation at oceanic ridges. Continental drift is no longer a proud theory in its own right; it has become a passive consequence of our new orthodoxy – plate tectonics.

We now have a new, mobilist orthodoxy, as definite and uncompromising as the staticism it replaced. In its light, the classical data for drift have been exhumed and proclaimed as proof positive. Yet these data played no role in validating the notion of wandering continents; drift triumphed only when it became the necessary consequence of a new theory.

The new orthodoxy colors our vision of all data; there are no 'pure facts' in our complex world. About five years ago, paleontologists found on Antarctica a fossil reptile named *Lystrosaurus*. It also lived in South Africa, and probably in South America as well (rocks of the appropriate age have not been found in South America). If anyone had floated such an argument for drift in the presence of Willis and Schuchert, he would have been howled down – and quite correctly. For Antarctica and South America are almost joined today by a string of islands, and they were certainly connected by a land bridge at various times in the past (a minor lowering of sea level would produce such a land bridge today). *Lystrosaurus* may well have walked in comfort, on a rather short journey at that. Yet the *New York Times* wrote an editorial proclaiming, on this basis alone, that continental drift had been proved.

Many readers may be disturbed by my argument for the primacy of theory. Does it not lead to dogmatism and disrespect for fact? It can, of course, but it need not. The lesson of history holds that theories are overthrown by rival theories, not that orthodoxies are unshakable. In the meantime, I am not distressed by the crusading zeal of plate tectonics, for two reasons. My intuition, culturally bound to be sure, tells me that it is basically true. My guts tell me that it's damned exciting – more than enough to show that conventional science can be twice as interesting as anything invented by all the von Dänikens and in all the Bermuda triangles of this and previous ages of human gullibility.

PHYLETIC SIZE DECREASE IN HERSHEY BARS

THE solace of my youth was a miserable concoction of something sweet and gooey, liberally studded with peanuts and surrounded by chocolate – real chocolate, at least. It was called 'Whizz' and it cost a nickel. Emblazoned on the wrapper stood its proud motto in rhyme – 'the best nickel candy there izz.' Sometime after the war, candy bars went up to six cents for a time, and the motto changed without fanfare – 'the best candy bar there izz.' Little did I suspect that an evolutionary process, persistent in direction and constantly accelerating, had commenced.

I am a paleontologist – one of those oddballs who parlayed his childhood fascination for dinosaurs into a career. We search the history of life for repeated patterns, mostly without success. One generality that works more often than it fails is called 'Cope's rule of phyletic size increase.' For reasons yet poorly specified, body size tends to increase fairly steadily within evolutionary lineages. Some have cited general advantages of larger bodies – greater foraging range, higher reproductive output, greater intelligence associated with larger brains. Others claim that founders of long lineages tend to be small, and that increasing size is more a drift away from diminutive stature than a positive achievement of greater bulk.

The opposite phenomenon of gradual size decrease is surpassingly rare. There is a famous foram (a single-celled marine creature) that got smaller and smaller before disappearing entirely. An extinct, but once major group, the graptolites (floating, colonial marine organisms, perhaps related to vertebrates) began life with a large number of stipes (branches bearing a row of individuals). The number of stipes then declined progressively in several lineages, to eight, four, and two, until finally all surviving graptolites possessed but a single stipe. Then they disappeared. Did they, like the *Incredible Shrinking Man* simply decline to invisibility – for he, having decreased enough to make his final exit through the mesh of a screen in his movie debut, must now be down to the size of a muon, but still, I suspect, hanging in there. Or did they snuff it entirely, like the legendary Foo-Bird who coursed in ever smaller circles until he flew up his own you-know-what and disappeared. What would a zero-stiped graptolite look like? In any case, they are no longer part of our world.

The rarities of nature are often commonplaces of culture; and phyletic size decrease surrounds us in products of human manufacture. Remember the come-on, once emblazoned on the covers of comic books – '52 pages, all comics.' And they only cost a dime. And remember when large meant large, rather than the smallest size in a sequence of detergent or cereal boxes going from large to gigantic to enormous.

Consider the Hershey Bar – a most worthy standard bearer for the general phenomenon of phyletic size decrease in manufactured goods. It is the unadvertised symbol of American quality. It shares with Band-Aids, Kleenex, Jell-o and the Fridge that rare distinction of attaching its brand name to the generic product. It has also been shrinking fast.

I have been monitoring informally, and with distress, this process for more than a decade. Obviously, others have followed it as well. The subject has become sufficiently sensitive that an official memo emanated in December 1978 from corporate headquarters at 19 East Chocolate Avenue – in Hershey, Pa. of course. Hershey chose the unmodified hangout and spilled all the beans, to coin an appropriate metaphor. This three page document is titled 'Remember the nickel bar?' (I do indeed, and ever so fondly, for I started to chomp them avidly in an age of youthful innocence, ever so long before I first heard of the nickel bag.) Hershey defends its shrinking bars and rising prices as a strictly average (or even slightly better than average) response to general inflation. I do not challenge this assertion since I use the bar as a synecdoche for general malaise – as an average, not an egregious, example.

I have constructed the accompanying graph from tabular data in the Hershey memo, including all information from mid-1965 to now. As a paleontologist used to interpreting evolutionary sequences, I spy two general phenomena: gradual phyletic size decrease within each price lineage, and occasional sudden mutation to larger size (and price)

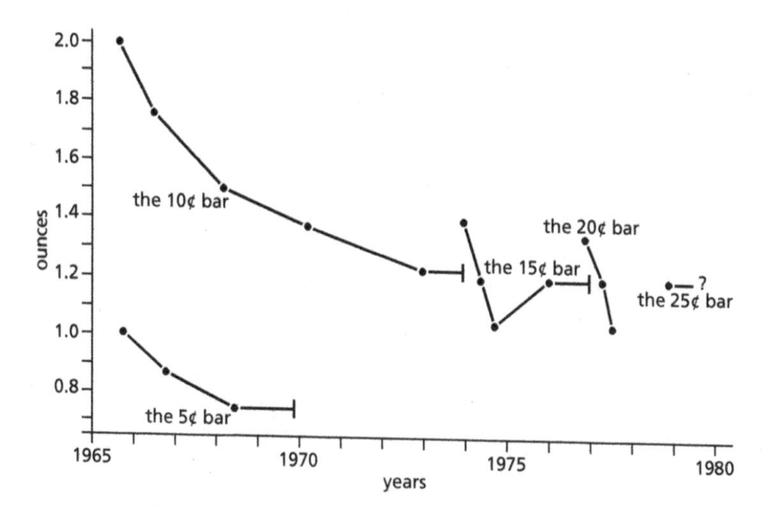

Figure 5. Hershey Bars bite the dust: a quantitative assessment.

following previous decline to dangerous levels. I am utterly innocent of economics, the dismal science. For me, bulls and bears have four legs and are called *Bos taurus* and *Ursus arctos*. But I think I finally understand what an evolutionist would call the 'adaptive significance' of inflation. Inflation is a necessary spin-off, or by-product, of a lineage's successful struggle for existence. For this radical explanation of inflation, you need grant me only one premise – that the manufactured products of culture, as fundamentally unnatural, tend to follow life's course in reverse. If organic lineages obey Cope's rule and increase in size, then manufactured lineages have an equally strong propensity for decreasing in size. Therefore, they either follow the fate of the Foo-Bird and we know them no longer, or they periodically restore themselves by sudden mutation to larger size – and, incidentally, fancier prices.

We may defend this thesis by extrapolating the tendencies of each price lineage on the graph. The nickel bar weighed an ounce in 1949. And it still weighed an ounce (following some temporary dips to 7/8 oz.) when our story began in September 1965. But it could delay its natural tendency no longer and decline began, to 7/8 oz. in September

1966 and finally to 3/4 oz. in May 1968 until its discontinuation on November 24, 1969, a day that will live in infamy. But just as well, for if you extrapolate its average rate of decline (1/4 ounce in thirty-two months), it would have become extinct naturally in May 1976. The dime bar followed a similar course, but beginning larger, it held on longer. It went steadily down from 2 oz. in August 1965 to 1.26 oz. in January 1973. It was officially discontinued on January 1, 1974, though I calculate that it would have become extinct on August 17, 1986. The fifteen-cent bar started hopefully at 1.4 oz. in January 1974, but then declined at an alarming rate far in excess of any predecessor. Unexpectedly, it then rallied, displaying the only (though minor) reverse toward larger size within a price lineage since 1965. Nonetheless, it died on December 31, 1976 – and why not, for it could only have lasted until December 31, 1988, and who would have paid fifteen cents for a crumb during its dotage? The twenty-cent bar (I do hope I'm not boring you) arose at 1.35 oz. in December 1976 and immediately experienced the most rapid and unreversed decline of any price lineage. It will die on July 15, 1979. The twenty-five-cent bar, now but a few months old, began at 1.2 oz. in December 1978. *Ave atque vale.*

The graph shows another alarming trend. Each time the Hershey Bar mutates to a new price lineage, it gets larger, but never as large as the founding member of the previous price lineage. The law of phyletic size decrease for manufactured goods must operate across related lineages as well as within them – thus ultimately frustrating the strategy of restoration by mutational jump. The ten-cent bar began at 2 oz. and was still holding firm when our story began in late 1965. The fifteen-cent bar arose at 1.4 oz., the twenty-cent bar at 1.35 oz., and the quarter bar at 1.2 oz. We can also extrapolate this rate of decrease across lineages to its final solution. We have seen a decrease of 0.8 oz. in three steps over thirteen years and four months. At this rate, the remaining four and a half steps will take another twenty years. And that ultimate wonder of wonders, the weightless bar, will be introduced in December 1998. It will cost forty-seven and a half cents.

The publicity people at Hershey's mentioned something about a ten-pound free sample. But I guess I've blown it. Still, I would remind everyone of Mark Twain's comment that there are 'lies, damned lies and statistics.' And I will say this for the good folks in Hershey, Pa. It's still the same damned good chocolate, what's left of it. A replacement of whole by broken almonds is the only compromise with quality

I've noticed, while I shudder to think what the 'creme' inside a Devil Dog is made of these days.

Still, I guess I've blown it. Too bad. A ten-pound bar titillates my wildest fancy. It would be as good as the 1949 Joe DiMaggio card that I never got (I don't think there was one in the series). And did I ever end up with a stack of pink bubble gum sheets for the effort. But that's another tale, to be told through false teeth at another time.

Postscript

I wrote this article (as anyone can tell from internal evidence) early in 1979. Since then, two interesting events have occurred. The first matched my predictions with uncanny accuracy. For the second, that specter of all science, the Great Exception (capital G, capital E), intervened and I have been temporarily foiled. And – as an avid Hershey bar chomper – am I ever glad for it.

The twenty-five-cent bar did just about what I said it would. It started at 1.2 oz. in December 1978, where I left it, and then plummeted to 1.05 oz. in March 1980 before becoming extinct in March 1982. But Hershey then added a twist to necessity when it replaced its lamented two-bit bar with the inevitable thirty-cent concoction. Previously, all new introductions had begun (despite their fancier prices) at lower weights than the proud first item of the previous price lineage. (I based my extrapolation to the weightless bar on this pattern.) But, wonder of wonders and salaam to the Great Exception, the thirty-cent bar began at a whopping 1.45 oz., larger than anything we've seen since the ten-cent bar of my long-lost boyhood.

As cynical readers might expect, a tale lies behind this peculiar move. In the *Washington Post* for July 11, 1982 (and with thanks to Ellis Yochelson for sending the article), Randolph E. Bucklin explains all under the title: 'Candy Wars: Price Tactic Fails Hershey.'

It seems that the good folks at (not on) Mars, manufacturers of Three Musketeers, Snickers, and M&M's, and Hershey's chief competitor, had made the unprecedented move of increasing the size of their quarter bars without raising prices. After a while, they snuck the price up to thirty cents but kept the new size. Hershey tried to hold the line with its shrinking quarter bars. But thousands of mom and pop stores couldn't be bothered charging a quarter for some bars and thirty cents for others (and couldn't remember which were Hershey's and which Mars's anyway) – and therefore charged thirty cents for both Mars's large bars and Hershey's minuscule offerings. Hershey's sales plummeted; finally, they capitulated to Mars's tactics, raising prices to thirty cents and beefing up sizes to Mars's level and above predictions of the natural trend.

As a scientist trained in special pleading, I have a ready explanation for the Great Exception. General trends have an intrinsic character; they continue when external conditions retain their constancy. An unanticipated and unpredictable catastrophe, like the late Cretaceous asteroid, or the sneaky sales tactic of Mars and Co., resets the system, and all bets are off. Still, the greater inevitability prevails. The thirty-cent bar will diminish and restitutions at higher prices will shrink as well. The weightless bar may come a few years later than I predicted (even a bit past the millennium) – but I still bet ya it'll cost about four bits.

PART IV

SIZE, FORM AND SHAPE

NATURALISTS begin by observing the living world around us; from the colours of flowers and the regular forms of pine cones or snail shells to the multitudinous shapes of fish and the pentadactyl limbs shared by so many mammals. How do these regularities come about? Evolution is above all the art of the possible. Natural selection does not have the possibility of an à la carte choice of the forms and structures available on which to work. Its raw materials are limited by chemistry and physics. So one of the abiding fascinations amongst both evolutionary and developmental biologists is the emergence of pattern in living forms. Steve was as fascinated as any by this problem, from his first collector's love of snails to the variations in skull size amongst human ancestors.

His hundredth essay in the *Natural History* series describes this love affair, with the Bahamian land snail *Cerion,* and reveals something of the combination of passion and probing question which together go to make a great researcher. The second essay in this short section addresses the question of size and shape in broader terms, and the third takes as its theme pattern generation. What could be more striking than the regularities of the black and white stripes of the zebra's coat, making the animal instantly recognizable even in a young child's colouring book? How, during development, do such regularities emerge? Their adaptive significance is clear, but the processes that generate them at first sight seem mysterious. Yet quite simple morphological forces are at work, as the great biologist and Gouldian precursor D'Arcy Thompson showed in his seminal work early in the last century. The stripes emerge in a process that a little mathematical analysis shows is both predictable and simple.

Not all such issues of size and shape are so easily resolved. Zoologists have long been concerned with the study of allometry – the preservation and transformation of form both during development and in evolutionary progressions between related species, members of which may look characteristically similar but vary markedly in size. Are there simple mathematical rules to explain these changes? Gould approached this question in collaboration with David Pilbeam in the early 1970s, and we conclude this section with an extract from their paper on size and scaling in human evolution from *Science*.

OPUS 100*

THROUGHOUT a long decade of essays I have never, and for definite reasons, written about the biological subject closest to me. Yet for this, my hundredth effort, I ask your indulgence and foist upon you the Bahamian land snail *Cerion*, mainstay of my own personal research and fieldwork. I love *Cerion* with all my heart and intellect but have consciously avoided it in this forum because the line between general interest and personal passion cannot be drawn from a perspective of total immersion – the image of doting parents driving friends and neighbors to somnolent distraction with family movies comes too easily to mind. These essays must follow two unbreakable rules: I never lie to you, and I strive mightily not to bore you. But, for this one time in a hundred, I will risk the second for personal pleasure alone.

Cerion is the most prominent land snail of West Indian islands. It ranges from the Florida Keys to the small islands of Aruba, Bonaire, and Curaçao, just off the Venezuelan coast, but the vast majority of species inhabit two principal centers – Cuba and the Bahamas. *Cerion*'s life includes little excitement by our standards. Most species inhabit rocks and sparse vegetation abutting the seashore. They may live for five to ten years, but they spend most of this time in the warm weather equivalent of hibernation (called estivation), hanging upside down from vegetation or affixed to rocks. After a rain or sometimes in the relative

* This is the fourth volume of essays compiled from my monthly column in *Natural History* magazine. I marked ten years of work, and never a deadline missed (I won't tell you about the numerous close calls), with this act of self-indulgence as my treat to myself for the one hundredth effort.

cool and damp of night, they descend from their twigs and stones, nibble at the fungi on decaying vegetation, and perhaps even copulate. We have marked and mapped the movement of individual snails and many can be found on the same few square yards of turf, year after year.

Why pick *Cerion?* Why, indeed, spend so much time on any detailed particular when all the giddy generalities of evolutionary theory beg for study in a lifetime too short to manage but a few? Iconoclast that I am, I would not abandon the central wisdom of natural history from its inception – that concepts without percepts are empty (as Kant said), and that no scientist can develop an adequate 'feel' for nature (that undefinable prerequisite of true understanding) without probing deeply into minute empirical details of some well-chosen group of organisms. Thus, Aristotle dissected squids and proclaimed the world's eternity, while Darwin wrote four volumes on barnacles and one on the origin of species. America's greatest evolutionists and natural historians, G. G. Simpson, T. Dobzhansky, and E. Mayr, began their careers as, respectively, leading experts on Mesozoic mammals, ladybird beetles, and the birds of New Guinea.

Scientists don't immerse themselves in particulars only for the grandiose (or self-serving) reason that such studies may lead to important generalities. We do it for fun. The pure joy of discovery transcends import. And we do it for adventure and for expansion. As drama, Bahamian field trips may seem risible compared with Darwin on the *Beagle*, Bates on the Amazon, and Wallace in the Malay Archipelago – although I would not care to repeat my only close brush with death, caught in a shoot-out among drug runners on North Andros. So much more do I value the quiet times in different worlds: an evening's discussion of bush medicine on Mayaguana, an exploration of ornamental carvings that adorn roofs on Long Island and South Andros, and the finest meal I have ever eaten – a campfire pot of fresh conch stewed with sweet potatoes from Jimmy Nixon's garden on Inagua, after a hot and hard day's work.

If all good naturalists must choose a group of organisms for detailed immersion, we do not select mindlessly or randomly (or even, as some cynics have suggested, because the Bahamas beat the Yukon as a field area). I am interested primarily in the evolution of form and have concentrated on how the varying shapes of an individual's growth can serve as a source of evolutionary change (see my technical book, *Ontogeny and Phylogeny*). An invertebrate paleontologist with these interests would naturally be led to snails, since their shells preserve a complete record of growth from egg to adult.

A student of form with a penchant for gastropods could not avoid *Cerion*, for this genus exhibits, among its several hundred species, a range of form unmatched by any other group of snails. Some *Cerions* are tall and pencil thin; others are shaped like golf balls. When a colleague ventured 'square snails' as an example of impossible animals at a public meeting, I was able to show him a peculiar quadrate *Cerion*. Five years ago, I discovered the largest *Cerion*, a thin and parallel-sided fossil giant from Mayaguana more than 70 mm tall. The smallest is a virtual sphere, scarcely 5 mm in diameter, from Little Inagua.

Cerion's mystery and special interest do not lie in its exuberant diversity alone; many groups of animals include some members with unusual propensities for speciation and consequent variation of form. Species are the fundamental units of biological diversity, distinct populations permanently isolated one from the other by an absence of interbreeding in nature. We should not be surprised that groups producing large numbers of species may become quite diverse in form, since more distinct units provide more opportunities for evolving a wide range of morphologies.

Faced with such a riotous array of shapes, older naturalists did name species aplenty in *Cerion*, some 600 of them. But few are biologically valid as distinct noninterbreeding populations. In ten years of fieldwork on all major Bahamian islands, we have only once found two distinct *Cerion* populations living in the same place and not interbreeding – true species, therefore. These included a giant and a dwarf – thus recalling various bad jokes about Chihuahuas and Great Danes. In all other cases, two forms, no matter how distinct in size and shape, interbreed and produce hybrids at their point of geographic contact. Somehow, *Cerion* manages to generate its unparalleled diversity of form without parceling its populations into true species. How can this happen? Moreover, if such different forms hybridize so readily, then the genetic differences between them cannot be great. How can such diversity of size and shape arise in the absence of extensive genetic change?

In a related and second mystery, distinct forms of *Cerion* often inhabit widely separated islands. The simplest explanation would propose that these far-flung colonies represent the same species and that hurricanes can blow snails great distances, producing haphazard distributions, or that colonies once inhabiting intermediate islands have become extinct, leaving large distances between survivors. Yet, all *Cerion* experts have developed the feeling (which I share) that these separated colonies, despite their detailed similarity for long lists of traits, have evolved independently *in situ*. If this unconventional interpretation is correct, how can such complex suites of associated traits evolve again and again?

Cerion thus presents two outstanding peculiarities amidst its unparalleled diversity: Its most distinct forms interbreed and are not true species, while these same forms, for all their complexity, may have evolved several times independently. Any scientist who can explain these odd phenomena for *Cerion* will make an important contribution to the understanding of form and its evolution in general. I shall try to describe the few preliminary and faltering steps we have made towards such a resolution.

Cerion has attracted the attention of several prominent naturalists, from Linnaeus, who named its first species in 1758, to Ernst Mayr, who pioneered the study of natural populations 200 years later. Still, despite the efforts of a tiny group of aficionados, *Cerion* has not received the renown it deserves in the light of its curious biology and its promise as an exemplar for the evolution of form. Its relative obscurity can be traced directly to past biological practice. Older naturalists buried *Cerion*'s unusual biology under such an impenetrable thicket of names (for invalid species) that colleagues interested in evolutionary theory have been unable to recover the pattern and interest from utter chaos.

The worst offender was C. J. Maynard, a fine amateur biologist who named hundreds of *Cerion* species from the 1880s through the 1920s. He imagined that he was performing a great service, proclaiming in 1889:

> Conchologists may take exception to some of my new species, thinking, perhaps, that I have used too trivial characters in separating them. Believing, however, as I do, that it is the imperative duty of naturalists today, to record minute points of differences among animals . . . I have not hesitated so to designate them, if for no other reason than for the benefit of coming generations.

I trust that I shall not be accused of undue cynicism in recognizing another reason. Maynard financed his Bahamian trips by selling shells, and more species meant more items to flog. *Caveat emptor.*

Professional colleagues were harsh on Maynard's overly fine splitting. H. A. Pilsbry, America's greatest conchologist, declared in uncharacteristically forceful prose that 'gods and men may well stand aghast at the naming of individual colonies from every sisal field and potato patch in the Bahamas.' W. H. Dall labeled Maynard's efforts as 'noxious and stupefying.' Yet, when tested in the crucible of practice, neither Pilsbry nor Dall lived up to his brave words. Each recognized at least half the species Maynard advocated, still sufficiently overinflated to bury any pattern in the forest of invalid names.

So rich was *Cerion*'s diversity, and so numerous its species, that G. B. Sowerby, the outstanding English conchologist, who fancied himself (with little justification) a poet, wrote this doggerel in introducing his monograph on the genus:

> Things that were not, at thy command,
> In perfect form before Thee stand;
> And all to their Creator raise
> A wondrous harmony of praise.

Sowerby then proceeded to list quite a chorus. And this quatrain dates from 1875, before Maynard ever named a *Cerion*!

In the light of this existing chaos, and before we can even ask the general questions about form that I posed above, we must pursue a much more basic and humble task. We must find out whether any pattern can be found in the ecological and geographic distribution of *Cerion*'s morphology. If we detect no correlation at all with geography or environment, then what can we explain? Fortunately, in a decade of work, we have reduced the chaos of existing names to predictable patterns and have thereby established the prerequisite for deeper explanation. Of the nature of that deeper explanation, we have intuitions and indications, but neither definite information nor even the tools to provide it (for we are stuck in an area of biology – the genetics of development – that is itself woefully undeveloped). Still, I think we have made a promising start.

I say 'we' because I realized right away that I could not do this work alone. I felt competent to analyze the growth and form of shells, but

I have no expertise in two areas that must be united with morphology in any comprehensive study: genetics and ecology. So I teamed up with David Woodruff, a biologist from the University of California at San Diego. For a decade we have done everything together, from blisters on Long Island to bullets on Andros.

(I must stop at this point, for I suddenly realize that I have almost broken my first rule. Scientists have a terrible tendency to present their work as a logical package, as if they thought everything out in careful and rigorous planning beforehand and then merely proceeded according to their good designs. It never works that way, if only because anyone who can think and see makes unanticipated discoveries and must fundamentally alter any preconceived strategy. Also, people get into problems for the damnedest of peculiar and accidental reasons. Projects grow like organisms, with serendipity and supple adjustment, not like the foreordained steps of a high school proof in plane geometry. Let me confess. I was first drawn to *Cerion* because I wanted to compare its fossils with snails I had studied on Bermuda. I studiously avoided all modern *Cerions* because I was petrified by the thicket of available names and considered them intractable. Woodruff first went to Inagua because he wanted to study color banding in another genus of snails. But he went at the height of mosquito season and lasted two days. We took our first trip together to Grand Bahama Island: I to study fossils, he to try the other genus again. But I soon discovered that Grand Bahama has no (or very few) rocks of terrestrial origin, hence no fossil land snails. The other genus wasn't much more common. We were stuck there for a week. So we studied the living *Cerions* and found a pattern behind the plethora of names. Since then, following Satchel Paige's advice, we have never looked back.)

About fifteen names had been proposed for the *Cerions* of Grand Bahama and neighboring Abaco Island. After a week, Woodruff and I recognized that only two distinct populations inhabited these islands, each restricted to a definite and different environment.

Abaco and Grand Bahama protrude above a shallow platform called Little Bahama Bank. When sea level was lower during the last ice age, the entire platform emerged and the islands were connected by land. Little Bahama Bank is separated by deep ocean from the larger Great Bahama Bank, source of the more familiar Bahamian islands (New Providence, with its capital city of Nassau, Bimini, Andros, Eleuthera, Cat, the Exuma chain, and many others). All these islands were also connected during glacial times of low sea level. As Woodruff and I moved from island to island on Great Bahama Bank, we found the same pattern of two different populations, always in the same distinctive environments. On Little Bahama Bank, a dozen invalid names had fallen into this pattern. On Great Bahama Bank, they collapsed, literally by the hundred. About one-third of all *Cerion* 'species' (close to 200 in all) turned out to be invalid names based on minor variants within this single pattern. We had reduced a chaos of improper names to a single, ecologically based order. (This reduction applies only to the islands of Little and Great Bahama Bank. Islands on other banks in the southeastern Bahamas, including Long Island, the southeasternmost island of Great Bahama Bank, contain truly different *Cerions*. These *Cerions* can also be reduced to coherent patterns based on few true species. But one essay can treat just so much, and I confine myself here to the northern Bahamas.)

Bahamian islands have two different kinds of coastlines. Major islands lie at the edge of their banks. The banks themselves are very shallow across their tops but plunge precipitously into deep ocean at their edges. Thus, bank-edge coasts abut the open ocean and tend to be raw and windy. Dunes build along windy coasts and solidify eventually into rock (often mistakenly called 'coral' by tourists). Bank-edge coasts are, therefore, usually rocky as well. By contrast, coastlines that border the interior parts of banks – I will call them bank-interior coasts – are surrounded by calm, shallow waters that extend for miles and do not promote the building of dunes. Bank-interior coasts, therefore, tend to be vegetated, low, and calm.

Woodruff and I found that bank-edge coasts in the northern Bahamas are invariably inhabited by thick-shelled, strongly ribbed, uniformly colored (white to darkish brown), relatively wide, and parallel-sided *Cerions*. To avoid writing most of the rest of this column in Latin, I will skip the formal names and refer to these forms as the 'ribby populations'. Bank-interior coasts are the home of thin-shelled, ribless or weakly ribbed, variegated (usually with alternating blotches of white and brown), narrow, and barrel-shaped *Cerions* – the 'mottled populations.' (Mottled *Cerions* also live away from coasts in the centers of islands, while ribby *Cerions* are confined exclusively to bank-edge coasts.)

This pattern is so consistent and invariable that we can 'map' hybrid zones even before we visit an island, simply by looking at a chart of bathymetry. Hybrid zones occur where bank-edge coasts meet bank-interior coasts.

This pattern might seem worthy of little more than an indulgent ho-hum. Perhaps mottled and ribby shells are not very different. Maybe the two environments elicit their differing forms directly from the same basic genetic stock, much as good and plentiful food can make a man fat and paltry fare eventually convert the same gent to a scarecrow. The very precision and predictability of the correlation between form and environment might suggest this biologically uninteresting solution. Two arguments, however, seem to stand conclusively against this interpretation and to indicate that mottled and ribby *Cerions* are different biological entities.

First, the ribby snails are not merely mottled forms with thicker and ribbier shells. As my technical contribution to our joint work, I measure each shell in twenty different ways. This effort permits me to characterize both growth and final adult form in mathematical terms. I have been able to show that the differences between ribby and mottled involve several independently varying determinants of form.

Second, an analysis of hybrid zones proves that they mark a mixture of two different entities, not a smooth blending of populations only superficially separate. My morphological analysis shows, in many cases, the anomalies of form, and the increased variation, that so often occur when two different developmental programs are mixed in offspring. Woodruff's genetic analysis also proves that the hybrids combine two substantially different systems, for he finds both generally increased genetic variability in hybrid samples, and genes detected in neither parental population.

We can demonstrate that ribby and mottled represent populations with substantial biological differences, but we cannot specify the cause of separation since we have been unable to distinguish between two hypotheses. First, ecological: ribby and mottled forms may be recent and immediate adaptations to their differing local environments. White or light-colored shells are inconspicuous against the bank-edge background of dune rocks, while thick and ribby shells protect their bearers on these windy and rocky coasts. Mottled shells are equally inconspicuous (indeed remarkably camouflaged) when dappled sunlight filters through the vegetation that houses *Cerion* on most bank-interior coasts, while thin and light shells are also well suited for hanging from thin twigs and grass blades. Second, historical: the pattern may be substantially older (although still probably adaptive for the reasons cited above). When sea level was much lower and the banks lay exposed during glacial periods, perhaps ribby populations inhabited all coasts (since all were then bank-edge), while mottled populations evolved for life in island interiors. As sea level rose, ribby and mottled snails simply kept their positions and preferences. The new bank-interior coasts were the interiors of previously larger islands and they continue as homes of mottled snails.

The distinction of mottled and ribby resolved nearly all the two hundred names previously given to *Cerions* from the northern Bahamas. But one problem (involving about ten more names) remained. A third kind of *Cerion*, bearing a thick, but smooth, pure white, and triangularly shaped shell, had been found on Eleuthera and Cat Island. Previous reports indicated nothing about their ecology or habits, but we found these thick white snails in two disjunct areas of southern Eleuthera and in southeastern Cat Island. They prefer island interiors and fit *Cerion*'s general pattern with gratifying predictability – that is, they hybridize with mottled populations as we approach bank-interior coasts and with ribby populations as we move toward bank-edge coasts. But what are they? Just as ecology and genetics resolved the basic pattern of mottled and ribby, we must call upon paleontology to explain our remaining source of diversity.

Fossil dunes of the Bahamas formed at times of high sea level during warmer periods between episodes of glaciation (ice ages). Three major sets of dunes built New Providence, the only Bahamian island with a documented geological pedigree. These include, from youngest to oldest, a few small dunes less than 10,000 years old and deposited since the last glaciers melted; an extensive set (forming the island's backbone), representing the high sea levels of 120,000 years ago, before the last glaciers formed; and a smaller set (situated near the island's center) built more than 200,000 years ago, before a previous glacial period. The oldest dunes contain a fossil *Cerion* now unknown in the Bahamas. The second and most extensive set includes two species of *Cerion*, a dwarf form now extinct and a large, smooth white species called *Cerion agassizi* (named for Alexander Agassiz, son of Louis, and a pioneer of scientific oceanography in the West Indies). The most recent set, as expected, contains either ribby or mottled *Cerions*, as in the modern fauna. We compared the large white snails of Eleuthera and Cat with *C. agassizi* and found no substantial differences. The small populations on these islands are surviving remnants of a species that once lived in abundance on all the islands of Great Bahama Bank.

The two hundred 'species' of northern Bahamian *Cerion* therefore reduce to three basic types with a sensible and ordered distribution. Geographic pattern identified ribby and mottled populations, but we needed an assist from history to understand the smooth white shells of Eleuthera and Cat. It is an awfully long stride from this taxonomic exercise in natural history to our ultimate goal – an understanding of how *Cerion*'s unparalleled diversity of form evolved – but we have taken the first step along the only pathway I know.

As an example of how this pattern illuminates the larger question, we have used our distinction of mottled and ribby to prove, for the first time, that the unconventional hypothesis advanced by most *Cerion* experts is indeed valid: the complex suite of characters defining such basic forms as mottled and ribby can evolve independently many times. We find the same distinction of mottled and ribby on both Little and Great Bahama banks. Conventional wisdom would hold that the mottled snails of both banks represent one stock, while the ribby snails of both banks form a different genealogical group. But Daniel Chung, a student of Woodruff's, and Simon Tillier, a leading anatomist of land snails at the Paris Museum, have studied the genital anatomy of these snails for us, and have made the following surprising discovery: *both* mottled and ribby snails of Little Bahama Bank share the *same* anatomy, while both mottled and ribby on Great Bahama Bank share a distinctly different set of genital structures. (Genital anatomy is the standard tool for establishing genealogical affinities among land snails. The differences are sufficiently profound and complex to indicate that shared anatomy reflects common descent while shared shell morphology must evolve independently.) Thus, the complex of traits defining mottled and ribby can evolve again and again. We would not have been able to reach this conclusion had we not extracted the pattern of mottled and ribby from a previous chaos of names.

At this point, I think we begin to peer through a glass darkly at the deeper mystery of form. We have shown that a complex set of independent traits can evolve in virtually the same way more than once. I do not see how this can happen if each trait must develop separately, following its own genetic pathway, each time. The traits must somehow be coordinated in *Cerion*'s genetic program; they must be evoked, or 'called forth,' together. Some genetic releaser must coordinate the joint appearance of these characters. Does the master genetic program of all *Cerions* encode alternate pathways representing the several basic forms that evolve again and again? The homeotic mutations of insects indicate that some such hierarchical system must regulate development, for the production of well-formed parts in the wrong places (legs where antennae should be, for example) indicates that some master switch must regulate all the genes that produce legs, and that higher controls can turn the master switch on in the wrong place or at the wrong time. Likewise, some master switch within *Cerion*'s program might evoke any one of its basic developmental pathways and evolve the set of traits that marks its fundamental forms again and again.

In this way, *Cerion* provides insight into what may be the most difficult and important problem in evolutionary theory: How can new and complex forms (not merely single features of obvious adaptive benefit) arise if each requires thousands of separate changes, and if intermediate stages make little sense as functioning organisms? If genetic and developmental programs are organized hierarchically, as homeotic mutations and the multiple evolution of basic forms in *Cerion* suggest, then new designs need not arise piecemeal (with all the intractable problems attached to such a view), but in a coordinated way by manipulating the master switches (or 'regulators') of developmental programs. Yet so deep is our present ignorance about the nature of development and embryology that we must look at final products (an adult *Drosophila* with a leg where its antenna should be or mottled *Cerions* evolved several times) to make uncertain inferences about underlying mechanisms.

I chose *Cerion* because I thought it might illuminate these large and woolly issues. Yet, although they lurk always in the back of my mind, they are not the source of my daily joy. Little predictions affirmed or small guesses proved wrong and exchanged for more interesting ideas are the food of continual satisfaction. *Cerion*, or any good field project, provides unending stimulation, so long as little puzzles remain as intensely absorbing, fascinating, and frustrating as big questions. Fieldwork is not like the one-hundred-thousandth essay on Shakespeare's sonnets; it always presents something truly new, not a gloss on previous commentaries.

I remember when we found the first population of living *Cerion agassizi* in central Eleuthera. Our hypothesis of *Cerion*'s general pattern required that two predictions be affirmed (or else we were in trouble):

this population must disappear by hybridization with mottled shells toward bank-interior coasts and with ribby snails toward the bank-edge. We hiked west toward the bank-interior and easily found hybrids right on the verge of the airport road. We then moved east toward the bank-edge along a disused road with vegetation rising to five feet in the center between the tire paths. We should have found our hybrids, but we did not. The *Cerion agassizi* simply stopped about two hundred yards north of our first ribby *Cerion*. Then we realized that a pond lay just to our east and that ribby forms, with their coastal preferences, might not favor the western side of the pond. We forded the pond and found a classic hybrid zone between *Cerion agassizi* and ribby *Cerions*. (Ribby *Cerion* had just managed to round the south end of the pond, but had not moved sufficiently north along the west side to establish contact with *C. agassizi* populations.) I wanted to shout for joy. Then I thought, 'But who can I tell; who cares?' And I answered myself, 'I don't have to tell anyone. We have just seen and understood something that no one has ever seen and understood before. What more does a man need?'

An eminent colleague, a fine theoretician who has paid his dues in the field, once said to me, only partly facetiously, that fieldwork is one hell of a way to get information. All that time, effort, and money, often for comparatively little when measured against the hours invested. True enough, especially when I count the hours spent drinking Cuban coffee, the one pleasure of my least favorite place, Miami airport. But all the frustration and dull, repetitive effort vanish to insignificance before the unalloyed joy of finding something new – and this pleasure can be savored nearly every day if one loves the little things as well. To say, 'We have discovered it; we understand it; we have made some sense and order of nature's confusion.' Can any reward be greater?

SIZE AND SHAPE

> Who could believe an ant in theory?
> A giraffe in blueprint?
> Ten thousand doctors of what's possible
> Could reason half the jungle out of being.

JOHN Ciardi's lines reflect a belief that the exuberant diversity of life will forever frustrate our arrogant claims to omniscience. Yet, however much we celebrate diversity and revel in the peculiarities of animals, we must also acknowledge a striking 'lawfulness' in the basic design of organisms. This regularity is most strongly evident in the correlation of size and shape.

Animals are physical objects. They are shaped to their advantage by natural selection. Consequently, they must assume forms best adapted to their size. The relative strength of many fundamental forces (gravity, for example) varies with size in a regular way, and animals respond by systematically altering their shapes.

The geometry of space itself is the major reason for correlations between size and shape. *Simply by growing larger*, any object will suffer continual decrease in relative surface area when its shape remains unchanged. This decrease occurs because volume increases as the cube of length (length x length x length), while surface increases only as the square (length x length): in other words, volume grows more rapidly than surface.

Why is this important to animals? Many functions that depend upon surfaces must serve the entire volume of the body. Digested food passes to the body through surfaces; oxygen is absorbed through surfaces in respiration; the strength of a leg bone depends upon the area of its cross section, but the legs must hold up a body increasing in weight by the cube of its length. Galileo first recognized this principle in his *Discorsi* of 1638, the masterpiece he wrote while under house arrest by the Inquisition. He argued that the bone of a large animal must thicken disproportionately to provide the same relative strength as the slender bone of a small creature.

One solution to decreasing surface has been particularly important in the progressive evolution of large and complex organisms: the development of internal organs. The lung is, essentially, a richly convoluted bag of surface area for the exchange of gases; the circulatory system distributes material to an internal space that cannot be reached by direct diffusion from the external surface of large organisms; the villi of our small intestine increase the surface area available for absorption of food (small mammals neither have nor need them).

Some simpler animals have never evolved internal organs; if they become large, they must alter their entire shape in ways so drastic that plasticity for further evolutionary change is sacrificed to extreme specialization. Thus, a tapeworm may be 20 feet long, but its thickness cannot exceed a fraction of an inch because food and oxygen must penetrate directly from the external surface to all parts of the body.

Other animals are constrained to remain small. Insects breathe through invaginations of their external surface. Oxygen must pass through these surfaces to reach the entire volume of the body. Since these invaginations must be more numerous and convoluted in larger bodies, they impose a limit upon insect size: at the size of even a small mammal, an insect would be 'all invagination' and have no room for internal parts.

We are prisoners of the perceptions of our size, and rarely recognize how different the world must appear to small animals. Since our relative surface area is so small at our large size, we are ruled by gravitational forces acting upon our weight. But gravity is negligible to very small animals with high surface to volume ratios; they live in a world dominated by surface forces and judge the pleasures and dangers of their surroundings in ways foreign to our experience.

An insect performs no miracle in walking up a wall or upon the surface of a pond; the small gravitational force pulling it down or under is easily counteracted by surface adhesion. Throw an insect off the roof and it floats gently down as frictional forces acting upon its surface overcome the weak influence of gravity.

The relative weakness of gravitational forces also permits a mode of growth that large animals could not maintain. Insects have an external skeleton and can only grow by discarding it and secreting a new one to accommodate the enlarged body. For a period between shedding and regrowth, the body must remain soft. A large mammal without any supporting structures would collapse to a formless mass under the influence of gravitational forces; a small insect can maintain its cohesion (related lobsters and crabs can grow much larger because they pass their 'soft' stage in the nearly weightless buoyancy of water). We have here another reason for the small size of insects.

The creators of horror and science-fiction movies seem to have no inkling of the relationship between size and shape. These 'expanders of the possible' cannot break free from the prejudices of their perceptions. The small people of *Dr Cyclops, The Bride of Frankenstein, The Incredible Shrinking Man*, and *Fantastic Voyage* behave just like their counterparts of normal dimensions. They fall off cliffs or down stairs with resounding thuds; they wield weapons and swim with Olympic agility. The large insects of films too numerous to name continue to walk up walls or fly even at dinosaurian dimensions. When the kindly entomologist of *Them* discovered that the giant queen ants had left for their nuptial flight, he quickly calculated this simple ratio: a normal ant is a fraction of an inch long and can fly hundreds of feet; these ants are many feet long and must be able to fly as much as 1,000 miles. Why, they could be as far away as Los Angeles! (Where, indeed, they were, lurking in the sewers.) But the ability to fly depends upon the surface area of wings, while the weight that must be borne aloft increases as the cube of length. We may be sure that even if the giant ants had somehow circumvented the problems of breathing and growth by molting, their sheer bulk would have grounded them permanently.

Other essential features of organisms change even more rapidly with increasing size than the ratio of surface to volume. Kinetic energy, in some situations, increases as length raised to the fifth power. If a child half your height falls down, its head will hit with not half, but only 1/32 the energy of yours in a similar fall. A child is protected more by its size than by a 'soft' head. In return, we are protected from the physical force of its tantrums, for the child can strike with, not half, but only 1/32 of the energy we can muster. I have long had a special sympathy for the poor dwarfs who suffer under the whip of cruel Alberich in Wagner's *Das Rheingold*. At their diminutive size, they haven't a chance of extracting, with mining picks, the precious minerals that Alberich demands, despite the industrious and incessant leitmotif of their futile attempt.*

This simple principle of differential scaling with increasing size may well be the most important determinant of organic shape. J. B. S. Haldane once wrote that 'comparative anatomy is largely the story of the struggle to increase surface in proportion to volume.' Yet its generality extends beyond life, for the geometry of space constrains ships, buildings, and machines, as well as animals.

Medieval churches present a good testing ground for the effects of size and shape, for they were built in an enormous range of sizes before the invention of steel girders, internal lighting, and air conditioning permitted modern architects to challenge the laws of size. The small, twelfth-century parish church of Little Tey, Essex, England, is a broad, simple rectangular building with a semicircular apse. Light reaches the interior through windows in the outer walls. If we were to build a cathedral simply by enlarging this design, then the area of outer walls and windows would increase as length squared, while the volume that light must reach would increase as length cubed. In other words, the area of the windows would increase far more slowly than the volume that requires illumination. Candles have limitations; the inside of such a cathedral would have been darker than the deed of Judas. Medieval churches, like tapeworms, lack internal systems and must alter their shape to produce more external surface as they are made larger. In addition, large churches had to be relatively narrow because ceilings were vaulted in stone and large widths could not be spanned without intermediate supports. The chapter house at Batalha, Portugal – one of the widest stone vaults in medieval architecture – collapsed twice during construction and was finally built by prisoners condemned to death.

Consider the large cathedral of Norwich, as it appeared in the twelfth century. In comparison with Little Tey, the rectangle of the nave has become much narrower; chapels have been added to the apse, and a transept runs perpendicular to the main axis. All these 'adaptations' increase the ratio of external wall and window to internal volume. It

* A friend has since pointed out that Alberich, a rather small man himself, would only wield the whip with a fraction of the force we could exert – so things might not have been quite so bad for his underlings.

is often stated that transepts were added to produce the form of a Latin cross. Theological motives may have dictated the position of such 'outpouchings,' but the laws of size required their presence. Very few small churches have transepts. Medieval architects had their rules of thumb, but they had, so far as we know, no explicit knowledge of the laws of size.

Large organisms, like large churches, have very few options open to them. Above a certain size, large terrestrial animals look basically alike – they have thick legs and relatively short, stout bodies. Large medieval churches are relatively long and have abundant outpouchings. The 'invention' of internal organs allowed animals to retain the highly successful shape of a simple exterior enclosing a large internal volume; the invention of internal lighting and structural steel has permitted modern architects to design large buildings of essentially cubic form. The limits are expanded, but the laws still operate. No large Gothic church is wider than long; no large animal has a sagging middle like a dachshund.

I once overheard a children's conversation in a New York playground. Two young girls were discussing the size of dogs. One asked: 'Can a dog be as large as an elephant?' Her friend responded: 'No, if it were as big as an elephant, it would look like an elephant.' How truly she spoke.

HOW THE ZEBRA GETS ITS STRIPES

SOME persistent, unanswered questions about nature possess a kind of majestic intractability. Does the universe have a beginning? How far does it extend? Others refuse to go away because they excite a pedestrian curiosity but seem calculated, in their very formulation, to arouse argument rather than inspire resolution. As a prototype for the second category, I nominate: Is a zebra a white animal with black stripes or a black animal with white stripes? I once learned that the zebra's white underbelly had decided the question in favor of black stripes on a blanched torso. But, to illustrate once again that 'facts' cannot be divorced from cultural contexts, I discovered recently that most African people regard zebras as black animals with white stripes.

In a poem about monkeys, Marianne Moore discussed some compatriots at the zoo and contrasted elephants and their 'strictly practical appendages' with zebras 'supreme in their abnormality.' Yet it has been established that the three species of zebras may not form a group of closest relatives – and that stripes either evolved more than once or represent an ancestral pattern in the progenitors of true horses and zebras. If stripes are not the markers of a few related oddballs but a basic pattern within a large group of animals, then the problems of their construction and meaning acquire more general interest. J. B. L. Bard, an embryologist from Edinburgh, has recently analyzed zebra stripes in the broad context of models for color in all mammals. He detected a developmental unity underlying the different patterns of adult striping among our three species of zebras and, *inter alia*, even proposed an answer to the great black-and-white issue in favor of the African viewpoint.

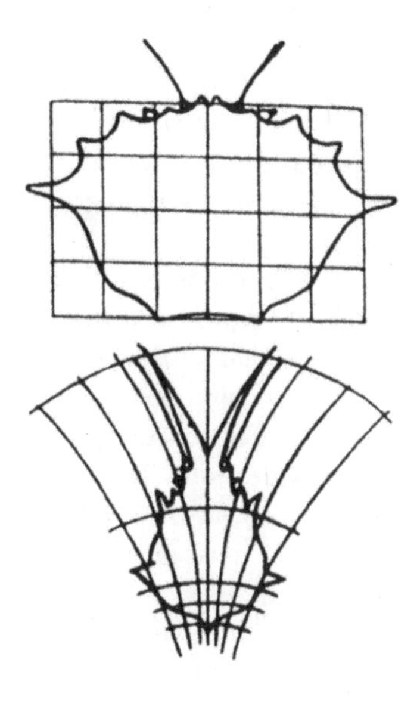

Figure 6. 'Transformed coordinates' on the carapaces of crabs from two different genera demonstrate unity of form. From *On Growth and Form* by D'Arcy Wentworth Thompson, 1917.

Biologists follow a number of intellectual styles. Some delight in diversity for its own sake and spend a lifetime describing intricate variations on common themes. Others strive to discover an underlying unity behind the differences that sort these few common themes into more than a million species. Among searchers for unity, the Scottish biologist and classical scholar D'Arcy Wentworth Thompson (1860–1948) occupies a special place. D'Arcy Thompson spent his life outside the mainstream, pursuing his own brand of Platonism and packing insights into his thousand-page classic, *On Growth and Form* – a book so broad in appeal that it won him an honorary degree at Oxford and, thirty years later, entered the *Whole Earth Catalog* as 'a paradigm classic.' D'Arcy Thompson struggled to reduce diverse expressions to common generating patterns. He believed that the basic patterns themselves had a kind of Platonic immutability as ideal designs, and that the shapes of organisms could only include a set of constrained variations upon the basic patterns. He developed a theory of 'transformed coordinates' to depict variations as expressions of a single pattern, stretched and distorted in various ways. But he worked before computing machines could express such transformations in numerical terms, and his theory achieved little impact because it never progressed much beyond the production of pretty pictures.

As a subtle thinker, D'Arcy Thompson understood that emphases on diversity and unity do not represent different theories of biology, but different aesthetic styles that profoundly influence the practice of science. No student of diversity denies that common generating patterns exist, and no searcher for unity fails to appreciate the uniqueness of particular expressions. But allegiance to one or the other style dictates, often subtly, how biologists view organisms and what they choose to study. We must reverse the maxim of the reprobate father teaching his son morality – do what I say, not what I do – and recognize that biological allegiances lie not so much in words but in actions and subjects chosen for research. Note what I do, not what I say. Of the 'pure taxonomist' – the describer of diversity – D'Arcy Thompson wrote:

> When comparing one organism with another, he describes the differences between them point by point and 'character' by 'character.' If he is from time to time constrained to admit the existence of 'correlation' between characters . . . he recognizes this fact of correlation somewhat vaguely, as a phenomenon due to causes which, except in rare instances, he can hardly hope to trace; and he falls readily into the habit of thinking and talking of evolution as though it had proceeded on the lines of his own description, point by point and character by character.

D'Arcy Thompson recognized, with sadness, that the theme of underlying unity had received much lip service, but little application. Differences between the striping patterns of our three zebra species had been described minutely and much energy had been invested in speculations about the adaptive significance of differences. But few had asked whether all the patterns might be reduced to a single system of generating forces. And few seemed to sense what significance such a proof of unity might possess for the science of organic form.

The vulgar version of Darwinism (not Darwin's) holds that natural selection is so powerful and pervasive in scrutinizing every variation and constructing optimal designs that organisms become collections of perfect parts, each minutely crafted for its special role. While not denying correlation in development or underlying unity in design, the vulgar Darwinian does relegate these concepts to unimportance because natural selection can always break a correlation or remold an inherited design.

The pure vulgar Darwinian may be a fiction; no one could be quite so foolish. But evolutionary biologists have often slipped into the practice of vulgar Darwinism (while denying its precepts) by following the reductionistic research strategy of analyzing organisms part by part and invoking natural selection as a preferred explanation for all forms and functions – the point of D'Arcy Thompson's profound statement cited above. Only in this way can I make sense of the curious fact that unity of design has received so little attention in the practice of research – although much lip service in textbooks – during the past forty years, while evolutionary biologists have generally preferred a rather strict construction of Darwinism in their explanations of nature.

For many reasons, ranging from the probable neutrality of much genetic variation to the non-adaptive nature of many evolutionary trends, this strict construction is breaking down, and themes of unity are receiving renewed attention. Old ideas are being rediscovered; D'Arcy Thompson, although never out of print, is now often out of bookstores (and in personal libraries). One old and promising theme emphasizes the correlated effects of changes in the timing of events in embryonic development. A small change in timing, perhaps the result of a minor genetic modification, may have profound effects on a suite of adult characters if the change occurs early in embryology and its effects accumulate thereafter.

The theory of human neoteny, often discussed in my essays, is an expression of this theme. It holds that a slowdown in maturation and rates of development has led to the expression in adult humans of many features generally found in embryos or juvenile stages of other primates. Not all these features need be viewed as direct adaptations built by natural selection. Many, like the 'embryonic' distribution of body hair on heads, armpits, and pubic regions, or the preservation of an embryonic membrane, the hymen, through puberty, may be non-adaptive consequences of a basic neoteny that is adaptive for other reasons – the value of slow maturation in a learning animal, for example.

Bard's proposal for 'a unity underlying the different zebra striping patterns' follows D'Arcy Thompson's theme of a basic motif stretched and pulled in different ways by varying forces of embryonic growth. These varying forces arise because the basic pattern develops at *different times* in the embryology of the three species. Bard thus combines the theme of transformed coordinates with the insight that substantial evolution can proceed by changes in the timing of development.

The basic pattern is simplicity itself: a series of parallel stripes deposited perpendicular to a line running along the embryonic zebra's back from head to tail – hang a sheet over a taut wire and paint vertical stripes on each side of it. These stripes are initially laid down at a constant size, no matter how big the embryo that forms them. They are 0.4 mm., or approximately twenty cell diameters, apart. The bigger the embryo, the greater the initial number of stripes. (I should point out that Bard's argument is a provocative model for testing, not a set of observations; no one has ever traced the embryology of zebra striping directly.)

The three zebra species differ in both number and configuration of stripes. In Bard's hypothesis, these complex variations arise only because the same basic pattern – the parallel stripes of constant spacing – develops during the fifth week of embryonic growth in one species, during the fourth week in another, and during the third week in the third species. Since the embryo undergoes complex changes in form during these weeks, the basic pattern is stretched and distorted in varying ways, leading to all the major differences in adult striping.

The three species differ most notably in patterns of striping on the rump and hind quarters. Grevy's zebra (*Equus grevyi*) has numerous fine and basically parallel stripes in these rear regions. On Bard's model, the stripes must have formed when the back part of the embryo was relatively large. (The larger the part, the more stripes it receives, since stripes are initially formed at constant size and spacing.) In the embryology of horses, the tail and hind regions expand markedly during the fifth week *in utero*. If adults possess numerous, fine posterior stripes, they must form after this embryonic expansion of the rear quarters. (Unfortunately, no one has ever studied the early embryology of zebras directly, and Bard assumes that the intrauterine growth patterns of true horses are followed by their striped relatives as well. Since basic features of early embryology tend to be highly conservative in evolution, true horses are probably fair models for zebras.)

The mountain zebra, *Equus zebra*, looks much like *E. grevyi* until we reach the haunch, where three broad stripes substitute for the numerous fine stripes of Grevy's zebra. Broad stripes on adults indicate initial formation on a small piece of embryo (where few stripes could fit), and later rapid growth of the piece (widening the stripes as the general area expands). If an embryo forms stripes in its fourth week, just before the posterior expansion that provides room for the many fine marks of Grevy's zebra, it will build the pattern of a mountain zebra during later embryonic growth.

Burchell's zebra, *Equus burchelli*, also has just a few broad stripes on its haunch. But, while the mountain zebra has fine stripes over most of its back and broad stripes only over the haunch, the broad stripes of Burchell's zebra begin in the middle of the belly and sweep back over the haunch. This pattern suggests an initial formation of stripes during the third week of embryonic growth. At this early stage, the embryo has a short, compact back, which later expands toward the rear in a broad, arching curve while the belly remains short. A stripe that initially ran vertically from belly to spine would be pulled toward the rear as the embryo's top surface expanded backward while its belly grew little. An adult stripe, subject to such deformation in its embryonic life, would be broad and would run from the belly up and over the haunch – as in Burchell's zebra.

Thus, Bard can render differences in rear striping of all three species as the results of deforming the same initial pattern at different times during normal embryonic growth. His hypothesis receives striking support from another source: the total number of stripes itself. Remember that Bard assumes a common size and spacing for stripes at their initial formation. Thus, the larger the embryo when stripes first form, the greater the number of stripes. Grevy's zebra, presumably forming its stripes as an embryo of five weeks and about 32 mm. in length, has eighty or so stripes as an adult – or about 0.4 mm. per stripe. Mountain zebras with a fourth-week embryo of some 14 to 19 mm. have about forty-three stripes – again about 0.4 mm. per stripe. Burchell's zebra has twenty-five to thirty stripes; if they form in a third-week embryo some 11 mm. long, we get the same value – about 0.4 mm. per stripe.

As additional support, and a lovely example of the difference between superficial appearance and knowledge of underlying causes, consider an old paradox involving hybrid offspring between zebras and true horses. These animals almost always have more stripes than their zebra parent. 'Common sense,' based on superficial appearance, declares this result puzzling. After all, the state between stripes and no stripes is few stripes. But if Bard is right about the underlying causes of striping, then this paradoxical result makes sense. The intermediate state between stripes and no stripes might well be a *delay* in the embryonic formation of stripes. If stripes then begin at their common size and spacing upon a larger embryo, the resulting adult will have *more* stripes.

If a unity of basic architecture underlies the diversity of zebra striping, then we must suspect that we are confronting a general pattern in nature, not just the 'supreme abnormality' that Marianne Moore described. Darwin viewed horses in this light and recognized that the capacity for striping in all horses constituted a powerful argument for evolution itself. If zebras are odd and perfect adaptations for camouflage, God might have made them as we find them. But if zebras merely actuate and exaggerate a potential property of all horses, then the occasional realization of striping in other horses – where it cannot be viewed as a perfected adaptation ordained by God – must indicate a community of evolutionary descent.

Darwin devoted much space in chapter 5 of the *Origin of Species* to an exhaustive tabulation of occasional striping in other horses. Asses, he found, often have 'very distinct transverse bars . . . like those on the legs of a zebra.' True horses often possess a spinal stripe, and some also have transverse leg bars. Darwin found a Welsh pony with three parallel stripes on each shoulder. And he noted that hybrids (with no zebra parents) were often strongly striped – an example of the common, and still mysterious, observation that hybrids often display ancestral reminiscences present in neither parent. 'I once saw a mule,' Darwin wrote, 'with its legs so much striped that any one at first would have thought that it must have been the product of a zebra.'

From this illustration of common, and often non-adaptive, patterns in all horses, Darwin drew one of his most powerful and passionate arguments for evolution – well worth quoting *in extenso*:

> He who believes that each equine species was independently created, will, I presume, assert that each species has been created with a tendency to vary, both under nature and under domestication, in this particular manner, so as often to become striped like other species of the genus; and that each has been created with a strong tendency, when crossed with species inhabiting distant quarters of the world, to produce hybrids resembling in their stripes, not their own parents, but other species of the genus. To admit this view is, as it seems to me, to reject a real for an unreal, or at least for an unknown, cause. It makes the works of God a mere mockery and deception; I would almost as soon believe with the old and ignorant cosmogonists, that fossil shells had never lived, but had been created in stone so as to mock the shells now living on the seashore.

The same theme also suggests an answer to the title of my previous essay 'What, If Anything, Is a Zebra?' I advanced the argument that zebras may not form a group of closest relatives but a set of different horses that had either evolved stripes independently or inherited them from a common ancestor (while asses and true horses lost them). Bard's hypothesis lends support to this conjecture because it suggests that the underlying pattern of zebra striping may be so simple that all horses include it in their repertoire of development. Zebras, then, may be the realization of a potential possessed by all horses.

Finally, moving from the sublime to the merely interesting, Bard proposes a solution to the primal dilemma and argues that zebras are black animals with white stripes after all. The white underbelly, he points out, is a lousy argument because many fully colored mammals are white underneath. Color may be generally inhibited in this region for reasons at present unknown. Mammals do not have their colors painted on a white background. The basic issue may then be rephrased: Does striping result from an inhibition or a deposition of melanin? If the first, zebras are black animals; if the second, they are white with black stripes.

Biologists often look to teratologies, or abnormalities of development, to solve such issues. Bard has uncovered an abnormal zebra whose 'stripes' are rows of dots and discontinuous blotches, rather than coherent lines of color. The dots and blotches are white on a black background. Bard writes: 'It is only possible to understand this pattern if the white stripes had failed to form properly and that therefore the 'default' color is black. The role of the striping mechanism is thus to inhibit natural pigment formation rather than to stimulate it.' The zebra, in other words, is a black animal with white stripes.

SIZE AND SCALING IN HUMAN EVOLUTION

(*with David Pilbeam*)

HUMAN paleontology shares a peculiar trait with such disparate subjects as theology and extraterrestrial biology: it contains more practitioners than objects for study. This abundance of specialists has assured the careful scrutiny of every bump on every bone. In this context, it is remarkable that the most general character of all – body size (the difference in absolute size among fossil hominids, and the clear phyletic trend toward larger bodies) – has been rather widely neglected.

Increase in body size has played an especially important role in evolution for two reasons.

1. It is so common. Several evolutionary phenomena are encountered so frequently that their canonization as 'law' has been widely accepted. 'Cope's law' of phyletic size increase is the best known and most widely touted of these statistical generalizations.[1]

2. It has such important and ineluctable consequences. Galileo[2] recognized that a large organism must change its shape in order to function in the same way as a smaller prototype. The primary law of size and shape involves unequal scaling of surfaces and volumes,[3] but other differential increases have their potent effect as well.[4] As a terrestrial vertebrate evolves to larger size, its limb bones become relatively thicker, the ratio of brain weight to body weight decreases, and digestive and respiratory surfaces become more complex.

We cannot begin to assess the nature of adaptation in lineages obeying Cope's rule until we establish 'criteria of subtraction' for recognizing the changes in shape that larger size requires. A simple description of changing shape will not suffice, for some changes merely compensate for increased size and reproduce the 'same' animal at a larger scale, while others represent special adaptations for particular conditions. Yet such a separation is rarely attempted.

We make such an attempt in this article and use it to argue a simple thesis about human evolution. We try to demonstrate that the three generally accepted species of australopithecines[5] represent the 'same' animal expressed over a wide range of size. In other words, size increase may be the only independent adaptation of these animals, changes in shape simply preserving the function of the smaller prototype at larger sizes. In evolving toward modern man, on the other hand, hominids also increased steadily in size, but they developed adaptations of brain and dentition that cannot be attributed to the mechanical requirements of larger bodies. In other words, the extinct branch of australopithecines did little more than increase in size during its evolution; the thriving branch of hominids increased in size and developed special adaptations as well.

Hominid Phylogeny

The hominids we discuss, known informally as Plio-Pleistocene hominids, come from African deposits ranging in age from a little less than 6 million to perhaps less than 1 million years.[6] The first early hominid from South Africa, an infant specimen of *Australopithecus africanus*, was discovered in 1914 at Taung. Many more specimens were recovered in South Africa during the decades that followed, and collecting continues today. Large to moderate samples are known from Swartkrans, Sterkfontein and Makapansgat; a handful of hominids come from Kromdraai; Taung is still represented by the original specimen alone. Those from Sterkfontein, Makapansgat, and Taung are frequently described as gracile, because they were apparently lighter in average body weight than the 'robust' forms from Kromdraai and Swartkrans.[7]

Many workers have emphasized the similarities between South African forms by classifying them in one genus, generally as two species: *Australopithecus africanus* (Taung, Sterkfontein, Makapansgat) and *A. robustus* (Kromdraai, Swartkrans).[8] Apart from probable differences in body size, there are other contrasts between these two forms, including differences in tooth size and shape, in cranial proportions, and (possibly) in postcranial anatomy. Brace[9] has argued for some time that the larger *A. robustus* was in many respects an allometric variant of the smaller form (although he has not quantified his argument).

Robinson,[10] however, has long advocated a very different explanation of variability in the South African hominids. He placed the robust material from Kromdraai and Swartkrans in a genus, *Paranthropus*, distinct from *Australopithecus*, in order to emphasize what he saw as important morphological differences between robust and gracile specimens. *Paranthropus* had larger cheek teeth and smaller anterior teeth than *Australopithecus*, which was therefore more like later *Homo* in dental size and proportions. Robinson[11] cited these contrasts as evidence for a major dietary difference between the two forms, with *Australopithecus* being an omnivore and carnivore and *Paranthropus* a herbivore. Cranial morphology also differed, the graciles having higher, more rounded calvariae, with more vaulted frontals. In these features, again, Robinson saw *Australopithecus* as significantly more like *Homo* than *Paranthropus*. Postcranially, according to Robinson and others, the two forms differed, the graciles being more like *Homo*. Recently, Robinson[12] has reclassified *A. africanus* as *Homo africanus* to emphasize his view that this gracile form is ancestral to later *Homo*.

Early hominids were little known in East Africa until 1959, when Leakey[13] described *Zinjanthropus boisei*, a very large (cheek) toothed form, from Olduvai in Tanzania. Subsequently, more dental, cranial, and postcranial remains of this 'hyper-robust,' large-bodied animal have been recovered elsewhere in East Africa (Olduvai and Lake Natron in Tanzania, East Rudolf in Kenya, Omo in Ethiopia) from well-dated sites.[14] Most workers now prefer to regard *A. boisei* an an *Australopithecus*;[15] however, Robinson[16] allies it to his *Paranthropus*, calling it *P. boisei*. These disagreements have centered around interpretations of dental and cranial anatomy, Robinson seeing *A. boisei*, with its relatively enormous cheek teeth, small brain, and low vaulted skull with massive brow-ridges, as a form far removed from the smaller and more manlike *A. africanus*.

In 1964, Leakey et al.[17] described another hominid from Olduvai Gorge, *Homo habilis*. The species is now reasonably well represented by cranial and dental remains from Olduvai, East Rudolf, Omo, and a few other sites.[18] *Homo habilis* differs from the *Australopithecus* species in having smaller cheek teeth and larger anterior teeth, an enlarged brain, and a postcranial skeleton (in the parts preserved) more like that of

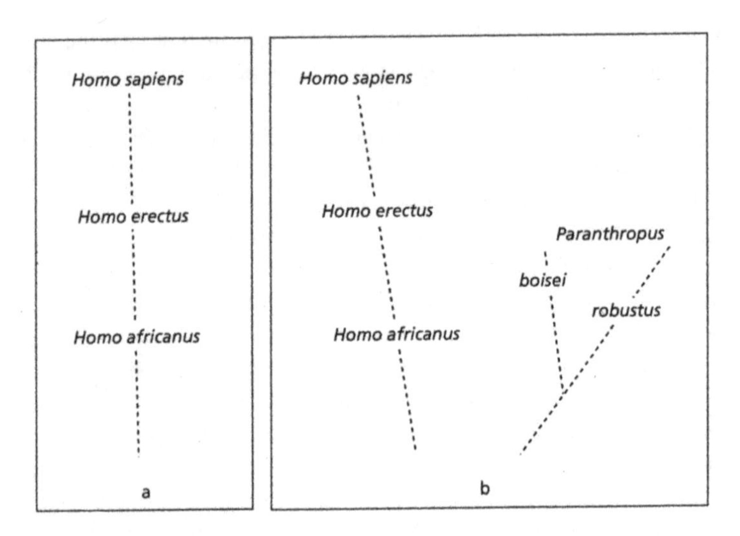

Figure 7. Three hypothetical schemes for Pliocene and Pleistocene human evolution. (a) The single-species hypothesis;[31] *H. africanus* includes all Plio-Pleistocene hominids. (b) Views of Robinson[32] and many others;[33] *H. africanus* includes gracile South and East African forms only. (c) Our current best estimate; a suitable hypothetical common ancestor would be *A. africanus*.

Homo in a number of features.[19] The two East African hominids, *H. habilis* and *A. boisei*, are less similar to each other than are the South African species *A. africanus* and *A. robustus*. Most material from East Africa can be assigned to *A. boisei* or *H. habilis*; a few specimens may, however, be closer to the smaller *Australopithecus* species.

These East African hominids are, in the main, well dated radiometrically, especially at Olduvai, East Rudolf, and Omo, where they span the period between 3 million and about 1 million years.[20] A couple of older hominid sites, Kanapoi and Lothagam in Kenya, go back to about 4 million and 5.5 million years, respectively.[21]

Unfortunately, there are no radiometric age determinations for South African sites. Some relative age estimates have been made, based mainly on faunal comparisons.[22] There is general agreement that the site of Swartkrans is younger than Sterkfontein and Makapansgat, possibly much younger, although there is disagreement about whether it is older

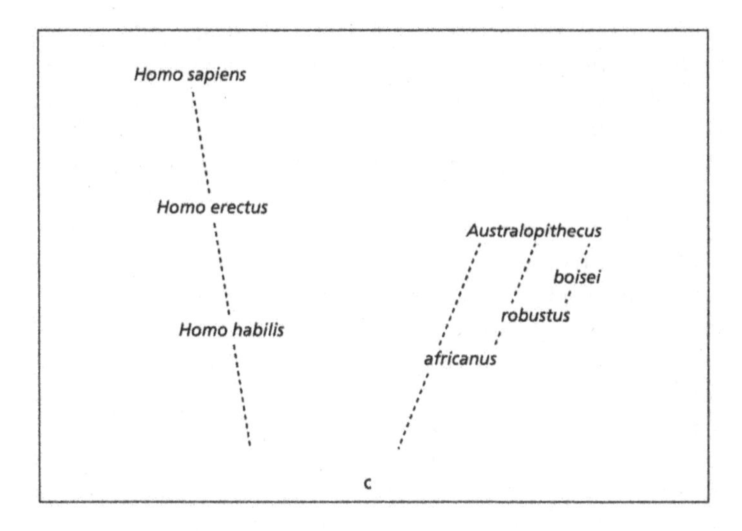

or younger than 2 million years; the Kromdraai hominids may well be undatable. The type specimen of *A. africanus*, the infant from Taung, has long been considered as equivalent in age to Sterkfontein, although Butzer[23] believes it to be younger.

Most workers regard *A. robustus* and *A. boisei* as closely related subspecies or species, or ancestor and descendant species.[24] Postcranially, they seem to be similar, the main difference being the greater mean body size of *A. boisei*, with its consequence of increased dental and cranial robustness.[25]

The relationship between *A. africanus* and *H. habilis* has been vigorously debated. Many see them as closely similar, even conspecific.[26] Others would separate them at the species level, on the basis of differences in tooth size and proportions, endocranial volume, and postcranial anatomy.[27] Whether they are different genera (that is, whether *habilis* is a *Homo* or an *Australopithecus*) is largely a matter of taste. In tooth proportions and brain volume, *H. habilis* is intermediate between *A. africanus* and *Homo erectus*.[28] The postcranial evidence is sparse, and equivocal, but what there is suggests that *H. habilis* is closer to *H. erectus*.

There certainly are resemblances between *A. africanus* and *Homo*. The question is, to what extent are these due to the small body size of *A. africanus*? For our purposes, we will follow most workers in placing

A. africanus, A. robustus, and *A. boisei* in one genus, and retaining *H. habilis* in *Homo*, a scheme that we believe is further justified by the work reported here. [Brace,[29] Wolpoff,[30] and others have included all these forms in a single lineage.]

Various views, including our own, on the most probable phylogeny for hominids are outlined in Figure 7. Precise temporal relationships are uncertain for the South African sites, but our general thesis is that Plio-Pleistocene hominids can be clustered into two major groups. One consists of *A. africanus, A. robustus*, and *A. boisei*, animals that are, in a number of characters, allometric variants of each other. Since they are scaled versions of the 'same' animal, precise temporal sequence becomes less important in evaluating evolutionary relationships. The other group consists of *H. habilis, H. erectus*, and *H. sapiens*, species that form a reasonable ancestor-descendant sequence showing increase in body weight through time. It seems probable that the ancestor of *H. habilis* resembled *A. africanus*, and so this form can be included in our second group as well, as a (perhaps hypothetical) ancestor. However, we believe that *A. africanus* exhibits 'advanced' dental and cranial features primarily because it is small. Thus, resemblances between it and other *Australopithecus* species have often been obscured because the size-related differences were not seen as such. We believe that *H. habilis* was probably the first animal to exhibit a shift away from the basic australopithecine adaptive pattern.

These conclusions have important consequences. *Australopithecus africanus* has been popularly advertised as a carnivorous, hunting form, the 'killer ape'.[34] On the contrary, we believe there are no good reasons for assuming that *A. africanus* was any less a vegetarian than *A. robustus* or *A. boisei*. However, *H. habilis* does show a shift in the direction of later hominids; yet, the evidence for large-scale bloodletting on its part is meager, to say the least.

Table 2 lists our estimates for body weight, cranial capacity, and tooth size in pongids and hominids.[35] Tooth areas are calculated by summing the products of lengths and breadths of individual teeth. Although this is not a very satisfactory measure of masticatory function, we prefer it to length, breadth, or module, and it is easy to calculate. Mandibular dentitions are used for hominids to increase sample size; maxillary areas are calculated for pongids (because their third lower premolar is sectorial) and considered equivalent to mandibular areas for comparison with the hominid data.

Animal	Body weight (g)			Cranial capacity (cm³)			Tooth area (mm²)		Area ratio x 100
	Male	Female	Combined	Male	Female	Combined	Posterior	Anterior	
				Hominids					
A. africanus			32,000			450	860	170	19.8
A. robustus			40,500			500	970	150	15.5
A. boisei			47,500			510	1140	140	12.3
H. habilis			43,000			725	750	185	24.7
H. erectus (Choukoutien)			53,000			1050	630	170	27.0
H. sapiens (Australian aborigines)	60,000	54,000	57,000			1230	560	145	25.9
				Great apes					
Pygmy chimp	38,500	32,000	35,250	355	330	343	415		
Chimp	49,000	41,000	45,000	410	380	395	480	145	30.2
Gorilla (lowland)	140,000	70,000	105,000	550	460	505	1030	155	15.0
Orangutan	69,000	37,000	53,000	415	370	393	730	175	24.0

Table 2. Estimates of mean body weights, cranial capacities, and tooth areas for great apes and hominids. Body weight estimates for fossils are considered accurate within about 20 percent. For hominids, tooth areas are for third lower premolar to third lower molar (posterior) and first lower incisor to lower canine (anterior). For great apes they are for third upper premolar to third upper molar (posterior) and first and second lower incisors (anterior). The area ratio is calculated as anterior/posterior.

Some of the estimates are quite accurate, while others are little more than educated guesses. Most have been rounded to avoid spurious appearances of high accuracy. Those for apes come from a variety of sources.[36] *Homo sapiens* is represented by a group of hunter-gatherers, the Australian aborigines. Body weights are from Martin and Saller,[37] cranial capacity estimates from Ashton and Spence,[38] and tooth areas from Campbell.[39] The *H. erectus* weight estimate is very approximate and is based on work by Weidenreich,[40] as are estimates of endocranial volume and dental size. We have selected this sample because it is the most complete for *H. erectus*. Of particular interest is the fact that body weight estimates are lower than samples from most *H. sapiens* groups.

For earlier hominids, there are many more problems in estimating population parameters.[41] Body weights for *A. africanus* apparently ranged from some 22 kilograms to no more than about 40 kilograms.[42] Holloway[43] has estimated the *A. africanus* endocranial volume at about 440 cubic centimeters; this may be an underestimate if a disproportionate number of the better specimens are females. Accordingly, we suggest a mean volume of 450 cm^3. Estimates of tooth area are approximate, but we believe not too inaccurate.[44] [Since complete dentitions are hard to come by, we have taken the mean area of each tooth and summed the averages. Where there is reason to suspect that a disproportionate number of large (male?) or small (female?) specimens have been preserved, we adjusted our estimates accordingly. We doubt that the values have more than about 5 percent error, and we are satisfied with their relative magnitudes.]

Estimates of body weight for *A. boisei* range from more than 70 kg to a little more than 20 kg, based mainly on published postcranial material.[45] An average of 45 to 50 kg seems acceptable. Brain volume estimates are from Holloway.[46] *Australopithecus robustus* body weights almost certainly fall between those of *A. africanus*, and *A. boisei* and probably overlapped both;[47] our best estimate is around 40 kg. The only complete *A. robustus* brain cast (from Swartkrans) yields volume of 530 cm^3.[48] However crushed and fragmentary crania from Swartkrans are clearly smaller than the larger *A. boisei* crania,[49] we think that our estimate of 500 cm^3 is reasonable; it may even be a little high. Tooth areas are based mainly on material from Swartkrans.[50]

Body weight estimates for *H. habilis* are difficult to make. It seems unlikely, on the basis of presumed *H. habilis* femora, that this species was, on average, lighter than *A. robustus* or heavier than *A. boisei.*[51] An estimate of between 40 and 45 kg seems reasonable. Brain volume estimates for *H. habilis* specimens have ranged between 600 and 800 cm^3[52] although the smaller values come from specimens that are fragmentary or that may not represent *H. habilis*. The complete braincase of hominid 1470 from East Rudolf has a volume of 775 cm^3. Hominid 7 (the type specimen) and hominid 16, both from Olduvai, probably had volumes of 700 cm^3 or more.[53] We have used a mean of 725 cm^3, which may be conservative, realizing that this estimate is subject to error. Tooth area estimates are fairly reliable.[54]

Scaling of Cranial Capacity

Cuvier[55] recognized that large animals have relatively small brains – that brain weight, in other words, increases more slowly than body weight as we progress from small to large species in a coherent taxonomic group. The treatment of relative brain weight with the equation

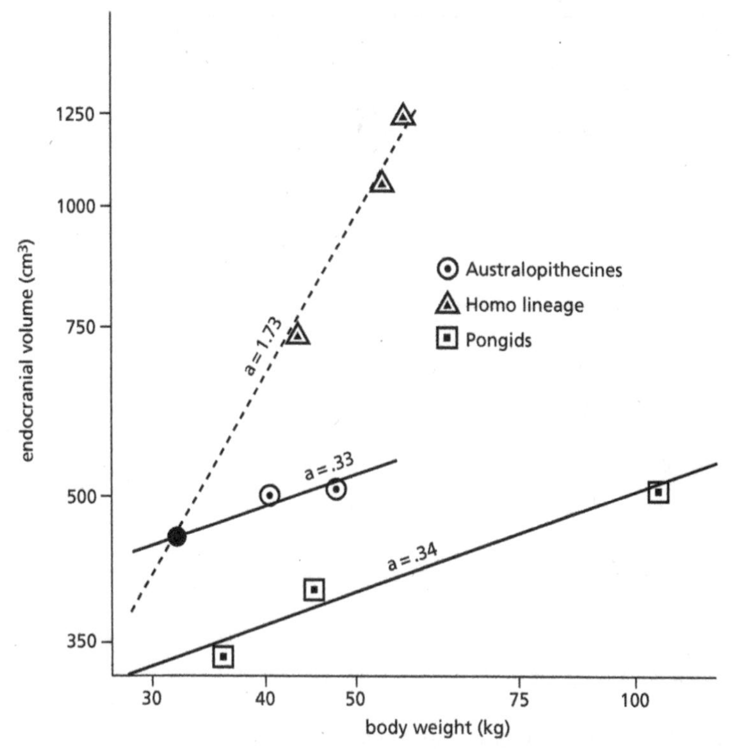

Figure 8. Scaling of endocranial volume in pongids and hominids. We believe that estimates for hominid body weights are accurate within about 20 per cent. The criterion for intraspecific functional equivalence is a = 0.2 to 0.4.

$$\text{brain weight} = b(\text{body weight})^a$$

(where a is the slope and b the y-intercept of a log-log plot) dates to work of Snell[56] and Dubois[57] in the 1890s and anticipates by more than thirty years the generalization of power functions in the allometric method of Huxley.[58] Jerison[59] has amassed all previous data and gathered much new information into an impressive synthesis of allometric studies on relative brain size. He proves that the slope of the power function for mammals – and for all vertebrate classes for that matter – is very close to 0.66 in static, mouse-to-elephant plots, where each point represents an average adult of a single species. (A slope of 2/3 implies that brain

Sequence	Correlation coefficient	Least squares slope	Major axis slope	Intercept of major axis
	Great apes			
Pygmy chimp-chimp-gorilla				
Combined	.988	0.338	0.339	10.16
Males only	.988	0.321	0.322	12.30
Females only	.993	0.412	0.415	4.55
Pigmy chimp-chimp-orangutan-gorilla Combined	.984	0.339	0.340	9.94
	Australopithecines			
A. africanus-A. robustus-A. boisei	.965	0.327	0.329	14.93
	Lineage of Homo sapiens			
A. africanus-H. habilis-H. erectus-H. sapiens	.999	1.73	1.73	7.3×10^{-6}

Table 3. Scaling of cranial capacity of great apes and hominids.

weight does not keep pace with body weight but increases only as fast as nonallometric body surfaces.)

The 0.66 slope for static, interspecific plots of major taxonomic groups does not exhaust the variety of brain-body relationships. Less than ten years after Snell and Dubois determined the interspecific values, Lapicque[60] (for birds and mammals) and Duboi[61] himself (for man) found that intraspecific plots (for adults within a species, races of a species, or very closely related species displaying the 'same' body plan over a wide range of size) produced slopes ranging between 0.2 and 0.4. (Lapicque, in fact, referred to this observation as a law, citing a 'universal' value of 5/18 or 0.28). In Lapicque's argument, 5/18 represents a brain that enlarges old neurons without adding new ones; 2/3 would then mark a brain that adds just enough neurons to serve the increased body bulk. Any scaling in excess of 2/3 must indicate an increase in cephalization. Confirmation of the range 0.2 to 0.4 has been abundant; these values for very closely related adults are as firmly established as the 2/3 slope for larger taxonomic groupings. Scholl,[62] for example, calculated 0.18 for several species of macaques; Bertalanffy and Pirozynski[63] cited 0.20 for adult rats; and we have computed 0.23 for Lapicque's data on races of domestic dogs[64] and 0.33 for Pilleri and Busnel's[65] data on adults of the cetacean *Delphinus delphis*. (Further citations can be found in Rohrs,[66] Frick,[67] and Bahrens[68].)

These generalities of scaling lead to a set of predictions: a sequence of closely related animals differing in size but not in function or 'grade' of evolution should yield a brain-body plot with a slope between 0.2 and 0.4. The highest slope that can be justified in making a claim of functional equivalence is 0.66. If a mammal evolves to larger sizes with cranial capacity scaling at greater than 0.66, we must affirm an increase in cephalization, for descendants will have larger brains than mammals of the same size living at the time of their ancestors. Ironically, a phyletic slope of 1 (signifying no change in shape) reflects a pronounced increase in cephalization, since growth of the brain along a slope of 0.66 marks the maximum rate of increase for functional equivalence.

Figure 8 and Table 3 present the data for cranial capacity and body size of pongids and hominids listed in Table 2. We have used the standard power function and report the parameters of both least squares (y on x) and major axis fits. Least squares is the oldest and most popular technique, but it is inappropriate in cases (such as these) where both variables are subject to error and extrapolative prediction is not the purpose of plotting. Jolicoeur[69] has argued persuasively that the major axis should be preferred among lines that consider errors in both variates. (Readers who prefer the reduced major axis may calculate it by dividing the least squares slope by the correlation coefficient.) When correlation coefficients get much below .95, different techniques yield markedly different slopes; the choice of a proper method is no trivial affair. We shall use the major axis fits as our primary reference. Zar[70] has argued that statistical fits should be made directly, but Jolicoeur[71] and Sacher[72] have defended the conventional logarithmic transformation used here.

Giles[73] and others have advanced the idea that gorillas and chimpanzees are different expressions of the 'same' body design for different sizes – or, to put it crudely, that gorillas are allometrically enlarged chimps. The addition of *Pan paniscus*, the pygmy chimpanzee, increases the size range for a claim of similar brain design. The slope of the power function is 0.339, comfortably within the range 0.2 to 0.4 for closely related forms. Addition of the orangutan changes the slope insignificantly ($a = 0.340$).

When australopithecines are plotted, we obtain a line of the same slope ($a = 0.329$), lying above the pongid regression. The australopithecines occupy a higher level of cephalization than pongids, for the ratio (australopithecine brain/pongid brain) is nearly constant for any common body weight. But, as with pongids, australopithecines scale among themselves with the predicted value for a series of animals differing only in size, but not in design. Large *A. boisei* is neither more nor less cephalized than its small prototype; it is, in fact, the very creature that we would predict if asked to build *A. africanus* at 1.5 times its average body weight. No one appreciates more than we the biometrical hazards of basing claims on three points; in a field as important yet as bereft of data as this, one must work with what one has. We do draw some comfort from the fact that very different estimates for brains and bodies of hominids yield the same form of scaling. Using Tobias's estimates,[74] the brain-body slope for australopithecines is 0.26 ($r = .984$) [while that for the sequence *A. africanus* to *H. sapiens* is 1.17 ($r = .994$)].

The story of our own lineage is quite different. The sequence *A. africanus–H. habilis–H. erectus–H. sapiens* yields a slope of 1.73. This value is far in excess of any that could be justified in asserting a claim of similar cephalization with increasing size. As we argued previously, even a slope of 1 (with the preservation of a constant brain-body ratio) would mark a pronounced improvement over the strong negative allometry of functional equivalence ($a = 0.66$). In fact, the brain increased with marked positive allometry during the evolution of our species.

Holloway[75] has noted, in a qualitative way, the tight correlation of cranial capacity and body weight during the evolution of *H. sapiens* (r, from our data, is .999). This strong correlation, he suggests, might mean that an essentially human level of brain function had been attained by *A. africanus*, and that further increase in brain size could be ascribed to increasing body size. We point out, however, that the strength of a correlation is a very different issue from the form of a regression. It is the parameters of regression that must determine functional claims in allometric studies. A slope greater than 1, no matter how tight the correlation, indicates a remarkable rise in cephalization with increasing size. The tightness of correlation need only reflect the strength of selection for mental traits grounded on increasing brain size.

These results show that all australopithecines had brains equally expanded beyond the ape grade. As Holloway[76] demonstrated, robust australopithecines had brains that were as much (or more) like those of later hominids in external morphology as those of gracile forms. According to Holloway, the brains of *A. africanus* and *A. robustus* differ from each other in external morphology no more than those of chimp and gorilla do. Thus, Robinson[77] is incorrect in stating that the gracile brain was 'significantly larger' than that of robusts, and that the robust forms had not 'embarked upon the hominid brain expansion.'

In the lineage leading to *H. sapiens*, brain volume does increase dramatically, but *A. africanus* was not the ancestor that first showed brain expansion beyond the australopithecine level; that honor must go to *H. habilis*, one reason for placing this species in *Homo*.

Cranial Allometry

Cranial anatomy differs among the australopithecines. *Australopithecus africanus* has a relatively small face and large, rounded braincase, while the robust forms have larger faces with lower calvaria, flatter frontals, bigger browridges, more robust zygomatic arches, better developed ectocranial structures (crests, ridges), and so forth.[78] Robinson[79] has argued that the more spherical braincase and smaller face of *A. africanus* is an 'advanced' feature, indicating an elusive relationship with later hominids; again, he sees the larger australopithecines as aberrant.

Many others[80] have suggested that these cranial differences are allometric. From our foregoing discussion of brain and tooth scaling, it seems highly likely that this is so. Again, skulls of African pongids – ranging from the delicately built, rounded cranium of the pygmy chimpanzee to that of the gorilla, large-faced and massively constructed[81] – provide us with an analog. An exact quantitative solution is hard to derive since sufficient complete and undistorted material has not been found for African hominids. Robinson[82] has argued that allometry cannot be invoked to explain the difference in cranial morphology between, for example, the Sterkfontein and Swartkrans samples because a single robust cranium from Swartkrans (SK 48) is supposedly smaller than a gracile one from Sterkfontein (Sts 5). Apart from the fact that comparisons should be between samples, and that it is clear from the data [SK 46[83] and SK 1585[84]] that there are larger crania at Swartkrans than at Sterkfontein, it should be emphasized that SK 48 is a damaged specimen.

The nearly universal trend both of primate ontogeny and of static series of closely related adults is toward negative allometry of the brain and positive allometry of the face.[85] Most features enumerated above are simply the consequences of one of these primary allometries – lower calvaria and flatter frontals reflect the relatively smaller brain of large primates, while a sagittal crest supports the massive musculature that a relatively large face and jaw requires. Both the australopithecine and chimpanzee-gorilla sequence display this set of allometric consequences;

larger forms are scaled-up replicas of their smaller prototypes. *Homo sapiens* provides the outstanding exception to this trend among primates, for we have evolved a relatively large brain and small face, in opposition to functional expectations at our size. We retain, as large adults, the cranial proportions that characterize juvenile or even fetal stages of other large primates; partial neoteny has probably played a major role in human evolution. *Australopithecus africanus* has a rounded braincase because it is a relatively small animal; *H. sapiens* displays this feature because we have evolved a large brain and circumvented the expectations of negative allometry. The resemblance is fortuitous; it offers no evidence of genetic similarity.

We have not analyzed postcranial remains, although this is a potentially fruitful area. Although Napier[86] and Robinson[87] see major differences between South African graciles and robusts in postcranial anatomy, other workers[88] have pointed out the similarities between the samples. Robinson has emphasized strong similarities between *A. africanus* and later hominids, although these resemblances are undetected by others.[89]

Conclusions

Our general conclusion is simply stated: many lineages display phyletic size increase; allometric changes almost always accompany increase in body size. We cannot judge adaptation until we separate such changes into those required by increasing size and those serving as special adaptations to changing environments.

In our view, the three australopithecines are, in a number of features, scaled variants of the 'same' animal. In these characters, *A. africanus* is no more 'advanced' than the larger, more robust forms. The one early hominid to show a significant departure from this adaptive pattern toward later hominids – cranially, dentally, and postcranially – is *H. habilis* from East Africa. The australopithecines, one of which was probably a precursor of the *Homo* lineage, were apparently a successful group of basically vegetarian hominids, more advanced behaviorally than apes,[90] but not hunter-gatherers.

The fossil hominids of Africa fall into two major groupings. One probable lineage, the australopithecines, apparently became extinct without issue; the other evolved to modern man. Both groups displayed steady increase in body size. We consider quantitatively two key characters of the hominid skull: cranial capacity and cheek tooth size. The variables are allometrically related to body size in both lineages. In australopithecines, the manner of relative growth neatly meets the predictions for functional equivalence over a wide range of sizes (negative allometry of cranial capacity with a slope against body weight of 0.2 to 0.4 and positive allometry of postcanine area with a slope near 0.75). In the *A. africanus* to *H. sapiens* lineage, cranial capacity increases with positive allometry (slope 1.73) while cheek teeth decrease absolutely (slope – 0.725). Clearly, these are special adaptations unrelated to the physical requirements of increasing body size. We examined qualitatively other features, which also seem to vary allometrically. Of course, many characters should be studied quantitatively, but we think that the scheme outlined here should be treated as the null hypothesis to be disproved.

Notes

1. N. D. Newell, *Evolution*, 3 (1949): 103; J. T. Bonner, *Journal of Paleontology*, 42, No 5, Suppl. (1968): 1; S. M. Stanley, *Evolution*, 27 (1973): 1.
2. This is in the 'Dialogue of the Second Day' in the *Discorsi* of 1638, the work Galileo wrote under house arrest by the Inquisition. It was translated as *Dialogues Concerning Two New Sciences* by H. Crew and A. De Salvio in 1914 (reprinted, New York: Dover, 1952).
3. D. W. Thompson, *On Growth and Form* (Cambridge: Cambridge University Press, 1942); S. J. Gould, *Biological Reviews*, 41 (1966): 587.
4. F. W. Went, *American Scientist*, 56 (1968): 400.
5. D. Pilbeam, *The Ascent of Man* (New York: Macmillan, 1972).
6. Ibid.
7. J. T. Robinson, *Early Hominid Posture and Locomotion* (Chicago: University of Chicago Press, 1972).
8. W. E. LeGros Clark, *The Fossil Evidence for Human Evolution* (Chicago: University of Chicago Press, 1955); P. V. Tobias, *Olduvai Gorge*, vol. 2 (Cambridge: Cambridge University Press, 1967).
9. C. L. Brace, *American Journal of Physical Anthropology*, 21 (1963): 87; *idem*, *Human Biology*, 35 (1963): 545; *idem*, *Yearbook of Physical Anthropology*, 16 (1972): 31.
10. Robinson, *Early Hominid Posture*.
11. *Idem*, in G. Kurth (ed.), *Evolution und Hominisation* (Stuttgart: Fischer, 1962), p. 120.
12. *Idem*, *Early Hominid Posture*; *idem*, in Th. Dobzhansky, M. K. Hecht and W. Steere (eds), *Evolutionary Biology* (New York: Appleton-Century-Crofts, 1967), p. 69.

13. L. S. B. Leakey, *Nature*, 184 (1959): 491.
14. M. D. Leakey, *Olduvai Gorge*, vol. 3 (Cambridge: Cambridge University Press, 1971); *idem, Nature*, 223 (1969): 756; P. V. Tobias, *Nature*, 149 (1965): 22; M. H. Day, *Nature*, 221 (1969): 230; W. W. Bishop and J. A. Miller (eds), *Calibration of Hominoid Evolution* (Edinburgh: Scottish Academic Press, 1972); Y. Coppens, *Comtes Rendus Hebdomadaire des Seances de l'Academie des Sciences. Serie D: Sciences Naturelles*, 274 (1972): 181; L. S. B. Leakey and M. D. Leakey, *Nature*, 202 (1964): 5; F. C. Howell, *Nature*, 223 (1969): 1234; Y. Coppens, *Comtes Rendus Hebdomadaire des Seances de l'Academie des Sciences. Serie D: Sciences Naturelles*, 271 (1970): 1968 and 2286; Y. Coppens, *Comtes Rendus Hebdomadaire des Seances de l'Academie des Sciences. Serie D: Sciences Naturelles*, 272 (1971): 36; F. C. Howell and Y. Coppens, *American Journal of Physical Anthropology*, 40 (1974): 1; V. J. Maglio, *Nature*, 239 (1972): 379; B. E. Bowen and C. F. Vondra, *Nature*, 242 (1973): 391; R. E. F. Leakey, ibid., 170; V. J. Maglio, *Transactions of the American Philosophical Society*, 63 (1973): 5; R. E. F. Leakey, *Nature*, 231 (1971): 241; R. E. F. Leakey, *Nature*, 237 (1972): 264; R. E. F. Leakey, *Nature*, 242 (1973): 447.
15. LeGros Clark, *Fossil Evidence*; Tobias, *Olduvai Gorge*, vol. 2.
16. Robinson, *Early Hominid Posture*.
17. L. S. B. Leakey, P. V. Tobias and J. R. Napier, *Nature*, 202 (1964): 7.
18. See Note 14.
19. P. V. Tobias, *Nature*, 209 (1966): 953; R. L. Holloway, *Nature*, 243 (1973): 98; R. E. F. Leakey, *Soc. Biol.*, 19 (1972): 99.
20. See Note 14.
21. V. J. Maglio, *Transactions of the American Philosophical Society*, 63 (1973): 5; B. Patterson, A. K. Behrensmeyer and W. D. Sill, *Nature*, 226 (1970): 918.
22. V. J. Maglio, *Transactions of the American Philosophical Society*, 63 (1973): 5.
23. K. W. Butzer, *Current Anthropology*, in press.
24. P. V. Tobias, *Nature*, 246 (1973): 79.
25. Pilbeam, *Ascent of Man*; A. C. Walker, *Journal of Human Evolution*, 2 (1973): 545.
26. Robinson, *Early Hominid Posture*.
27. Pilbeam, *Ascent of Man*.
28. See Note 19.
29. C. L. Brace, *Yearbook of Physical Anthropology*, 16 (1972): 31.
30. M. H. Wolpoff, *Man*, 6 (1971): 601.
31. C. L. Brace, *Yearbook of Physical Anthropology*, 16 (1972): 31; M. H. Wolpoff, *Man*, 6 (1971): 601.
32. J. T. Robinson, *Early Hominid Posture and Locomotion* (Chicago: University of Chicago Press, 1972).
33. For example, D. Pilbeam, *The Ascent of Man* (New York: Macmillan, 1972).
34. R. Ardrey, *African Genesis* (London: Collins, 1961).
35. C. O. Lovejoy and K. G. Heiple, *American Journal of Physical*

Anthropology, 32 (1970): 33. Body weight estimates, which are of crucial importance, are especially difficult to make because of the incomplete nature of the remains. Body build, body proportions, the relationship of relative joint size to overall body weight, are all unknown factors. We have tried to estimate possible ranges and pick up a central figure as 'average', realising fully that such an estimate can be accurate only to within 15 to 20 percent. Our results suggest that some of the estimates may not be far wrong. We thank A. C. Walker for fruitful discussion on this point, and emphasize that he does not quite share the confidence with which these estimates are made.

36. C. L. Brace, *Yearbook of Physical Anthropology*, 16 (1972): 31; J. R. Napier and P. Napier, *A Handbook of Living Primates* (New York: Academic Press, 1967); D. Pilbeam, *Peabody Museum of Natural History at Yale University Bulletin*, 31 (1969); D. J. Hooijer, *Zool. Mededelingen Rijksmuseum van Natuurlijke Historie Leiden*, 29 (1948): 175; D. W. Frayer, *American Journal of Physical Anthropology*, 39 (973): 413; E. H. Ashton and S. Zuckerman, *Proceedings of the Royal Society of London. Series B: Biological Sciences*, 234 (1950): 59. Pygmy chimp estimates are from a variety of sources, and are less reliable. Body weights are compiled by us and E. L. Simons, based on data from zoological gardens and from various workers in Zaire (particularly A. D. Horn); cranial volumes are from A. H. Schultz (in *Homenaje a Juan Comas en su 65 Aniversario*, University of Mexico, 1965, p. 337) and H. C. Coolidge, *American Journal of Physical Anthropology*, 18 (1933): 1. Tooth measures were taken by David Pilbeam. E. H. Ashton and T. F. Spence, *Proceedings of the Zoological Society of London*, 130 (1958): 169.
37. R. Martin and K. Saller, *Lehrbuch der Anthropologie* (Stuttgart: Fischer, 1959).
38. E. H. Ashton and T. F. Spence, *Proceedings of the Zoological Society of London*, 130 (1958): 169.
39. T. D. Campbell, *Dentition and Palate of the Australian Aboriginal* (Adelaide: Hassell, 1925).
40. F. Weidenreich, *Palaeontol. Sin. Ser. D*, 7 (1937); *idem, Palaeontol. Sin. Ser. D*, 116 (1941); *idem, Palaeontol. Sin. Ser. D*, 127 (1943).
41. See Note 32.
42. See Note 32 and Robinson, *Early Hominid Posture.*
43. R. L. Holloway, op. cit.
44. The estimates are mainly from J. T. Robinson, *Transvaal Museum Mem.*, 6 (1952), and data collected by David Pilbeam.
45. A. C. Walker, op. cit.
46. R. L. Holloway, op. cit.
47. A. C. Walker, op. cit.
48. R. L. Holloway, *American Journal of Physical Anthropology*, 37 (1972): 173.
49. Tobias, *Olduvai Gorge*, vol. 2; E. H. Ashton and T. F. Spence, op. cit.; R. Broom and J. T. Robinson, *Transvaal Museum Mem.*, 6 (1952).
50. See Note 44; C. K. Brain, *Nature*, 225 (1970): 1112.
51. R. E. F. Leakey, *Nature*, 242 (1973): 447; R. E. F. Leakey, *Soc. Biol.*, 19 (1972): 99.

52. R. E. F. Leakey, *Nature*, 242 (1973): 447; R. L. Holloway, *Nature*, 243 (1973): 98.
53. R. L. Holloway and A. C. Walker, personal communications.
54. V. J. Maglio, *Transactions of the American Philosophical Society*, 63 (1973): 5; R. E. F. Leakey, *Nature*, 237 (1972): 264; P. V. Tobias, *Nature*, 209 (1966): 953; P. V. Tobias and G. H. R. von Koenigswald, *Nature*, 204 (1064): 515; M. H. Day and R. E. F. Leakey, *American Journal of Physical Anthropology*, 39 (1973): 341; R. E. F. Leakey and B. A. Wood, ibid., 355.
55. G. Cuvier, *Recherches sur les ossemens fossils* (Paris, 1812).
56. O. Snell, *Archiv für Psychiatrie und Nervenkrankheiten* 23 (1891): 436.
57. E. Dubois, *Bulletin et Memoires de la Société d'Anthropologique de Paris*, 8 (1897): 337.
58. J. S. Huxley, *Problems of Relative Growth* (London: MacVeagh, 1932).
59. H. Jerison, *Evolution of the Brain and Intelligence* (New York: Academic Press, 1973).
60. L. Lapicque, *Comptes Rendus des Seances de la Société de Biologie et de ses Filiales*, 50 (1898): 62.
61. E. Dubois, *Arch. Anthropol.* 25 (1897): 423.
62. D. A. Scholl, *Proceedings of the Royal Society of London. Series B: Biological Sciences*, 135 (1948): 243.
63. L. von Bertalanffy and W. J. Pirozynski, *Evolution*, 4 (1952): 387.
64. Tabulated in F. Weidenreich, *Transactions of the American Philosophical Society*, 31 (1941): 321.
65. G. Pilleri and R. G. Busnel, *Acta Anatomica*, 73 (1969): 92.
66. M. Rohrs, *Zeitschrift für wissenschaftliche Zoologie*, 162 (1959): 1; *Zeitschrift für Morphologie und Anthropologie*, 51 (1961): 289.
67. H. Frick, *Zeitschrift für Säugetierkunde*, 26 (1961): 138.
68. D. Bahrens, *Morphologisches Jahrbuch*, 101 (1960): 279.
69. P. Jolicoeur, *Biometrics*, 24 (1968): 679; P. Jolicoeur and J. E. Mosimann, *Biometrie-Praximetrie*, 9 (1968): 121.
70. J. H. Zar, *BioScience*, 18 (1968): 1118.
71. P. Jolicoeur, personal communication.
72. G. A. Sacher, in C. R. Noback and W. Montagna (eds), *The Primate Brain* (New York: Appleton-Century-Crofts, 1970), p. 245.
73. E. Giles, *Human Biology*, 28 (1956): 43.
74. P. V. Tobias, *The Brain in Hominid Evolution* (New York: Columbia University Press, 1971).
75. R. L. Holloway, in C. R. Noback and W. Montagna, op. cit., p. 209.
76. *Idem*, in R. H. Tuttle (ed.), *The Functional and Evolutionary Biology of Primates* (Chicago: Aldine, 1972), p. 185.
77. Robinson, *Early Hominid Posture*, p. 220.
78. Pilbeam, *Ascent of Man*.
79. R. E. F. Leakey, *Nature*, 237 (1972): 264.
80. Pilbeam, *Ascent of Man*; P. V. Tobias, *Olduvai Gorge*, vol. 2; C. L. Brace, *American Journal of Physical Anthropology*, 21 (1963): 87; *idem*, *Human Biology*, 35 (1963): 545; *idem*, *Yearbook of Physical Anthropology*, 16 (1972): 31.

81. See Note 64.
82. J. T. Robinson, *American Journal of Physical Anthropology*, 21 (1963): 595.
83. R. Broom and J. T. Robinson, op. cit.
84. R. L. Holloway, *American Journal of Physical Anthropology*, 37 (1972): 173; C. K. Brain, op. cit.
85. A. A. Abbie, *Proceedings of the Linnaean Society of New South Wales*, 83 (1958): 197; S. J. Gould, *Ontogeny and Phylogeny* (Cambridge, Mass.: Harvard University Press, 1977).
86. J. R. Napier, *Arch. Biol.*, 75 (1964): 673.
87. Robinson, *Early Hominid Posture.*
88. See Note 35; A. C. Walker, op. cit.; A. L. Zihlman, *Proceedings of the 3rd International Congress of Primatology*, 1 (1971): 54; C. O. Lovejoy, K. G. Heiple and A. H. Burstein, *American Journal of Physical Anthropology*, 38 (1973): 757.
89. Zihlman, op. cit.; O. J. Lewis, *Journal of Human Evolution*, 2 (1973): 1; C. E. Oxnard, in M. H. Day (ed.), *Human Evolution* (London: Taylor and Francis, 1973), p. 103.
90. A. E. Mann, *The Paleodemography of Australopithecus* (Ann Arbor, Mich.: University Microfilms, 1968).

PART V

STAGES AND SEQUENCES

EVER since Darwin – to coin a Gouldian phrase – evolutionary theorists have been fascinated by the idea of progression, of stages. Steve shared these concerns but added a typical twist of his own: his fascination with the iconography used to demonstrate this progression. By your icons, by your metaphors, you shall know them, revealing as they are of ideology as much as of methodology. Of course, he wasn't above a little iconographic persuasion himself, notably in his use of church architecture – of which more later. Some of this fascination has appeared in previous sections, but here we show it in its full flowering in the early pages of his paean to the contingency and variety of living forms: *Wonderful Life*. The thesis of that book, that if you were to wind the tape of history back and then replay it you would be in the highest degree unlikely to find humans reappearing, is a challenge to the more deterministically minded evolutionists and in particular to one of the researchers, Simon Conway Morris, on whose fieldwork some of the arguments in *Wonderful Life* depend. Conway Morris was one of the researchers studying the extraordinarily rich fossil deposits of the Burgess Shale in the Canadian Rockies, revealing a host of surprising organisms, animals that failed to fit readily into the standard account of sequential evolution. Instead, they seemed to come from quite different phyla, leaving no recognisable successor species. Gould uses the example of the fossils of the Burgess Shale to point to the contingency, the at least partially accidental quality, of evolution, of which species become extinct and which survive in newly evolved forms. Regrettably, to my mind, Conway Morris has repaid Gould's praise with distinct negativity, but the extract we have chosen sidesteps this particular controversy.

Stage theory appears not just in biological evolution but also in the theories some anthropologists and cultural historians apply to the chronology of human artifacts also – a theme within evolutionary psychology to which we return shortly. The second extract in this short section replays the theory no longer in biological evolution but in the theorizing of the development of prehistoric cave art by the early discoverers of the wonders of the painted caves of south-west France.

THE LADDER AND THE CONE: ICONOGRAPHIES OF PROGRESS

FAMILIARITY has been breeding overtime in our mottoes, producing everything from contempt (according to Aesop) to children (as Mark Twain observed). Polonius, amidst his loquacious wanderings, urged Laertes to seek friends who were tried and true, and then, having chosen well, to 'grapple them' to his 'soul with hoops of steel.'

Yet, as Polonius's eventual murderer stated in the most famous soliloquy of all time, 'there's the rub.' Those hoops of steel are not easily unbound, and the comfortably familiar becomes a prison of thought.

Words are our favored means of enforcing consensus; nothing inspires orthodoxy and purposeful unanimity of action so well as a finely crafted motto – Win one for the Gipper, and God shed his grace on thee. But our recent invention of speech cannot entirely bury an earlier heritage. Primates are visual animals par excellence, and the iconography of persuasion strikes even closer than words to the core of our being. Every demagogue, every humorist, every advertising executive, has known and exploited the evocative power of a well-chosen picture.

Scientists lost this insight somewhere along the way. To be sure, we use pictures more than most scholars, art historians excepted. *Next slide please* surpasses even *It seems to me that* as the most common phrase in professional talks at scientific meetings. But we view our pictures only as ancillary illustrations of what we defend by words. Few scientists would view an image itself as intrinsically ideological in content. Pictures, as accurate mirrors of nature, just are.

I can understand such an attitude directed toward photographs of objects – though opportunities for subtle manipulation are legion even here. But many of our pictures are incarnations of concepts masquerading as neutral descriptions of nature. These are the most potent sources of conformity, since ideas passing as descriptions lead us to equate the tentative with the unambiguously factual. Suggestions for the organization of thought are transformed to established patterns in nature. Guesses and hunches become things.

The familiar iconographies of evolution are all directed – sometimes crudely, sometimes subtly – toward reinforcing a comfortable view of human inevitability and superiority. The starkest version, the chain of being or ladder of linear progress, has an ancient, pre-evolutionary pedigree (see A. O. Lovejoy's classic, *The Great Chain of Being*, 1936). Consider, for example, Alexander Pope's *Essay on Man*, written early in the eighteenth century:

> Far as creation's ample range extends,
> The scale of sensual, mental powers ascends:
> Mark how it mounts, to man's imperial race,
> From the green myriads in the peopled grass.

And note a famous version from the very end of that century. In his *Regular Gradation in Man* (1799), British physician Charles White shoehorned all the ramifying diversity of vertebrate life into a single motley sequence running from birds through crocodiles and dogs, past apes, and up the conventional racist ladder of human groups to a Caucasian paragon, described with the rococo flourish of White's dying century:

> Where shall we find, unless in the European, that nobly arched head, containing such a quantity of brain . . . ? Where the perpendicular face, the prominent nose, and round projecting chin? Where that variety of features, and fullness of expression, . . . those rosy cheeks and coral lips?

This tradition never vanished, even in our more enlightened age. In 1915, Henry Fairfield Osborn celebrated the linear accretion of cognition in a figure full of illuminating errors (Fig. 9). Chimps are not ancestors but modern cousins, equally distant in evolutionary terms from the unknown forebear of African great apes and humans. *Pithecanthropus* (*Homo erectus* in modern terms) is a potential ancestor, and the only legitimate member of the sequence. The inclusion of

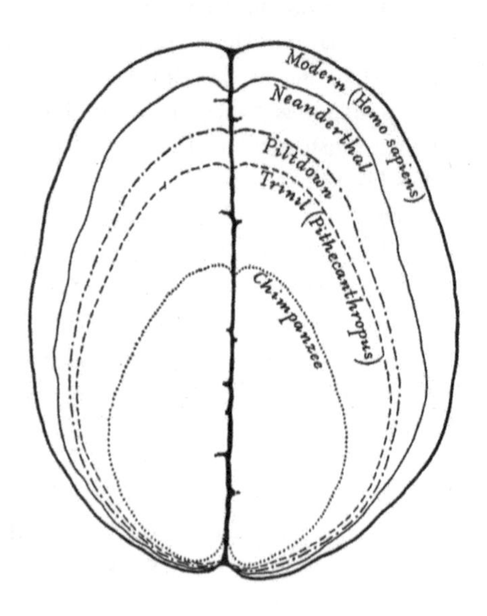

Figure 9. Progress in the evolution of the human brain as illustrated by Henry Fairfield Osborn in 1915.

Piltdown is especially revealing. We now know that Piltdown was a fraud composed of a modern human cranium and an ape's jaw. As a contemporary cranium, Piltdown possessed a brain of modern size; yet so convinced were Osborn's colleagues that human fossils must show intermediate values on a ladder of progress, that they reconstructed Piltdown's brain according to their expectations. As for Neanderthal, these creatures were probably close cousins belonging to a separate species, not ancestors. In any case, they had brains as large as ours, or larger, Osborn's ladder notwithstanding.

The march of progress is *the* canonical representation of evolution – the one picture immediately grasped and viscerally understood by all. This may best be appreciated by its prominent use in humor and in advertising. These professions provide our best test of public perceptions. Jokes and ads must click in the fleeting second that our attention grants them. Consider Figure 10, a cartoon drawn by

Figure 10. A 'scientific creationist' takes his appropriate place in the march of progress. By Bill Day, in the *Detroit Free Press*.

Bill Day, on 'scientific creationism.' Or Figure 11, by my friend Mike Peters, on the social possibilities traditionally open to men and to women. For advertising, consider the evolution of rental television (Fig. 12).*

The straitjacket of linear advance goes beyond iconography to the definition of evolution: the word itself becomes a synonym for *progress*. The makers of Doral cigarettes once presented a linear sequence of 'improved' products through the years, under the heading 'Doral's theory of evolution.'† (Perhaps they are now embarrassed by this

* Invoking another aspect of the same image – the equation of old and extinct with inadequate – Granada exhorts us to rent rather than buy because 'today's latest models could be obsolete before you can say brontosaurus.'

† Wonderfully ironic, since the sequence showed, basically, more effective filters. Evolution, to professionals, is adaptation to changing environments, not progress. Since the filters were responses to new conditions – public knowledge of health dangers – Doral did use the term *evolution* properly. Surely, however, they intended 'absolutely better' rather than 'punting to maintain profit' – a rather grisly claim in the light of several million deaths attributable to cigarette smoking.

Figure 11. More mileage from the iconography of the ladder. By Mike Peters, in the *Dayton Daily News*.

Figure 12. The march of progress as portrayed in another advertisement.

misguided claim, since they refused me permission to reprint the ad.)

Life is a copiously branching bush, continually pruned by the grim reaper of extinction, not a ladder of predictable progress. Most people may know this as a phrase to be uttered, but not as a concept brought into the deep interior of understanding. Hence we continually make errors inspired by unconscious allegiance to the ladder of progress, even when we explicitly deny such a superannuated view of life. For example, consider two errors, the second providing a key to our conventional misunderstanding of the Burgess Shale.

First, in an error that I call 'life's little joke', we are virtually compelled to the stunning mistake of citing unsuccessful lineages as classic 'textbook cases' of 'evolution.' We do this because we try to extract a single line of advance from the true topology of copious branching. In this misguided effort, we are inevitably drawn to bushes so near the brink of total annihilation that they retain only one surviving twig. We then view this twig as the acme of upward achievement, rather than the probable last gasp of a richer ancestry.

Consider the great warhorse of tradition – the evolutionary ladder of horses themselves (Fig. 13). To be sure, an unbroken evolutionary connection does link *Hyracotherium* (formerly called *Eohippus*) to modern *Equus*. And, yes again, modern horses are bigger, with fewer toes and higher crowned teeth. But *Hyracotherium–Equus* is not a ladder, or even a central lineage. This sequence is but one labyrinthine pathway among thousands on a complex bush. This particular route has achieved prominence for just one ironic reason – because all other twigs are extinct. *Equus* is the only twig left, and hence the tip of a ladder in our false iconography. Horses have become the classic example of progressive evolution because their bush has been so unsuccessful. We never grant proper acclaim to the real triumphs of mammalian evolution. Who ever hears a story about the evolution of bats, antelopes, or rodents – the current champions of mammalian life? We tell no such tales because we cannot linearize the bounteous success of these creatures into our favored ladder. They present us with thousands of twigs on a vigorous bush.

Need I remind everyone that at least one other lineage of mammals, especially dear to our hearts for parochial reasons, shares with horses both the topology of a bush with one surviving twig, and the false iconography of a march to progress?

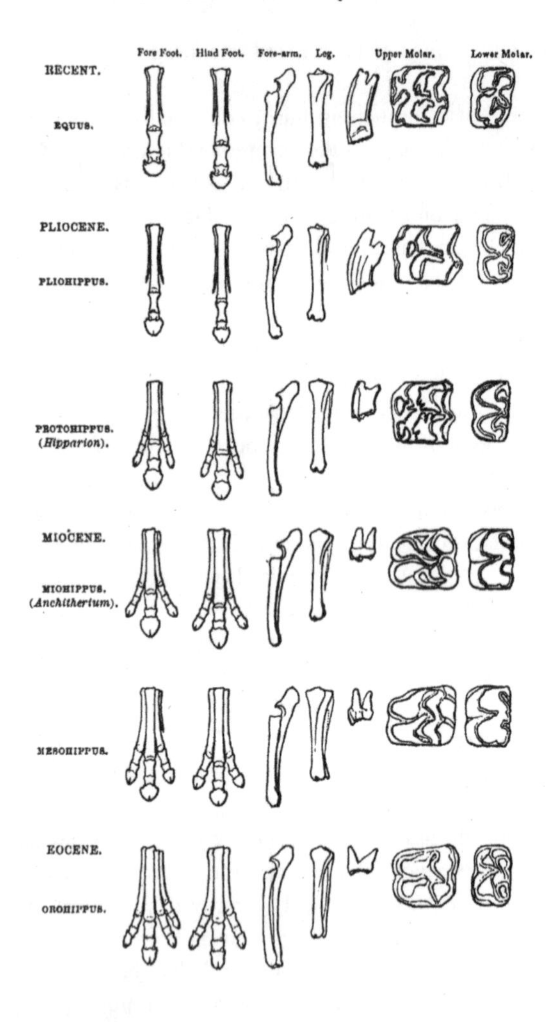

Figure 13. The original version of the ladder of progress for horses, drawn by the American paleontologist O. C. Marsh for Thomas Henry Huxley after Marsh had shown his recently collected Western fossils to Huxley on his only visit to the United States. Marsh convinced his English visitor about this sequence, thus compelling Huxley to revamp his lecture on the evolution of horses given in New York in 1876. Note the steady decrease in number of toes and increase in height of teeth. Since Marsh drew all his specimens the same size, we do not see the other classical trend of increase in stature.

In a second great error, we may abandon the ladder and acknowledge the branching character of evolutionary lineages, yet still portray the tree of life in a conventional manner chosen to validate our hopes for predictable progress.

The tree of life grows with a few crucial constraints upon its form. First, since any well-defined taxonomic group can trace its origin to a single common ancestor, an evolutionary tree must have a unique basal trunk.* Second, all branches of the tree either die or ramify further. Separation is irrevocable; distinct branches do not join.†

Yet, within these constraints of *monophyly* and *divergence*, the geometric possibilities for evolutionary trees are nearly endless. A bush may quickly expand to maximal width and then taper continuously, like a Christmas tree. Or it may diversify rapidly, but then maintain its full width by a continuing balance of innovation and death. Or it may, like a tumbleweed, branch helter-skelter in a confusing jumble of shapes and sizes.

Ignoring these multifarious possibilities, conventional iconography has fastened upon a primary model, the 'cone of increasing diversity,'

* A properly defined group with a single common ancestor is called monophyletic. Taxonomists insist upon monophyly in formal classification. However, many vernacular names do not correspond to well-constituted evolutionary groups because they include creatures with disparate ancestries – 'polyphyletic' groups in technical parlance. For example, folk classifications that include bats among birds, or whales among fishes, are polyphyletic. The vernacular term *animal* itself probably denotes a polyphyletic group, since sponges (almost surely), and probably corals and their allies as well, arose separately from unicellular ancestors – while all other animals of our ordinary definitions belong to a third distinct group. The Burgess Shale contains numerous sponges, and probably some members of the coral phylum as well, but this book will treat only the third great group – the coelomates, or animals with a body cavity. The coelomates include all vertebrates and all common invertebrates except sponges, corals, and their allies. Since the coelomates are clearly monophyletic, the subjects of this book form a proper evolutionary group.

† This fundamental principle, while true for the complex multicellular animals treated in this book, does not apply to all life. Hybridization between distant lineages occurs frequently in plants, producing a 'tree of life' that often looks more like a network than a conventional bush. (I find it amusing that the classic metaphor of the tree of life, used as a picture of evolution ever since Darwin and so beautifully accurate for animals, may not apply well to plants, the source of the image.) In addition, we now know that genes can be transferred laterally, usually by viruses, across species boundaries. This process may be important in the evolution of some unicellular creatures, but probably plays only a small role in the phylogeny of complex animals, if only because two embryological systems based upon intricately different developmental pathways cannot mesh, films about flies and humans notwithstanding.

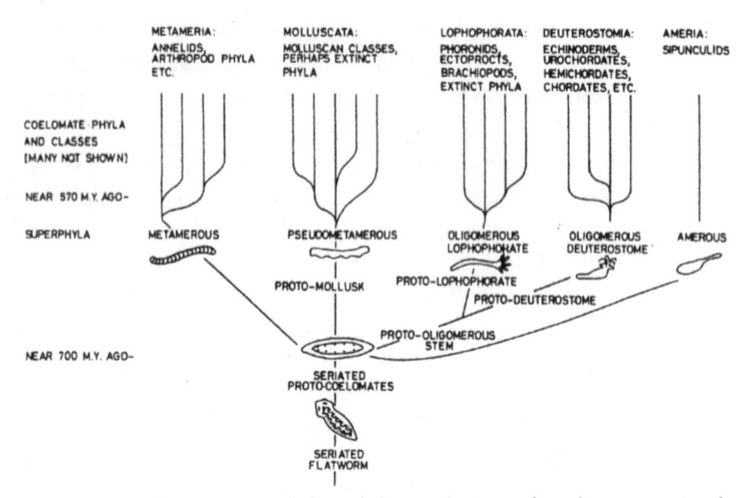

Figure 14. A recent iconography of the evolution of coelomate animals, drawn according to the convention of the cone of increasing diversity.

an upside-down Christmas tree. Life begins with the restricted and simple, and progresses ever upward to more and more and, by implication, better and better. Figure 14 on the evolution of coelomates (animals with a body cavity, the subjects of this book), shows the orderly origin of everything from a simple flatworm. The stem splits to a few basic stocks; none becomes extinct; and each diversifies further, into a continually increasing number of subgroups.

Figure 15 presents three examples of cones drawn from popular modern textbooks. All these trees show the same pattern: branches grow ever upward and outward, splitting from time to time. If some early lineages die, later gains soon overbalance these losses. Early deaths can eliminate only small branches near the central trunk. Evolution unfolds as though the tree were growing up a funnel, always filling the continually expanding cone of possibilities.

In its conventional interpretation, the cone of diversity propagates an interesting conflation of meanings. The horizontal dimension shows diversity – fishes plus insects plus snails plus starfishes at the top take up much more lateral room than just flatworms at the bottom. But what does the vertical dimension represent? In a literal reading, up and down should record only younger and older in geological time: organisms at the neck of the funnel are ancient; those at the lip, recent. But we also

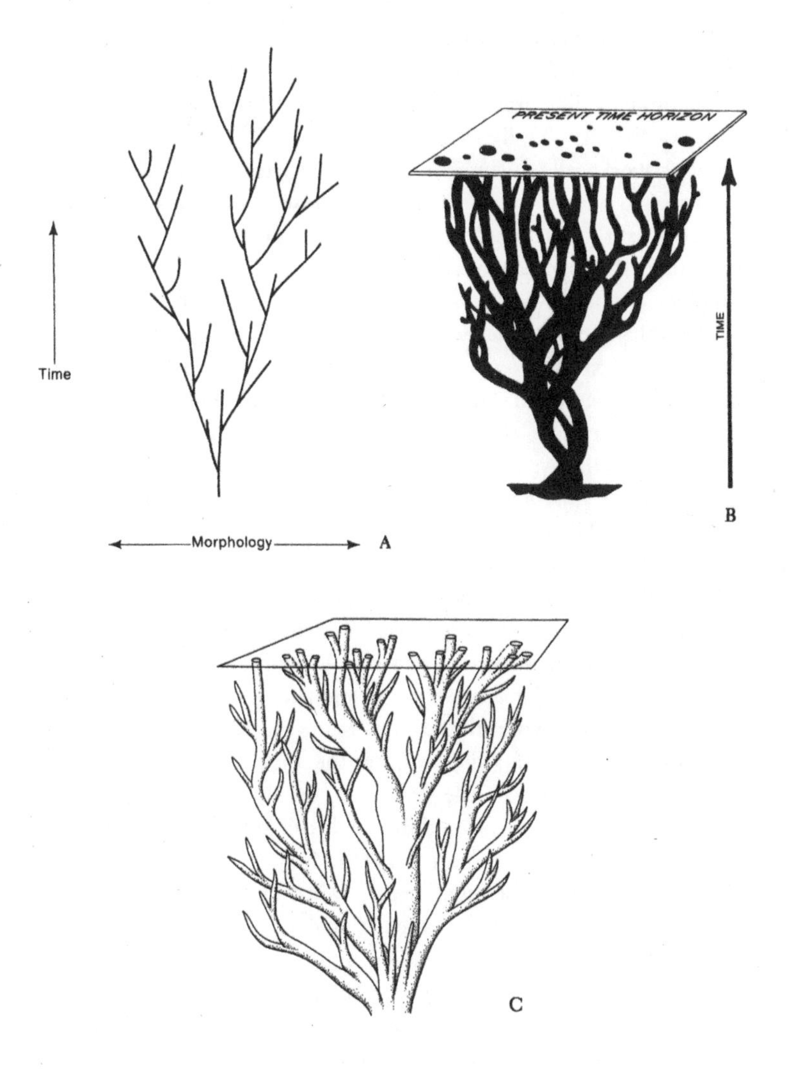

Figure 15. The iconography of the cone of increasing diversity, as seen in examples from textbooks. All these diagrams are presented as simple objective portrayals of evolution; none are explicit representations of diversification as opposed to some other evolutionary process. The data of the Burgess Shale falsify this central view of arthropod evolution as a continuous process of increasing diversification.

read upward movement as simple to complex, or primitive to advanced. *Placement in time is conflated with judgment of worth.*

Our ordinary discourse about animals follows this iconography. Nature's theme is diversity. We live surrounded by coeval twigs of life's tree. In Darwin's world, all (as survivors in a tough game) have some claim to equal status. Why, then, do we usually choose to construct a ranking of implied worth (by assumed complexity, or relative nearness to humans, for example)? In a review of a book on courtship in the animal kingdom, Jonathan Weiner (*New York Times Book Review*, March 27, 1988) describes the author's scheme of organization: 'Working in loosely evolutionary order, Mr Walters begins with horseshoe crabs, which have been meeting and mating on dark beaches in synchrony with tide and moon for 200 million years.' Later chapters make the 'long evolutionary leap to the antics of the pygmy chimpanzee.' Why is this sequence called 'evolutionary order'? Anatomically complex horseshoe crabs are not ancestral to vertebrates; the two phyla, Arthropoda and Chordata, have been separate from the very first records of multicellular life.

In another recent example, showing that this error infests technical as well as lay discourse, an editorial in *Science*, the leading scientific journal in America, constructs an order every bit as motley and senseless as White's 'regular gradation'. Commenting on species commonly used for laboratory work, the editors discuss the 'middle range' between unicellular creatures and guess who at the apex: 'Higher on the evolutionary ladder,' we learn, 'the nematode, the fly and the frog have the advantage of complexity beyond the single cell, but represent far simpler species than mammals' (June 10, 1988).

The fatuous idea of a single order amidst the multifarious diversity of modern life flows from our conventional iconographies and the prejudices that nurture them – the ladder of life and the cone of increasing diversity. By the ladder, horseshoe crabs are judged as simple; by the cone, they are deemed old.* And one implies the other under the grand conflation discussed above – down on the ladder also means old, while

* Another factual irony: despite the usual picture of horseshoe crabs as 'living fossils,' *Limulus polyphemus* (our American East Coast species) has no fossil record whatever. The genus *Limulus* ranges back only some 20 million years, not 200 million. We mistakenly regard horseshoe crabs as 'living fossils' because the group has never produced many species, and therefore never developed much evolutionary potential for diversification; consequently, modern species are morphologically similar to early forms. But the species themselves are not notably old.

low on the cone denotes simple.

I don't think that any particular secret, mystery, or inordinate subtlety underlies the reasons for our allegiance to these false iconographies of ladder and cone. They are adopted because they nurture our hopes for a universe of intrinsic meaning defined in our terms. We simply cannot bear the implications of Omar Khayyám's honesty:

> Into this Universe, and Why not knowing,
> Nor whence, like Water willy-nilly flowing:
> And out of it, as Wind along the Waste
> I know not Whither, willy-nilly blowing.

A later quatrain of the *Rubáiyát* proposes a counteracting strategy, but acknowledges its status as a vain hope:

> Ah Love! could you and I with Fate conspire
> To grasp this sorry Scheme of Things entire,
> Would we not shatter it to bits – and then
> Re-mold it nearer to the Heart's Desire!

Most myths and early scientific explanations of Western culture pay homage to this 'heart's desire.' Consider the primal tale of Genesis, presenting a world but a few thousand years old, inhabited by humans for all but the first five days, and populated by creatures made for our benefit and subordinate to our needs. Such a geological background could inspire Alexander Pope's confidence, in the *Essay on Man*, about the deeper meaning of immediate appearances:

> All Nature is but art, unknown to thee;
> All chance, direction, which thou canst not see;
> All discord, harmony not understood;
> All partial evil, universal good.

But, as Freud observed, our relationship with science must be paradoxical because we are forced to pay an almost intolerable price for each major gain in knowledge and power – the psychological cost of progressive dethronement from the center of things, and increasing marginality in an uncaring universe. Thus, physics and astronomy relegated our world to a corner of the cosmos, and biology shifted our status from a simulacrum of God to a naked, upright ape.

To this cosmic redefinition, my profession contributed its own special shock – geology's most frightening fact, we might say. By the turn of the last century, we knew that the earth had endured for millions of years, and that human existence occupied but the last geological milli-microsecond of this history – the last inch of the cosmic mile, or the last second of the geological year, in our standard pedagogical metaphors.

We cannot bear the central implication of this brave new world. If humanity arose just yesterday as a small twig on one branch of a flourishing tree, then life may not, in any genuine sense, exist for us or because of us. Perhaps we are only an afterthought, a kind of cosmic accident, just one bauble on the Christmas tree of evolution.

What options are left in the face of geology's most frightening fact? Only two, really. We may, as this book advocates, accept the implications and learn to seek the meaning of human life, including the source of morality, in other, more appropriate, domains – either stoically with a sense of loss, or with joy in the challenge if our temperament be optimistic. Or we may continue to seek cosmic comfort in nature by reading life's history in a distorted light.

If we elect the second strategy, our maneuvers are severely restricted by our geological history. When we infested all but the first five days of time, the history of life could easily be rendered in our terms. But if we wish to assert human centrality in a world that functioned without us until the last moment, we must somehow grasp all that came before as a grand preparation, a foreshadowing of our eventual origin.

The old chain of being would provide the greatest comfort, but we now know that the vast majority of 'simpler' creatures are not human ancestors or even prototypes, but only collateral branches on life's tree. The cone of increasing progress and diversity therefore becomes our iconography of choice. The cone implies predictable development from simple to complex, from less to more. *Homo sapiens* may form only a twig, but if life moves, even fitfully, toward greater complexity and higher mental powers, then the eventual origin of self-conscious intelligence may be implicit in all that came before. In short, I cannot understand our continued allegiance to the manifestly false iconographies of ladder and cone except as a desperate finger in the dike of cosmically justified hope and arrogance.

I leave the last word on this subject to Mark Twain, who grasped so graphically, when the Eiffel Tower was the world's tallest building, the implications of geology's most frightening fact:

> Man has been here 32,000 years. That it took a hundred million years to prepare the world for him* is proof that that is what it was done for. I suppose it is. I dunno. If the Eiffel Tower were now representing the world's age, the skin of paint on the pinnacle knob at its summit would represent man's share of that age; and anybody would perceive that the skin was what the tower was built for. I reckon they would, I dunno.

* Twain used Lord Kelvin's estimate, then current, for the age of the earth. The estimated ages have lengthened substantially since then, but Twain's proportions are just about right. He took human existence as about 1/30,000 of the earth's age. At current estimates of 250,000 years for the origin of our species, *Homo sapiens*, the earth would be 7.5 billion years old if our span were 1/30,000 of totality. By best current estimates, the earth is 4.5 billion years old.

UP AGAINST A WALL

WE are, above all, a contentious lot, unable to agree on much of anything. Alexander Pope caught the essence of our discord in a couplet (though modern technology has vitiated the force of his simile):

> 'Tis with our judgments as our watches, none
> Go just alike, yet each believes his own.

Most proclamations of unanimity therefore convey a fishy odor – arising either from imposed restraint ('elections' in dictatorial one-party states), or comedic invention to underscore an opposite reality (as when Ko-Ko, in Gilbert and Sullivan's *Mikado*, reads a document signed by the Attorney-General, the Lord Chief Justice, the Master of the Rolls, the Judge Ordinary, and the Lord Chancellor – and then proclaims, 'Never knew such unanimity on a point of law in my life.' But the document has been endorsed by only one signatory – for Pooh-Bah holds all the aforementioned titles!).

Paleontologists probably match the average among human groups for levels of contentiousness among individuals (while students of human prehistory surely rank near the top, for this field contains more practitioners than objects for study, thus breeding a high level of acquisitiveness and territoriality). Yet one subject – and only one – elicits absolute unanimity of judgment among students of ancient life, though for reasons more visceral than intellectual. Every last mother's son and daughter among us stands in reverent awe and amazement before the great cave paintings done by our ancestors in southern and central Europe between roughly thirty thousand and ten thousand years ago.

If this wonderment stands as our only point of consensus (not confined, by the way, to scientific professionals, but shared with any member of *Homo sapiens* possessing the merest modicum of curiosity about our past), please don't regard me as a Scrooge or a Grinch if I point out that our usual rationale for such awe arises from a pairing of reasons – one entirely appropriate, and the other completely invalid. For I don't impart this news to suggest any diminution of wonder, but rather to clear away some conceptual baggage that, once discarded, might free us to appreciate even more fully this amazing beginning of our most worthy institution.

For the good reason, we look at the best and most powerful examples of this art, and we just know that we have fixed a Michelangelo in our gaze. Comparisons of this sort seem so obvious, and so just, that they have become a virtual cliché for anyone's description of a first reaction to a wonderfully painted cave wall. For example, in describing his emotional reaction to the newly discovered cave of Chauvet – the source of eventual denouement for this essay as well – a noted expert wrote: 'Looking closely at the splendid heads of the four horses, I was suddenly overcome with emotion. I felt a deep and clear certainty that here was the work of one of the great masters, a Leonardo da Vinci of the Solutrean revealed to us for the first time. It was both humbling and exhilarating.'

For the bad argument, our amazement also arises for a conceptual reason added to our simple (and entirely appropriate) visceral awe. We are, in short, surprised, even stunned, to discover that something so old could be so sophisticated. Old should mean rudimentary – either primitive by greater evolutionary regress toward an apish past, or infantile by closer approach to the first steps on our path toward modernity. (These metaphors of grunting coarseness or babbling juvenility probably hold about equal sway in the formation of our prejudices.) As we travel in time down our own evolutionary tree, we should encounter ever-older ancestors of ever-decreasing mental capacity. The first known expressions of representational art should therefore be crude and primitive. Instead, we see the work of a primal Picasso – and we are dumbstruck.

I dedicate this essay to tracing the prevalence of this view in the lifework of the two greatest scholars of Paleolithic cave art. I shall then argue that this equation of older with more rudimentary both violates the expectations of evolutionary theory when properly construed, and has now also been empirically disproven by discoveries at Chauvet and elsewhere. I shall then suggest that the more appropriate expectation of maximal sophistication for this earliest art should only increase our appreciation – for we trade a false (if heroic) view of ever-expanding triumph for a deeply satisfying feeling of oneness with people who were, biologically, fully us in circumstances of maximal distance, both temporal and cultural, from our current lives.

(No species of punditry deserves more ridicule than the art form known as Monday-morning quarterbacking or backseat driving – the 'I told you so' of the nonparticipant. This essay veers dangerously toward such an unworthy activity. I am, after all, a paleontologist and expert on land snails, not an art historian or a student of human culture. What right do I have to criticize the monumental and lifelong efforts of the Abbé Henri Breuil and André Leroi-Gourhan, the most learned and prolific of true devotees? In defense, I would say, first, that honorable errors do not count as failures in science, but as seeds for progress in the quintessential activity of correction. No great and new study has ever developed without substantial error, and we need only cite a famous line from Darwin: 'False facts are highly injurious to the progress of science, for they often endure long; but false views, if supported by some evidence, do little harm, for every one takes a salutary pleasure in proving their falseness.' In reverse of Mark Antony, I have come to praise Breuil and Leroi-Gourhan, not to bury them. Second, the perspective of cognate fields can often bring light to neighboring disciplines too set in favored ways. I therefore speak from my close vicinity of evolutionary theory to point out – as many have before me, and on the same basis – that conventional expectations have no sanction in our current understanding of evolution, and therefore represent lingering prejudices that we might wish to reassess and then choose to discard.)

The general title of 'ice-age' or 'Paleolithic' (literally, Old Stone Age) art has been applied to the great variety and geographic spread of works in two major categories – smaller and movable objects usually called 'portable' (the so-called Venus figurines, the deer, horses, and other animals carved in bone or ivory on disks, plaques, and spear-throwers, for example); and the engravings and paintings on cave walls (and now from a few open sites as well), dubbed 'parietal.' (A *paries* is a Latin wall; if you belong to my generation, you will remember, now with amusement but then with utmost frustration, the parietal hours of college dorms, when members of the opposite sex had to return within the walls of their own rooms, and not remain in yours.) European portable art extends from Spain to Siberia; parietal art has been found mostly in France and northern Spain, with a few Italian sites, and perhaps others even more distant. (Decorated caves in other genres, but perhaps of equal or even greater age, have been found in many other areas of the world, from Africa to Australia.)

Current radiocarbon dates (from charcoal) of paintings in parietal art span a range from 32,410 years B.P. (before the present) at Chauvet, to 11,600 years B.P. at Le Portel. This period corresponds to the occupation of Europe by our own species, *Homo sapiens* (often called 'Cro-Magnon' in this incarnation to honor a French site of first discovery. Remember that the Cro-Magnon people are us – by both bodily anatomy and parietal art – not some stooped and grunting distant ancestor). The immediately earlier inhabitants of Europe, the famous Neanderthal people, did not (so far as we know) produce any representational art. Neanderthals overlapped Cro-Magnon in Europe, probably into the time of Cro-Magnon's early parietal art. This striking cultural difference reinforces the opinion that Neanderthal and Cro-Magnon were two separate, albeit closely related, species, and not end-points of a smooth evolutionary continuum. On this view, Neanderthal died out, while Cro-Magnon continues as us.

Two subjects have long dominated the theoretical discussion on parietal art: function and chronology. The two greatest scholars in this field – the Abbé Henri Breuil and André Leroi-Gourhan – differed profoundly in their views about function, but (somewhat paradoxically) agreed substantially in their proposals about chronology.

Of the several French priests who have become distinguished students of ancient life (in a land where both Catholic traditions and general intellectual commitments favor harmony between the different domains of science and religion, Pierre Teilhard de Chardin surely won the most fame, while the Abbé Henri Breuil may well have done the best work. Breuil, a talented artist, spent nearly sixty years copying figures from cave walls (at a time when photography yielded poor images in such subterranean conditions), and then comparing the results in his compendia of drawings. He traced directly from cave walls whenever possible, and with utmost care at all times (drawings are not inherently more subjective than photographs). But he sometimes had to work by difficult and indirect methods. He could not, for example, press paper against the famous painted ceiling of Altamira because any direct contact detached the pasty pigment used by Paleolithic artists. Therefore, positioned like Michelangelo under the Sistine Chapel ceiling, he lay on his back, cushioned on soft sacks of ferns, while holding his paper as close to the roof as possible, and making imperfect sketches.

As he drew the animals one by one, Breuil tended to read their meaning in the same piecemeal fashion – that is, as individuals rather than parts of integrated compositions. He held that the paintings functioned as a kind of 'hunting magic' to make game plentiful (if you draw it, it will come), or to ensure success in the kill (game animals are often painted with wounds and spear holes). Breuil wrote in his summary book of 1952:

> Here, for the first time, men dreamed of great art and, by the mystical contemplation of their works, gave to their contemporaries the assurance of success in their hunting expeditions, of triumphs in the struggle against the enormous pachyderms and grazing animals.

In the next generation, André Leroi-Gourhan, director of the Musée de l'Homme in Paris, approached the same subject of meaning from the maximally different perspective of his 'card-carrying' membership in one of the major intellectual movements of our century – French structuralism as embodied in the work of anthropologist Claude Lévi-Strauss. This form of structuralism searches for timeless and integrative themes based on dichotomous divisions that may record much of nature's reality, but mostly reflect the brain's basic mode of operation. Thus, we separate nature from culture (the raw and the cooked, in Lévi-Strauss's terminology), light from darkness, and, above all, male from female.

Leroi-Gourhan therefore viewed each cave as an integrated composition, a sanctuary in which the numbers and positions of animals bore unified meaning within a scheme set by the primary duality of male and female. Each animal became a symbol, with a primary division between horses as male and bisons as female. He also interpreted abstract signs and artifacts as sexually labeled, with spears (for example) as male, and wounds as female. He viewed the cave itself as fundamentally female, thus requiring a definite positioning and grouping of male symbols. Led by this theory, Leroi-Gourhan treated each cave as a unity and compiled extensive statistical tables of numbers and locations – in maximal contrast with Breuil's concentration on each animal in and for itself. Leroi-Gourhan wrote:

> Clearly, the core of the system rests upon the alternation, complementarity, or antagonism between male and female values, and one might think of 'a fertility cult.' . . . There are few religions, primitive or evolved, that do not somewhere involve a confrontation of the same values, whether divine couples such as Jupiter and Juno are concerned, or principles such as yang and yin. There is little doubt that Paleolithic men were familiar with the division of the animal and human world into two opposite halves, or that they supposed the union of these two halves to govern the economy of living beings . . . Paleolithic people represented in the caves the two great categories of living creatures, the corresponding male and female symbols, and the symbols of death on which the hunters fed. In the central area of the cave, the system is expressed by groups of male symbols placed around the main female figures, whereas in the other parts of the sanctuary we find exclusively male representations, the complements, it seems, to the underground cavity itself.

And yet, despite their maximal ideological difference on the function and meaning of cave art, Breuil and Leroi-Gourhan maintained agreement on the second great subject of chronology. To be sure (and as we shall see), these two great scholars diverged on many particulars, but they shared an unswerving and defining conviction – a kind of central and unshakable faith – that the chronology of cave art must record a progression from crude and simple beginnings to ever more refined and sophisticated expression. In so doing, these scholars could assimilate the earliest known history of representational art to the classic myths and sagas of Western culture – the hero's birth, his first faltering steps, his rise to maturity, his triumph and dominion, and, ultimately, his tragic fall (for both Breuil and Leroi-Gourhan included a final stage of degeneration after the ice retreated and the game dispersed).

In regarding a progressive chronology as a consummation so devoutly to be wished, Breuil and Leroi-Gourhan had complex motives. They were, no doubt and in large part, simply caught up in conventional modes of thinking, so deep and so automatic in our culture that such views rarely bubble to a conscious surface where they might be questioned. But an important technical reason also drove both scholars to such a hope. Layers of sediment can often be dated by various means, now conventional. But a cave is a hole in the ground; how can you specify the age of a cavity? (You might ascertain the age of rocks forming the cave wall, but these dates bear no relationship at all to the age of the cave as a hole.) How, then, can you know the time of a prehistoric painting or engraving on a cave wall? (Today we can date the pigments by carbon-14 and other methods, particularly the charcoal used to draw black lines, but Breuil had no access to such techniques at all, and the carbon-14 methods of Leroi-Gourhan's time required so much material to obtain a date that entire paintings would have been sacrificed – a procedure that no one, quite properly, would ever sanction.)

The only hope for dating therefore inhered in the paintings themselves – in the search for an internal criterion that could order this earliest art into a chronological sequence. Breuil struggled mightily to establish such an order by superposition – that is, by studying paintings drawn over earlier paintings. He succeeded to some extent, but technical problems proved too daunting for a general solution. You cannot always specify the sequence of overlap on an essentially flat surface; moreover, even if you can, the painting on top could have been executed the next day, or a thousand years after, the one below. Leroi-Gourhan contrasted the ease of dating portable objects found in strata with difficulties for paintings on cave walls: 'A reindeer incised on a small plaque, found in a layer that also yielded hundreds of flints, is often easy to date, but a mammoth painted on a cave wall three feet or more above the ground is cut off from all chronological clues.'

Both scholars therefore turned to the venerable technique of art historians of later times – the analysis of styles. But a problem of circular reasoning now intrudes, for we need a source of evidence separate from the paintings themselves. We can place Michelangelo's style in the sixteenth century, and Picasso's in our own, because we have independent evidence about dates from a known historical record. But nothing either in abstract logic or pictorial necessity dictates that one form of mannerism must be four hundred years old, while another style of cubism could only emerge much later. If we had absolutely no other evidence but Michelangelo's *Last Judgment* and Picasso's *Guernica* –

no texts, no contexts, no witnesses – we could not know their temporal order.

In such a context of abysmally limited information – the situation faced by Breuil and Leroi-Gourhan – we must try to construct a theory of stylistic change that might establish a chronological sequence by internal evidence. (If we could say, for example, that realism must precede abstraction, then we could place Michelangelo before Picasso by internal criteria alone.) I don't fault these scholars for seeking such a theory of stylistic change – for how else could they have proceeded, given the limitations? But I am intrigued that they fell back so easily and so uncritically – almost automatically, it might seem – upon the most conventional form of progressivist mythology: a chronology ordered by simple to complex, or rude to sophisticated.

I can better grasp Breuil's attraction to the legend of progress. He was, after all, a child of the late nineteenth century – the great age of maximal faith in human advance, especially in Western nations at the height of their imperial and industrial expansion (the ravages of World War I ended this illusion for many, though not, apparently, for Breuil). But Leroi-Gourhan's assent is more puzzling, for his philosophical commitment to structuralism led him to view the symbolic ensemble of each cave as the expression of an unvarying human psyche, with its dualistic contrasts of male and female, danger and safety, and so forth.

In fact, and in several interesting passages, Leroi-Gourhan addressed this issue directly. He acknowledges that structuralism does lead to a hypothesis of unvarying form and function for caves as sanctuaries throughout the history of Paleolithic parietal art. But given this constancy of structure, he then argues, how could we untangle chronology except by the hope (and expectation) that styles used to paint the constant symbols will change in a systematic way through time? A bison may always represent the female moiety, and may always occupy the same position within a cave – but artists may learn to paint bisons better through time. Leroi-Gourhan writes:

> The same content persists from first to last. The pairing of animal species with signs appears in the Aurignacian [the first period of cave art] and disappears in the terminal Magdalenian [the last period]. Consequently, the ideological unity of cave art rules out the guideposts that it might provide for us had there been changes in the basic themes. Only variations in

> the representation of this uniform subject matter are discernible in the course of a stylistic study.

Parietal art includes a complex array of figures and signs. The figures mostly depict the large mammals of ice-age Europe (various deer, horses, bison, mammoths, rhinos, lions, and several others), but we also find occasional humans (and the more frequent, and wonderful, handprints, often stenciled by placing a hand against the wall and blowing paint around it with some kind of Paleolithic spray can). In another category, rarely given as much attention but surely surpassing the animal figures in number (and perhaps in interpretive importance), a large variety of signs and symbols festoon the walls – some identifiable as pictures of weapons or body parts (often sexual), others as geometric forms, and still others quite mysterious.

In the progressivist chronologies of Breuil and Leroi-Gourhan, figures and signs show a superficially opposite directionality. Figures begin with crude and simple outlines and progress to more supple and complex realism, complete with dimensionality and perspective. Signs, on the other hand, become simpler and more symbolic, with identifiable pictures (of vulvas, for example) evolving to less variable, more symbolic, and often highly simplified geometric representation. Leroi-Gourhan wrote: 'The animal figures . . . show a development in form towards a more and more precise analysis. The geometricization of the signs in contrast with the character of the animal figures is one of the interesting aspects of research into the meaning of the designs.'

But these apparently opposite directions of change for figures and symbols really represent – as Breuil and Leroi-Gourhan repeatedly emphasized – different facets of the same overall theme of progress as the basis of chronology. In painting figures, the artists were trying to do better in representing the animals themselves – and the supposed sequence of styles marks their continual improvement. But, in drawing signs, the same artists were knowingly developing a system of symbols – and symbols gain universality and meaning by becoming more abstract and reduced to a geometric essence. After all – and the analogy was not lost on these scholars – most alphabets derived their letters as simplified pictures of objects (while the same argument applies with even more force to the evolution of such character systems as Chinese).

Breuil initially proposed a system of five stages in a single sequence of greater realism and complexity for figures (his papers of the early

1900s make fascinating reading). He later developed his famous theory of two successive cycles, each with a complete history of progress to a pinnacle, followed by late decline. (Breuil continued to hold, despite mounting evidence to the contrary, that the first cycle could be recognized by drawings of animals in 'twisted perspective' – that is, with more than one plane represented, as in a bison with a body seen from the side, but with a face pointing forward.)

Breuil's two schemes are not so different as they appear – for his first notion of five sequential steps included a period of decline in the middle. I was particularly struck by his adjectives of judgment in supposedly objective descriptions of the stages. In an early article of 1906, he marks the animals of Stage 3 (later to become the decline at the end of the first cycle) as 'of a deplorable design, and with a disconcerting lack of proportion.' He then praises the recovery of Stage 4 as one might describe a Renaissance artist trying to re-create the lost glories of an ancient Greece: 'The artists sought to rediscover the model lost in the preceding stage. They obtained this result by polychromy [figures of more than one color]. These paintings are timid at first . . .' Breuil concluded his paper by stating: 'Paleolithic art, after an almost infantile beginning, rapidly developed a lively way of depicting animal forms, but didn't perfect its painting techniques until an advanced stage.'

Leroi-Gourhan, in contrast, developed a theory of four successive stages in a single series. But his sequence of progress scarcely differed from Breuil's – though the older scholar wanted to run the story twice. Both schemes began with immobile animals stiffly carved in crude outlines with no interior coloration, and moved on to ever more accurate images, drawn with a much better feel for mobility, rendered in better perspective, and more richly colored. (The later artists, Leroi-Gourhan believed, reached such a state of perfection that their art stagnated a bit at the end, becoming rather academic in replication of excellence.)

Mario Ruspoli, a disciple of Leroi-Gourhan, epitomized the theory well in his 1986 book *The Cave of Lascaux*. 'From the earliest images onward, one has the impression of being in the presence of a system refined by time . . . The development of Paleolithic cave art may be summed up as 15,000 years of apprenticeship followed by 8,000 years of academicism.'

Leroi-Gourhan recognized the essential similarity of his view with the earlier theories of Breuil. After a detailed (though respectful) critique of Breuil, and an extensive compendium of their particular differences, Leroi-Gourhan acknowledges the fundamental similarity in their common concept of progress as the key to a chronology of Paleolithic art:

> The theory . . . is logical and rational: art apparently began with simple outlines, then developed more elaborate forms to achieve modeling, and then developed polychrome or bichrome painting before it eventually fell into decadence.

This progressivist theory of increasingly complex and supple realism in Paleolithic painting dominated the field for decades. Writing of Leroi-Gourhan's four-stage theory, Brigitte and Gilles Delluc (in Ruspoli's book), state simply: 'The classification was fairly soon adopted by everyone.' And yet, I think everyone now realizes that the hypothesis of progressivism in Paleolithic art cannot hold. The march to greater and more complex realism doesn't make any sense theoretically, and has now been disproven empirically at Chauvet and elsewhere.

Theoretical dubiety. I don't want to use this essay as one more rehearsal for my favorite theme that Darwinian evolution cannot be read as a theory of progress, but only as a mechanism for building better adaptation to changing local environments – and that the equation of evolution with progress represents our strongest cultural impediment to a proper understanding of this greatest biological revolution in the history of human thought. Still, I can't help pointing out that this prejudice must underlie the ready proposition and acceptance of such a manifestly improbable notion as linear progress for the history of parietal art from thirty thousand to ten thousand years ago.

But why do I label the progressivist hypothesis of Breuil and Leroi-Gourhan as 'manifestly improbable'? After all, humans did evolve from apish ancestors with smaller brains and presumably more limited mental capacities, artistic and otherwise. So why shouldn't we see progress through time?

The answer to this query requires a consideration of proper scale. The twenty-thousand-year span of known parietal art does not reach deep into our apish ancestry (where a notion of general mental advance could be defended). The earliest parietal art lies well within the range of our current species, *Homo sapiens*. (By best estimates, *Homo sapiens* evolved in Africa some 200,000 years ago, and had probably migrated into the Levant [if not into Europe proper] by about ninety thousand years ago.) Therefore, the painters of the first known parietal art were far closer in time to folks living today than to the original *Homo sapiens*.

But a progressivist critic might still retort: 'Okay, I now understand that we are only discussing a sliver of human history, not most of the whole story since our split from the common ancestor of chimpanzees and us. But the trend of the whole should also be manifest within the shorter history of individual species, for evolution should move slowly and steadily to higher levels of mentality.' Herein lies the key prejudice underlying our uncritical acceptance of the progressivist paradigm for the history of art. It just feels 'right' to us that the very earliest art should be primitive. Older in time should mean more and more rudimentary in mental accomplishment.

And here, I think, we make a simple (but deep and widespread) error. Apparently similar phenomena of different scale do not become automatically comparable, but often (I would say usually) differ profoundly. Changes *between* species in an evolutionary sequence represent a completely different phenomenon from variation (spatial or temporal) *within* a single species. Humans have bigger brains than ancestral monkeys; these monkeys have bigger brains than distantly ancestral fishes. This increase in brain size does record a great gain in mental complexity. But a correlation of size and smarts across species does not imply that variation in brain size among modern humans also correlates with intelligence. In fact, normal adults differ in brain size by as much as 1,000 cubic centimeters, and no correlation has ever been found between size and intelligence (the average human brain occupies about 1,300 cc of volume).

Similarly, while evolution obviously produces change between one species and the next in a sequence of descent, most individual species don't alter much during their geological lifetimes. Large, widespread, and successful species tend to be especially stable. Humans fall into this category, and the historical record supports such a prediction. Human bodily form has not altered appreciably in 100,000 years. As I stated earlier, the Cro-Magnon cave painters are us – so why should their mental capacity differ from ours? We don't regard Plato or King Tut as dumb, even though they lived a long time ago. Remember that the distance from Plato to the parietal painters spans far less time than the interval separating these painters from the first *Homo sapiens*.

But defenders of progressivism in parietal art might still fall back upon one potentially promising argument: cultural change differs profoundly from biological evolution. We can admit biological stability and still expect an accumulative and progressive history of art or invention. The road has been both long and upward from Jericho and some scratch farming to New York City and the World Wide Web.

Fair enough in principle – but, again, the known timing precludes such an argument in practice. I will admit that if we happen to catch art at the very beginning, we would not expect full sophistication right away. But the oldest known parietal art, at thirty thousand years ago, lies well into the history of *Homo sapiens* in Europe – far closer to us today than to the first invasion from Africa. I don't know why earlier art hasn't been found (perhaps we just haven't made the discovery yet; perhaps people only moved into areas with caves at a much later time). I doubt that Ugh, the first Cro-Magnon orator, spoke in truly dulcet tones. But we surely don't regard Pericles as worse than Martin Luther King, Jr, just because he lived a few thousand years ago. Phidias doesn't pale before Picasso, and no modern composer beats Bach by mere virtue of residence in the twentieth century. Please remember that the first known Cro-Magnon artist, at thirty thousand years ago, stands closer to Pericles and Phidias than to Ugh, the orator, and Ur, the very first painter. So why should parietal art be any more primitive than the great statue of Athena that once graced the Parthenon?

As a final point, why should areas as distant as southern Spain, northeastern France, and southeastern Italy go through a series of progressive stages in lockstep over twenty thousand years? Regional and individual variation can swamp general trends, even today in a world of airplanes and televisions. Why did we ever think that evolution should imply a primary signal of uniform advance?

This general line of criticism has been well articulated by Paul G. Bahn and Jean Vertut in their 1988 book *Images of the Ice Age*. (I am pleased that they found our paleontological theory of punctuated equilibrium useful in constructing their critique.)

> The development of Paleolithic art was probably akin to evolution itself: not a straight line or ladder, but a much more circuitous path – a complex growth like a bush, with parallel shoots and a mass of offshoots; not a slow, gradual

> change, but a 'punctuated equilibrium,' with occasional flashes of brilliance . . . Each period of the Upper Paleolithic almost certainly saw the coexistence and fluctuating importance of a number of styles and techniques, . . . as well as a wide range of talent and ability . . . Consequently, not every apparently 'primitive' or 'archaic' figure is necessarily old (Leroi-Gourhan fully admitted this point), and some of the earliest art will probably look quite sophisticated.

Empirical disproof. Theoretical arguments may be dazzling, but give me a good old fact any time. The linear schemes of Breuil and Leroi-Gourhan had been weakening for many years as new information accumulated and old certainties evaporated. But one technical advance truly opened the floodgates. Thanks to a new method of radiocarbon dating – called AMS, for Accelerator Mass Spectrometry – only tiny amounts of charcoal need now be used, and paintings may therefore be analyzed without removing significant material.

In late 1994, three French explorers discovered a wonderful new site, now called Chauvet Cave. The animals at Chauvet, particularly the magnificent horses and lions, match anything else in Paleolithic art for sophistication and accuracy. But the radiocarbon dates, multiply repeated and presumably accurate, give ages in excess of thirty thousand years – making Chauvet the oldest of all known caves with parietal art. If the very oldest includes the very best, then our previous theories of linear advance must yield. In his epilogue to a gorgeous book, published in 1996, on this new site, Jean Clottes, a leading expert on Paleolithic art, writes:

> The subdivision of Paleolithic art proposed by Leroi-Gourhan, in successive styles, must be revised. His Style I, in which Chauvet Cave should be placed, was defined as archaic and very crude without any definite mural depictions, and is obviously no longer adequate. We now know that sophisticated techniques for wall art were invented . . . at an early date. The rendering of perspective through various methods, the generalized use of shading, the outlining of animals, the reproduction of movement and reliefs, all date back more than 30,000 years . . . This means that the Aurignacians, who coexisted with the last Neanderthals before replacing them,

> had artistic capabilities identical to those of their successors. Art did not have a linear evolution from clumsy and crude beginnings, as had been believed since the work of the Abbé Henri Breuil.

Let us not lament any lost pleasure in abandoning the notion that we now reside on an ever-rising pinnacle of continuous mental advance, looking back upon benighted beginnings. Consider instead the great satisfaction in grasping our true fellowship with the first known Paleolithic artists. There but for the grace of thirty thousand additional years go I. These paintings speak so powerfully to us today because we know the people who did them; they are us.

In a famous paradox, Francis Bacon wrote: *antiquitas saeculi, juventus mundi* (or, roughly, 'the old days were the world's youth'). In other words, don't think of the Paleolithic as a time of ancient primitivity, but as a period of vigorous youth for our species (while we today must represent the graybeards). Paleolithic art records our own early age, and we feel a visceral union with the paintings of Chauvet because, as Wordsworth wrote, 'the child is father of the man.' But we should also note the less frequently cited first verse of his poem:

> My heart leaps up when I behold
> A rainbow in the sky;
> So was it when my life began;
> So is it now I am a man.

We have loved the rainbow for thirty thousand unbroken years and more. We have struggled to depict the beauty and power of nature across all these ages. The art of Chauvet – and Lascaux, and Altamira, and a hundred other sites – makes our heart leap because we see our own beginnings on these walls, and know that we were, even then, worthy of greatness.

PART VI

SOCIOBIOLOGY AND EVOLUTIONARY PSYCHOLOGY

FROM their earliest appearances in intellectual debate, evolutionary theories have both reflected and been employed in arguments over the human social and political condition. In both Europe and the United States, social Darwinism, eugenics and many other forms of political prescription preceded the debate that erupted famously, and most fiercely, around the publication, in 1975, of E. O. Wilson's book *Sociobiology*. Gould describes the 'pervasive influence' of such pseudo-Darwinism in fields as far apart as cultural anthropology and child development in his history of the relationship between evolutionary and developmental thinking in biology, *Ontogeny and Phylogeny*, a brief extract of which survey forms the first of our pieces in this section. One such pervasive influence has been the glib phrase, first heard around the end of the nineteenth century, that 'ontogeny recapitulates phylogeny'. The rationale for this belief comes from studying the development of the human foetus during pregnancy. From conception to birth, the foetus seems to pass through a variety of forms – fish, amphibian etc – that resemble those of our evolutionary ancestors. These resemblances are not accidental, reflecting as they do some of the exigencies of development and the fact that evolutionary pressures can only work on existing forms. However the 'recapitulation' theory, once popular, is now thoroughly discredited. In this extract, Gould shows how its metaphorical – ideological if you like – use has permeated many fields of human enquiry.

The conflicts that followed the publication of Wilson's book have been documented – though scarcely neutrally – by those involved and, sometimes, by their acolytes amongst historians of science. Steve was one of those centrally involved in the counterattack on the Wilsonian theses, notably the assumption that central features of United States society – its class, race and gender structure, its inequalities of status wealth and power – were adaptive, evolved features of the human condition, deducible from Darwinian principles. One of the central assaults on sociobiological assumptions was on what Gould and his co-author, Richard Lewontin, christened 'the Panglossian paradigm' –

the assumption that the features we observe in the living world around us are there because they have an adaptive significance in this best of all possible worlds. As we describe in the Introduction to this collection, Gould delivered the key paper at a Darwin symposium in the prestigious halls of London's Royal Society in a characteristic bravura performance, and in tribute we here reprint 'The Spandrels of San Marco and the Panglossian Paradigm' in full.

During the 1990s sociobiology metamorphosed into evolutionary psychology, and once again Gould was in the thick of the critical debate. 'More Things in Heaven and Earth' originated as a critique of the philosopher Daniel Dennett in the *New York Review of Books*, and was then edited for *Alas, Poor Darwin*, the collection of critiques of evolutionary psychology by Hilary Rose and one of the current editors. Dennett's prior attack on Gould's exaptationist spandrels certainly helped set the tone of this robust assault on fundamentalist ultra-Darwinism.

Two final extracts in this section show the pervasive influence of pseudo-Darwinian ideas outside sociobiology. Thus both Engels and Freud were enthusiastic Darwinians. 'Posture Maketh the Man' explores some of Engels' views. Freud combined then-fashionable ideas on the 'primitive horde' in early human society and the notion that ontogeny recapitulated phylogeny in a heady mix which still embarrasses even the most devoted of Freudians – and gives full scope to Gould's amused dismissal.

ONTOGENY AND PHYLOGENY

ERNST Haeckel described the revolutionary power of Darwinian thought with a characteristic flourish:

> Dogma and authority, mutually dedicated to the suppression of all free thought and unfettered knowledge of nature, have erected a barrier of prejudice two or three times stronger than the Great Wall of China about the fortress of organic morphology – a citadel into which all kinds of distorted superstition have withdrawn as their last outpost. Nevertheless, we go into the battle without fear and sure of victory.[1]

We may smile at the exaggeration or wince at a vigor rarely encountered in scientific treatises; still, we cannot deny that evolutionary theory was the most upsetting idea that later nineteenth-century science introduced to a world steadily retreating from traditional notions of static order. The influence of evolution in fields far removed from biology has been documented almost to exhaustion (at least to mine), but the impact of Haeckel's favorite weapon has not been widely noted by historians. Nonetheless, the theory of recapitulation played a fundamental role in a host of diverse disciplines; I suspect that its influence as an import from evolutionary theory into other fields was exceeded only by natural selection itself during the nineteenth century.

The historical chapters of this book deal almost exclusively with theories and debates about the mechanisms of relationships between ontogeny and phylogeny. I would need another volume even to begin an adequate treatment of how biologists used recapitulation in their daily work. This omission is unfortunate because we cannot judge a theory's impact merely by documenting lengthy debates about its mechanism. We grasp the importance of recapitulation only when we understand that it served as the organizing idea for generations of work in comparative embryology, physiology, and morphology. In pooh-poohing the biogenetic law, E. G. Conklin gave an excellent summary of its beguiling appeal: 'Here was a method which promised to reveal more important secrets of the past than would the unearthing of all the buried monuments of antiquity – in fact nothing less than a complete genealogical tree of all the diversified forms of life which inhabit the earth. It promised to reveal not only the animal ancestry of man and the line of his descent but also the method of origin of his mental, social and ethical faculties'.[2]

In my own field of paleontology, for example, it governed most studies in phyletic reconstruction from Haeckel's day right through the 1930s. At the turn of the century, the classification of almost every invertebrate phylum relied upon morphological criteria chosen for their ontogenetic value in constructing phylogeny from the biogenetic law (facial sutures of trilobites, suture patterns of ammonites, for example). As late as 1957, Jesse James Galloway wrote: 'Ideally, a classification is built on the basis of comparative structure, and the application of the Law of Recapitulation, checked by the known geologic range of each taxonomic group'.[3] Faith in recapitulation was unbounded among paleontologists. In 1898, James Perrin Smith echoed a common, if extreme, claim:

> One can even prophesy concerning the occurrence of unknown genera in certain horizons when he finds their counterparts in youthful stages of later forms; in fact he could often furnish just as exact a description as if he had the adult genus before him.[4]

With ample justification, a recent and popular textbook of invertebrate paleontology states: 'It is no exaggeration to say that the theory of recapitulation has had more effect upon paleontologic thought than has any doctrine aside from that of organic evolution itself'.[5] My colleague Bernhard Kummel tells of an argument he had in the 1940s with R. C. Moore, the greatest 'classical' paleontologist of our century. Kummel, as a bright young Turk, was expressing some gentle doubts about recapitulation over dinner one evening when Moore, his patience stretched to the limit, brought down his fist and exclaimed: 'Bernie, do you deny the law of gravity!'

Though I have retreated shamefully before the task of documenting how biologists worked with the biogenetic law in their empirical studies, I cannot abandon the historical treatment of recapitulation without some discussion of its actual use. But I shall adopt a different tactic and explore the impact of recapitulation in fields far removed from biology. As a criterion for the importance of an idea, widespread and influential exportation to other disciplines must rank as highly as dominance within a field. With such an *embarras de richesses*, this chapter can be little more than a potpourri of citations. My files are bulging and my diligent scouts send me more material every week.

I shall present five essays on subjects strongly influenced by recapitulation: criminal anthropology, racism, child development, primary education, and psychoanalysis. Many other areas would have furnished equally impressive proof of influence. I have, for example, considered none of the arts, though recapitulation played a strong role in each branch. References abound in nineteenth-century literature. Tennyson, for example, wrote in the epithalamium of *In Memoriam* (1850) about a child who, in gestation, will be

> moved through life of lower phase
> Result in man, be born and think.

William Blake, in his *First Book of Urizen* (1794), invoked a similar image while his scientific counterparts were advocating recapitulation as an essential ingredient of romantic biology:

> Many sorrows and dismal throes,
> Many forms of fish, bird and beast
> Brought forth an Infant form
> Where was a worm before.[6]

Writers on the history and ethnology of art delighted in comparing the scribbles of 'civilized' children with the finished products of ancient civilizations and modern 'primitives'.[7] John Wesley Powell, America's premier ethnologist, compared child development with human history in tracing the evolution of music 'from dance to symphony'.[8]

Recapitulation intruded itself into every subject that offered even the remotest possibility of a connection between children of 'higher' races and the persistent habits of adult 'savages.' One need only consult the 500-page compilation of A. F. Chamberlain's *The Child: A Study in the Evolution of Man* (1900)[9] – an uncritical, but copious compendium of recapitulation. Consider the following:

The 'counting out' rhymes used by children to begin games and make decisions: 'a notable example of the survival in the usage of children of the serious practices of adults in primitive stages of culture . . . Children now select their leader or partner as once men selected victims for sacrifice'.[10]

A boy's practice of hunting and fishing with his hands, a reminiscence of prehistoric life before the evolution of tools: 'The old proverb, "a bird in the *hand* is worth two in the bush," grew up quite naturally it would seem'.[11]

Name changing: 'The child, the savage and the paranoiac love many names, like to change them, conceal them from strangers, etc.'[12]

Indiscriminate eating: 'A prominent abdomen is a noticeable characteristic alike of children, women, and many primitive races; a "pot-bellied" child and a "pot-bellied" savage are common enough'.[13]

'Orophily,' believe it or not: 'the delight in being upon a mound, a height, a hill, and commanding the universe around or merging oneself into it'.[14]

Belief in the reality of dreams: 'The child and the savage meet on this ground, some young boys and girls being as firmly impressed with the reality of their dreams as are the Brazilian Indians'.[15]

Love of adornment: 'The child is one with the savage in picking up the pebble from the beach or the bright feather from the ground'.[16]

Finally, while still in the domain of the ridiculous, consider Havelock Ellis on posture: 'The apes are but imperfect bipeds, with tendencies towards the quadrupedal attitude; the human infant is as imperfect a biped as the ape; savage races do not stand so erect as civilized races. Country people . . . tend to bend forward, and the aristocrat is more erect than the plebeian. In this respect women appear to be nearer to the infantile condition than men'.[17]

Perhaps an even stronger testimony to the beguiling appeal of recapitulation can be found in its tenacious survival in casual references of modern humanists, more than half a century after scientists ditched it. 'Ontogeny recapitulates phylogeny' is a literary epitome too appealing to resist, whatever its truth value.

Consider the following from a recent popular science-fiction novel (Edgar Rice Burroughs's *Out of Time's Abyss* should also be consulted):

> The brief span of an individual life is misleading. Each one of us is as old as the entire biological kingdom, and our bloodstreams are tributaries of the great sea of its total memory. The uterine odyssey of the growing foetus recapitulates the entire evolutionary past, and its central nervous system is a coded time-scale, each nexus of neurones and each spinal level marking a symbolic station, a unit of neuronic time.[18]

Or this from a work in literary criticism:

> The earliest age of mankind is associated with the verdure of springtime, with the spontaneity of childhood, and often with the awakening of love . . . Novelists, discovering for themselves the principle that ontogeny recapitulates phylogeny, have concentrated more and more intensively on the joys and pangs of adolescence.[19]

Or W. H. Auden writing on the death of Stravinsky:

> Stravinsky's life as a composer is as good a demonstration as any I know of the difference between a major and a minor artist . . . The minor artist, that is to say, once he has reached maturity and found himself, ceases to have a history. A major artist, on the other hand, is always re-finding himself, so that the history of his works recapitulates or mirrors the history of art.[20]

Or the Jungian poetry of Theodore Roethke:

> By snails, by leaps of frog, I came here, spirit.
> Tell me, body without skin, does a fish sweat?
> I can't crawl back through those veins,
> I ache for another choice.
>
> (from 'Praise to the End!' 1951)

Or this, from a nuclear physicist:

> The Fermilab synchrotron employs a four-stage acceleration process. Protons, obtained by ionizing hydrogen in a discharge tube, receive their first push in the electric field produced by the 750,000 volts of a Cockcroft-Walton generator, a direct descendant of the first particle accelerator. Ontogeny recapitulates phylogeny for accelerators too![21]

Or even Dr Spock setting the child-rearing habits of a nation:

> Each child as he develops is retracing the whole history of mankind, physically and spiritually, step by step. A baby starts off in the womb as a single tiny cell, just the way the first living thing appeared in the ocean. Weeks later, as he lies in the amniotic fluid in the womb, he has gills like a fish. Towards the end of his first year of life, when he learns to clamber to his feet, he's celebrating that period millions of years ago when man's ancestors got up off all fours . . . The child in the years after six gives up part of his dependence on his parents. He makes it his business to find out how to fit into the world outside his family. He takes seriously the rules of the game. He is probably reliving that stage of human history when our wild ancestors found it was better not to roam the forest in independent family groups but to form larger communities.[22]

Criminal Anthropology

Evolutionary theory quickly became the primary weapon for many efforts in social change. Reformers argued that the social and legal systems of Western Europe had been founded on antiquated notions of natural reason or Christian morality; and did not face squarely the irrevocable biology of human nature. Proposals for change might shock traditional ethics, but if they brought social procedure into harmony with human biology, we might establish the beginning of a rational and scientific order freed from ancient superstition and therefore, in the long run, humane in the literal sense.

The late nineteenth-century school of 'criminal anthropology' pursued this argument relentlessly. Previous systems of criminal law relied on social and ethical ideas of justice, fairness, protection, and retribution. They made no attempt to judge the biology of criminals – to learn if any recognizable peculiarity of their heredity might predispose them to lawlessness. Previous systems studied the crime; modern science would study the criminal. 'Criminal anthropology,' wrote Sergi, 'studies the delinquent in his natural place – that is to say, in the field of biology and pathology'.[23]

Criminal anthropology had its roots in Italy where Cesare Lombroso, its founding father, published the first edition of *L'uomo delinquente* ('Criminal Man') in 1876. It spread widely and became one of the most important scientific and social movements of the late nineteenth century.

Lombroso argued that many criminals were born with an almost irrevocable predisposition to lawlessness. These born criminals could be recognized by definite physical signs; they were indeed 'a well characterized anthropological variety'.[24] At this point Lombroso's argument takes a phyletic turn. The stigmata of the born criminal are not anomalous marks of disease or hereditary disorder; they are the atavistic features of an evolutionary past. The born criminal pursues his destructive ways because he is, literally, a savage in our midst – and we can recognize him because he carries the morphological signs of an apish past. Lombroso records his moment of truth:

> In 1870 I was carrying on for several months researches in the prisons and asylums of Pavia upon cadavers and living persons, in order to determine upon substantial differences between the insane and criminals, without succeeding very well. Suddenly, the morning of a gloomy day in December, I found in the skull of a brigand a very long series of atavistic anomalies . . . The problem of the nature and of the origin of the criminal seemed to me resolved; the characters of primitive men and of inferior animals must be reproduced in our times.[25]

Since ontogeny recapitulates phylogeny, a perfectly normal child must pass through a savage phase as well. The child, too, is a natural criminal at one stage of his development. The normal adult passes on to civilization as he mounts the phyletic scale in his own growth; the born criminal remains mired in his brutish past. This argument rings today with such patent absurdity (and viciousness) that I hasten to add a few disclaimers to explain why it attracted such a following during the late nineteenth century.

Lombroso and his school never attributed all criminal acts to innate depravity. In fact, their argument depended upon a strict separation between born criminals and those induced to commit their acts for social reasons such as poverty or extreme anger. Lombroso believed that no more than 40 percent of criminals bore the anthropometric stigmata of an inherent criminal disposition. His theories and recommendations applied to this group alone. E. Ferri[26] presented a graded classification, ranging from born criminals to criminals of passion via such intermediate categories as the criminal of contracted habit who transgresses by his own free choice but does it so assiduously that his children, through Lamarckian inheritance, become born criminals.

Although these distinctions might seem to mitigate the force of Lombroso's claims, they actually served to make the theory invincible to disproof. Lombroso did not tie specific criminal acts to criminal types – the types were defined by inferred biology, not by what they did. Murder might be the work of the most incorrigible born criminal or the most law-abiding cuckold. Hence, Lombroso's theory was invincible. Take the perpetrator of any criminal act: if he possesses the stigmata, he performed it by his biological nature; if not, by his social circumstance. All cases are covered. You build a category to incorporate exceptions within your theory; all cases can then be allocated.

Furthermore, the criminal anthropologists were, for the most part, neither petty sadists nor proto-fascists. The major figures in the Italian school were socialists who viewed their theory as the spearhead for a rational, scientific society based on human realities. They wished, for example, to reform criminal law by adapting punishment to the nature of criminals rather than to the severity of crimes. Social criminals should be jailed for the time needed to secure their amendment, but born criminals offered little hope for permanent cure. Lombroso and Ferri did not recommend the death penalty (though they did not oppose it);[27] human sensibility required some retreat from what might be biologically preferable. They tended to recommend irrevocable detention for life (in pleasant, but isolated surroundings) for any recidivist with the telltale stigmata.[28] In addition, born criminals were not hopelessly doomed to a life of wrongdoing. Eyeglasses can cure inherited problems of vision, but they must be worn for life. If born criminals were identified early in childhood, they might be sent to bucolic retreats or shipped out to sea as cabin boys; they might be isolated and supervised in perpetuity, but they could not be cured: 'Theoretical ethics passes over these diseased brains, as oil does over marble, without penetrating it'.[29]

The personal motivations of the Italian school may have been scientific, even humane, but their primary impact lay in another direction. And so it has always been with extreme versions of biological determinism – witness the conversion of some well-intentioned eugenics into a rationale for Hitler. Innate determination is a dangerous argument, for it is easily seized by men in power as a 'scientific' excuse for preserving a status quo or eliminating any unfavored group as biologically perverse. Once Lombroso claimed a rationale for capital punishment in the operation of natural selection, he could not distance himself from Taine's implication: 'You have shown us fierce and lubricious orang-utans with human faces. It is evident that as such they cannot act otherwise. If they ravish, steal, and kill it is by virtue of their own nature and their past, but there is all the more reason for destroying them when it has been proved that they will always remain orang-utans.' (Quoted, not unfavorably, in Lombroso;[30] in fact, Lombroso adds: 'The fact that there exist such beings as born criminals, organically fitted for evil, atavistic reproductions, not simply of savage men but even of the fiercest animals, far from making us more compassionate towards them, as has been maintained, steels us against all pity.'[31]) And once Ferri recommended that 'tattooing, anthropometry, physiognomy . . . reflex activity, vaso-motor reactions and the range of sight' be used as criteria of judgment by magistrates,[32] it was not a big step to Hitler's 'final solution' for 'undesirable' groups.

Lombroso, in the 1887 edition of *L'uomo delinquente*, presents his argument in a strict phyletic mode. Part 1, on the 'embryology of crime,' devotes its three chapters to demonstrating that what we call crime among civilized adults is normal behavior in animals (and even plants), adult savages, and children of civilized cultures – the three-fold parallelism of classical recapitulation theory. Lombroso begins with a collection of animal anecdotes in the anthropomorphic tradition. Animals may kill from rage (an ant, made impatient by a recalcitrant aphid, killed and devoured it);[33] they murder to suppress revolts (some ant slaves, tired of being pushed around, seized the queen's leg and tried to pull her out of the nest; the queen seized the rebel by the head and killed her);[34] they do away with sexual rivals (an adulterous male stork and his lover found the rightful husband chasing frogs on a mud flat and killed him with their beaks);[35] they violate the young and undeveloped (male ants, deprived of sexual females, attempt to copulate with workers; but the workers, with their atrophied sexual organs, suffer greatly and often die); they form criminal associations (three communal beavers shared an area with one solitary individual; the three communals visited the solitary and were well-treated; the solitary returned the visit in a neighborly way and was killed for his solicitude);[36] even the voracious habits of insectivorous plants are considered as an 'equivalent of crime,' though I fail to see how this interspecific action differs from any other form of eating.

In part 2, on the 'Pathological Anatomy and Anthropometry of Crime,' Lombroso discusses the atavistic stigmata of born criminals – their physical links to a past represented today by living savages and our own children. The list of apish or primitive human features includes: relatively long arms, prehensile foot with mobile big toe, low and narrow forehead, large ears, thick skull, large and prognathous jaw, copious hair on the male chest, browner skin, and such physiological characters as diminished sensitivity to pain and absence of vascular reaction (criminals and savages do not blush). Atavisms do not stop at the primate level. Large canine teeth and a flat palate recall a distant mammalian past. The median occipital fossette of many criminals looks like that of rodents (and, by recapitulation, of three month old fetuses).[37] Lombroso even compares the common facial asymmetry of criminals with the normal condition of flatfishes![38] 'The atavism of the criminal, when he lacks absolutely every trace of shame and pity, may go back far beyond the savage, even to the brutes themselves'.[39]

But the stigmata are not only physical. The social behavior of the born criminal also allies him with savages and children. Lombroso and his school placed special emphasis on tattooing: 'Tattooing is one of the essential characters of primitive man – one that still survives in the savage state'.[40] Lombroso analyzed the content of tattoos upon born criminals and found them lewd, lawless, or exculpating ('born under an unlucky star,' 'no luck,' and 'vengeance,' for example, though one read, he had to admit, 'long live France and french fried potatoes').[41] Criminal slang is a language of its own, markedly similar in such features as onomatopoeia and personification to the talk of children and savages: 'Atavism contributes to it more than anything else. They speak differently because they feel differently; they speak like savages, because they are true savages in the midst of our brilliant European civilization'.[42] Criminals are often as insensible to pain as savages (Ellis tells the tale of some Maoris who cut off their toes to wear European boots[43]), and Lombroso notes that 'Their physical insensibility well recalls that of savage peoples who can bear in rites of puberty, tortures that a white man could never endure. All travellers know the indifference of Negroes and American savages to pain: the former cut their hands and laugh in order to avoid work; the latter, tied to the torture post, gaily sing the praises of their tribe while they are slowly burnt'.[44]

In later years, Lombroso retreated steadily before criticism of his atavistic theory. While maintaining his strict adherence to the congenital nature of crime, he broadened both his criteria and his notion of cause. To the category of atavisms, he added the criterion of developmental arrest. Of course, under the biogenetic law, many arrests had the same phyletic significance as atavisms, since they brought the ancient phyletic states of embryology and childhood forward to the adult. But others could only be classed as anomalies and illnesses (Lombroso lays great stress on epilepsy). Stigmata are still to be found, but they need not have phyletic significance. We may therefore 'see in the criminal the savage man and, at the same time, the sick man'.[45]

The recapitulatory argument for natural criminality of children is one of the two or three central themes in Lombroso's fabric – not a mere collateral point. He devoted much of his work to cataloguing the criminal nature of child behavior in general and the statistics of legally criminal acts by children in particular. 'One of the most important discoveries of my school is that in the child up to a certain age are manifested the saddest tendencies of the criminal man. The germs of delinquency and of criminality are found normally even in the first periods of human life'.[46] Havelock Ellis proclaimed that 'the child is naturally, by his organization, nearer to the animal, to the savage, to the criminal, than the adult'.[47] An American follower added: 'It is proved by voluminous evidence, easily accessible, that children are born criminals'.[48]

Lombroso cites the following traits as normal in children and criminal (or disposing towards it) in adults: anger, vengeance, jealousy, lying, lack of moral sense, lack of affection, cruelty, laziness, use of slang, vanity, alcoholism (if alcohol is made available),[49] predisposition to obscenity, imitation, and lack of foresight.[50] Many members of his school made special studies to compare the expression of these traits in children, savages, and criminals. A. F. B. Crofton, for example, considered the use of onomatopoeia in criminal slang, children's talk, and primitive language.[51] The thief calls a train a 'rattler,' the child a 'choo-choo.'

But nature and education conspire to bring good from natural evil. The child mounts through his phyletic history and reaches the promise of his species as a young adult: 'Now when the child becomes a youth, largely through the training of his parents and of the school, but more so by nature itself, when inclined to the good, all this criminality disappears, just as in the fully developed fetus the traces of the lower animals gradually disappear which are so conspicuous in the first months of the fetal life'.[52] In a nature not 'inclined to the good,' however, these traits persist and the arrested features of childhood produce a criminal adult. When Lombroso shifted his emphasis from atavisms to developmental arrests as criteria for the stigmata of criminality, he greatly increased the importance of a recapitulatory argument based upon children (as Havelock Ellis emphasized so strongly).[53] When the physical signs of an apish ancestry served as primary markers of the born criminal, children found no place in the argument (though they always played a major role in all discussions of behavioral signs from tattooing to slang) – humans are neotenous, and our babies resemble adult ancestors in physical appearance even less than we do. In developmental arrest, however, children become the focus of inquiry. Crime is still a phyletic question, because children, by the biogenetic law, are close to ancestors.

The classical argument for recapitulation involves a threefold parallelism of paleontology, comparative anatomy, and ontogeny. Morphologists occasionally added a fourth source of evidence – teratology and the phyletic explanation of abnormalities as developmental arrests. This fourth criterion – the abnormal individual as an arrested juvenile – forms an important part of the usage made by other disciplines of the biogenetic law. We have seen how Lombroso invoked it in his theory of criminality. We will encounter it again in Freud's theory of neurosis.

Racism

Vietnam reminds me of the development of a child.
Gen. William Westmoreland

Few connections are more intimate and pervasive than that between racism and statements by scientists about human diversity (I do not say scientific statements) made before the Second World War. Consequently, we should not be surprised that the very first sustained argument for recapitulation in morphology was cast in a racist mold. H. F. Autenrieth receives traditional credit for a work published in 1797 (C. F. Kielmeyer's article of 1793 spoke only of physiology, and earlier statements are either analogistic or incidental). After arguing that the completed forms of lower animals are merely stages in the ontogeny of higher forms, Autenrieth speaks of 'certain traits which seem, in the adult African, to be less changed from the embryonic condition than in the adult European' ('quaedam, quae in adulto Afro minus, quam adulto Europaeo ex reliquiis embryonis mutata videntur'[54]).

For anyone who wishes to affirm the innate inequality of races, few biological arguments can have more appeal than recapitulation, with its insistence that children of higher races (invariably one's own) are passing through and beyond the permanent conditions of adults in lower races. If adults of lower races are like white children, then they may be treated as such – subdued, disciplined, and managed (or, in the paternalistic tradition, educated but equally subdued). The 'primitive-as-child' argument stood second to none in the arsenal of racist arguments supplied by science to justify slavery and imperialism. (I do not think that most scientists who upheld the primitive-as-child argument consciously intended to promote racism. They merely expressed their allegiance to the prevailing views of white intellectuals and leaders of European society. Still, the arguments were used by politicians and I can find no evidence that any recapitulationist ever objected.)

Biological arguments based on innate inferiority spread rapidly after evolutionary theory permitted a literal equation of modern 'lower' races with ancestral stages of higher forms. But similar arguments were far from unknown before 1859. Several of the leading pre-evolutionary recapitulationists ranked human races by the primitive-as-child argument.

Schiller, a godfather of Naturphilosophie, wrote: 'The discoveries which our European sailors have made in foreign seas . . . show us that different people are distributed around us . . . just as children of different ages may surround a grown-up man'.[55]

Chambers, from an evolutionary perspective, wrote in his *Vestiges* of 1844 that 'the varieties of his race are represented in the progressive development of an individual of the highest, before we see the adult Caucasian, the highest point yet attained in the animal scale.' Agassiz endeared himself to the proponents of slavery when, as Europe's leading natural historian, he chose to settle in America and to maintain that Blacks represent a separate and lower species. 'The brain of the Negro,' he claimed, 'is that of the imperfect brain of a seven month's infant in the womb of the White'.[56]

Étienne Serres approached the subject from a more liberal perspective. Serres separated humans from other animals on the criterion of perfectibility. Animals are fixed in their created status; human races, however, can advance by evolution in both the physical and intellectual realms.[57] The polygenist belief that human races are separate species relies on dead anatomy which refutes the senses and disgusts the spirit by the aridity of its considerations . . . It replaces with multiple species the sublime idea of human unity – a tendency even more dangerous because it seems to lend scientific support to the enslavement of races less advanced in civilization than the Caucasian'.[58] Nonetheless, however flexible in future movement, the scale of human races could still be ranked from lower to higher – and recapitulation provided the major criterion for ranking. Serres wrote: 'May we not say in a general manner that, in its progress, the Ethiopian race has stopped at the beginning of Caucasian adolescence . . . and the Mongolian race at [Caucasian] adolescence itself – times of arrest which form degrees of development within the unity of the human species'.[59] Serres was hard pressed to find many morphological signs of a parallel between Caucasian ontogeny and the adult sequence Negro–Mongolian–Caucasian. Difficult though it be to take seriously today, Serres emphasized the relative position of the belly button.[60] The relative distance between penis and navel increases during Caucasian ontogeny. Adult blacks have a navel as low slung as that of a Caucasian child; the adult Mongolian navel is a bit higher.

Biological arguments for racism may have been common before

1859, but they increased by orders of magnitude following the acceptance of evolutionary theory. The litany is familiar: cold, dispassionate, objective, modern science shows us that races can be ranked on a scale of superiority. If this offends Christian morality or a sentimental belief in human unity, so be it; science must be free to proclaim unpleasant truths. But the data were worthless. We never have had, and still do not have, any unambiguous data on the innate mental capacities of different human groups – a meaningless notion anyway since environments cannot be standardized. If the chorus of racist arguments did not follow a constraint of data, it must have reflected social prejudice pure and simple – anything from an a priori belief in universal progress among apolitical but chauvinistic scientists to an explicit desire to construct a rationale for imperialism.

Some recapitulationists ranked races by physical traits. E. D. Cope searched the human body to find a mere handful of characters that would affirm white superiority on the primitive-as-child argument:

> Let it be particularly observed that two of the most prominent characters of the negro [*sic*] are those of immature stages of the Indo European race in its characteristic types. The deficient calf is the character of infants at a very early stage; but, what is more important, the flattened bridge of the nose and shortened nasal cartilages are universally immature conditions of the same parts in the Indo-European.[61]

D. G. Brinton made a more comprehensive claim:

> The adult who retains the more numerous fetal, infantile or simian traits, is unquestionably inferior to him whose development has progressed beyond them . . . Measured by these criteria, the European or white race stands at the head of the list, the African or negro at its foot . . . All parts of the body have been minutely scanned, measured and weighed, in order to erect a science of the comparative anatomy of the races.[62]

But arguments about the brain provided more direct ammunition for ranking by evolutionary status. C. Vogt wrote:

> In the brain of the Negro the central gyri are like those in a foetus of seven months, the secondary are still less marked. By its rounded apex and less developed posterior lobe the Negro brain resembles that of our children, and by the protuberance of the parietal lobe, that of our females.[63]

If the primitive-as-child argument worked for the material ground of intelligence, it would perform even better for the mental traits themselves. Most racist recapitulationists relied more on the products of mind than on physical criteria for ranking. Cope argued that 'some of these features have a purely physical significance, but the majority of them are . . . intimately connected with the development of the mind'.[64] 'The intellectual traits of the uncivilized,' claimed Herbert Spencer 'are traits recurring in the children of the civilized'.[65] Lord Avebury compared 'modern savage mentality to that of a child',[66] while the English leader of child study stated: 'As we all know, the lowest races of mankind stand in close proximity to the animal world. The same is true for the infants of civilized races'.[67] And the American leader of child study, G. Stanley Hall, maintained that 'most savages in most respects are children, or, because of sexual maturity, more properly, adolescents of adult size'.[68]

My catalogue of specific examples is far too long to relate, but a sampling would include:

1. On the animism of children:

> Is it too fanciful to suppose that the belief of the savage in the occasional visits of the real spirit god to his idol has for its psychological motive the impulse which prompts the child ever and again to identify his toys and even his pictures with the realities which they represent.[69]

> The child who says, 'I am an engine' or 'I am a tiger,' making appropriate movements and sounds the while, is enjoining an imaginative identification of a genuinely primitive character. Is it not a justifiable distinction to say that the child is playing, while the savage is intensely serious at his rites.[70]

2. On aesthetic sensibility:

> In much of this first crude utterance of the aesthetic sense of the child we have points of contact with the first manifestations of taste in the race. Delight in bright, glistening things, in gay things, in strong contrasts of color, as well as in certain forms of movement, as that of feathers – the favorite personal adornment – this is known to be characteristic of the savage and gives to his taste in the eyes of civilized man the look of childishness. On the other hand, it is doubtful whether the savage attains to the sentiment of the child for the beauty of flowers.[71]

3. On art:

> If we look at representation by drawing or sculpture, we find that the efforts of the earliest races of which we have any knowledge were quite similar to those which the untaught hand of infancy traces on its slate or the savage depicts on the rocky faces of hills.[72]

The recapitulatory argument for ranking extended beyond races to any set of categories for which wealthy, Nordic males wished to assert their superiority. Lower classes within any society were a favorite target. Cope, for example, listed several simian characters among 'the lower classes of the Irish'.[73]

Women fitted the argument especially well for two reasons – the social observation that men wrote all the textbooks and the morphological fact that skulls of adult women are more childlike than those of men. Since the child is a living primitive, the adult woman must be as well. In 1821, J. F. Meckel noted the lesser differentiation of women from a common (and primitive) embryonic type; he also suspected that women, with their smaller brains, were innately inferior in intelligence.[74] Again, recapitulationists quickly moved from morphology to mental traits. Cope, for example, discoursed on the 'metaphysical characteristics' of women:

> The gentler sex is characterized by a great impressibility . . . , warmth of emotion, submission to its influence rather than that of logic, timidity and irregularity of action in the outer world. All these qualities belong to the male sex, as

> a general rule, at some period of life, though different individuals lose them at very various periods . . . perhaps all men can recall a period of youth when they were hero-worshippers – when they felt the need of a stronger arm, and loved to look up to the powerful friend who could sympathize with and aid them. This is the 'woman stage' of character.[75]

G. Stanley Hall argued that women's greater propensity for suicide expresses the primitive stage of submission to elemental forces.

> This is one expression of a profound psychic difference between the sexes. Woman's body and soul is phyletically older and more primitive, while man is more modern, variable, and less conservative. Women are always inclined to preserve old customs and ways of thinking. Women prefer passive methods; to give themselves up to the power of elemental forces, as gravity, when they throw themselves from heights or take poison, in which methods of suicide they surpass man. [Havelock] Ellis thinks drowning is becoming more frequent, and that therein women are becoming more womanly.[76]

At this point, I hasten to add that I am not selecting the crackpot statements of a bygone age. I am quoting the major works of recognized leaders. The sway of biological determinism, the lack of sensitivity to environmental influence, and the blatant desire to crown one's own group as biologically superior are quite characteristic of the time – and scarcely extinct today.

But can we prove that these assertions emerged from the scientific literature to influence social and political life? They surely did, as two lines of evidence attest. First of all, many scientists drew explicit political conclusions in their widely read books. Consider Vogt's justification for colonialism:

> The grown-up Negro partakes, as regards his intellectual faculties, of the nature of the child, the female, and the senile white . . . Some tribes have founded states, possessing a peculiar organization; but, as to the rest, we may boldly assert

> that the whole race has, neither in the past nor in the present, performed anything tending to the progress of humanity or worthy of preservation.[77]

Tarde added that many criminals (equivalent to primitives in Lombroso's theory) 'would have been the ornament and moral aristocracy of a tribe of Red Indians'.[78] The occasional anti-imperialists agreed completely with the primitive-as-child argument and only disputed the implied right of conquest:

> We are laboring to prevent the 'big fist' of adults from breaking in upon childhood and its evolutional activities; we ought also to labor to prevent the bigger fist of 'civilized races' from breaking in upon the like evolutional activities of primitive peoples with even more disastrous results . . . We ought to be as fair to the 'naughty race' abroad as we are to the 'naughty boy' at home.[79]

Second, many politicians and statesmen borrowed the primitive-as-child argument with an explicit bow to science and recapitulation. The Rev. Josiah Strong, in his plea for an imperial America, noted that modern science had provided the rationale justifying colonialism as an ethical venture. In pre-evolutionary days, Henry Clay had voiced religious doubts about the concept of racial superiority. 'I contend,' Clay argued, 'that it is to arraign the disposition of Providence Himself to suppose that He has created beings incapable of governing themselves'.[80] Strong replied that recapitulation, with its primitive-as-child argument, had married imperialism with scientific respectability. We had not only the right, but also the duty, to annex the Philippines:

> Clay's conception was formed . . . before modern science had shown that races develop in the course of centuries as individuals do in years, and that an undeveloped race, which is incapable of self-government is no more of a reflection on the Almighty than is an undeveloped child who is incapable of self-government. The opinions of men who in this enlightened day believe that the Filipinos are capable of self-government because everybody is, are not worth considering.[81]

In a similar vein, B. Kidd used recapitulation to justify the conquest of tropical Africa:

> The evolution in character which the race has undergone has been northwards from the tropics. The first step to the solution of the problem before us is simply to acquire the principle that [we are] dealing with peoples who represent the same stage in the history of the development of the race that the child does in the history of the development of the individual. The tropics will not, therefore, be developed by the natives themselves.[82]

The argument even turned up in the first verse of Kipling's most famous paean for colonialism:

> Take up the White Man's Burden
> Send forth the best ye breed
> Go, bind your sons to exile
> To serve the captives' need:
> To wait in heavy harness,
> On fluttered folk and wild –
> Your new-caught, sullen peoples,
> Half-devil and half-child.

Theodore Roosevelt, who had received an advance copy, wrote to Henry Cabot Lodge that it 'was very poor poetry but made good sense from the expansion point of view'.[83]

There should, in a view of history that motivates many scientists, be a happy ending to this sorry tale. By 1920, the theory of recapitulation was in disarray. By the 1930s Haeckel's insistence on universal acceleration and the consequent pushing back of ancestral adult characters to the juvenile stages of descendants had yielded to an expanded version granting equal orthodoxy to the opposite process: the juvenile traits of ancestors may, by retardation of development, become the adult stages of descendants. This appearance of ancestral juvenile traits in adult descendants is called paedomorphosis (child-shaped). It demands a conclusion exactly opposite to the primitive-as-child argument – for the child is now a harbinger of things to come, not a storehouse for the adult traits of ancestors. The child of a 'lower' race should be like the adult of a 'higher' group. In short, for racist arguments, two contradictory claims are necessary:

1. Under recapitulation, whites, as children, reach the level of black adults; whites then continue on to higher things during their ontogeny.

2. Under paedomorphosis, white adults retain the characteristics of black children, while blacks continue to develop (or rather devolve) during their ontogeny.

The irony of this conceptual change lies in the fact that our own species represents a most impressive case of paedomorphosis. In feature after feature, we resemble the juvenile stages of other primates – and this includes such markers of intellectual status as our bulbous cranium and relatively large brain. Even the nineteenth-century recapitulationists knew this in their heart of hearts, though they labored mightily to explain it away. Cope, for example, wrote at length about human features that display retarded development. In fact, Cope tried to have it both ways by arguing that whites are superior in both recapitulated and paedomorphic traits:

> The Indo-European race is then the highest by virtue of the acceleration of growth in the development of the muscles by which the body is maintained in the erect position (extensors of the leg), and in those important elements of beauty, a well-developed nose and beard. It is also superior in those points in which it is more embryonic than the other races, viz., the want of prominence of the jaws and cheek-bones, since these are associated with a greater predominance of the cerebral part of the skull, increased size of cerebral hemispheres, and greater intellectual power.[84]

When this shift from recapitulation to paedomorphosis occurred, during the 1920s and 1930s, the available data on human evolution included fifty years of accumulated facts, virtually all supporting the claim that black (and other 'primitive') adults were like white children. The men who collected these data – and it was always men – claimed that they had done so in the spirit of objective science, caring only for truth and untrammelled by political constraint. In fact, they often argued that their inegalitarian conclusions proved that hard science had triumphed over liberal or Christian sentimentalism. If their motivations were so simple and unsullied, then the replacement of recapitulation by paedomorphosis as an explanation for human evolution should have led to the following honest admission: hard facts prove that white children are like black adults; under paedomorphosis, children of primitives are like adult stages of advanced forms; therefore, blacks are superior to whites.

Needless to say, nothing of the sort happened. I presented the suggestion that it might have occurred not as a plausible hypothesis but merely as a bit of rhetoric illustrating the absurdity of any claim that scientists act 'objectively' on matters so vital to the interests of their patrons (and their own privileged position). In fact, the founders of paedomorphosis quietly forgot all the old data on adult-blacks-as-white-children and set out to find some opposing information that would reaffirm racism on the opposite, paedomorphic model.

Louis Bolk, Dutch anatomist and primary proponent of human paedomorphosis, reversed the catechism of recapitulation and proclaimed: 'In his fetal development the negro [*sic*] passes through a stage that has already become the final stage for the white man'.[85] Bolk made no attempt to hide his general interpretation: 'It is obvious that I am, on the basis of my theory, a convinced believer in the inequality of races. All races have not moved the same distance forward on the path of human evolution [*Menschwerdung*]'.[86] Or, more explicitly:

> The question is of great importance from an anthropological as well as from a sociological point of view. For it need hardly be emphasized that a different degree of fetalization [Bolk's term for paedomorphosis] means a more or less advanced state of hominization. The farther a race has been somatically fetalized and physiologically retarded, the further it has grown away from the pithecoid ancestor of man. Quantitative differences in fetalization and retardation are the base of racial inequivalence. Looked at from this point of view, the division of mankind into higher and lower races is fully justified.[87]

Bolk then scoured the human body for a selected list of traits to affirm the greater paedomorphosis of whites. He cites a more rounded skull, lesser prognathism of the jaw, slower somatic development, and longer life (without considering any environmental influence). 'The white race,' he concludes, 'appears to be the most progressive, as being the most retarded'.[88]

The argument is by no means extinct today, despite the efforts of Ashley Montagu[89] and other antiracists (scientific antiracism is largely a post-Hitler phenomenon). H. J. Eysenck[90] notes that African and black American babies exhibit faster sensorimotor development than whites. By age three, however, whites surpass blacks in IQ. There is also a slightly negative correlation between first year sensorimotor development and later IQ. Eysenck invokes paedomorphosis to link these facts and imply an innate mental inferiority among blacks: 'These findings are important because of a very general view in biology according to which the more prolonged the infancy the greater in general are the cognitive or intellectual abilities of the species. This law appears to work even within a given species'.[91] (We have here a classic example of a potentially meaningless, noncausal correlation. Suppose that differences in IQ are completely determined by environment; then, rapid motor development does not cause low IQ – it is merely another measure of racial identification, and a poorer one than skin color at that.)

As a final proof of extrascientific motivation, I note the conspiracy of silence that has surrounded two aspects of the paedomorphic argument that are very uncomfortable for white males anxious to retain their exalted status. First, it is hard to deny that Mongoloids – not Caucasians – are the most paedomorphic of human groups. Bolk performed a song and dance about the facts he listed and ended up by arguing that Caucasian and Mongoloid differences were too close to call.[92] But the iconoclastic Havelock Ellis, an early supporter of human paedomorphosis, faced the issue squarely in 1894: 'On the whole, it may be said that the yellow races are nearest to the infantile conditions; negroes and Australians are farthest removed from it, often although not always in the direction of the Ape; while the white races occupy an intermediate position'.[93] His generosity toward yellow skins did not extend to black, although it is easy to list an impressive set of features for which Africans are the most strongly paedomorphic of human groups.[94] Ellis continues, presaging Bolk's argument: 'The child of many African races is scarcely if at all less intelligent than the European child, but while the African as he grows up becomes stupid and obtuse, and his whole social life falls into a state of hide-bound routine, the European retains much of his childlike vivacity'.[95]

Second, women are clearly more paedomorphic than men. Again, Bolk chose to ignore the issue and Ellis met it directly with an admission of inferiority: 'The infant ape is very much nearer to Man than the adult ape. This means that the infant ape is higher in the line of evolution than the adult, and the female ape, by approximating to the infant type, is somewhat higher than the male'.[96] Women, Ellis affirms, are leading the direction of human evolution:

> She bears the special characteristics of humanity in a higher degree than man . . . Her conservatism is thus compensated and justified by the fact that she represents more nearly than man the human type to which man is approximating. This is true of physical characters: the large-headed, delicate-faced, small-boned man of urban civilization is much nearer to the typical woman than is the savage. Not only by his large brain, but by his large pelvis, the modern man is following a path first marked out by woman.[97]

Notes

1. E. Haeckel, *Generelle Morphologie der Organismen: Allgemeine Grundzüge der organischen Formen-Wissenschaft, mechanisch begründet durch die von Charles Darwin reformirte Descendenz-Theorie*, vol. 1 (Berlin: Georg Reimer, 1866), p. xv.
2. E. G. Conklin, 'Embryology and Evolution', in F. Mason (ed.), *Creation by Evolution* (New York: Macmillan, 1928), p. 70.
3. J. J. Galloway, 'Structure and Classification of the Stromatoporoidea', *Bulletin of American Paleontology*, 37 (1957): 395.
4. J. P. Smith, 'Evolution of Fossil Cephalopoda', in D. S. Jordan (ed.), *Footnotes to Evolution* (New York: D. Appleton, 1898), p. 122.
5. W. H. Easton, *Invertebrate Paleontology* (New York: Harper & Row, 1960), p. 33.
6. Blake, cited in J. Oppenheimer, 'Recapitulation', in P. P. Weiner (ed.), *Dictionary of the History of Ideas*, vol. 4 (New York: Charles Scribner's Sons, 1973), pp. 56–9.
7. J. Sully, 'Studies of Childhood, XIV: The Child as Artist', *Popular Science*, 48 (1895): 385–95; *idem, Studies of Childhood* (New York: D. Appleton, 1896).
8. J. W. Powell, 'Evolution of Music from Dance to Symphony', *Proceedings of the American Association for the Advancement of Science*, (1889): 1–21.
9. A. F. Chamberlain, *The Child: A Study of the Evolution of Man* (London: Walter Scott, 1900).

10. Ibid., p. 277.
11. Ibid., p. 279.
12. Ibid., p. 302.
13. Ibid., p. 315.
14. Ibid., p. 320.
15. Ibid., p. 336.
16. Ibid., p. 453.
17. H. Ellis, *Man and Woman* (New York: Charles Scribner's Sons, 1894), p. 59.
18. J. G. Ballard, *The Drowned World* (London: Penguin, 1965), p. 43.
19. H. Levin, *The Myth of the Golden Age in the Renaissance* (Bloomington: Indiana University Press, 1969).
20. W. H. Auden, 'Craftsman, Artist, Genius', *Observer*, 11 April 1971.
21. R. R. Wilson, 'High-Energy Physics at Fermilab', *Bulletin of the American Academy of Arts and Sciences*, 29 (1975): 20.
22. B. Spock, *Baby and Child Care*, revised edition (New York: Pocket Books, 1968), p. 229.
23. Sergi, cited in H. Zimmern, 'Criminal Anthropology in Italy', *Popular Science*, 52 (1898): 744.
24. Ferri to Fourth International Conference on Criminal Anthropology, 1896, cited in M. Parmelee, *The Principles of Anthropology and Sociology in Their Relations to Criminal Procedure* (New York: Macmillan, 1912), p. 80.
25. Lombroso, cited in ibid., p. 25.
26. E. Ferri, *Criminal Sociology* (New York: D. Appleton, undated), p. 45.
27. Ibid., p. 240; Lombroso in letter to Zimmern, cited in H. Zimmern, 'Reformatory Prisons and Lombroso's Theories', *Popular Science*, 43 (1893): 600–1.
28. C. Lombroso, *L'homme criminel: Criminel-né – fou moral – épileptique. Etude anthropologique et médico-legale* (Paris: Germer Bailliere, 1887), p. xvii.
29. *Idem*, 'Criminal Anthropology Applied to Pedagogy', *The Monist*, 6 (1895): 58.
30. *Idem*, *Crime: Its Causes and Remedies* (Boston: Little, Brown and Co., 1911), p. 428.
31. Ibid., p. 427.
32. Ferri, op. cit., p. 166.
33. Lombroso, *L'homme criminel*, p. 13.
34. Ibid.
35. Ibid., p. 16.
36. Ibid., p. 18.
37. Ibid., p. 181.
38. Lombroso, *Crime*, p. 373.
39. Ibid., p. 368.
40. Lombroso, *L'homme criminel*, p. 284; see also Zimmern, 'Criminal Anthropology', 746.
41. Lombroso, *L'homme criminel*, p. 267.

42 Ibid., p. 467.
43. H. Ellis, *The Criminal* (New York: Charles Scribner's Sons, 1911), p. 116.
44. Lombroso, *L'homme criminel*, p. 319. The typical heads-I-win-tails-you-lose argument of an incontrovertible racism. The very behaviour that would be regarded as heroic for a white man – think of how many great Western heroes died with courage in excruciating pain – demands a different interpretation when the victim is an Indian. In this case, he is no hero because he does not feel the pain of his martyrdom.
45. Ibid., p. 651.
46. Lombroso, 'Criminal Anthropology', 53.
47. Ellis, *The Criminal*, p. 211.
48. E. S. Morse, 'Natural Selection and Crime', *Popular Science*, 41 (1892): 438.
49. Lombroso argues that children, despite a common impression to the contrary, have a natural penchant for alcohol. That scientists fail to note it merely reflects a class bias; for middle- and upper-class parents never give their children the opportunity to indulge this natural vice. 'One who lives among the upper classes has no idea of the passion babies have for alcoholic liquor, but among the lower classes it is only too common a thing to see even suckling babies drink wine and liquors with wonderful delight.' (Lombroso, 'Criminal Anthropology', 56).
50. Lombroso, *L'homme criminel*, p. 99.
51. A. F. B. Crofton, 'The Language of Crime', *Popular Science*, 50 (1897): 831–5.
52. Lombroso, 'Criminal Anthropology', 56.
53. Ellis, *The Criminal*.
54. H. F. Autenrieth, *Observationum ad historiam embryonic facientium, pars prima* (Tübingen, 1797), cited in O. Temkin, 'German Concepts of Ontogeny and History around 1800', *Bulletin of the History of Medicine*, 24 (1950): 227–46.
55. Schiller, cited in H. Schmidt, *Das biogenetische Grundgesetz Ernst Haeckels und seine Gegner* (Frankfurt: Neuer Frankfurter Verlag, 1909), p. 156.
56. Agassiz, cited in W. Stanton, *The Leopard's Spots: Scientific Attitudes toward Race in America, 1815–1859* (Chicago: University of Chicago Press, 1960), p. 100.
57. E. R. A. Serres, 'Principes d'embryogénie, de zoogénie et de teratogénie', *Mém. Acad. Sci.*, 25 (1860): 771.
58. Ibid., p. ii. This remarkable passage is the only attack I can find levelled by a recapitulationist against the pseudoscientific justification of political positions based on racial rank. It is, of course, an attack on the polygenist theory, not on the concept or use of recapitulation.
59. Ibid., p. 765.
60. Ibid., pp. 763–5.
61. Cope, 1883, in E. D. Cope, *The Origin of the Fittest* (New York: Macmillan, 1887), p. 146.
62. D. G. Brinton, *Races and Peoples* (New York: N. D. C. Hodges, 1890), p. 48.
63. C. Vogt, *Lectures on Man* (London: Longman, Green, Longman & Roberts, 1864), p. 183.

64. Cope, 1883, in Cope, op. cit., p. 293.
65. H. Spencer, *The Principles of Sociology*, third edition (New York: D. Appleton, 1895), p. 89.
66. Lord Avebury (John Lubbock), *The Origin of Civilization and the Primitive Condition of Man* (London: Longmans, 1870), p. 4.
67. Sully, op. cit., 5.
68. G. S. Hall, *Adolescence: Its Psychology and Its Relations to Physiology, Anthropology, Sociology, Sex, Crime, Religion, and Education*, vol. 2 (New York: D. Appleton, 1904), p. 649.
69. Sully, op. cit., 392.
70. J. Murphy, *Primitive Man* (London: Oxford University Press, 1927), p. 109.
71. Sully, op. cit., 386.
72. Cope, 1870, in Cope, op. cit., p. 153. Note the full threefold parallelism of recapitulation: paleontology ('fossil' record of historical ancestors), comparative anatomy (modern 'savages'), and ontogeny.
73. Cope, 1883, in ibid., p. 291.
74. J. F. Meckel, *System der vergleichenden Anatomie*, 7 vols (Halle: Rengersche Buchhandlung, 1821), p. 416–17.
75. Cope, 1870, in Cope, op. cit., p. 159.
76. G. S. Hall, op. cit., p. 194.
77. Vogt, op. cit., p. 192.
78. Lombroso, cited in Ellis, op. cit., p. 254.
79. A. F. Chamberlain, *The Contact of 'Higher' and 'Lower' Races* (Worcester, MA: Clark University, undated), p. 1.
80. Clay, cited in J. Strong, *Expansion under New World-Conditions* (New York: Baker and Taylor, 1900), p. 289.
81. Ibid, pp. 289–90.
82. B. Kidd, *The Control of the Tropics* (New York: Macmillan, 1898), p. 51.
83. Roosevelt, cited in R. F. Weston, *Racism in US Imperialism: The Influence of Racial Assumptions on American Foreign Policy, 1893–1946* (Columbia: University of South Carolina Press, 1972), p. 35.
84. Cope, 1883, in Cope, op. cit., pp. 288–90.
85. L. Bolk, 'On the Problem of Anthropogenesis', *Proc. Section Sciences Kon. Akad. Wetens. Amsterdam*, 29 (1926): 473.
86. L. Bolk, *Das Problem der Menschwerdung* (Jena: Gustav Fischer, 1926), p. 38.
87. L. Bolk, 'Origin of Racial Characteristics in Man', *American Journal of Physical Anthropology*, 13 (1929): 26–7.
88. Ibid., 75.
89. M. F. A. Montagu, 'Time, Morphology, and Neoteny in the Evolution of Man', in *idem* (ed.), *Culture and the Evolution of Man* (New York: Oxford University Press, 1962), pp. 324–42.
90. H. J. Eysenck, *The IQ Argument: Race, Intelligence and Education* (New York: Library Press, 1971).
91. Ibid., p. 79.
92. Bolk, 'Origin of Racial Characteristics', 28.

93. H. Ellis, *Man and Woman* (New York: Charles Scribner's Sons, 1894), p. 28.
94. Montagu, op. cit., p. 331.
95. Ellis, *Man and Woman*, p. 518.
96. Ibid., p. 517.
97. Ibid., p. 519.

THE SPANDRELS OF SAN MARCO AND THE PANGLOSSIAN PARADIGM: A CRITIQUE OF THE ADAPTATIONIST PROGRAMME

(*with Richard Lewontin*)

1. Introduction

THE great central dome of St Mark's Cathedral in Venice presents in its mosaic design a detailed iconography expressing the mainstays of Christian faith. Three circles of figures radiate out from a central image of Christ: angels, disciples, and virtues. Each circle is divided into quadrants, even though the dome itself is radially symmetrical in structure. Each quadrant meets one of the four spandrels in the arches below the dome. Spandrels – the tapering triangular spaces formed by the intersection of two rounded arches at right angles – are necessary architectural by-products of mounting a dome on rounded arches. Each spandrel contains a design admirably fitted into its tapering space. An evangelist sits in the upper part flanked by the heavenly cities. Below, a man representing one of the four Biblical rivers (Tigris, Euphrates, Indus and Nile) pours water from a pitcher into the narrowing space below his feet.

The design is so elaborate, harmonious and purposeful that we are tempted to view it as the starting point of any analysis, as the cause in some sense of the surrounding architecture. But this would invert the proper path of analysis. The system begins with an architectural constraint: the necessary four spandrels and their tapering triangular form. They provide a space in which the mosaicists worked; they set the quadripartite symmetry of the dome above.

Such architectural constraints abound and we find them easy to understand because we do not impose our biological biases upon them. Every fan vaulted ceiling must have a series of open spaces along the midline of the vault, where the sides of the fans intersect between the pillars. Since the spaces must exist, they are often used for ingenious ornamental effect. In King's College Chapel in Cambridge, for example, the spaces contain bosses alternately embellished with the Tudor rose and portcullis. In a sense, this design represents an 'adaptation', but the architectural constraint is clearly primary. The spaces arise as a necessary by-product of fan vaulting; their appropriate use is a secondary effect. Anyone who tried to argue that the structure exists because the alternation of rose and portcullis makes so much sense in a Tudor chapel would be inviting the same ridicule that Voltaire heaped on Dr Pangloss: 'Things cannot be other than they are . . . Everything is made for the best purpose. Our noses were made to carry spectacles, so we have spectacles. Legs were clearly intended for breeches, and we wear them.' Yet evolutionary biologists, in their tendency to focus exclusively on immediate adaptation to local conditions, do tend to ignore architectural constraints and perform just such an inversion of explanation.

As a closer example, recently featured in some important biological literature on adaptation, anthropologist Michael Harner has proposed that Aztec human sacrifice arose as a solution to chronic shortage of meat (limbs of victims were often consumed, but only by people of high status).[1] E. O. Wilson has used this explanation as a primary illustration of an adaptive, genetic predisposition for carnivory in humans.[2] Harner and Wilson ask us to view an elaborate social system and a complex set of explicit justifications involving myth, symbol, and tradition as mere epiphenomena generated by the Aztecs as an unconscious rationalization masking the 'real' reason for it all: need for protein. But M. Sahlins has argued that human sacrifice represented just one part of an elaborate cultural fabric that, in its entirety, not only represented the material expression of Aztec cosmology, but also performed such utilitarian functions as the maintenance of social ranks and systems of tribute among cities.[3]

We strongly suspect that Aztec cannibalism was an 'adaptation' much like evangelists and rivers in spandrels, or ornamented bosses in ceiling spaces: a secondary epiphenomenon representing a fruitful use of available parts, not a cause of the entire system. To put it crudely: a system developed for other reasons generated an increasing number of fresh bodies; use might as well be made of them. Why invert the whole system in such a curious fashion and view an entire culture as the epiphenomenon of an unusual way to beef up the meat supply? Spandrels do not exist to house the evangelists. (Moreover, as Sahlins argues, it is not even clear that human sacrifice was an adaptation at all. Human cultural practices can be orthogenetic and drive towards extinction in ways that Darwinian processes, based on genetic selection, cannot. Since each new monarch had to outdo his predecessor in even more elaborate and copious sacrifice, the practice was beginning to stretch resources to the breaking point. It would not have been the first time that a human culture did itself in. And, finally, many experts doubt Harner's premise in the first place.[4] They argue that other sources of protein were not in short supply, and that a practice awarding meat only to privileged people who had enough anyway, and who used bodies so inefficiently (only the limbs were consumed, and partially at that) represents a mighty poor way to run a butchery.)

We deliberately chose nonbiological examples in a sequence running from remote to more familiar: architecture to anthropology. We did this because the primacy of architectural constraint and the epiphenomenal nature of adaptation are not obscured by our biological prejudices in these examples. But we trust that the message for biologists will not go unheeded: if these had been biological systems, would we not, by force of habit, have regarded the epiphenomenal adaptation as primary and tried to build the whole structural system from it?

2. The Adaptationist Programme

We wish to question a deeply engrained habit of thinking among students of evolution. We call it the adaptationist programme, or the Panglossian paradigm. It is rooted in a notion popularized by A. R. Wallace and A. Weismann (but not, as we shall see, by Darwin) towards the end of the nineteenth century: the near omnipotence of natural selection in forging organic design and fashioning the best among possible worlds. This programme regards natural selection as so powerful and the constraints upon it so few that direct production of adaptation through its operation becomes the primary cause of nearly all organic form, function, and behaviour. Constraints upon the pervasive power of natural selection are recognized of course (phyletic inertia primarily among them, although immediate architectural constraints, as discussed in the last section, are rarely acknowledged). But they are usually dimissed as unimportant or else, and more frustratingly, simply acknowledged and then not taken to heart and invoked.

Studies under the adaptationist programme generally proceed in two steps:

(1) An organism is atomized into 'traits' and these traits are explained as structures optimally designed by natural selection for their functions. For lack of space, we must omit an extended discussion of the vital issue: 'what is a trait?' Some evolutionists may regard this as a trivial, or merely a semantic problem. It is not. Organisms are integrated entities, not collections of discrete objects. Evolutionists have often been led astray by inappropriate atomization, as D'Arcy Thompson loved to point out.[5] Our favourite example involves the human chin.[6] If we regard the chin as a 'thing', rather than as a product of interaction between two growth fields (alveolar and mandibular), then we are led to an interpretation of its origin (recapitulatory) exactly opposite to the one now generally favoured (neotenic).

(2) After the failure of part-by-part optimization, interaction is acknowledged via the dictum that an organism cannot optimize each part without imposing expenses on others. The notion of 'trade-off' is introduced, and organisms are interpreted as best compromises among competing demands. Thus, interaction among parts is retained completely within the adaptationist programme. Any suboptimality of a part is explained as its contribution to the best possible design for the whole. The notion that suboptimality might represent anything other than the immediate work of natural selection is usually not entertained. As Dr Pangloss said in explaining to Candide why he suffered from venereal disease: 'It is indispensable in this best of worlds. For if Columbus, when visiting the West Indies, had not caught this disease, which poisons the source of generation, which frequently even hinders generation, and is clearly opposed to the great end of Nature, we should have neither chocolate nor cochineal.' The adaptationist programme is truly Panglossian. Our world may not be good in an abstract sense, but it is the very best we could have. Each trait plays its part and must be as it is.

At this point, some evolutionists will protest that we are caricaturing their view of adaptation. After all, do they not admit genetic drift, allometry, and a variety of reasons for non-adaptive evolution? They do, to be sure, but we make a different point. In natural history, all possible things happen sometimes; you generally do not support your favoured phenomenon by declaring rivals impossible in theory. Rather, you acknowledge the rival, but circumscribe its domain of action so narrowly that it cannot have any importance in the affairs of nature. Then, you often congratulate yourself for being such an undogmatic and ecumenical chap. We maintain that alternatives to selection for best overall design have generally been relegated to unimportance by this mode of argument. Have we not all heard the catechism about genetic drift: it can only be important in populations so small that they are likely to become extinct before playing any sustained evolutionary role.[7]

The admission of alternatives in principle does not imply their serious consideration in daily practice. We all say that not everything is adaptive; yet, faced with an organism, we tend to break it into parts and tell adaptive stories as if trade-offs among competing, well designed parts were the only constraint upon perfection for each trait. It is an old habit. As G. J. Romanes complained about A. R. Wallace in 1900: 'Mr Wallace does not expressly maintain the abstract impossibility of laws and causes other than those of utility and natural selection . . . Nevertheless, as he nowhere recognizes any other law or cause . . . , he practically concludes that, on inductive or empirical grounds, there *is* no such other law or cause to be entertained.'[8]

The adaptationist programme can be traced through common styles of argument. We illustrate just a few; we trust they will be recognized by all:

1. If one adaptive argument fails, try another. Zig-zag commissures of clams and brachiopods, once widely regarded as devices for strengthening the shell, become sieves for restricting particles above a given size.[9] A suite of external structures (horns, antlers, tusks) once viewed as weapons against predators, become symbols of intraspecific competition among males.[10] The eskimo face, once depicted as 'cold engineered',[11] becomes an adaptation to generate and withstand large masticatory forces.[12] We do not attack these newer interpretations; they may all be right. We do wonder, though, whether the failure of one adaptive explanation should always simply inspire a search for another of the same general form, rather than a consideration of alternatives to the proposition that each part is 'for' some specific purpose.

2. If one adaptive argument fails, assume that another must exist; a weaker version of the first argument. R. Costa and P. M. Bisol, for example, hoped to find a correlation between genetic polymorphism and stability of environment in the deep sea, but they failed.[13] They conclude: 'The degree of genetic polymorphism found would seem to indicate absence of correlation with the particular environmental factors

which characterize the sampled area. The results suggest that the adaptive strategies of organisms belonging to different phyla are different.'[14]

3. In the absence of a good adaptive argument in the first place, attribute failure to imperfect understanding of where an organism lives and what it does. This is again an old argument. Consider A. R. Wallace on why all details of colour and form in land snails must be adaptive, even if different animals seem to inhabit the same environment: 'The exact proportions of the various species of plants, the numbers of each kind of insect or of bird, the peculiarities of more or less exposure to sunshine or to wind at certain critical epochs, and other slight differences which to us are absolutely immaterial and unrecognizable, may be of the highest significance to these humble creatures, and be quite sufficient to require some slight adjustments of size, form, or colour, which natural selection will bring about.'[15]

4. Emphasize immediate utility and exclude other attributes of form. Fully half the explanatory information accompanying the full-scale fibreglass *Tyrannosaurus* at Boston's Museum of Science reads: 'Front legs a puzzle: how *Tyrannosaurus* used its tiny front legs is a scientific puzzle; they were too short even to reach the mouth. They may have been used to help the animal rise from a lying position.' (We purposely choose an example based on public impact of science to show how widely habits of the adaptationist programme extend. We are not using glass beasts as straw men; similar arguments and relative emphases, framed in different words, appear regularly in the professional literature.) We don't doubt that *Tyrannosaurus* used its diminutive front legs for something. If they had arisen *de novo*, we would encourage the search for some immediate adaptive reason. But they are, after all, the reduced product of conventionally functional homologues in ancestors (longer limbs of allosaurs, for example). As such, we do not need an explicitly adaptive explanation for the reduction itself. It is likely to be a developmental correlate of allometric fields for relative increase in head and hindlimb size. This non-adaptive hypothesis can be tested by conventional allometric methods[16] and seems to us both more interesting and fruitful than untestable speculations based on secondary utility in the best of possible worlds. One must not confuse the fact that a structure is used in some way (consider again the spandrels, ceiling spaces and Aztec bodies) with the primary evolutionary reason for its existence and conformation.

3. Telling Stories

> All this is a manifestation of the rightness of things, since if there is a volcano at Lisbon it could not be anywhere else. For it is impossible for things not to be where they are, because everything is for the best
>
> Dr Pangloss on the great Lisbon earthquake of 1755 in which up to 50,000 people lost their lives.

We would not object so strenuously to the adaptationist programme if its invocation, in any particular case, could lead in principle to its rejection for want of evidence. We might still view it as restrictive and object to its status as an argument of first choice. But if it could be dismissed after failing some explicit test, then alternatives would get their chance. Unfortunately, a common procedure among evolutionists does not allow such definable rejection for two reasons. First, the rejection of one adaptive story usually leads to its replacement by another, rather than to a suspicion that a different kind of explanation might be required. Since the range of adaptive stories is as wide as our minds are fertile, new stories can always be postulated. And if a story is not immediately available, one can always plead temporary ignorance and trust that it will be forthcoming, as did Costa and Bisol.[17] Secondly, the criteria for acceptance of a story are so loose that many pass without proper confirmation. Often, evolutionists use *consistency* with natural selection as the sole criterion and consider their work done when they concoct a plausible story. But plausible stories can always be told. The key to historical research lies in devising criteria to identify proper explanations among the substantial set of plausible pathways to any modern result.

We have, for example,[18] criticized D. P. Barash's work on aggression in mountain bluebirds[19] for this reason. Barash mounted a stuffed male near the nests of two pairs of bluebirds while the male was out foraging. He did this at the same nests on three occasions at ten-day intervals: the first before eggs were laid, the last two afterwards. He then counted aggressive approaches of the returning male towards both the model and the female. At time one, aggression was high towards the model and lower towards females but substantial in both nests. Aggression towards the model declined steadily for times two and three and plummeted to near zero towards females. Barash reasoned that this made evolutionary sense since males would be more sensitive to intruders before eggs were laid than afterwards (when they can have some confidence that their genes are inside). Having devised this plausible story, he considered his work as completed:

> The results are consistent with the expectations of evolutionary theory. Thus aggression toward an intruding male (the model) would clearly be especially advantageous early in the breeding season, when territories and nests are normally defended . . . The initial aggressive response to the mated female is also adaptive in that, given a situation suggesting a high probability of adultery (i.e. the presence of the model near the female) and assuming that replacement females are available, obtaining a new mate would enhance the fitness of males . . . The decline in male-female aggressiveness during incubation and fledgling stages could be attributed to the impossibility of being cuckolded after the eggs have been laid . . . The results are consistent with an evolutionary interpretation[20]

They are indeed consistent, but what about an obvious alternative, dismissed without test by Barash? Male returns at times two and three, approaches the model, tests it a bit, recognizes it as the same phoney he saw before, and doesn't bother his female. Why not at least perform the obvious test for this alternative to a conventional adaptive story: expose a male to the model for the *first* time after the eggs are laid.

Since we criticized Barash's work, E. S. Morton, M. S. Geitgey and S. McGrath repeated it,[21] with some variations (including the introduction of a female model), in the closely related eastern bluebird *Sialia sialis*. 'We hoped to confirm', they wrote, that Barash's conclusions represent 'a widespread evolutionary reality, at least within the genus *Sialia*. Unfortunately, we were unable to do so.' They found no 'anti-cuckoldry' behaviour at all: males never approached their females aggressively after testing the model at any nesting stage. Instead, females often approached the male model and, in any case, attacked female models more than males attacked male models. 'This violent response resulted in the near destruction of the female model after presentations and its complete demise on the third, as a female flew off with the model's head early in the experiment to lose it for us in the brush'.[22] Yet, instead of calling Barash's selected story into question, they merely devise one of their own to render both results in the adaptationist mode. Perhaps, they conjecture, replacement females are scarce in their species and abundant in Barash's. Since Barash's males can replace a potentially 'unfaithful' female, they can afford to be choosy and possessive. Eastern bluebird males are stuck with uncommon mates and had best be respectful. They conclude: 'If we did not support Barash's suggestion that male bluebirds show anticuckoldry adaptations, we suggest that both studies still had "results that are consistent with the expectations of evolutionary theory",[23] as we presume any careful study would.' But what good is a theory that cannot fail in careful study (since by 'evolutionary theory', they clearly mean the action of natural selection applied to particular cases, rather than the fact of transmutation itself).

4. The Master's Voice Re-examined

Since Darwin has attained sainthood (if not divinity) among evolutionary biologists, and since all sides invoke God's allegiance, Darwin has often been depicted as a radical selectionist at heart who invoked other mechanisms only in retreat, and only as a result of his age's own lamented ignorance about the mechanisms of heredity. This view is false. Although Darwin regarded selection as the most important of evolutionary mechanisms (as do we), no argument from opponents angered him more than the common attempt to caricature and trivialize his theory by stating that it relied exclusively upon natural selection. In the last edition of the *Origin*, he wrote:

> As my conclusions have lately been much misrepresented, and it has been stated that I attribute the modification of species exclusively to natural selection, I may be permitted to remark that in the first edition of this work, and subsequently, I placed in a most conspicuous position – namely at the close of the Introduction – the following words: 'I am convinced that natural selection has been the main, but not the exclusive means of modification.' This has been of no avail. Great is the power of steady misrepresentation.[24]

G. J. Romanes, whose once famous essay on Darwin's pluralism versus the panselectionism of Wallace and Weismann deserves a resurrection, noted of this passage: 'In the whole range of Darwin's writings there cannot be found a passage so strongly worded as this: it presents the only note of bitterness in all the thousands of pages which he has published.'[25] Apparently, Romanes did not know the letter Darwin wrote to *Nature* in 1880, in which he castigated Sir Wyville Thomson for caricaturing his theory as panselectionist:

> I am sorry to find that Sir Wyville Thomson does not understand the principle of natural selection . . . If he had done so, he could not have written the following sentence in the Introduction to the Voyage of the Challenger: 'The character of the abyssal fauna refuses to give the least support to the theory which refers the evolution of species to extreme variation guided only by natural selection.' This is a standard of criticism not uncommonly reached by theologians and metaphysicians when they write on scientific subjects, but is something new as coming from a naturalist . . . Can Sir Wyville Thomson name any one who has said that the evolution of species depends only on natural selection? As far as concerns myself, I believe that no one has brought forward so many observations on the effects of the use and disuse of parts, as I have done in my 'Variation of Animals and Plants under Domestication'; and these observations were made for this special object. I have likewise there adduced a considerable body of facts, showing the direct action of external conditions on organisms.[26]

We do not now regard all of Darwin's subsidiary mechanisms as significant or even valid, though many, including direct modification and correlation of growth, are very important. But we should cherish his consistent attitude of pluralism in attempting to explain Nature's complexity.

5. A Partial Typology of Alternatives to the Adaptationist Programme

In Darwin's pluralistic spirit, we present an incomplete hierarchy of alternatives to immediate adaptation for the explanation of form, function, and behaviour.

1. No adaptation and no selection at all. At present, population geneticists are sharply divided on the question of how much genetic polymorphism within populations and how much of the genetic differences between species is, in fact, the result of natural selection as opposed to purely random factors. Populations are finite in size and the isolated populations that form the first step in the speciation process are often founded by a very small number of individuals. As a result of this restriction in population size, frequencies of alleles change by *genetic drift*, a kind of random genetic sampling error. The stochastic process of change in gene frequency by random genetic drift, including the very strong sampling process that goes on when a new isolated population is formed from a few immigrants, has several important consequences. First, populations and species will become genetically differentiated, and even fixed for different alleles at a locus in the complete absence of any selective force at all.

Secondly, alleles can become fixed in a population *in spite of natural selection*. Even if an allele is favoured by natural selection, some proportion of population, depending upon the product of population size N and selection intensity s, will become homozygous for the less fit allele because of genetic drift. If Ns is large this random fixation for unfavourable alleles is a rare phenomenon, but if selection coefficients are on the order of the reciprocal of population size ($Ns = 1$) or smaller, fixation for deleterious alleles is common. If many genes are involved in influencing a metric character like shape, metabolism or behaviour, then the intensity of selection on each locus will be small and Ns per locus may be small. As a result, many of the loci may be fixed for nonoptimal alleles.

Thirdly, new mutations have a small chance of being incorporated into a population, even when selectively favoured. Genetic drift causes the immediate loss of most new mutations after their introduction. With a selection intensity s, a new favourable mutation has a probability of only $2s$ of ever being incorporated. Thus, one cannot claim that, eventually, a new mutation of just the right sort for some adaptive argument will occur and spread. 'Eventually' becomes a very long time if only one in 1,000 or one in 10,000 of the 'right' mutations that do occur ever get incorporated in a population.

2. No adaptation and no selection on the part at issue; form of the part is a correlated consequence of selection directed elsewhere. Under this important category, Darwin ranked his 'mysterious' laws of the

'correlation of growth'. Today, we speak of pleiotrophy, allometry, 'material compensation',[27] and mechanically forced correlations in D'Arcy Thompson's sense.[28] Here we come face to face with organisms as integrated wholes, fundamentally not decomposable into independent and separately optimized parts.

Although allometric patterns are as subject to selection as static morphology itself,[29] some regularities in relative growth are probably not under immediate adaptive control. For example, we do not doubt that the famous 0.66 interspecific allometry of brain size in all major vertebrate groups represents a selected 'design criterion,' though its significance remains elusive.[30] It is too repeatable across too wide a taxonomic range to represent much else than a series of creatures similarly well designed for their different sizes. But another common allometry, the 0.2 to 0.4 intraspecific scaling among homeothermic adults differing in body size, or among races within a species, probably does not require a selectionist story though many, including one of us, have tried to provide one.[31] R. Lande[32] (personal communication) has used the experiments of Falconer[33] to show that selection upon *body size alone* yields a brain–body slope across generations of 0.35 in mice.

More compelling examples abound in the literature on selection for altering the timing of maturation.[34] At least three times in the evolution of arthropods (mites, flies and beetles), the same complex adaptation has evolved, apparently for rapid turnover of generations in strongly *r*-selected feeders on superabundant but ephemeral fungal resources: females reproduce as larvae and grow the next generation within their bodies. Offspring eat their mother from inside and emerge from her hollow shell, only to be devoured a few days later by their own progeny. It would be foolish to seek adaptive significance in paedomorphic morphology per se; it is primarily a by-product of selection for rapid cycling of generations. In more interesting cases, selection for small size (as in animals of the interstitial fauna) or rapid maturation (dwarf males of many crustaceans) has occurred by progenesis,[35] and descendant adults contain a mixture of ancestral juvenile and adult features. Many biologists have been tempted to find primary adaptive meaning for the mixture, but it probably arises as a by-product of truncated maturation, leaving some features 'behind' in the larval state, while allowing others, more strongly correlated with sexual maturation, to retain the adult configuration of ancestors.

3. The decoupling of selection and adaptation.

i) Selection without adaptation. Lewontin has presented the following hypothetical example: 'A mutation which doubles the fecundity of individuals will sweep through a population rapidly. If there has been no change in efficiency of resource utilization, the individuals will leave no more offspring than before, but simply lay twice as many eggs, the excess dying because of resource limitation. In what sense are the individuals or the population as a whole better adapted than before? Indeed, if a predator on immature stages is led to switch to the species now that immatures are more plentiful, the population size may actually decrease as a consequence, yet natural selection at all times will favour individuals with higher fecundity.'[36]

ii) Adaptation without selection. Many sedentary marine organisms, sponges and corals in particular, are well adapted to the flow regimes in which they live. A wide spectrum of 'good design' may be purely phenotypic in origin, largely induced by the current itself. (We may be sure of this in numerous cases, when genetically identical individuals of a colony assume different shapes in different microhabitats.) Larger patterns of geographic variation are often adaptive and purely phenotypic as well. B. W. Sweeney and R. L. Vannote, for example, showed that many hemimetabolous aquatic insects reach smaller adult size with reduced fecundity when they grow at temperatures above and below their optima.[37] Coherent, climatically correlated patterns in geographic distribution for these insects – so often taken as a priori signs of genetic adaptation – may simply reflect this phenotypic plasticity.

'Adaptation' – the good fit of organisms to their environment – can occur at three hierarchical levels with different causes. It is unfortunate that our language has focused on the common result and called all three phenomena 'adaptation': the differences in process have been obscured and evolutionists have often been misled to extend the Darwinian mode to the other two levels as well. First, we have what physiologists call 'adaptation': the phenotypic plasticity that permits organisms to mould their form to prevailing circumstances during ontogeny. Human 'adaptations' to high altitude fall into this category (while others, like resistance of sickling heterozygotes to malaria, are genetic and Darwinian). Physiological adaptations are not heritable, though the capacity to develop them presumably is. Secondly, we have a 'heritable' form of non-Darwinian adaptation in humans (and, in rudimentary ways, in a few other advanced social species): cultural adaptation (with heritability imposed by learning). Much confused thinking in human sociobiology arises from a failure to distinguish this mode from Darwinian adaptation based on genetic variation. Finally, we have adaptation arising from the conventional Darwinian mechanism of selection upon genetic variation. The mere existence of a good fit between organism and environment is insufficient evidence for inferring the action of natural selection.

4. Adaptation and selection but no selective basis for differences among adaptations. Species of related organisms, or subpopulations within a species, often develop different adaptations as solutions to the same problem. When 'multiple adaptive peaks' are occupied, we usually have no basis for asserting that one solution is better than another. The solution followed in any spot is a result of history; the first steps went in one direction, though others would have led to adequate prosperity as well. Every naturalist has his favourite illustration. In the West Indian land snail *Cerion*, for example, populations living on rocky and windy coasts almost always develop white, thick and relatively squat shells for conventional adaptive reasons. We can identify at least two different developmental pathways to whiteness from the mottling of early whorls in all *Cerion*, two paths to thickened shells and three styles of allometry leading to squat shells. All twelve combinations can be identified in Bahamian populations, but would it be fruitful to ask why – in the sense of optimal design rather than historical contingency – *Cerion* from eastern Long Island evolved one solution, and *Cerion* from Acklins Island another?

5. Adaptation and selection, but the adaptation is a secondary utilization of parts present for reasons of architecture, development or history. We have already discussed this neglected subject in the first section on spandrels, spaces and cannibalism. If blushing turns out to be an adaptation affected by sexual selection in humans, it will not help us to understand why blood is red. The immediate utility of an organic structure often says nothing at all about the reason for its being.

6. Another, and Unfairly Maligned, Approach to Evolution

In continental Europe, evolutionists have never been much attracted to the Anglo-American penchant for atomizing organisms into parts and trying to explain each as a direct adaptation. Their general alternative exists in both a strong and a weak form. In the strong form, as advocated by such major theorists as O. H. Schindewolf,[38] A. Remane,[39] and P.-P. Grassé[40] natural selection under the adaptationist programme can explain superficial modifications of the *Bauplan* that fit structure to environment: why moles are blind, giraffes have long necks, and ducks webbed feet, for example. But the important steps of evolution, the construction of the *Bauplan* itself and the transition between *Baupläne*, must involve some other unknown, and perhaps 'internal', mechanism. We believe that English biologists have been right in rejecting this strong form as close to an appeal to mysticism.

But the argument has a weaker – and paradoxically powerful – form that has not been appreciated, but deserves to be. It also acknowledges conventional selection for superficial modifications of the *Bauplan*. It also denies that the adaptationist programme (atomization plus optimizing selection on parts) can do much to explain *Baupläne* and the transitions between them. But it does not therefore resort to a fundamentally unknown process. It holds instead that the basic body plans of organisms are so integrated and so replete with constraints upon adaptation (categories 2 and 5 of our typology) that conventional styles of selective arguments can explain little of interest about them. It does not deny that change, when it occurs, may be mediated by natural selection, but it holds that constraints restrict possible paths and modes of change so strongly that the constraints themselves become much the most interesting aspect of evolution.

Rupert Riedl, the Austrian zoologist who has tried to develop this thesis for English audiences,[41] writes:

> The living world happens to be crowded by universal patterns of organization which, most obviously, find no direct explanation through environmental conditions or adaptive radiation, but exist primarily through universal requirements which can only be expected under the systems conditions of complex organization itself . . . This is not self-evident, for the whole of the huge and profound thought collected in the field of morphology, from Goethe to Remane, has virtually been cut off from modern biology. It is not taught in most American universities. Even the teachers who could teach it have disappeared.

Constraints upon evolutionary change may be ordered into at least two categories. All evolutionists are familiar with *phyletic* constraints, as embodied in W. K. Gregory's classic distinction between habitus and heritage.[42] We acknowledge a kind of phyletic inertia in recognizing, for example, that humans are not optimally designed for upright posture because so much of our *Bauplan* evolved for quadrupedal life. We also invoke phyletic constraint in explaining why no molluscs fly in air and no insects are as large as elephants.

Developmental constraints, a subcategory of phyletic restrictions, may hold the most powerful rein of all over possible evolutionary pathways. In complex organisms, early stages of ontogeny are remarkably refractory to evolutionary change, presumably because the differentiation of organ systems and their integration into a functioning body is such a delicate process, so easily derailed by early errors with accumulating effects. K. E. von Baer's fundamental embryological laws[43] represent little more than a recognition that early stages are both highly conservative and strongly restrictive of later development. Haeckel's biogenetic law, the primary subject of late nineteenth century evolutionary biology, rested upon a misreading of the same data.[44] If development occurs in integrated packages, and cannot be pulled apart piece by piece in evolution, then the adaptationist programme cannot explain the alteration of developmental programmes underlying nearly all changes of *Bauplan*.

The German paleontologist A. Seilacher, whose work deserves far more attention than it has received, has emphasized what he calls '*bautechnischer*', or *architectural*, constraints.[45] These arise not from former adaptations retained in a new ecological setting (phyletic constraints as usually understood), but as architectural restrictions that never were adaptations, but rather the necessary consequences of materials and designs selected to build basic *Baupläne*. We devoted the first section of this paper to nonbiological examples in this category. Spandrels must exist once a blueprint specifies that a dome shall rest on rounded arches. Architectural constraints can exert a far-ranging influence upon organisms as well. The subject is full of potential insight because it has rarely been acknowledged at all.

In a fascinating example, A. Seilacher has shown that the divaricate form of architecture occurs again and again in all groups of molluscs, and in brachiopods as well.[46] This basic form expresses itself in a wide variety of structures: raised ornamental lines (not growth lines because they do not conform to the mantle margin at any time), patterns of coloration, internal structures in the mineralization of calcite, and incised grooves. He does not know what generates this pattern and feels that traditional and nearly exclusive focus on the adaptive value of each manifestation has diverted attention from questions of its genesis in growth and also prevented its recognition as a general phenomenon. It must arise from some characteristic pattern of inhomogeneity in the growing mantle, probably from the generation of interference patterns around regularly spaced centres; simple computer simulations can generate the form in this manner.[47] The general pattern may not be a direct adaptation at all.

Seilacher then argues that most manifestations of the pattern are probably nonadaptive. His reasons vary, but seem generally sound to us. Some are based on field observations: colour patterns that remain invisible because clams possessing them either live buried in sediments or remain covered with a periostracum so thick that the colours cannot be seen. Others rely on more general principles: presence only in odd and pathological individuals, rarity as a developmental anomaly, excessive variability compared with much reduced variability when the same general structure assumes a form judged functional on engineering grounds.

In a distinct minority of cases, the divaricate pattern becomes functional in each of the four categories. Divaricate ribs may act as scoops and anchors in burrowing,[48] but they are not properly arranged for such function in most clams. The colour chevrons are mimetic in one species (*Pteria zebra*) that lives on hydrozoan branches; here the variability is strongly reduced. The mineralization chevrons are probably adaptive in only one remarkable creature, the peculiar bivalve *Corculum cardissa* (in other species, they either appear in odd specimens or only as post-mortem products of shell erosion). This clam is uniquely flattened in an anterio-posterior direction. It lies on the substrate, posterior up. Distributed over its rear end are divaricate triangles of mineralization. They are translucent, while the rest of the shell is opaque. Under these windows dwell endosymbiotic algae!

All previous literature on divaricate structure has focused on its adaptive significance (and failed to find any in most cases). But Seilacher is probably right in representing this case as the spandrels, ceiling holes and sacrificed bodies of our first section. The divaricate pattern is a fundamental architectural constraint. Occasionally, since it is there, it is used to beneficial effect. But we cannot understand the pattern or its evolutionary meaning by viewing these infrequent and secondary adaptations as a reason for the pattern itself.

F. Galton contrasted the adaptationist programme with a focus on constraints and modes of development by citing a telling anecdote about Herbert Spencer's fingerprints:

> Much has been written, but the last word has not been said, on the rationale of these curious papillary ridges; why in one man and in one finger they form whorls and in another loops. I may mention a characteristic anecdote of Herbert Spencer in connection with this. He asked me to show him my Laboratory and to take his prints, which I did. Then I spoke of the failure to discover the origin of these patterns, and how the fingers of unborn children had been dissected to ascertain their earliest stages, and so forth. Spencer remarked that this was beginning in the wrong way; that I ought to consider the purpose the ridges had to fulfil, and to work backwards. Here, he said, it was obvious that the delicate mouths of the sudorific glands required the protection given to them by the ridges on either side of them, and therefrom he elaborated a consistent and ingenious hypothesis at great length. I replied that his arguments were beautiful and deserved to be true, but it happened that the mouths of the ducts did not run in the valleys between the crests, but along the crests of the ridges themselves.[49]

We feel that the potential rewards of abandoning exclusive focus on the adaptationist programme are very great indeed. We do not offer a council of despair, as adaptationists have charged; for nonadaptive does not mean nonintelligible. We welcome the richness that a pluralistic approach, so akin to Darwin's spirit, can provide. Under the adaptationist programme, the great historic themes of developmental morphology and *Bauplan* were largely abandoned; for if selection can break any correlation and optimize parts separately, then an organism's integration counts for little. Too often, the adaptationist programme gave us an evolutionary biology of parts and genes, but not of organisms. It assumed that all transitions could occur step by step and underrated the importance of integrated developmental blocks and pervasive constraints of history and architecture. A pluralistic view could put organisms, with all their recalcitrant, yet intelligible, complexity, back into evolutionary theory.

Notes

1. M. Harner, 'The Ecological Basis for Aztec Sacrifice', *American Ethnologist*, 4 (1977): 117–35.
2. E. O. Wilson, *On Human Nature* (Cambridge, MA: Harvard University Press, 1978).
3. M. Sahlins, 'Culture as Protein and Profit', *New York Review of Books*, 23 November 1978, pp. 45–53.
4. B. R. Ortiz de Montellano, 'Aztec Cannibalism: An Ecological Necessity?', *Science*, 200 (1978): 611–17.
5. D. W. Thompson, *Growth and Form* (New York: Macmillan, 1942).
6. S. J. Gould, *Ontogeny and Phylogeny* (Cambridge, MA: Harvard University Press, 1977), pp. 381–2; R. C. Lewontin, 'Adaptation', *Scientific American*, 239, 3 (1978): 156–69.
7. But see R. Lande, 'Natural Selection and Random Genetic Drift in Phenotypic Evolution', *Evolution*, 30 (1976): 314–34.
8. G. J. Romanes, 'The Darwinism of Darwin and of the Post-Darwinian Schools', in *idem, Darwin, and after Darwin*, new edition, vol. 2 (London: Longman, Green & Co: 1900).
9. M. J. S. Rudwick, 'The Function of the Zig-Zag Deflections in the Commissures of Fossil Brachiopods', *Palaeontology*, 7 (1964): 135–71.
10. L. S. Davitashvili, *Teoriya polovogo otbora* [Theory of Sexual Selection] (Moscow: Akademii Nauk, 1961).
11. C. S. Coon, S. M. Garn and J. B. Birdsell, *Races* (Springfield, Ohio: C. Thomas, 1950).
12. B. T. Shea, 'Eskimo Craniofacial Morphology, Cold Stress and the Maxillary Sinus', *American Journal of Physical Anthropology*, 47 (1977): 289–300.
13. R. Costa and P. M. Bisol, 'Genetic Variability in Deep-Sea Organisms', *Biological Bulletin*, 155 (1978): 125–33.
14. Ibid., 132, 133.
15. R. Wallace, *Darwinism* (London: Macmillan, 1899), p. 148.
16. In general, S. J. Gould, 'Allometry in Primates, with Emphasis on Scaling and the Evolution of the Brain', *Approaches to Primate Paleobiology. Contrib. Primatol.*, 5 (1974): 244–92; on limb reductions, R. Lande, 'Evolutionary Mechanisms of Limb Loss in Tetrapods', *Evolution*, 32 (1978): 73–92.
17. Costa and Bisol, op. cit.
18. S. J. Gould, 'Sociobiology: The Art of Storytelling', *New Scientist*, 80 (1978): 530–3.
19. D. P. Barash, 'Male Response to Apparent Female Adultery in the Mountain

Bluebird: An Evolutionary Interpretation', *American Naturalist*, 110 (1976): 1097–1101.
20. Ibid., 1099, 1100.
21. E. S. Morton, M. S. Geitgey and S. McGrath, 'On Bluebird "Responses to Apparent Female Adultery"', *American Naturalist*, 112 (1978): 968–71.
22. Ibid., 969.
23. Barash, op. cit., 1099.
24. C. Darwin, *The Origin of Species* (London: John Murray, 1872), p. 395.
25. Romanes, op. cit.
26. C. Darwin, 'Sir Wyville Thomson and Natural Selection', *Nature*, 23 (1880): 32.
27. B. Rensch, *Evolution above the Species Level* (New York: Columbia University Press, 1959), pp. 179–87.
28. D. W. Thompson, op. cit.; S. J. Gould, 'D'Arcy Thompson and the Science of Form', *New Literary History*, 2, 2 (1971): 229–58.
29. S. J. Gould, 'Allometry and Size in Ontogeny and Phylogeny', *Biological Reviews*, 41 (1966): 587–640.
30. H. J. Jerison, *Evolution of the Brain and Intelligence* (New York: Academic Press, 1973).
31. Gould, 'Allometry in Primates'.
32. R. Lande, personal communication.
33. D. S. Falconer, 'Replicated Selection for Body Weight in Mice', *Genetical Research*, 22 (1973): 291–321.
34. Gould, *Ontogeny and Phylogeny*.
35. Ibid., pp. 324–36.
36. R. C. Lewontin, 'Sociobiology as an Adaptationist Program', *Behavioral Science*, 1979 (in press).
37. B. W. Sweeney and R. L. Vannote, 'Size Variation and the Distribution of Hemimetabolous Aquatic Insects: Two Thermal Equilibrium Hypotheses', *Science*, 200 (1978): 444–6.
38. O. H. Schindewolf, *Grundfragen der Paläontologie* (Stuttgart: Schweizerbart, 1950).
39. A. Remane, *Die Grundlagen des natürlichen Systems der vergleichenden Anatomie und der Phylogenetik* (Königstein/Taunus: Koeltz, 1971).
40. P.-P. Grassé, *Evolution of Living Organisms* (New York: Academic Press, 1977).
41. R. Riedl, 'A Systems-analytical Approach to Macro-evolutionary Phenomena', *Quarterly Review of Biology*, 52 (1977): 351–70; *idem*, *Die Ordnung des Lebendigen* (Hamburg: Paul Parey, 1975), now being translated into English by R. Jefferies.
42. W. K. Gregory, 'Habitus Factors in the Skeleton of Fossil and Recent Mammals', *Proceedings of the American Philosophical Society*, 76 (1936): 429–44.
43. K. E. von Baer, *Entwicklungsgeschichte der Tiere* (Königsberg: Bornträger, 1828).
44. Gould, *Ontogeny and Phylogeny*.

45. A. Seilacher, 'Arbeitskonzept zur Konstruktionsmorphologie', *Lethaia*, 3 (1970): 393–6.
46. *Idem*, 'Divaricate Patterns in Pelecypod Shells', *Lethaia*, 5 (1972): 325–43.
47. C. H. Waddington and J. R. Cowe, 'Computer Simulation of a Molluscan Pigmentation Pattern', *Journal of Theoretical Biology*, 25 (1969): 219–25.
48. S. M. Stanley, 'Relation of Shell Form to Life Habits in the Bivalvia (Mollusca)', *Memo. Geological Society of America*, 125 (1970).
49. F. Galton, *Memories of My Life* (London: Methuen, 1909), p. 257.

MORE THINGS IN HEAVEN AND EARTH

Darwinian Fundamentalism

With copious evidence ranging from Plato's haughtiness to Beethoven's tirades, we may conclude that most brilliant people of history tend to be a prickly lot. But Charles Darwin must have been the most genial of geniuses. He was kind to a fault, even to the undeserving, and he never uttered a harsh word – or hardly ever, as his countryman Captain Corcoran once said. Darwin's disciple, George Romanes, expressed surprise at the only sharply critical Darwinian statement he had even encountered: 'In the whole range of Darwin's writings there cannot be found a passage so strongly worded as this: it presents the only note of bitterness in all the thousands of pages which he has published.' Darwin directed the passage which Romanes found so striking against people who would simplify and caricature his theory as claiming that natural selection, and only natural selection, caused all evolutionary changes. He wrote in the last (1872) edition of *The Origin of Species*:

> As my conclusions have lately been much misrepresented, and it has been stated that I attribute the modification of species exclusively to natural selection, I may be permitted to remark that in the first edition of this work and subsequently, I place in a most conspicuous position – namely at the close of the Introduction – the following words: 'I am convinced that natural selection has been the main but not the exclusive means of modification.' This has been of no avail. Great is the power of steady misrepresentation.

Darwin clearly loved his distinctive theory of natural selection – the powerful ideas that he often identified in letters as his dear 'child'. But, like any good parent, he understood limits and imposed discipline. He knew that the complex and comprehensive phenomena of evolution could not be fully rendered by any single cause, even one so ubiquitous and powerful as his own brainchild.

In this light, especially given history's tendency to recycle great issues, I am amused by an irony that has recently ensnared evolutionary theory. A movement of strict constructionism, a self-styled form of Darwinian fundamentalism, has risen to some prominence in a variety of fields, from the English biological heartland of John Maynard Smith to the uncompromising ideology (albeit in graceful prose) of his compatriot Richard Dawkins, to the equally narrow and more ponderous writing of the American philosopher Daniel Dennett.[1] Moreover, a larger group of strict constructionists are now engaged in an almost mordantly self-conscious effort to 'revolutionize' the study of human behaviour along a Darwinian straight and narrow under the name of 'evolutionary psychology'.

Some of these ideas have filtered into the general press, but the uniting theme of Darwinian fundamentalism has not been adequately stressed or identified. Professionals, on the other hand, are well aware of the connections. My colleague Niles Eldredge, for example, speaks of this co-ordinated movement as ultra-Darwinism.[2] Amid the variety of their subject matter, the ultra-Darwinists share a conviction that natural selection regulates everything of any importance in evolution, and that adaptation emerges as a universal result and ultimate test of selection's ubiquity.

The irony of this situation is twofold. First, as illustrated by the quotation above, Darwin himself strongly opposed the ultras of his own day. (In one sense, this nicety of history should not be relevant to modern concerns: maybe Darwin was overcautious, and modern ultras therefore out-Darwin Darwin for good reason. But since the modern ultras push their line with an almost theological fervour, and since the views of founding fathers do matter in religion, though supposedly not in science, Darwin's own fierce opposition does become a factor in judgement.) Second, the invigoration of modern evolutionary biology with exciting non-selectionist and non-adaptationist data from the three central disciplines of population genetics, developmental biology and paleontology (see examples below) makes our pre-millennial decade an especially unpropitious time for Darwinian fundamentalism – and seems only to reconfirm Darwin's own eminently sensible pluralism.

Charles Darwin often remarked that his revolutionary work had two distinct aims: first, to demonstrate the fact of evolution (the genealogical connection of all organisms and a history of life regulated by 'descent with modification'); second, to advance the theory of natural selection as the most important mechanism of evolution. Darwin triumphed in his first aim (American creationism of the Christian far right notwithstanding). Virtually all thinking people accept the factuality of evolution and no conclusion in science enjoys better documentation. Darwin also succeeded substantially in his second aim. Natural selection, an immensely powerful idea with radical philosophical implications, is surely a major cause of evolution as validated in theory and demonstrated by countless experiments. But is natural selection as ubiquitous and effectively exclusive as the ultras propose?

The radicalism of natural selection lies in its power to dethrone some of the deepest and most traditional comforts of Western thought, particularly the notion that nature's benevolence, order, and good design, with humans at a sensible summit of power and excellence, prove the existence of an omnipotent and benevolent creator who loves us most of all (the old-style theological version), or at least that nature has meaningful directions and that humans fit into a sensible and predictable pattern regulating the totality (the modern and more secular version).

To these beliefs Darwinian natural selection presents the most contrary position imaginable. Only one causal force produces evolutionary change in Darwin's world: the unconscious struggle among individual organisms to promote their own personal reproductive success – nothing else, and nothing higher (no force, for example, works explicitly for the good of species or the harmony of ecosystems). Richard Dawkins would narrow the focus of explanation even one step further – to genes struggling for reproductive success within passive bodies (organisms) under the control of genes – a hyper-Darwinian idea that I regard as a logically flawed and basically foolish caricature of Darwin's genuinely radical intent.

The very phenomena that traditional views cite as proof of benevolence and intentional order – the good design or organisms and the harmony of ecosystems – arise by Darwin's process of natural selection only as side consequences of a singular causal principle of apparently opposite meaning: organisms struggling for themselves alone. (Good design becomes one pathway to reproductive success, while the harmony of ecosystems records a competitive balance among victors.) Darwin's system should be viewed as morally liberating, not cosmically depressing. The answers to moral questions cannot be found in nature's factuality in any case, so why not take the 'cold bath' of recognizing nature as nonmoral, and not constructed to match our hopes? After all, life existed on earth for 3.5 billion years before we arrived; why should life's causal ways match our prescriptions for human meaning or decency?

We now reach the technical and practical point that sets the ultra-Darwinian research agenda. Natural selection can be observed directly, but only in the unusual circumstances of controlled experiments in laboratories (on organisms with very short generations such as fruit flies) or within simplified and closely monitored systems in nature. Since evolution, in any substantial sense, takes so much time (more than the entire potential history of human observing!), we cannot, except in special circumstances, watch the process in action, and must therefore try to infer causes from results – the standard procedure in any historical science, by the way, and not a special impediment facing evolutionists.

The generally accepted result of natural selection is adaptation – the shaping of an organism's form, function and behaviour to achieve the Darwinian *summum bonum* of enhanced reproductive success. We must therefore study natural selection primarily from its results – that is, by concentrating on the putative adaptations of organisms. If we can interpret all relevant attributes of organisms as adaptations for reproductive success, then we may infer that natural selection has been the cause of evolutionary change. This strategy of research – the so-called adaptationist programme – is the heart of Darwinian biology and the fervent, singular credo of the ultras.

Since the ultras are fundamentalists at heart, and since fundamentalists generally try to stigmatise their opponents by depicting them as apostates from the one true way, may I state for the record that I (along with all other Darwinian pluralists) do not deny either the existence and central importance of adaptation, or the production of adaptation by natural selection. Yes, eyes are for seeing and feet are for moving. And, yes again, I know of no scientific mechanism other than natural selection with the proven power to build structures of such eminently workable design.

But does all the rest of evolution – all the phenomena of organic diversity, embryological architecture and genetic structure, for example – flow by simple extrapolation from selection's power to create the good design of organisms? Does the force that makes a functional eye also explain why the world houses more than 500,000 species of beetles and fewer than fifty species of priapulid worms? Or why most nucleotides in multicellular creatures do not code for any enzyme or protein involved in the construction of an organism? Or why ruling dinosaurs died and subordinate mammals survived to flourish and, along one oddly contingent pathway, to evolve a creature capable of building cities and understanding natural selection?

I do not deny that natural selection has helped us to explain phenomena at scales very distant from individual organisms, from the behaviour of an ant colony to the survival of a redwood forest. But selection cannot suffice as a full explanation for many aspects of evolution: for other types and styles of causes become relevant, or even prevalent, in domains both far above and far below the traditional Darwinian locus of the organism. These other causes are not, as the ultras often claim, the product of thinly veiled attempts to smuggle purpose back into biology. These additional principles are as directionless, nonteleological and materialistic as natural selection itself – but they operate differently from Darwin's central mechanism. In other words, I agree with Darwin that natural selection is 'not the exclusive means of modification'.

What an odd time to be a fundamentalist about adaptation and natural selection – when each major subdiscipline of evolutionary biology has been discovering other mechanisms as adjuncts to selection's centrality. Population genetics has worked out in theory, and validated in practice, an elegant, mathematical account of the large role that neutral, and therefore non-adaptive, changes play in the evolution of nucleotides, or individual units of DNA programmes. Eyes may be adaptations, but most substitutions of one nucleotide for another within populations may not be adaptive.

In the most stunning evolutionary discoveries of the past decade, developmental biologists have documented an astonishing 'conservation', or close similarity, of basic pathways of development among phyla that have been evolving independently for at least 500 million years, and that seem so different in basic anatomy (insects and vertebrates, for example). The famous homeotic genes of fruit flies – responsible for odd mutations that disturb the order of parts along the main body axis, placing legs, for example, where antennae or mouth parts should be – are also present (and repeated four times on four separate chromosomes) in vertebrates, where they function in effectively the same way. The major developmental pathway for eyes is conserved and mediated by the same gene in squids, flies and vertebrates, though the end products differ substantially (our single-lens eye versus the multiple facets of insects). The same genes regulate the formation of top and bottom surfaces in insects and vertebrates, though with inverted order – as our back, with the spinal cord running above the gut, is anatomically equivalent to an insect's belly, where the main nerve cords run along the bottom surface, with the gut above.

One could argue, I suppose, that these instances of conservation only record adaptation, unchanged through all of life's vicissitudes because their optimality can't be improved. But most biologists feel that such stability acts primarily as a constraint upon the range and potentiality of adaptation, for if organisms of such different function and ecology must build bodies along the same basic pathways, then limitation of possibilities rather than adaptive honing to perfection becomes a dominant theme in evolution. At a minimum, in explaining evolutionary pathways through time, the constraints imposed by history rise to equal prominence with the immediate advantages of adaptation.

My own field of paleontology has strongly challenged the Darwinian premise that life's major transformations can be explained by adding up, through the immensity of geological time, the successive tiny changes produced generation after generation by natural selection. The extended stability of most species, and the branching off of new species in geological moments (however slow by the irrelevant scale of a human life) – the pattern known as punctuated equilibrium – require that long-term evolutionary trends be explained as the distinctive success of some species versus others, and not as a gradual accumulation of adaptations generated by organisms within a continuously evolving population. A trend may be set by high rates of branching in certain species within a larger group. But individual organisms do not branch; only populations do – and the causes of a population's branching can rarely be reduced to the adaptive improvement of its individual organisms.

The study of mass extinction has also disturbed the ultra-Darwinian consensus. We now know, at least for the terminal Cretaceous event some 65 million years ago which wiped out dinosaurs along with about 50 per cent of marine invertebrate species, that some episodes of mass extinction are both truly catastrophic and triggered by extraterrestrial impact. The death of some groups (like dinosaurs) in mass extinctions and the survival of others (like mammals), while surely not random, probably bears little relationship to the evolved, adaptive reasons for success of lineages in normal Darwinian times dominated by competition. Perhaps mammals survived (and humans ultimately evolved) because small creatures are more resistant to catastrophic extinction. And perhaps Cretaceous mammals were small primarily because they could not compete successfully in the larger size ranges of dominant dinosaurs. Immediate adaptation may bear no relationship to success over immensely long periods of geological change.

Why then should Darwinian fundamentalism be expressing itself so stridently when most evolutionary biologists have become more pluralistic in the light of these new discoveries and theories? I am no psychologist, but I suppose that the devotees of any superficially attractive cult must dig in when a general threat arises. 'That old-time religion; it's good enough for me.' There is something immensely beguiling about strict adaptationism – the dream of an underpinning simplicity for an enormously complex and various world. If evolution were powered by a single force producing one kind of result, and if life's long messy history could therefore be explained by extending small or orderly increments of adaptation through the immensity of geological time, then an explanatory simplicity might descend upon evolution's overt richness. Evolution then might become 'algorithmic', a surefire logical procedure, as in Daniel Dennett's reverie. But what is wrong with messy richness, so long as we can construct an equally rich texture of satisfying explanation?

Daniel Dennett's 1995 book, *Darwin's Dangerous Idea*, presents itself as the ultras' philosophical manifesto of pure adaptationism. Dennett explains the strict adaptationist view well enough, but he defends a blinkered picture of evolution in assuming that all important phenomena can be explained thereby.

Dennett bases his argument on three images or metaphors, all sharing the common error of assuming that conventional natural selection, working in the adaptationist mode, can account for all evolution by extension – so that the entire history of life becomes one grand solution to problems in design. 'Biology is engineering,' Dennett tells us again and again. In a devastating review, published in the leading professional journal *Evolution*, and titled 'Dennett's Dangerous Idea', H. Allen Orr notes:

> His review of attempts by biologists to circumscribe the role of natural selection borders on a zealous defense of panselectionism. It is also absurdly unfair . . . Dennett fundamentally misunderstands biologists' worries about adaptationism. Evolutionists are essentially unanimous that – where there is 'intelligent Design' – it is caused by natural selection . . . Our problem is that, in many adaptive stories, the protagonist does not show dead-obvious signs of Design.

In his first metaphor Dennett describes Darwin's dangerous idea of natural selection as a 'universal acid' – to honour both its ubiquity and its power to corrode traditional Western beliefs. Speaking of adaptation, natural selection's main consequence, Dennett writes, 'It plays a crucial role in the analysis of every biological event at every scale from the creation of the first self-replicating macromolecule on up.' I certainly accept the acidic designation – for the power and influence of the idea of natural selection does lie in its radical philosophical content – but few biologists would defend the blithe claim for ubiquity. If Dennett chooses to restrict his personal interest to the engineering side of biology – the part that natural selection does construct – then he is welcome to do so. But he may not impose this limitation upon others, who know that the record of life contains many more evolutionary things than are dreamt of in Dennett's philosophy.

Natural selection does not explain why many evolutionary transitions from one nucleotide to another are neutral, and therefore non-adaptive. Natural selection does not explain why a meteor crashed into the earth 65 million years ago, setting in motion the extinction of half the world's species. As Orr points out, Dennett's disabling parochialism lies most clearly exposed in his failure to discuss the neutral theory of molecular evolution, or even to mention the name of its founder, the great Japanese geneticist Motoo Kimura – for few evolutionary biologists would deny that this theory ranks among the most interesting and powerful adjuncts to evolutionary explanation since Darwin's formulation of natural selection. You don't have to like the idea, but how can you possibly leave it out?

In a second metaphor Dennett continually invokes an image of cranes and skyhooks. In his reductionist account of evolution cranes build the good design of organisms upward from nature's physicochemical substrate. Cranes are good. Natural selection is evolution's basic crane; all other cranes (sexual reproduction, for example) act as mere auxiliaries to boost the speed or power of natural selection in constructing organisms of good design. Skyhooks, on the other hand, are spurious forms of special pleading that reach down from the numinous heavens and try to build organic complexity with ad hoc fallacies and speculations unlinked to other proven causes. Skyhooks, of course, are bad. Everything that isn't natural selection, or an aid to the operation of natural selection, is a skyhook.

But a third (and correct) option exists to Dennett's oddly dichotomous Hobson's choice: either accept the idea of one basic crane with auxiliaries, or believe in skyhooks. May I suggest that the platform of evolutionary explanation houses an assortment of basic cranes, all helping to build the edifice of life's history in its full grandeur (not only the architecture of well-engineered organisms). Natural selection may be the biggest crane with the largest set of auxiliaries, but Kimura's theory of neutralism is also a crane; so is punctuated equilibrium; so is the channeling of evolutionary change by developmental constraints. 'In my father's house are many mansions' – and you need a lot of cranes to build something so splendid and variegated.

For his third metaphor – though he would demur and falsely label the claim as a fundamental statement about causes – Dennett describes evolution as an 'algorithmic process'. Algorithms are abstract rules of calculation, fully general in making no reference to particular content. In Dennett's words, 'an algorithm is a certain sort of formal process that can be counted on – logically – to yield a certain sort of result whenever it is "run" or instantiated'. If evolution truly works by an algorithm, then all else in Dennett's simplistic system follows: we need only one kind of crane to supply the universal acid.

I am perfectly happy to allow – indeed I do not see how anyone could deny – that natural selection, operating by its bare-bones mechanics, is algorithmic: variation proposes and selection disposes. So if natural selection builds all of evolution without the interposition of auxiliary processes or intermediary complexities, then I suppose that evolution is algorithmic too. But – and here we encounter Dennett's disabling error once again – evolution includes so much in addition to natural selection that it cannot be algorithmic in Dennett's simple calculational sense.

Yet Dennett yearns to subsume all the phenomenology of nature under the limited aegis of adaptation as an algorithmic result of natural selection. He writes, 'Here, then, is Darwin's dangerous idea: the algorithmic level *is* the level that best accounts for the speed of the antelope, the wing of the eagle, the shape of the orchid, the diversity of species, and all the other occasions for wonder in the world of nature' (Dennett's italics). I will grant the antelope's run, the eagle's wing, and much of the orchid's shape – for these are adaptations, produced by natural selection, and therefore legitimately in the algorithmic domain. But can Dennett really believe his own imperialistic extensions? Is the diversity of species no more than a calculational consequence of natural selection? Can anyone really believe, beyond the hype of rhetoric, that '*all* the other occasions for wonder in the world of nature' flow from adaptation?

Perhaps Dennett only gets excited when he can observe adaptive design, the legitimate algorithmic domain; but such an attitude surely represents an impoverished view of nature's potential interest. I regard the neutral substitution of nucleotides as an 'occasion for wonder in the world of nature'. And I marvel at the probability that the impact of a meteor wiped out dinosaurs and gave mammals a chance. If this contingent event had not occurred, and imparted a distinctive pattern to the evolution of life, we would not be here to wonder about anything at all!

The Fallacies of 'Evolutionary Psychology' and the Pleasures of Pluralism

Darwin began the last paragraph of *The Origin of Species* (1859) with a famous metaphor about life's diversity and ecological complexity:

> It is interesting to contemplate an entangled bank, clothed with many plants of many kinds, with birds singing on the bushes, with various insects flitting about, and with worms

> crawling through the damp earth, and to reflect that these elaborately constructed forms, so different from each other, and dependent on each other in so complex a manner, have all been produced by laws acting around us.

He then begins the final sentence of the book with an equally famous statement: 'There is grandeur in this view of life . . .'

For Darwin, as for any scientist, a kind of ultimate satisfaction (Darwin's 'grandeur') must reside in the prospect that so much variety and complexity might be generated from natural regularities – the 'laws acting around us accessible to our intellect and empirical probing'. But what is the proper relationship between underlying laws and explicit results? The fundamentalists among evolutionary theorists revel in the belief that one overarching law – Darwin's central principle of natural selection – can render the full complexity of outcomes (by working in conjunction with auxiliary principles, like sexual reproduction, that enhance its rate and power).

The 'pluralists', on the other hand – a long line of thinkers including Darwin himself, however ironic this may seem since the fundamentalists use the cloak of his name for their distortion of his position – accept natural selection as a paramount principle (truly *primus inter pares*), but then argue that a set of additional laws, as well as a large role for history's unpredictable contingencies, must also be invoked to explain the basic patterns and regularities of the evolutionary pathways of life. Both sides locate the 'grandeur' of 'this view of life' in the explanation of complex and particular outcomes by general principles, but ultra-Darwinian fundamentalists pursue one true way, while pluralists seek to identify a set of interacting explanatory modes, all fully intelligible, although not reducible to a single grand principle like natural selection.

Daniel Dennett devotes the longest chapter in *Darwin's Dangerous Idea* to an excoriating caricature of my ideas, all in order to bolster his defence of Darwinian fundamentalism. If an argued case can be discerned at all amid the slurs and sneers, it would have to be described as an effort to claim that I have, thanks to some literary skill, tried to raise a few piddling, insignificant and basically conventional ideas to 'revolutionary' status, challenging what he takes to be the true Darwinian scripture. Dennett claims that I have promulgated three 'false alarms' as supposed revolutions against the version of Darwinism that he and his fellow defenders of evolutionary orthodoxy continue to espouse.

Dennett first attacks my view that punctuated equilibrium is the dominant pattern of evolutionary change in the history of living organisms. This theory, formulated by Niles Eldredge and me in 1972, proposes that the two most general observations made by paleontologists form a genuine and primary pattern of evolution and do not arise as artefacts of an imperfect fossil record. The first observation notes that most new species originate in a geological 'moment'. The second holds that species generally do not change in any substantial or directional way during their geological lifetimes – usually a long period averaging five to ten million years for fossil invertebrate species. Punctuated equilibrium does not challenge accepted genetic ideas about the rates at which species emerge (for the geological 'moment' of a single rock layer may represent many thousand years of accumulation). But the theory does contravene conventional Darwinian expectations for gradual change over geological periods and does suggest a substantial revision of standard views about the causes of long-term evolutionary trends. For such trends must now be explained by the higher rates at which some species branch off from others, and the greater durations of some stable species as distinguished from others, and not as the slow and continuous transformation of entire populations.

In his second attack, Dennett denigrates the importance of non-adaptive side consequences ('spandrels' in my terminology) as sources for later and fruitful reuse. In principle, spandrels define the major category of important evolutionary features that do not arise as adaptations. Since organisms are complex and highly integrated entities, any adaptive change must automatically 'throw off' a series of structural by-products – like the mold marks on an old bottle or, in the case of an architectural spandrel itself, the triangular space 'left over' between a rounded arch and the rectangular frame of wall and ceiling. Such by-products may later be co-opted for useful purposes, but they didn't arise as adaptations. Reading and writing are now highly adaptive for humans, but the mental machinery for these crucial capacities must have originated as spandrels that were co-opted later, for the brain reached its current size and conformation tens of thousands of years before any human invented reading or writing.

Third, and finally, Dennett denies theoretical importance to the roles of contingency and chance in the history of life, a history that has few predictable particulars and no inherent directionality, especially given the persistence of bacteria as the most common and dominant form of life on Earth ever since their origin as the first fossilised creatures some 3.5 billion years ago.[3] Bacteria are biochemically more diverse, and live in a wider range of environments (including near-boiling waters, and pore spaces in rocks up to two miles beneath the earth's surface), than all other living things combined. The number of *E. coli* cells in the gut of each human being exceeds the total number of human beings that have ever lived.

These three concepts work as pluralistic correctives to both the poverty and limited explanatory power of the ultra-Darwinian research programme. Punctuated equilibrium requires that substantial evolutionary trends over geological time, the primary phenomenon of macroevolution, be explained by the greater long-term success of some species versus others within a group of species descended from a common ancestor. Such trends cannot be explained, as Darwinian fundamentalists would prefer, as the adaptive success of individual organisms in conventional competition extrapolated through geological time as the slow and steady transformation of populations by natural selection. The principle of spandrels, discussed at greater length later in this chapter, stresses the role that nonadaptive side consequences play in structuring the directions and potentials of future evolutionary change. Taken together, punctuated equilibrium and spandrels invoke the operation of several important principles in addition (and sometimes even opposed) to conventional natural selection working in the engineering mode that Dennett sees as the only valid mechanism of evolution.

My third pluralistic corrective to traditional theory does not invoke other principles in addition to natural selection, but rather stresses the limits faced by *any* set of general principles in our quest to explain the actual patterns of life's history. Crank your algorithm of natural selection to your heart's content, and you cannot grind out the contingent patterns built during the earth's geological history. You will get predictable pieces here and there (convergent evolution of wings in flying creatures), but you will also encounter too much randomness from a plethora of sources, too many additional principles from within biological theory and too many unpredictable impacts from environmental histories beyond biology (including those occasional meteors) – all showing that the theory of natural selection must work in concert with several other principles of change to explain the observed pattern of evolution.

The fallacy of Dennett's argument undermines his other imperialist hope that the universal acid of natural selection might reduce human cultural change to the Darwinian algorithm as well. Dennett, following Dawkins, tries to identify human thoughts and actions as 'memes', thus viewing them as units that are subject to a form of selection analogous to natural selection of genes. Cultural change, working by memetic selection, then becomes as algorithmic as biological change operating by natural selection on genes – thus uniting the evolution of organisms and thoughts under a single ultra-Darwinian rubric:

> According to Darwin's dangerous idea . . . not only all your children and your children's children, but all your brain-children and your brainchildren's brainchildren must grow from the common stock of Design elements, genes and memes . . . Life and all its glories are thus united under a single perspective.

But, as Dennett himself correctly and repeatedly emphasises, the generality of an algorithm depends upon 'substrate neutrality'. That is, the various materials (substrates) subject to the mechanism (natural selection in this case) must all permit the mechanism to work in the same effective manner. If one kind of substrate tweaks the mechanism to operate differently (or, even worse, not to work at all), then the algorithm fails. To choose a somewhat silly example which actually played an important role in recent American foreign policy, the Cold War 'domino theory' held that communism must be stopped everywhere because if one country turned red, then others would do so as well, for countries are like dominoes standing on their ends and placed one behind the other – so that the toppling of one must propagate down the entire line to topple all. Now if you devised a general formula (an algorithm) to describe the necessary propagation of such toppling, and wanted to cite the algorithm as a general rule for all systems made of a series of separate objects, then the generality of your algorithm would depend upon substrate neutrality – that is, upon the algorithm's common working regardless of substrate (similarly for dominoes and nations in this case). The domino theory failed because differences in substrate affect the outcome, and such differences can even derail the operation of the algorithm. Dominoes must topple, but the second nation in a line might brace itself, stay upright upon impact, and therefore fail to propagate the collapse.

Natural selection does not enjoy this necessary substrate neutrality.

As the great evolutionist R. A. Fisher showed many years ago in the founding document of modern Darwinism (*The Genetical Theory of Natural Selection*, 1930), natural selection requires Mendelian inheritance to be effective. Genetic evolution works upon such a substrate and can therefore be Darwinian. Cultural (or memetic) change manifestly operates on the radically different substrate or Lamarckian inheritance, or the passage of acquired characters to subsequent generations. Whatever we invent in our lifetimes, we can pass on to our children by our writing and teaching. Evolutionists have long understood that Darwinism cannot operate effectively in systems of Lamarckian inheritance – for Lamarckian change has such a clear direction and permits evolution to proceed so rapidly that the much slower process of natural selection shrinks to insignificance before the Lamarckian juggernaut.

This crucial difference between biological and cultural evolution also undermines the self-proclaimed revolutionary pretensions of a much-publicized doctrine – 'evolutionary psychology' – that could be quite useful if proponents would change their propensity for cultism and ultra-Darwinian fealty for a healthy dose of modesty.[4] Humans are animals and the mind evolved; therefore, all curious people must support the quest for an evolutionary psychology. But the movement that has commandeered this name adopts a fatally restrictive view of the meaning and range of evolutionary explanation. 'Evolutionary psychology' has, in short, fallen into the same ultra-Darwinian trap that ensnared Daniel Dennett and his *confrères* – for disciples of this new art confine evolutionary accounts to the workings of natural selection and consequent adaptation for personal reproductive success.

Evolutionary psychology, as a putative science of human behaviour, itself evolved by 'descent with modification' from 1970s-style sociobiology. But the new species, like many children striving for independence, shuns its actual ancestry by taking a new name and exaggerating some genuine differences while ignoring the much larger amount of shared doctrine – all done, I assume, to avoid the odour of sociobiology's dubious political implications and speculative failures (amid some solid successes when based on interesting theory and firm data, mostly from nonhuman species).

Three major claims define the core commitments of evolutionary psychology; each embodies a considerable strength and a serious (in one case, fatal) weakness:

1. *Modularity*. Human behaviour and mental operations can be

divided into a relatively discrete set of items, or mental organs. (In one prominent study, for example, authors designate a 'cheater detector' as a mental organ, since the ability to discern infidelity and other forms of prevarication can be so vital to Darwinian success – the adaptationist rationale.) The argument for modularity flows, in part, from exciting work in neurobiology and cognitive science on localisation of function within the brain – as shown, for example, in the precise mapping, to different areas of the cerebral cortex, of mental operations formerly regarded as only arbitrarily divisible by social convention (production of vowels and consonants, for example, or the naming of animals and tools).

Ironically, though, neurobiology and evolutionary psychology employ the concept of modularity for opposite theoretical purposes. Neurobiologists do so to stress the complexity of an integrated organ. Evolutionary psychology uses modularity to atomise behaviour into a priori, subjectively defined and poorly separated items (not known modules empirically demonstrated by neurological study), so that selective value and adaptive significance can be postulated for individual items as the ultra-Darwinian approach requires.

2. *Universality*. Evolutionary psychologists generally restrict their study to universal aspects of human behaviour and mentality, thereby explicitly avoiding the study of differences among individuals or groups. They argue that variations among individuals, and such groups as races and social classes, only reflect the influence of diverse environments upon a common biological heritage. In this sense, they argue, evolutionary psychology adopts a 'liberal' position in contrast with the conservative implications of most previous evolutionary arguments about behaviour, which viewed variation among individuals and groups as results of different, and largely unalterable, genetic constitutions.

I welcome much of this change; but, in one important respect, this new approach to universals and differences continues to follow the old strategy of finding an adaptationist narrative (often in the purely speculative or storytelling mode) to account for genetic differences built by natural selection. For the most publicised work in evolutionary psychology has centred on the universality in all human societies of a particular kind of difference: the putative evolutionary reasons for supposedly universal behavioural differences between males and females.

3. *Adaptation*. Evolutionary psychologists claim that they have reformed the old adaptationism of sociobiology into a new and exciting

approach. They will no longer just assume, they now say, that all prominent and universal behaviours must, ipso facto, be adaptive to modern humans in boosting reproductive success. They recognize, instead, that many such behaviours may be tragically out of whack with the needs of modern life, and may even lead to our destruction – aggressivity in a nuclear age, for example.

Again, I applaud this development. If this principle were advanced in conjunction with the recognition that a putative evolutionary origin does not necessarily imply an adaptive value at all, then evolutionary psychology could make a substantial advance in applying Darwinian theory to human behaviour. But the advocates of evolutionary psychology proceed in the opposite direction by twisting the observation that the behaviour of modern humans may not necessarily have adaptive value into an even more dogmatic, and even less scientifically testable, panadaptionist claim. Evolutionary universals may not be adaptive now, they say, but such behaviours must have *arisen* as adaptations in the different ancestral environment of life as small bands of hunter-gatherers on the African savannahs – for evolutionary theory 'means' a search for adaptive origins.

The task of evolutionary psychology then turns into a speculative search for reasons why a behaviour that may harm us now must once have originated for adaptive purposes. To take an illustration proposed seriously by Robert Wright in *The Moral Animal*,[5] a sweet tooth leads to unhealthy obesity today but must have arisen as an adaptation. Wright therefore states, 'The classic example of an adaptation that has outlived its logic is the sweet tooth. Our fondness for sweetness was designed for an environment in which fruit existed but candy didn't.'

This statement ranks as pure guesswork in the cocktail party mode; Wright presents no neurological evidence of a brain module for sweetness and no paleontological data about ancestral feeding. This 'just-so story' therefore cannot stand as a 'classic example of an adaptation' in any sense deserving the name of science.

Much of evolutionary psychology therefore devolves into a search for the so-called EEA, or 'environment of evolutionary adaptation', which allegedly prevailed in prehistoric times. Evolutionary psychologists have gained some sophistication in recognizing that they need not postulate current utility to advance a Darwinian argument; but they have made their enterprise even less operational by placing their central postulate outside the primary definition of science – for claims about an EEA usually cannot be tested in principle but only subjected to speculation. At least an argument about modern utility can be tested by studying the current impact of a given feature upon reproductive success. Indeed, the disproof of many key sociobiological speculations about current utility pushed evolutionary psychology to the revised tactic of searching for an EEA instead.

But how can we possibly know in detail what small bands of hunter-gatherers did in Africa two million years ago? These ancestors left some tools and bones and paleoanthropologists can make some ingenious inferences from such evidence. But how can we possibly obtain the key information that would be required to show the validity of adaptive tales about an EEA: relations of kinship, social structures and sizes of groups, different activities of males and females, the roles of religion, symbolising, storytelling and a hundred other central aspects of human life that cannot be traced in fossils? We do not even know the original environment of our ancestors – did ancestral humans stay in one region or move about? How did environments vary through years and centuries?

In short, evolutionary psychology is as ultra-Darwinian as any previous behavioural theory in insisting upon adaptive reasons for origin as the key desideratum of the enterprise. But the chief strategy proposed by evolutionary psychologists for identifying adaptation is untestable and therefore unscientific. This central problem does not restrain leading disciples from indulging in reveries about the ubiquity of original adaptation as the source of revolutionary power for the putative new science. I detect not a shred of caution in this proclamation by Wright – embodying the three principal claims of evolutionary psychology as listed above:

> The thousands and thousands of genes that influence human behaviour – genes that build the brain and govern neurotransmitters and other hormones, thus defining our 'mental organs' [note the modularity claim] – are here for a reason. And the reason is that they goaded our ancestors into getting their genes into the next generation [the claim for adaptation in the EEA]. If the theory of natural selection is correct, then essentially everything about the human mind should be intelligible in these terms [the ultra-Darwinian faith in adaptationism]. The basic ways we feel about each other, the basic kinds of things we think about each other and say to each

> other [note the claim for universality], are with us today by virtue of their past contribution to genetic fitness.

Wright's closing sermon is more suitable to a Sunday pulpit than a work of science:

> The theory of natural selection is so elegant and powerful as to inspire a kind of faith in it – not *blind* faith, really . . . But faith nonetheless; there is a point after which one no longer entertains the possibility; of encountering some fact that would call the whole theory into question.
>
> I must admit to having reached this point. Natural selection has now been shown to plausibly account for so much about life in general and the human mind in particular that I have little doubt that it can account for the rest.

This adaptationist premise is the fatal flaw of evolutionary psychology in its current form. The premise also seriously compromises – by turning a useful principle into a central dogma with asserted powers for nearly universal explanation – the most promising theory of evolutionary psychology: the recognition that differing Darwinian requirements for males and females imply distinct adaptive behaviours centred upon male advantage in spreading sperm as widely as possible (since a male need invest no energy in reproduction beyond a single ejaculation) and female strategies for extracting additional time and attention from males (in the form of parental care or supply of provisions, etc.). In most sexually reproducing species, males generate large numbers of 'cheap' sperm, while females make relatively few 'energetically expensive' eggs and then must invest much time and many resources in nurturing the next generation.

This principle of differential 'parental investment' makes Darwinian sense and probably does underlie some different, and broadly general, emotional propensities of human males and females. But contrary to claims in a recent deluge of magazine articles, parental investment will not explain the full panoply of supposed sexual differences so dear to pop psychology. For example, I do not believe that members of my gender are willing to rear babies only because clever females beguile us. A man may feel love for a baby because the infant looks so darling and dependent and because a father sees a bit of himself in his progeny.

This feeling need not arise as a specifically selected Darwinian adaptation for my reproductive success, or as the result of a female ruse, culturally imposed. Direct adaptation represents only one mode of evolutionary origin. After all, I also have nipples not because I need them, but because women do and all humans share the same basic pathways of embryological development.

If evolutionary psychologists continue to push the theory of parental investment as a central dogma, they will eventually suffer the fate of the Freudians, who also had some good insights but failed spectacularly, and with serious harm imposed upon millions of people (women, for example, who were labelled as 'frigid' when they couldn't make an impossible physiological transition from clitoral to vaginal orgasm), because they elevated a limited guide into a rigid creed that became more of an untestable and unchangeable religion than a science.

Exclusive adaptationism suffers fatally from two broad classes of error, one external to Darwinian theory, the other internal. The external error arises from fundamental differences in principle and mechanism between, on the one hand, genetic Darwinian evolution and, on the other, human cultural change, which cannot be basically Darwinian at all. Since every participant in these debates, including Dennett and the evolutionary psychologists, agrees that much of human behaviour arises by culturally induced rather than genetically coded change, giving total authority to Darwinian explanation requires that culture also work in a Darwinian manner. (Dennett, as discussed earlier, makes such a claim for cultural change in arguing for the 'substrate neutrality' of natural selection.) But for two fundamental reasons (and a host of other factors), cultural change unfolds in virtual antithesis to Darwinian requirements.

First, topological: as the common metaphor proclaims, biological evolution builds a tree of life – a system based upon continuous diversification and separation. A lineage, after branching off from ancestors as a new species, attains an entirely independent evolutionary fate. Nature cannot make a new mammalian species by mixing 20 per cent dugong with 30 per cent rat and 50 per cent aardvark. But cultural change works largely by an opposite process of joining, or interconnection, of lineages.

Marco Polo visits China and returns with many of the customs and skills that later distinguish Italian culture. I speak English because my grandparents migrated to the United States. Moreover, this interdigitation implies that human cultural change needn't even follow genealogical lines

– the most basic requirement of a Darwinian evolutionary process – for even the most distant cultural lineages can borrow from each other with ease. If we want a biological metaphor for cultural change, we should probably invoke infection rather than evolution.

Second, casual: as argued above, human cultural change operates fundamentally in the Lamarckian mode, while genetic evolution remains firmly Darwinian. Lamarckian processes are so labile, so directional and so rapid that they overwhelm Darwinian rates of change. Since Lamarckian and Darwinian systems work so differently, cultural change will receive only limited (and metaphorical) illumination from Darwinism.

The internal error of adaptationism arises from a failure to recognize that even the strictest operation of pure natural selection builds organisms full of nonadaptive parts and behaviours. Non-adaptations arise for many reasons in Darwinian systems, but consider only my favourite principle of 'spandrels'.

All organisms evolve as complex and interconnected wholes, not as loose alliances of separate parts, each independently optimized by natural selection. Any adaptive change must also generate, in addition, a set of spandrels or non-adaptive by-products. These spandrels may later be 'co-opted' for a secondary use. But we would make an egregious logical error if we argued that these secondary uses explain the existence of a spandrel. I may realize some day that my favourite boomerang fits beautifully into the arched space of my living-room spandrel, but you would think me pretty silly if I argued that the spandrel exists to house the boomerang. Similarly, snails build their shells by winding a tube around an axis of coiling. This geometry of growth generates an empty cylindrical space, called an umbilicus, along the axis. A few species of snails use the umbilicus as a brooding chamber for storing eggs. But the umbilicus arose as a non-adaptive spandrel, not as an adaptation for reproduction. The overwhelming majority of snails do not use their umbilici for brooding, or for much of anything.

If any organ is, prima facie, replete with spandrels the human brain must be our finest candidate – thus making adaptationism a particularly dubious approach to human behaviour. I can adopt (indeed I do) the most conventional Darwinian argument for why the human brain evolved to a large size – and the non-adaptationist principle of spandrels may still dominate human nature. I am content to believe that the human brain became large by natural selection, and for adaptive reasons – that is, for some set of activities that our savannah ancestors could only perform with bigger brains.

Does this argument imply that all genetically and biologically based attributes of our universal human nature must therefore be adaptations? Of course not. Many, if not most, universal behaviours are probably spandrels, often co-opted later in human history for important secondary functions. The human brain is the most complicated device for reasoning and calculating, and for expressing emotion, ever evolved on earth. Natural selection made the human brain big, but most of our mental properties and potentials may be spandrels – that is, non-adaptive side consequences of building a device with such structural complexity. If I put a small computer (no match for a brain) in my factory, my adaptive reasons for so doing (to keep accounts and issue paychecks) represent a tiny subset of what the computer, by virtue of inherent structure, can do (factor-analyse my data on land snails, beat or tie anyone perpetually in tick-tack-toe). In pure numbers, the spandrels overwhelm the adaptations.

The human brain must be bursting with spandrels that establish central components of what we call human nature but that arose as non-adaptations, and therefore fall outside the compass of evolutionary psychology or any other ultra-Darwinian theory. The brain did not enlarge by natural selection so that we would be able to read or write. Even such an eminently functional and universal institution as religion arose largely as a spandrel if we accept Freud's old and sensible argument that humans devised religious beliefs largely to accommodate the most terrifying fact that our large brains forced us to acknowledge: the inevitability of personal mortality. We can scarcely argue that the brain got large so that we would know we must die!

In summary, Darwin cut to the heart of nature by insisting so forcefully that 'natural selection has been the main, but not the exclusive, means of modification' – and that hard-line adaptationism could only represent a simplistic caricature and distortion of his theory. We live in a world of enormous complexity in organic design and diversity – a world where some features of organisms evolved by an algorithmic form of natural selection, some by an equally algorithmic theory of unselected neutrality, some by the vagaries of history's contingency, and some as by-products of other processes. Why should such a complex and various world yield to one narrowly construed cause? Let us have a cast of cranes, some more important and general, others for particular things – but all subject to scientific understanding, and all working together in a comprehensible way. Why not admit for theory the same delight that Robert Louis Stevenson expressed for objects in his 'Happy Thought':

> The world is so full of a number of things,
> I'm sure we should all be as happy as kings.

Notes

1. Daniel Dennett, *Darwin's Dangerous Idea: Evolution and the Meanings of Life* (New York: Simon & Schuster, 1995).
2. Niles Eldredge, *Reinventing Darwin: The Great Debate at the High Table of Evolutionary Theory* (New York: John Wiley, 1995).
3. See my recent book *Life's Grandeur* (London: Cape, 1996; published in the USA as *Full House*, New York: Crown, 1996) for an account of this.
4. See such technical works as J. Barkow, L. Cosmides and J. Tooby (eds), *The Adapted Mind: Evolutionary Psychology and the Generation of Culture* (Oxford: Oxford University Press, 1992); D. M. Buss, *The Evolution of Desire* (New York: Basic Books, 1994); and, especially, for its impact by good writing and egregiously simplistic argument, the popular book of Robert Wright, *The Moral Animal: Why We Are the Way We Are: The New Science of Evolutionary Psychology* (New York: Random House, 1994).
5. Ibid., *The Moral Animal.*

POSTURE MAKETH THE MAN

NO event did more to establish the fame and prestige of The American Museum of Natural History than the Gobi Desert expeditions of the 1920s. The discoveries, including the first dinosaur eggs, were exciting and abundant, and the sheer romance fit Hollywood's most heroic mold. It is still hard to find a better adventure story than Roy Chapman Andrews's book (with its chauvinistic title): *The New Conquest of Central Asia*. Nonetheless, the expeditions utterly failed to achieve their stated purpose: to find in Central Asia the ancestors of man. And they failed for the most elementary of reasons – we evolved in Africa, as Darwin had surmised fifty years earlier.

Our African ancestors (or at least our nearest cousins) were discovered in cave deposits during the 1920s. But these australopithecines failed to fit preconceived notions of what a 'missing link' should look like, and many scientists refused to accept them as bona fide members of our lineage. Most anthropologists had imagined a fairly harmonious transformation from ape to human, propelled by increasing intelligence. A missing link should be intermediate in both body and brain – Alley Oop or the old (and false) representations of stoopshouldered Neanderthals. But the australopithecines refused to conform. To be sure, their brains were bigger than those of any ape with comparable body size, but not much bigger. Most of our evolutionary increase in brain size occurred after we reached the australopithecine level. Yet these small-brained australopithecines walked as erect as you or I. How could this be? If our evolution was propelled by an enlarging brain, how could upright posture – another 'hallmark of hominization,' not just an incidental feature – originate first? In a 1963 essay, George

Gaylord Simpson used this dilemma to illustrate

> the sometimes spectacular failure to predict discoveries even when there is a sound basis for such prediction. An evolutionary example is the failure to predict discovery of a 'missing link,' now known [*Australopithecus*], that was upright and tool-making but had the physiognomy and cranial capacity of an ape.

We must ascribe this 'spectacular failure' primarily to a subtle prejudice that led to the following, invalid extrapolation: We dominate other animals by brain power (and little else); therefore, an increasing brain must have propelled our own evolution at all stages. The tradition for subordinating upright posture to an enlarging brain can be traced throughout the history of anthropology. Karl Ernst von Baer, the greatest embryologist of the nineteenth century (and second only to Darwin in my personal pantheon of scientific heroes) wrote in 1828: 'Upright posture is only the consequence of the higher development of the brain . . . all differences between men and other animals depend upon construction of the brain.' One hundred years later, the English anthropologist G. E. Smith wrote: 'It was not the adoption of the erect attitude or the invention of articulate language that made man from an ape, but the gradual perfecting of a brain and the slow building of the mental structure, of which erectness of carriage and speech are some of the incidental manifestations.'

Against this chorus of emphasis upon the brain, very few scientists upheld the primacy of upright posture. Sigmund Freud based much of his highly idiosyncratic theory for the origin of civilization upon it. Beginning in his letters to Wilhelm Fliess in the 1890s and culminating in his 1930 essay on *Civilization and Its Discontents*, Freud argued that our assumption of upright posture had reoriented our primary sensation from smell to vision. This devaluation of olfaction shifted the object of sexual stimulation in males from cyclic odors of estrus to the continual visibility of female genitalia. Continual desire of males led to the evolution of continual receptivity in females. Most mammals copulate only around periods of ovulation; humans are sexually active at all times (a favorite theme of writers on sexuality). Continual sexuality has cemented the human family and made civilization possible; animals with strongly cyclic copulation have no strong impetus for stable family structure. 'The fateful process of civilization,' Freud concludes, 'would thus have set in with man's adoption of an erect posture.'

Although Freud's ideas gained no following among anthropologists, another minor tradition did arise to stress the primacy of upright posture. (It is, by the way, the argument we tend to accept today in explaining the morphology of australopithecines and the path of human evolution.) The brain cannot begin to increase in a vacuum. A primary impetus must be provided by an altered mode of life that would place a strong, selective premium upon intelligence. Upright posture frees the hands from locomotion and for manipulation (literally, from *manus* = 'hand'). For the first time, tools and weapons can be fashioned and used with ease. Increased intelligence is largely a response to the enormous potential inherent in free hands for manufacture – again, literally. (Needless to say, no anthropologist has ever been so naïve as to argue that brain and posture are completely independent in evolution, that one reached its fully human status before the other began to change at all. We are dealing with interaction and mutual reinforcement. Nevertheless, our early evolution did involve a more rapid change in posture than in brain size; complete freeing of our hands for using tools preceded most of the evolutionary enlargement of our brain.)

In another proof that sobriety does not make right, von Baer's mystical and oracular colleague Lorenz Oken hit upon the 'correct' argument in 1809, while von Baer was led astray a few years later. 'Man by the upright walk obtains his character,' writes Oken, 'the hands become free and can achieve all other offices. . . . With the freedom of the body has been granted also the freedom of the mind.' But the champion of upright posture during the nineteenth century was Darwin's German bulldog Ernst Haeckel. Without a scrap of direct evidence, Haeckel reconstructed our ancestor and even gave it a scientific name, *Pithecanthropus alalus*, the upright, speechless, small-brained ape-man. (*Pithecanthropus*, by the way, is probably the only scientific name ever given to an animal before it was discovered. When Du Bois discovered Java Man in the 1890s, he adopted Haeckel's generic name but he gave it the new specific designation *Pithecanthropus erectus*. We now usually include this creature in our own genus as *Homo erectus*.)

But why, despite Oken and Haeckel's demurral, did the idea of cerebral primary become so strongly entrenched? One thing is sure; it had nothing to do with direct evidence – for there was none for any position. With the exception of Neanderthal (a geographic variant of our own species according to most anthropologists), no fossil humans were discovered until the closing years of the nineteenth century, long after the dogma of cerebral primary was established. But debates based on no evidence are among the most revealing in the history of science, for in the absence of factual constraints, the cultural biases that affect all thought (and which scientists try so assiduously to deny) lie nakedly exposed.

Indeed, the nineteenth century produced a brilliant exposé from a source that will no doubt surprise most readers – Friedrich Engels. (A bit of reflection should diminish surprise. Engels had a keen interest in the natural sciences and sought to base his general philosophy of dialectical materialism upon a 'positive' foundation. He did not live to complete his 'dialectics of nature,' but he included long commentaries on science in such treatises as the *Anti-Dühring*.) In 1876, Engels wrote an essay entitled, *The Part Played by Labor in the Transition from Ape to Man*. It was published posthumously in 1896 and, unfortunately, had no visible impact upon Western science.

Engels considers three essential features of human evolution: speech, a large brain, and upright posture. He argues that the first step must have been a descent from the trees with subsequent evolution to upright posture by our ground-dwelling ancestors. 'These apes when moving on level ground began to drop the habit of using their hands and to adopt a more and more erect gait. This was the decisive step in the transition from ape to man.' Upright posture freed the hand for using tools (labor, in Engels's terminology); increased intelligence and speech came later.

> Thus the hand is not only the organ of labor, it is also the product of labor. Only by labor, by adaptation to ever new operations . . . by the ever-renewed employment of these inherited improvements in new, more and more complicated operations, has the human hand attained the high degree of perfection that has enabled it to conjure into being the pictures of Raphael, the statues of Thorwaldsen, the music of Paganini.

Engels presents his conclusions as though they followed deductively from the premises of his materialist philosophy, but I am confident that he cribbed them from Haeckel. The two formulations are almost identical, and Engels cites the relevant pages of Haeckel's work for other purposes in an earlier essay written in 1874. But no matter. The importance of Engels's essay lies, not in its substantive conclusions, but in its trenchant political analysis of why Western science was so hung up on the a priori assertion of cerebral primacy.

As humans learned to master their material surroundings, Engels argues, other skills were added to primitive hunting – agriculture, spinning, pottery, navigation, arts and sciences, law and politics, and finally, 'the fantastic reflection of human things in the human mind: religion.' As wealth accumulated, small groups of men seized power and forced others to work for them. Labor, the source of all wealth and the primary impetus for human evolution, assumed the same low status of those who labored for the rulers. Since rulers governed by their will (that is, by feats of mind), actions of the brain appeared to have a motive power of their own. The profession of philosophy followed no unsullied ideal of truth. Philosophers relied on state or religious patronage. Even if Plato did not consciously conspire to bolster the privileges of rulers with a supposedly abstract philosophy, his own class position encouraged an emphasis on thought as primary, dominating, and altogether more noble and important than the labor it supervised. This idealistic tradition dominated philosophy right through to Darwin's day. Its influence was so subtle and pervasive that even scientific, but apolitical, materialists like Darwin fell under its sway. A bias must be recognized before it can be challenged. Cerebral primacy seemed so obvious and natural that it was accepted as given, rather than recognized as a deep-seated social prejudice related to the class position of professional thinkers and their patrons. Engels writes:

> All merit for the swift advance of civilization was ascribed to the mind, to the development and activity of the brain. Men became accustomed to explain their actions from their thoughts, instead of from their needs. . . . And so there arose in the course of time that idealistic outlook on the world which, especially since the downfall of the ancient world, has dominated men's minds. It still rules them to such a degree that even the most materialistic natural scientists of the Darwinian school are still unable to form any clear idea of the origin of man, because under that ideological influence they do not recognize the part that has been played therein by labor.

The importance of Engels's essay does not lie in the happy result that *Australopithecus* confirmed a specific theory proposed by him – via Haeckel – but rather in his perceptive analysis of the political role of science and of the social biases that must affect all thought.

Indeed, Engels's theme of the separation of head and hand has done much to set and limit the course of science throughout history. Academic science, in particular, has been constrained by an ideal of 'pure' research, which in former days barred a scientist from extensive experimentation and empirical testing. Ancient Greek science labored under the restriction that patrician thinkers could not perform the manual work of plebeian artisans. Medieval barber-surgeons who had to deal with battlefield casualties did more to advance the practice of medicine than academic physicians who rarely examined patients and who based their treatment on a knowledge of Galen and other learned texts. Even today, 'pure' researchers tend to disparage the practical, and terms such as 'aggie school' and 'cow college' are heard with distressing frequency in academic circles. If we took Engels's message to heart and recognized our belief in the inherent superiority of pure research for what it is – namely social prejudice – then we might forge among scientists the union between theory and practice that a world teetering dangerously near the brink so desperately needs.

FREUD'S EVOLUTIONARY FANTASY

IN 1897, the public schools of Detroit carried out an extensive experiment with a new and supposedly ideal curriculum. In the first grade children would read *The Song of Hiawatha* because, at this age, they recapitulated the 'nomadic' and 'savage' stages of their evolutionary past and would therefore appreciate such a like-minded hero. During the same years, Rudyard Kipling wrote poetry's greatest paean to imperialism, 'The White Man's Burden.' Kipling admonished his countrymen to shoulder the arduous responsibility of serving these 'new-caught, sullen people, half-devil and half-child.' Teddy Roosevelt, who knew the value of a good line, wrote to Henry Cabot Lodge that Kipling's effort 'was very poor poetry but made good sense from the expansion point of view.'

These disparate incidents record the enormous influence upon popular culture of an evolutionary idea that ranks second only to natural selection itself for impact beyond biology. This theory held, mellifluously and perhaps with a tad of obfuscation in terminology, that 'ontogeny recapitulates phylogeny,' or that an organism, during the course of its embryonic growth, passes through a series of stages representing adult ancestors in their proper historical order. The gill slits of a human embryo record our distant past as a fish, while our later embryonic tail (subsequently resorbed) represents the reptilian stage of our ancestry.

Biology abandoned this idea some fifty years ago, for a variety of reasons chronicled in my book *Ontogeny and Phylogeny* (1977), but not before the theory of recapitulation – to cite just three examples of its widespread practical influence – had served as the basis for an influential proposal that 'born criminals' acted by necessity as the unfortunate result of a poor genetic shake as manifested by their retention of apish features successfully transcended in the ontogeny of normal people; buttressed a variety of racist claims by depicting adults in 'primitive' cultures as analogs of Caucasian children in need of both discipline and domination; and structured the primary-school curricula of many cities by treating young children as equivalent to grown men and women of a simpler past.

The theory of recapitulation also played a profound, but almost completely unrecognized, role in the formulation of one of the half-dozen most influential movements of the twentieth century: Freudian psychoanalysis. Although the legend surrounding Freud tends to downplay the continuity of his ideas with preexisting theories, and to view psychoanalysis as an abrupt and entirely novel contribution to human thought, Freud trained as a biologist in the heyday of evolution's first discovery, and his theory sank several deep roots in the leading ideas of Darwin's world. (See Frank J. Sulloway's biography *Freud, Biologist of the Mind*, 1979, with its argument that nearly all creative geniuses become surrounded by a mythology of absolute originality.)

The 'threefold parallelism' of classical recapitulation theory in biology equated the child of an advanced species both with an adult ancestor and with adults of any 'primitive' lineages that still survived (the human embryo with gill slits, for example, represents both an actual ancestral fish that lived some 300 million years ago and all surviving fishes as well; similarly, in a racist extension, white children might be compared both with fossils of adult *Homo erectus* and with modern adult Africans). Freud added a fourth parallel: the neurotic adult who, in important respects, represents a normal child, an adult ancestor, or a normal modern adult from a primitive culture. This fourth term for adult pathologies did not originate with Freud, but arose within many theories of the time – as in Lombroso's notion of *l'uomo delinquente* (criminal man), and in various interpretations of neonatal deformity or mental retardation as the retention of an embryonic stage once normal in adult ancestors.

Freud often expressed his convictions about recapitulation. He wrote in his *Introductory Lectures on Psychoanalysis* (1916), 'Each individual somehow recapitulates in an abbreviated form the entire development of the human race.' In a note penned in 1938, he evoked a graphic image for his fourth term: 'With neurotics it is as though we were in a prehistoric landscape – for instance in the Jurassic. The great saurians are still running around; the horsetails grow as high as palms.'

Moreover, these statements do not represent merely a passing fancy or a peripheral concern. Recapitulation occupied a central and pervasive place in Freud's intellectual development. Early in his career, before he formulated the theory of psychosexual stages (anal, oral, and genital), he wrote to Wilhelm Fliess, his chief friend and collaborator, that sexual repression of olfactory stimuli represented our phyletic transition to upright posture: 'Upright carriage was adopted, the nose was raised from the ground, and at the same time a number of what had formerly been interesting sensations connected with the earth became repellent' (letter of 1897). Freud based his later theory of psychosexual stages explicitly upon recapitulation: the anal and oral stages of childhood sexuality represent our quadrupedal past, when senses of taste, touch, and smell predominated. When we evolved upright posture, vision became our primary sense and reoriented sexual stimuli to the genital stage. Freud wrote in 1905 that oral and anal stages 'almost seem as though they were harking back to early animal forms of life.'

In his later career, Freud used recapitulation as the centerpiece for two major books. In *Totem and Taboo* (1913), subtitled *Some Points of Agreement Between the Mental Life of Savages and Neurotics*, Freud inferred a complex phyletic past from the existence of the Oedipus complex in modern children and its persistence in adult neurotics, and from the operation, in primitive cultures, of incest taboos and totemism (identification of a clan with a sacred animal that must be protected, but may be eaten once a year in a great totemic feast). Freud argued that early human society must have been organized as a patriarchal horde, ruled by a dominant father who excluded his sons from sexual contact with women of the clan. In frustration, the sons killed their domineering father, but then, in their guilt, could not possess the women (incest taboo). They expiated their remorse by identifying their slain father with a totemic animal, but celebrated their triumph by reenactment during the annual totemic feast. Modern children relive this act of primal parricide in the Oedipus complex. Freud's last book, *Moses and Monotheism* (1939), reiterates the same theme in a particular context. Moses, Freud argues, was an Egyptian who cast his lot with the Jews. Eventually his adopted people killed him and, in their overwhelming guilt, recast him as a prophet of a single, all-powerful God, thus also creating the ethical ideals that lie at the heart of Judeo-Christian civilization.

A new discovery, hailed as the most significant in many years by Freudian scholars, has now demonstrated an even more central role for recapitulation in Freud's theory than anyone had ever imagined or been willing to allow – although, again, almost every commentator has missed the connection because Freud's biological influences have been slighted by a taxonomy that locates him in another discipline, and because the eclipse of recapitulation has placed this formerly dominant theory outside the consciousness of most modern scholars. In 1915, in the shadow of war and as he began his sixtieth year, Freud labored with great enthusiasm on a book that would set forth the theoretical underpinnings of all his work – the 'metapsychology,' as he called the project. He wrote twelve papers for his work, but later abandoned his plans for unknown reasons much discussed by scholars. Five of the twelve papers were eventually published (with *Mourning and Melancholia* as the best known), but the other seven were presumed lost or destroyed. In 1983, Ilse Grubrich-Simitis discovered a copy, in Freud's hand, of the twelfth and most general paper. The document had resided in a trunk, formerly the property of Freud's daughter Anna (who died in 1983), and otherwise filled with the papers of Freud's Hungarian collaborator Sándor Ferenczi. Harvard University Press published this document in 1987 under the title *A Phylogenetic Fantasy* (translated by Axel and Peter T. Hoffer and edited and explicated by Dr Grubrich-Simitis).

The connection with Ferenczi reinforces the importance of recapitulation as a centerpiece of Freud's psychological theory. Freud had been deeply hurt by the estrangement and opposition of his leading associates, Alfred Adler and Carl Jung. But Ferenczi remained loyal, and Freud strengthened both personal and intellectual ties with him during this time of stress. 'You are now really the only one who still works beside me,' Freud wrote to Ferenczi on July 31, 1915. In preparing the metapsychological papers, Freud's interchange with Ferenczi became so intense that these works might almost be viewed as a joint effort. The twelfth paper, the phylogenetic fantasy, survived only because Freud sent a draft to Ferenczi for his criticism. Ferenczi had received the most extensive biological training of all Freud's associates, and no one else in the history of psychoanalysis maintained so strong a commitment to recapitulation. When Freud sent his phylogenetic fantasy to Ferenczi on July 12, 1915, he ended his letter by stating, 'Your priority in all this is evident.'

Ferenczi wrote a remarkable work titled *Thalassa: A Theory of Genitality* (1924), perhaps best known today in mild ridicule for claiming that much of human psychology records our unrecognized yearning to return to the comforting confines of the womb, 'where there is no such painful disharmony between ego and environment that characterizes existence in the external world.' By his own admission, Ferenczi wrote *Thalassa* 'as an adherent of Haeckel's recapitulation theory.'

Ferenczi viewed sexual intercourse as an act of reversion toward a phyletic past in the tranquillity of a timeless ocean – a 'thalassal regressive trend . . . striving towards the aquatic mode of existence abandoned in primeval time.' He interpreted the weariness of postcoital repose as symbolic of oceanic tranquillity. He also viewed the penis as a symbolic fish, so to speak, reaching toward the womb of the primeval ocean. Moreover, he pointed out, the fetus that arises from this union passes its embryonic life in an amniotic fluid, thus recalling the aquatic environment of our ancestors.

Ferenczi tried to locate even earlier events in our modern psychic lives. He also likened the repose following coitus to a striving further back toward the ultimate tranquillity of a Precambrian world before the origin of life. Ferenczi viewed the full sequence of a human life – from the coitus of parents to the final death of their offspring – as a recapitulation of the gigantic tableau of our entire evolutionary past (Freud would not proceed nearly this far into such a realm of conflating possible symbol with reality). Coitus, in the repose that strives for death, represents the early earth before life, while impregnation recapitulates the dawn of life. The fetus, in the womb of its symbolic ocean, then passes through all ancestral stages from the primal amoeba to a fully formed human. Birth recapitulates the colonization of land by reptiles and amphibians, while the period of latency, following youthful sexuality and before full maturation, repeats the torpor induced by ice ages.

With this recollection of human life during the ice ages, we can connect Ferenczi's thoughts with Freud's phylogenetic fantasy – for Freud, eschewing Ferenczi's overblown, if colorful, inferences about an earlier past, begins with the glacial epoch in trying to reconstruct human history from current psychic life. The basis for Freud's theory lay in his attempt to classify neuroses according to their order of appearance during human growth.

Theories inevitably impose themselves upon our perceptions; no exclusive, objective, or obvious way exists for describing nature. Why should we classify neuroses primarily by their *time* of appearance? Neuroses might be described and ordered in a hundred other ways (by social effect, by common actions or structure, by emotional impacts upon the psyche, by chemical changes that might cause or accompany them). Freud's decision stemmed directly from his commitment to an evolutionary explanation of neurosis – a scheme, moreover, that Freud chose to base upon the theory of recapitulation. In this view, sequential events of human history set the neuroses – for neurotic people become fixated at a stage of growth that normal people transcend. Since each stage of growth recapitulates a past episode in our evolutionary history, each neurosis fixates on a particular prehistoric stage in our ancestry. These behaviors may have been appropriate and adaptive then, but they now produce neuroses in our vastly different modern world. Therefore, if neuroses can be ordered by time of appearance, we will obtain a guide to their evolutionary meaning (and causation) as a series of major events in our phyletic history. Freud wrote to Ferenczi on July 12, 1915, 'What are now neuroses were once phases of human conditions.' In the *Phylogenetic Fantasy*, Freud asserts that 'the neuroses must also bear witness to the history of the mental development of mankind.'

Freud begins by acknowledging that his own theory of psychosexual stages, combined with Ferenczi's speculations, may capture some truly distant aspects of phylogeny by their appearance in the development of very young children. For the phylogenetic fantasy, however, he confines himself to more definite (and less symbolic) parts of history that lie recorded in two sets of neuroses developing later in growth – the transference neuroses and the narcissistic neuroses of his terminology. As the centerpiece of the phylogenetic fantasy, Freud orders these neuroses in six successive stages: the three transference neuroses (anxiety hysteria, conversion hysteria, and obsessional neurosis), followed by the three narcissistic neuroses (dementia praecox [schizophrenia], paranoia, and melancholia-mania [depression]).

> There exists a series to which one can attach various far-reaching ideas. It originates when one arranges the . . . neuroses . . . according to the point in time at which they customarily appear in the life of the individual. . . . Anxiety hysteria . . . is the earliest, closely followed by conversion

> hysteria (from about the fourth year); somewhat later in prepuberty (9–10) obsessional neuroses appear in children. The narcissistic neuroses are absent in childhood. Of these, dementia praecox in classic form is an illness of the puberty years, paranoia approaches the mature years, and melancholia-mania the same time period, otherwise not specifiable.

Freud interprets the transference neuroses as recapitulations of behaviors that we developed to cope with difficulties of human life during the ice ages: 'The temptation is very great to recognize in the three dispositions to anxiety hysteria, conversion hysteria, and obsessional neurosis regressions to phases that the whole human race had to go through at some time from the beginning to the end of the Ice Age, so that at that time all human beings were the way only some of them are today.' Anxiety hysteria represents our first reaction to these difficult times: 'Mankind, under the influence of the privations that the encroaching Ice Age imposed upon it, has become generally anxious. The hitherto predominantly friendly outside word, which bestowed every satisfaction, transformed itself into a mass of threatening perils.'

In these parlous times, large populations could not be supported, and limits to procreation became necessary. In a process adaptive for the time, humans learned to redirect their libidinal urges to other objects, and thereby to limit reproduction. The same behavior today, expressed as a phyletic memory, has become inappropriate and therefore represents the second neurosis – conversion hysteria: 'It became a social obligation to limit reproduction. Perverse satisfactions that did not lead to the propagation of children avoided this prohibition. . . . The whole situation obviously corresponds to the conditions of conversion hysteria.'

The third neurosis, obsession, records our mastery over these difficult conditions of the Ice Age. We needed to devote enormous resources of energy and thought to ordering our lives and overcoming the hostilities of the environment. This same intensely directed energy may now be expressed neurotically in obsessions to follow rules and to focus on meaningless details. This behavior, once so necessary, now 'leaves as compulsion, only the impulses that have been displaced to trivialities.'

Freud then locates the narcissistic neuroses of later life in the subsequent, postglacial events of human history that he had already identified in *Totem and Taboo*. Schizophrenia records the father's revenge as he castrates his challenging sons:

> We may imagine the effect of castration in that primeval time as an extinguishing of the libido and a standstill in individual development. Such a state seems to be recapitulated by dementia praecox which . . . leads to giving up every love-object, degeneration of all sublimations, and return to auto-erotism. The youthful individual behaves as though he had undergone castration.

(In *Totem and Taboo*, Freud had only charged the father with expelling his sons from the clan; now he opts for the harsher punishment of castration. Commentators have attributed this change to Freud's own anger at his 'sons' Adler and Jung for their break with his theories, and their foundation of rival schools. By castration, Freud could preclude the possibility of their future success. I am not much attracted to psychoanalytic speculations of this genre. Freud was, of course, not unaware that a charge of castration posed difficulty for his evolutionary explanation – for the mutilated sons could leave no offspring to remember the event in heredity. Freud speculates that younger sons were spared, thanks to the mother's intercession; these sons lived to reproduce but were psychically scarred by the fate that had befallen their brothers.)

The next neurosis, paranoia, records the struggle of exiled sons against the homosexual inclinations that must inevitably arise within their bonded and exiled group: 'It is very possible that the long-sought hereditary disposition of homosexuality can be glimpsed in the inheritance of this phase of the human condition. . . . Paranoia tries to ward off homosexuality, which was the basis for the organization of brothers, and in so doing must drive the victim out of society and destroy his social sublimations.'

The last neurosis of depression then records the murder of the father by his triumphant sons. The extreme swings in mood of the manic-depressive record both the exultation and the guilt of parricide: 'Triumph over his death, then mourning over the fact they all still revered him as a model.'

From our current standpoint, these speculations seem so farfetched that we may be tempted simply to dismiss them as absurd, even though they emanate from such a distinguished source. Freud's claims are, to be sure, quite wrong, based on knowledge gained in the past half-century. (In particular, Freud's theory is fatally and falsely Eurocentric. Human evolution was not shaped near the ice sheets of northern Europe, but in Africa. We can also cite no reason for supposing that European Neanderthals, who were probably not our ancestors in any case, suffered unduly during glacial times with their abundant game for hunting. Finally, we can offer not a shred of evidence that human social organizations once matched Freud's notion of a domineering father who castrated his sons and drove them away – an awfully precarious way to assure one's Darwinian patrimony.)

But the main reason that we must not dismiss Freud's theory as absurd lies in its consonance with biological ideas then current. Science has since abandoned the biological linchpins of Freud's theory, and most commentators don't know what these concepts entailed or that they ever even existed. Freud's theory therefore strikes us as a crazy speculation that makes absolutely no sense according to modern ideas of evolution. Well, Freud's phylogenetic fantasy *is* bold, wildly beyond data, speculative in the extreme, idiosyncratic – and wrong. But Freud's speculation does become comprehensible once one recognizes the two formerly respectable biological theories underpinning the argument.

The first theory, of course, is recapitulation itself, as discussed throughout this essay. Recapitulation must provide the primary warrant for Freud's fantasy, for recapitulation allowed Freud to interpret a normal feature of childhood (or a neurosis interpreted as fixation to some childhood stage) as necessarily representing an adult phase of our evolutionary past. But recapitulation does not suffice, for one also needs a mechanism to convert the experiences of adults into the heredity of their offspring. Conventional Darwinism could not provide such a mechanism in this case – and Freud understood that his fantasy demanded allegiance to a different version of heredity.

Freud's fantasy requires the passage to modern heredity of events that affected our ancestors only tens of thousands of years ago at most. But such events – anxiety at approaching ice sheets, castration of sons and murder of fathers – have no hereditary impact. However traumatic, such events do not affect the eggs and sperm of parents, and therefore cannot pass into heredity under Mendelian and Darwinian rules.

Freud, therefore, held firmly to his second biological linchpin – the Lamarckian idea, then already unfashionable but still advocated by some prominent biologists, that acquired characters will be inherited.

Under Lamarckism, all theoretical problems for Freud's mechanism disappear. Any important and adaptive behavior developed by adult ancestors can pass directly into the heredity of offspring – and quickly. A primal parricide that occurred just ten or twenty thousand years ago may well be encoded as the Oedipal complex of modern children.

I credit Freud for his firm allegiance to the logic of his argument. Unlike Ferenczi, who concocted an untenable melange of symbolism and causality in *Thalassa* (the placenta, for example, as a newly evolved adaptation of mammals, cannot, therefore, enclose a phyletic vestige of the primeval ocean). However, Freud's theory obeyed a rigidly consistent biological logic rooted in two notions since discredited – recapitulation and Lamarckian inheritance.

Freud understood that his theory depended upon the validity of Lamarckian inheritance. He wrote in the *Phylogenetic Fantasy*, 'One can justifiably claim that the inherited dispositions are residues of the acquisition of our ancestors.' He also recognized that Lamarckism had been falling from fashion since the rediscovery of Mendel's laws in 1900. In their collaboration, Freud and Ferenczi dwelt increasingly upon the necessary role of Lamarckism in psychoanalysis. They planned a joint book on the subject, and Freud dug in with enthusiasm, reading Lamarck's works in late 1916 and writing a paper on the subject (unfortunately never published and apparently not preserved) that he sent to Ferenczi in early 1917. But the project never came to fruition, as the privations of World War I made research and communication increasingly difficult. When Ferenczi nudged Freud one last time in 1918, Freud responded, 'Not disposed to work . . . too much interested in the end of the world drama.'

Illogic remains slippery and vacuous (*Thalassa* can never be proved or rejected; so the idea has simply been forgotten). But in a logical argument, one must live or die by the validity of required premises. Lamarckism has been firmly rejected, and Freud's evolutionary theory of neurosis falls with the validation of Mendel. Freud himself chronicled with great remorse the slippage of Lamarckism from respectability. In *Moses and Monotheism*, he continued to recognize his need for Lamarckism while acknowledging the usual view of its failure:

> This state of affairs is made more difficult, it is true, by the present attitude of biological science, which rejects the idea

> of acquired qualities being transmitted to descendants. I admit, in all modesty, that in spite of this I cannot picture biological development proceeding without taking this factor into account.

Since most commentators have not grasped the logic of Freud's theory because they have not recognized the roles of Lamarckism and recapitulation, they fall into a dilemma, particularly if they generally favor Freud. Without these two biological linchpins of recapitulation and Lamarckism, Freud's fantasy sounds crazy. Could Freud really mean that these events of recent history somehow entered the inheritance of modern children and the fixated behavior of neurotics? Consequently, a muted or kindly tradition has arisen for viewing Freud's claims as merely symbolic. He didn't really mean that exiled sons actually killed their father and that Oedipal complexes truly reenacted a specific event of our past. Freud's words should therefore be regarded sympathetically as colorful imagery providing insight into the psychological meaning of neurosis. Daniel Goleman, reporting on the discovery of *A Phylogenetic Fantasy* (in the *New York Times*, 10 February 1987), writes:

> In the manuscript, according to many scholars, Freud appeared to be turning to a literary mechanism he would use often in the explication of his ideas; he put forward a story that might or might not be grounded in reality but whose mythological content revealed what he saw as basic human conflicts.

I strongly reject this 'kindly' tradition of watering down Freud's well-formulated mechanism to myth or metaphor. In fact, I don't view this tradition as kindly at all, for in order to make Freud appear cogent in an inappropriate context of modern ideas, such interpretations sacrifice the sharp logic and consistency of Freud's actual argument. Freud's writing gives no indication that he intended his phylogenetic speculation as anything but a potentially true account of actual events. If Freud had meant these ideas only as metaphor, then why did he work out such consistency with biological theory based on Lamarckism and recapitulation? And why did he yearn so strongly for Lamarckism after its popularity had faded?

Freud recognized his fantasy as speculation, of course, but he meant every word as potential reality. In fact, the end of *Totem and Taboo* features an incisive discussion of this very subject, with a firm denial of any metaphorical intent. Freud writes:

> It is not accurate to say that obsessional neurotics, weighed down under the burden of an excessive morality, are defending themselves only against psychical reality and are punishing themselves for impulses which were merely felt. Historical reality has a share in the matter as well.

Freud's closing line then reiterates this argument with a literary fillip. He quotes the famous parody of the first line of John's Gospel ('In the beginning was the Word'), as spoken by Faust in Part 1 of Goethe's drama: *Im Anfang war die Tat* (In the beginning was the Deed).

Finally, in explicating Freud's belief in the reality of his story, and in recognizing the firm logic of his argument, I do not defend his method of speculation devoid of any actual evidence in the historical or archaeological record. I believe that such purely speculative reconstructions of history do more harm than good because they give the study of history a bad name. These speculative reveries often lead students of the 'hard' experimental sciences to dismiss the investigation of history as a 'soft' enterprise unworthy of the name science. But history, pursued in other ways, includes all the care and rigor of physics or chemistry at its best. I also deplore the overly adaptationist premise that any evolved feature not making sense in our present life must have arisen long ago for a good reason rooted in past conditions now altered. In our tough, complex, and partly random world, many features just don't make functional sense, period. We need not view schizophrenia, paranoia, and depression as postglacial adaptations gone awry: perhaps these illnesses are immediate pathologies, with remediable medical causes, pure and simple.

Freud, of course, recognized the speculative character of his theory. He called his work a phylogenetic 'fantasy,' and he ultimately abandoned any thought of publication, perhaps because he regarded the work as too outré and unsupported. He even referred playfully to the speculative character of his manuscript, begging that readers 'be patient if once in a while criticism retreats in the face of fantasy and unconfirmed things are presented, merely because they are stimulating and open up distant vistas.' He then wrote to Ferenczi that scientific creativity must be defined as a 'succession of daringly playful fantasy and relentlessly realistic criticism.' Perhaps the phase of relentless criticism intruded before Freud dared to publish his phylogenetic fantasy.

We are therefore left with a paradoxical, and at least mildly disturbing, thought. Freud's theory ranks as a wild speculation, based upon false biology and rooted in no direct data at all about phylogenetic history. Yet the manuscript has been published and analyzed with painstaking care more than half a century later. Hundreds of unknown visionaries develop equally far-fetched but interesting and coherent speculations every day – but we ignore them or, at best, laugh at such crazy ideas. Rewrite the *Phylogenetic Fantasy* to remove the literary hand of Freud's masterly prose, put Joe Blow's name on the title page, and no one will pay the slightest attention. We live in a world of privilege, and only great thinkers win a public right to fail greatly.

PART VII

RACISM, SCIENTIFIC AND OTHERWISE

THE use of Darwinian and pseudo-Darwinian ideas in support of a racist agenda has long been a source of concern to evolutionary biologists. Gould's *Mismeasure of Man* has become a standard source in the refutation of scientific racism, and this section begins with one of its classic demolition jobs – that of the claim that there are differences in head size between 'racial' groups that in some way are related to assumed differences in intelligence. Gould reanalyzes the data presented by some of the great craniologists of the nineteenth century to demonstrate the systematic biases and flaws in both method and measurement that their preconceptions generate. The whole chapter should be obligatory reading: we have included merely the first half.

Attacks on more overt forms of racism appear in the other extracts in this section. First the Nazi view of Jews and of 'racial mixing' in 'The Most Unkindest Cut of All', and, in 'A Tale of Two Work Sites', a reflection of the endemic racism of American treatment of immigrant east European, mainly Jewish, workers. White America's notorious eugenic policies, involving compulsory sterilization, are reflected in the story of 'Carrie Buck's Daughter', but a reader might wonder why an account of the tribulations of the biologist Ernest Everett Just should appear here. Only towards the end of Gould's tribute to this 'thoughtful embryologist' does the reason become clear.

MEASURING HEADS: PAUL BROCA AND THE HEYDAY OF CRANIOLOGY

> No rational man, cognisant of the facts, believes that the average negro is the equal, still less the superior, of the average white man. And, if this be true, it is simply incredible that, when all his disabilities are removed, and our prognathous relative has a fair field and no favor, as well as no oppressor, he will be able to compete successfully with his bigger-brained and smaller-jawed rival, in a contest which is to be carried on by thoughts and not by bites.
>
> T. H. Huxley

The Allure of Numbers

Introduction

Evolutionary theory swept away the creationist rug that had supported the intense debate between monogenists and polygenists, but it satisfied both sides by presenting an even better rationale for their shared racism. The monogenists continued to construct linear hierarchies of races according to mental and moral worth; the polygenists now admitted a common ancestry in the prehistoric mists, but affirmed that races had been separate long enough to evolve major inherited differences in talent and intelligence. As historian of anthropology George Stocking writes: 'The resulting intellectual tensions were resolved after 1859 by a comprehensive evolutionism which was at once monogenist and racist, which affirmed human unity even as it relegated the dark-skinned savage to a status very near the ape.'[1]

The second half of the nineteenth century was not only the era of evolution in anthropology. Another trend, equally irresistible, swept through the human sciences – the allure of numbers, the faith that rigorous measurement could guarantee irrefutable precision, and might mark the transition between subjective speculation and a true science as worthy as Newtonian physics. Evolution and quantification formed an unholy alliance; in a sense, their union forged the first powerful theory of 'scientific' racism – if we define 'science' as many do who misunderstand it most profoundly: as any claim apparently backed by copious numbers. Anthropologists had presented numbers before Darwin, but the crudity of Morton's analysis belies any claim to rigor. By the end of Darwin's century, standardized procedures and a developing body of statistical knowledge had generated a deluge of more truthworthy numerical data.

This chapter is the story of numbers once regarded as surpassing all others in importance – the data of craniometry, or measurement of the skull and its contents. The leaders of craniometry were not conscious political ideologues. They regarded themselves as servants of their numbers, apostles of objectivity. And they confirmed all the common prejudices of comfortable white males – that blacks, women, and poor people occupy their subordinate roles by the harsh dictates of nature.

Science is rooted in creative interpretation. Numbers suggest, constrain, and refute; they do not, by themselves, specify the content of scientific theories. Theories are built upon the interpretation of numbers, and interpreters are often trapped by their own rhetoric. They believe in their own objectivity, and fail to discern the prejudice that leads them to one interpretation among many consistent with their numbers. Paul Broca is now distant enough. We can stand back and show that he used numbers not to generate new theories but to illustrate a priori conclusions. Shall we believe that science is different today simply because we share the cultural context of most practicing scientists and mistake its influence for objective truth? Broca was an exemplary scientist; no one has ever surpassed him in meticulous care and accuracy of measurement. By what right, other than our own biases, can we identify his prejudice and hold that science now operates independently of culture and class?

Francis Galton – Apostle of Quantification

No man expressed his era's fascination with numbers so well as Darwin's celebrated cousin, Francis Galton (1822–1911). Independently wealthy, Galton had the rare freedom to devote his considerable energy and intelligence to his favorite subject of measurement. Galton, a pioneer of modern statistics, believed that, with sufficient labor and ingenuity, anything might be measured, and that measurement is the primary criterion of a scientific study. He even proposed and began to carry out a statistical inquiry into the efficacy of prayer! Galton coined the term 'eugenics' in 1883 and advocated the regulation of marriage and family size according to hereditary endowment of parents.

Galton backed his faith in measurement with all the ingenuity of his idiosyncratic methods. He sought, for example, to construct a 'beauty map' of the British Isles in the following manner:

> Whenever I have occasion to classify the persons I meet into three classes, 'good, medium, bad,' I use a needle mounted as a pricker, wherewith to prick holes, unseen, in a piece of paper, torn rudely into a cross with a long leg. I use its upper end for 'good,' the cross arm for 'medium,' the lower end for 'bad.' The prick holes keep distinct, and are easily read off at leisure. The object, place, and date are written on the paper. I used this plan for my beauty data, classifying the girls I passed in streets or elsewhere as attractive, indifferent, or repellent. Of course this was a purely individual estimate, but it was consistent, judging from the conformity of different attempts in the same population. I found London to rank highest for beauty; Aberdeen lowest.[2]

With good humor, he suggested the following method for quantifying boredom:

> Many mental processes admit of being roughly measured. For instance, the degree to which people are bored, by counting the number of their fidgets. I not infrequently tried this method at the meetings of the Royal Geographical Society, for even there dull memoirs are occasionally read. . . . The use of a watch attracts attention, so I reckon time by the

> number of my breathings, of which there are 15 in a minute. They are not counted mentally, but are punctuated by pressing with 15 fingers successively. The counting is reserved for the fidgets. These observations should be confined to persons of middle age. Children are rarely still, while elderly philosophers will sometimes remain rigid for minutes altogether.[3]

Quantification was Galton's god, and a strong belief in the inheritance of nearly everything he could measure stood at the right hand. Galton believed that even the most socially embedded behaviors had strong innate components: 'As many members of our House of Lords marry the daughters of millionaires,' he wrote, 'it is quite conceivable that our Senate may in time become characterized by a more than common share of shrewd business capacity, possibly also by a lower standard of commercial probity than at present.'[4] Constantly seeking new and ingenious ways to measure the relative worth of peoples, he proposed to rate blacks and whites by studying the history of encounters between black chiefs and white travelers:

> The latter, no doubt, bring with them the knowledge current in civilized lands, but that is an advantage of less importance than we are apt to suppose. A native chief has as good an education in the art of ruling men, as can be desired; he is continually exercised in personal government, and usually maintains his place by the ascendancy of his character shown every day over his subjects and rivals. A traveller in wild countries also fills, to a certain degree, the position of a commander, and has to confront native chiefs at every inhabited place. The result is familiar enough – the white traveller almost invariably holds his own in their presence. It is seldom that we hear of a white traveller meeting with a black chief whom he feels to be the better man.[5]

Galton's major work on the inheritance of intelligence, *Hereditary Genius* (1869), included anthropometry among its criteria, but his interest in measuring skulls and bodies peaked later when he established a laboratory at the International Exposition of 1884. There, for threepence, people moved through his assembly line of tests and measures, and received his assessment at the end. After the Exposition, he maintained the lab for six years at a London museum. The laboratory became famous and attracted many notables, including Gladstone:

> Mr Gladstone was amusingly insistent about the size of his head, saying that hatters often told him that he had an Aberdeenshire head – 'a fact which you may be sure I do not forget to tell my Scotch constituents.' It was a beautifully shaped head, though rather low, but after all it was not so very large in circumference.[6]

Lest this be mistaken for the harmless musings of some dotty Victorian eccentric, I point out that Sir Francis was taken quite seriously as a leading intellect of his time. The American hereditarian Lewis Terman, the man most responsible for instituting IQ tests in America, retrospectively calculated Galton's IQ at above 200, but accorded only 135 to Darwin and a mere 100–110 to Copernicus. Darwin, who approached hereditarian arguments with strong suspicion, wrote after reading *Hereditary Genius*: 'You have made a convert of an opponent in one sense, for I have always maintained that, excepting fools, men did not differ much in intellect, only in zeal and hard work'.[7] Galton responded: 'The rejoinder that might be made to his remark about hard work, is that character, including the aptitude for work, is heritable like every other faculty.'

A Curtain-raiser with a Moral: Numbers Do Not Guarantee Truth

In 1906, a Virginia physician, Robert Bennett Bean, published a long, technical article comparing the brains of American blacks and whites. With a kind of neurological green thumb, he found meaningful differences wherever he looked – meaningful, that is, in his favored sense of expressing black inferiority in hard numbers.

Bean took special pride in his data on the corpus callosum, a structure within the brain that contains fibers connecting the right and left hemispheres. Following a cardinal tenet of craniometry, that higher mental functions reside in the front of the brain and sensorimotor capacities toward the rear, Bean reasoned that he might rank races by the relative sizes of parts within the corpus callosum. So he measured the length of the genu, the front part of the corpus callosum, and

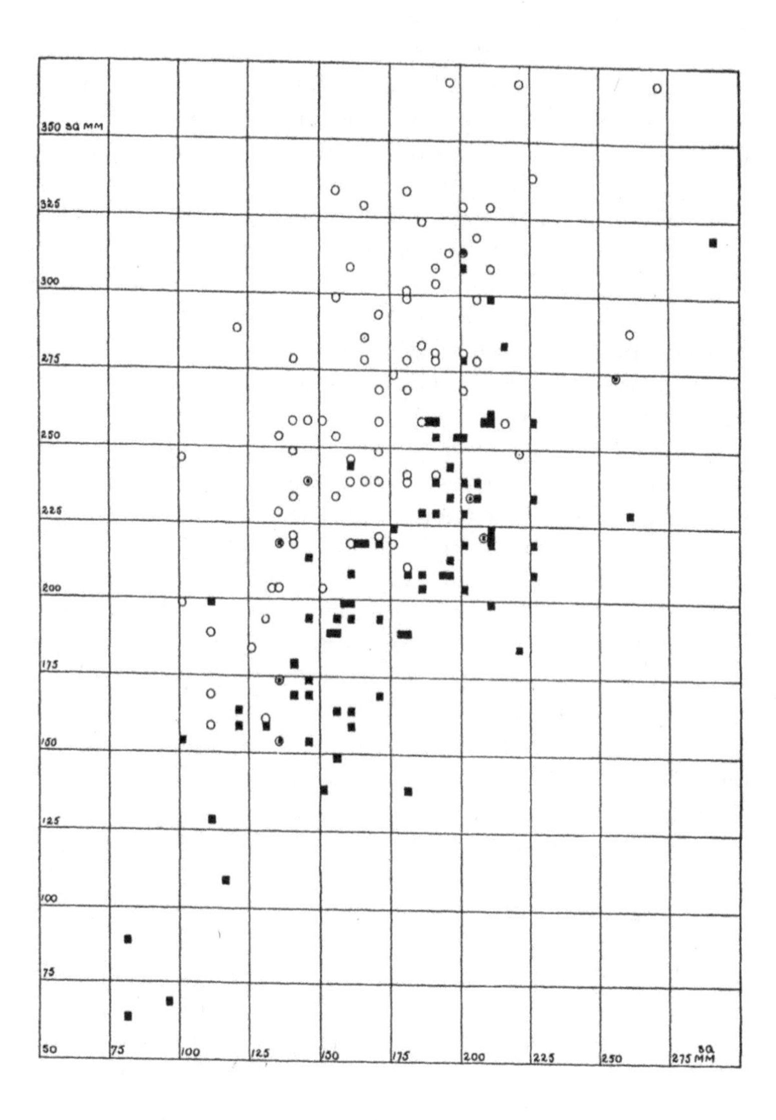

Figure 16. Bean's plot of the genu on the y-axis vs. the splenium on the x-axis. White circles are, unsurprisingly, for white brains; black squares for black brains. Whites seem to have a larger genu, hence more up front, and presumably more intelligence.

compared it with the length of the splenium, the back part. He plotted genu vs. splenium (Fig. 16) and obtained, for a respectably large sample, virtually complete separation between black and white brains. Whites have a relatively large genu, hence more brain up front in the seat of intelligence. All the more remarkable, Bean exclaimed,[8] because the genu contains fibers both for olfaction and for intelligence! Bean continued: We all know that blacks have a keener sense of smell than whites; hence we might have expected larger genus in blacks if intelligence did not differ substantially between races. Yet black genus are smaller despite their olfactory predominance; hence, blacks must really suffer from a paucity of intelligence. Moreover, Bean did not neglect to push the corresponding conclusion for sexes. Within each race, women have relatively smaller genus than men.

Bean then continued his discourse on the relatively greater size of frontal vs. parietal and occipital (side and back) parts of the brain in whites. In the relative size of their frontal areas, he proclaimed, blacks are intermediate between 'man [*sic*] and the ourang-outang'.[9]

Throughout this long monograph, one common measure is conspicuous by its absence: Bean says nothing about the size of the brain itself, the favored criterion of classical craniometry. The reason for this neglect lies buried in an addendum: black and white brains did not differ in overall size. Bean temporized: 'So many factors enter into brain weight that it is questionable whether discussion of the subject is profitable here.' Still, he found a way out. His brains came from unclaimed bodies given to medical schools. We all know that blacks have less respect for their dead than whites. Only the lowest classes of whites – prostitutes and the depraved – would be found among abandoned bodies, 'while among Negroes it is known that even the better classes neglect their dead'. Thus, even an absence of measured difference might indicate white superiority, for the data 'do perhaps show that the low class Caucasian has a larger brain than a better class Negro'.[10]

Bean's general conclusion, expressed in a summary paragraph before the troublesome addendum, proclaimed a common prejudice as the conclusion of science:

> The Negro is primarily affectionate, immensely emotional, then sensual and under stimulation passionate. There is love of ostentation, and capacity for melodious articulation; there is undeveloped artistic power and taste – Negroes make good

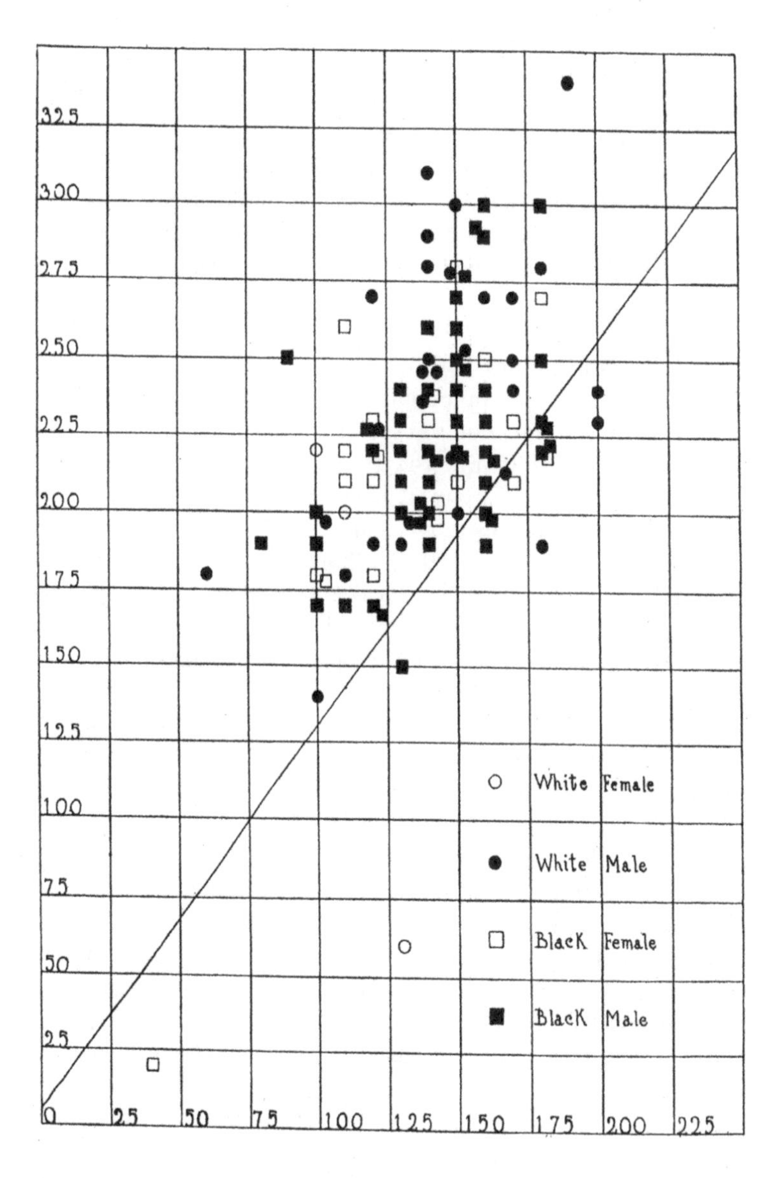

Figure 17. Mall's plot of genu vs. splenium. Mall measured the brains without knowing whether they came from whites or blacks. He found no difference between the races. The line represents Bean's separation between whites and blacks.

> artisans, handicraftsmen – and there is instability of character incident to lack of self-control, especially in connection with the sexual relation; and there is lack of orientation, or recognition of position and condition of self and environment, evidenced by a peculiar bumptiousness, so called, that is particularly noticeable. One would naturally expect some such character for the Negro, because the whole posterior part of the brain is large, and the whole anterior portion is small.

Bean did not confine his opinions to technical journals. He published two articles in popular magazines during 1906, and attracted sufficient attention to become the subject of an editorial in *American Medicine* for April 1907.[11] Bean had provided, the editorial proclaimed, 'the anatomical basis for the complete failure of the negro schools to impart the higher studies – the brain cannot comprehend them any more than a horse can understand the rule of three. . . . Leaders in all political parties now acknowledge the error of human equality. . . . It may be practicable to rectify the error and remove a menace to our prosperity – a large electorate without brains.'

But Franklin P. Mall, Bean's mentor at Johns Hopkins, became suspicious: Bean's data were too good. He repeated Bean's work, but with an important difference in procedure – he made sure that he did not know which brains were from blacks and which from whites until *after* he had measured them.[12] For a sample of 106 brains, using Bean's method of measurement; he found no difference between whites and blacks in the relative sizes of genu and splenium (Fig. 17). This sample included eighteen brains from Bean's original sample, ten from whites, eight from blacks. Bean's measure of the genu was larger than Mall's for seven whites, but for only a single black. Bean's measure of the splenium was larger than Mall's for seven of the eight blacks.

I use this small tale of zealotry as a curtain-raiser because it illustrates so well the major contentions of this chapter and book:

1. Scientific racists and sexists often confine their label of inferiority to a single disadvantaged group; but race, sex, and class go together, and each acts as a surrogate for the others. Individual studies may be limited in scope, but the general philosophy of bio-

logical determinism pervades – hierarchies of advantage and disadvantage follow the dictates of nature; stratification reflects biology. Bean studied races, but he extended his most important conclusion to women, and also invoked differences of social class to argue that equality of size between black and white brains really reflects the inferiority of blacks.

2. Prior prejudice, not copious numerical documentation, dictates conclusions. We can scarcely doubt that Bean's statement about black bumptiousness reflected a prior belief that he set out to objectify, not an induction from data about fronts and backs of brains. And the special pleading that yielded black inferiority from equality of brain size is ludicrous outside a shared context of a priori belief in the inferiority of blacks.

3. Numbers and graphs do not gain authority from increasing precision of measurement, sample size, or complexity in manipulation. Basic experimental designs may be flawed and not subject to correction by extended repetition. Prior commitment to one among many potential conclusions often guarantees a serious flaw in design.

4. Craniometry was not just a plaything of academicians, a subject confined to technical journals. Conclusions flooded the popular press. Once entrenched, they often embarked on a life of their own, endlessly copied from secondary source to secondary source, refractory to disproof because no one examined the fragility of primary documentation. In this case, Mall nipped a dogma in the bud, but not before a leading journal had recommended that blacks be barred from voting as a consequence of their innate stupidity.

But I also note an important difference between Bean and the great European craniometricians. Bean committed either conscious fraud or extraordinary self-delusion. He was a poor scientist following an absurd experimental design. The great craniometricians, on the other hand, were fine scientists by the criteria of their time. Their numbers, unlike Bean's, were generally sound. Their prejudices played a more subtle role in specifying interpretations and in suggesting what numbers might be gathered in the first place. Their work was more refractory to exposure, but equally invalid for the same reason: prejudices led through data in a circle back to the same prejudices – an unbeatable system that gained authority because it seemed to arise from meticulous measurement.

Bean's story has been told several times,[13] if not with all its details.

But Bean was a marginal figure on a temporary and provincial stage. I have found no modern analysis of the main drama, the data of Paul Broca and his school.

Masters of Craniometry: Paul Broca and His School

The Great Circle Route

In 1861 a fierce debate extended over several meetings of a young association still experiencing its birth pangs. Paul Broca (1824–1880), professor of clinical surgery in the faculty of medicine, had founded the Anthropological Society of Paris in 1859. At a meeting of the society two years later, Louis Pierre Gratiolet read a paper that challenged Broca's most precious belief: Gratiolet dared to argue that the size of a brain bore no relationship to its degree of intelligence.

Broca rose in his own defense, arguing that 'the study of the brains of human races would lose most of its interest and utility' if variation in size counted for nothing.[14] Why had anthropologists spent so much time measuring skulls, unless their results could delineate human groups and assess their relative worth?

> Among the questions heretofore discussed within the Anthropological Society, none is equal in interest and importance to the question before us now. . . . The great importance of craniology has struck anthropologists with such force that many among us have neglected the other parts of our science in order to devote ourselves almost exclusively to the study of skulls. . . . In such data, we hoped to find some information relevant to the intellectual value of the various human races.[15]

Broca then unleashed his data and poor Gratiolet was routed. His final contribution to the debate must rank among the most oblique, yet abject concession speeches ever offered by a scientist. He did not abjure his errors; he argued instead that no one had appreciated the subtlety of his position. (Gratiolet, by the way, was a royalist, not an egalitarian. He merely sought other measures to affirm the inferiority of blacks and women – earlier closure of the skull sutures, for example.)

Broca concluded triumphantly:

> In general, the brain is larger in mature adults than in the elderly, in men than in women, in eminent men than in men of mediocre talent, in superior races than in inferior races.[16] . . . Other things equal, there is a remarkable relationship between the development of intelligence and the volume of the brain.[17]

Five years later, in an encyclopedia article on anthropology, Broca expressed himself more forcefully:

> A prognathous [forward-jutting] face, more or less black color of the skin, woolly hair and intellectual and social inferiority are often associated, while more or less white skin, straight hair and an orthognathous [straight] face are the ordinary equipment of the highest groups in the human series.[18] . . . A group with black skin, woolly hair and a prognathous face has never been able to raise itself spontaneously to civilization.[19]

These are harsh words, and Broca himself regretted that nature had fashioned such a system.[20] But what could he do? Facts are facts. 'There is no faith, however respectable, no interest, however legitimate, which must not accommodate itself to the progress of human knowledge and bend before truth'.[21] Paul Topinard, Broca's leading disciple and successor, took as his motto: '*J'ai horreur des systèmes et surtout des systèmes a priori*' (I abhor systems, especially a priori systems).

Broca singled out the few egalitarian scientists of his century for particularly harsh treatment because they had debased their calling by allowing an ethical hope or political dream to cloud their judgment and distort objective truth. 'The intervention of political and social considerations has not been less injurious to anthropology than the religious element'.[22] The great German anatomist Friedrich Tiedemann, for example, had argued that blacks and whites did not differ in cranial capacity. Broca nailed Tiedemann for the same error I uncovered in Morton's work. When Morton used a subjective and imprecise method of reckoning, he calculated systematically lower capacities for blacks than when he measured the same skulls with a precise technique.

Tiedemann, using an even more imprecise method, calculated a black average 45 cc above the mean value recorded by other scientists. Yet his measures for white skulls were no larger than those reported by colleagues. (For all his delight in exposing Tiedemann, Broca apparently never checked Morton's figures, though Morton was his hero and model. Broca once published a one-hundred-page paper analyzing Morton's techniques in the most minute detail.)[23]

Why had Tiedemann gone astray? 'Unhappily,' Broca wrote, 'he was dominated by a preconceived idea. He set out to prove that the cranial capacity of all human races is the same.'[24] But 'it is an axiom of all observational sciences that facts must precede theories'.[25] Broca believed, sincerely I assume, that facts were his only constraint and that his success in affirming traditional rankings arose from the precision of his measures and his care in establishing repeatable procedures.

Indeed, one cannot read Broca without gaining enormous respect for his care in generating data. I believe his numbers and doubt that any better have ever been obtained. Broca made an exhaustive study of all previous methods used to determine cranial capacity. He decided that lead shot, as advocated by '*le célèbre* Morton',[26] gave the best results, but he spent months refining the technique, taking into account such factors as the form and height of the cylinder used to receive the shot after it is poured from the skull, the speed of pouring shot into the skull, and the mode of shaking and tapping the skull to pack the shot and to determine whether or not more will fit in.[27] Broca finally developed an objective method for measuring cranial capacity. In most of his work, however, he preferred to weigh the brain directly after autopsies performed by his own hands.

I spent a month reading all of Broca's major work, concentrating on his statistical procedures. I found a definite pattern in his methods. He traversed the gap between fact and conclusion by what may be the usual route – predominantly in reverse. Conclusions came first and Broca's conclusions were the shared assumptions of most successful white males during his time – themselves on top by the good fortune of nature, and women, blacks, and poor people below. His facts were reliable (unlike Morton's), but they were gathered selectively and then manipulated unconsciously in the service of prior conclusions. By this route, the conclusions achieved not only the blessing of science, but the prestige of numbers. Broca and his school used facts as illustrations, not as constraining documents. They began with conclusions, peered through their facts, and came back in a circle to the same conclusions. Their example repays a closer study, for unlike Morton (who manipulated data, however unconsciously), they reflected their prejudices by another, and probably more common, route: advocacy masquerading as objectivity.

Selecting Characters

When the 'Hottentot Venus' died in Paris, Georges Cuvier, the greatest scientist and, as Broca would later discover to his delight, the largest brain of France, remembered this African woman as he had seen her in the flesh.

> She had a way of pouting her lips exactly like what we have observed in the orang-utan. Her movements had something abrupt and fantastical about them, reminding one of those of the ape. Her lips were monstrously large [those of apes are thin and small as Cuvier apparently forgot]. Her ear was like that of many apes, being small, the tragus weak, and the external border almost obliterated behind. These are animal characters. I have never seen a human head more like an ape than that of this woman.[28]

The human body can be measured in a thousand ways. Any investigator, convinced beforehand of a group's inferiority, can select a small set of measures to illustrate its greater affinity with apes. (This procedure, of course, would work equally well for white males, though no one made the attempt. White people, for example, have thin lips – a property shared with chimpanzees – while most black Africans have thicker, consequently more 'human,' lips.)

Broca's cardinal bias lay in his assumption that human races could be ranked in a linear scale of mental worth. In enumerating the aims of ethnology, Broca included: 'to determine the relative position of races in the human series'.[29] It did not occur to him that human variation might be ramified and random, rather than linear and hierarchical. And since he knew the order beforehand, anthropometry became a search for characters that would display the correct ranking, not a numerical exercise in raw empiricism.

Thus Broca began his search for 'meaningful' characters – those that would display the established ranks. In 1862, for example, he tried the ratio of radius (lower arm bone) to humerus (upper arm bone), reasoning that a higher ratio marks a longer forearm – a character of apes. All began well: blacks yielded a ratio of .794, whites .739. But then Broca ran into trouble. An Eskimo skeleton yielded .703, an Australian aborigine .709, while the Hottentot Venus, Cuvier's near ape (her skeleton had been preserved in Paris), measured a mere .703. Broca now had two choices. He could either admit that, on this criterion, whites ranked lower than several dark-skinned groups, or he could abandon the criterion. Since he knew[30] that Hottentots, Eskimos, and Australian aborigines ranked below most African blacks, he chose the second course: 'After this, it seems difficult to me to continue to say that elongation of the forearm is a character of degradation or inferiority, because, on this account, the European occupies a place between Negroes on the one hand, and Hottentots, Australians, and eskimos on the other'.[31]

Later, he almost abandoned his cardinal criterion of brain size because inferior yellow people scored so well:

> A table on which races were arranged by order of their cranial capacities would not represent the degrees of their superiority or inferiority, because size represents only one element of the problem [of ranking races]. On such a table, Eskimos, Lapps, Malays, Tartars and several other peoples of the Mongolian type would surpass the most civilized people of Europe. A lowly race may therefore have a big brain.[32]

But Broca felt that he could salvage much of value from his crude measure of overall brain size. It may fail at the upper end because some inferior groups have big brains, but it works at the lower end because small brains belong exclusively to people of low intelligence. Broca continued:

> But this does not destroy the value of small brain size as a mark of inferiority. The table shows that West African blacks have a cranial capacity about 100 cc less than that of European races. To this figure, we may add the following: Caffirs, Nubians, Tasmanians, Hottentots, Australians. These

> examples are sufficient to prove that if the volume of the brain does not play a decisive role in the intellectual ranking of races, it nevertheless has a very real importance.[33]

An unbeatable argument. Deny it at one end where conclusions are uncongenial; affirm it by the same criterion at the other. Broca did not fudge numbers; he merely selected among them or interpreted his way around them to favored conclusions.

In choosing among measures, Broca did not just drift passively in the sway of a preconceived idea. He advocated selection among characters as a stated goal with explicit criteria. Topinard, his chief disciple, distinguished between 'empirical' characters 'having no apparent design,' and 'rational' characters 'related to some physiological opinion'.[34] How then to determine which characters are 'rational'? Topinard answered: 'Other characteristics are looked upon, whether rightly or wrongly, as dominant. They have an affinity in negroes to those which they exhibit in apes, and establish the transition between these and Europeans'.[35] Broca had also considered this issue in the midst of his debate with Gratiolet, and had reached the same conclusion:

> We surmount the problem easily by choosing, for our comparison of brains, races whose intellectual inequalities are completely clear. Thus, the superiority of Europeans compared with African Negroes, American Indians, Hottentots, Australians and the Negroes of Oceania, is sufficiently certain to serve as a point of departure for the comparison of brains.[36]

Particularly outrageous examples abound in the selection of individuals to represent groups in illustrations. Thirty years ago, when I was a child, the Hall of Man in the American Museum of Natural History still displayed the characters of human races by linear arrays running from apes to whites. Standard anatomical illustrations, until this generation, depicted a chimp, a Negro, and a white, part by part in that order – even though variation among whites and blacks is always large enough to generate a different order with other individuals: chimp, white, black. In 1903, for example, the American anatomist E. A. Spitzka published a long treatise on brain size and form in 'men of eminence.' He printed the following figure (Fig. 18) with a comment: 'The jump

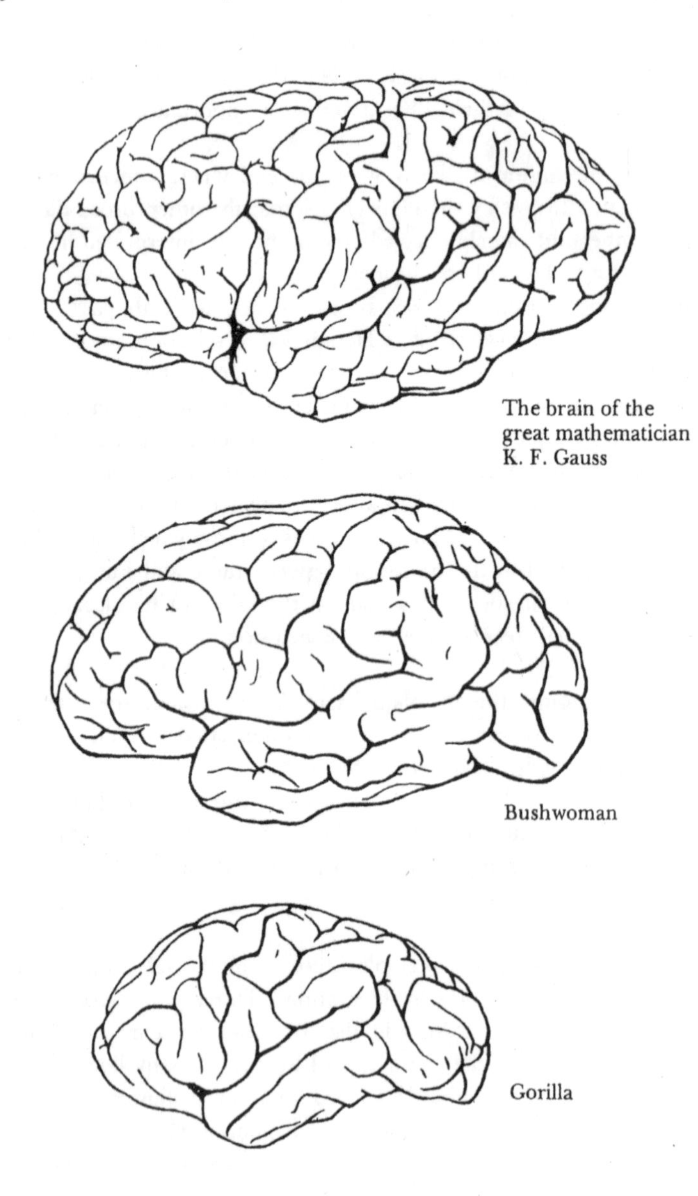

Figure 18. Spitzka's chain of being according to brain size.

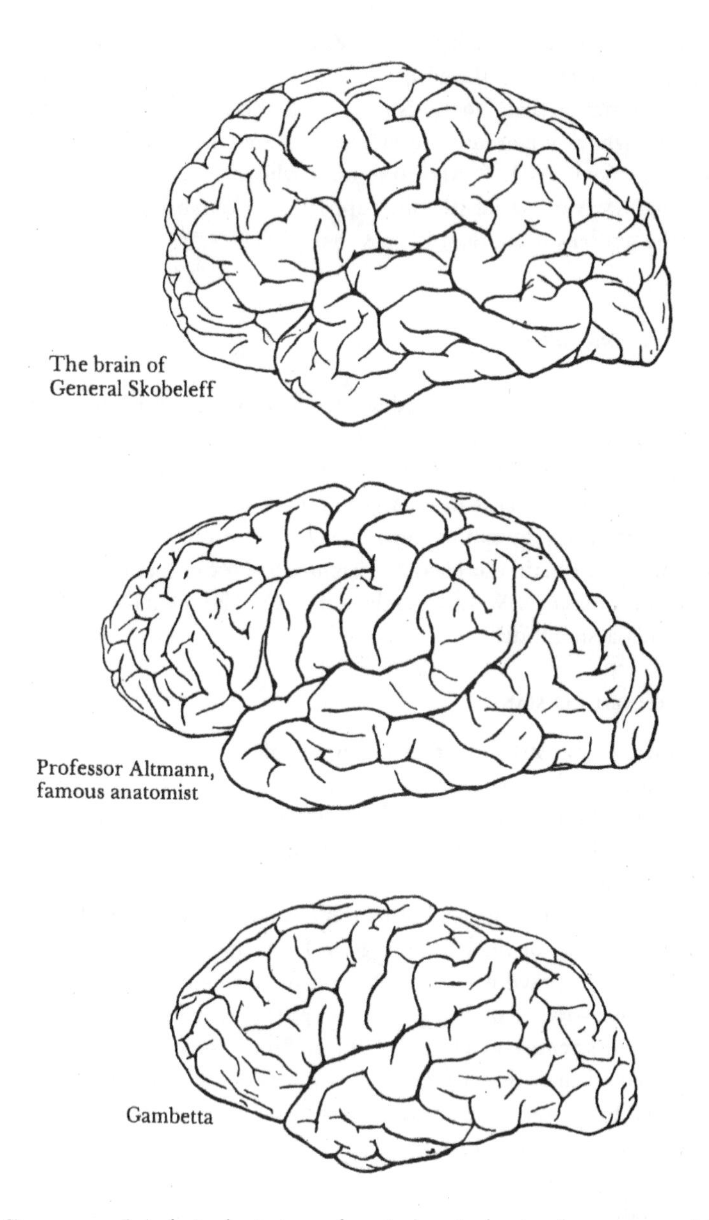

Figure 19. Spitzka's depiction of variation in brain size among white men of eminence.

from a Cuvier or a Thackeray to a Zulu or a Bushman is not greater than from the latter to the gorilla or the orang'.[37] But he also published a similar figure (Fig. 19) illustrating variation in brain size among eminent whites apparently never realizing that he had destroyed his own argument. As F. P. Mall, the man who exposed Bean, wrote of these figures: 'Comparing [them], it appears that Gambetta's brain resembles the gorilla's more than it does that of Gauss.'[38]

Averting Anomalies

Inevitably, since Broca amassed so much disparate and honest data, he generated numerous anomalies and apparent exceptions to his guiding generality – that size of brain records intelligence and that comfortable white males have larger brains than women, poor people, and lower races. In noting how he worked around each apparent exception, we obtain our clearest insight into Broca's methods of argument and inference. We also understand why data could never overthrow his assumptions.

BIG-BRAINED GERMANS

Gratiolet, in his last desperate attempt, pulled out all the stops. He dared to claim that, on average, German brains are 100 grams heavier than French brains. Clearly, Gratiolet argued, brain size has nothing to do with intelligence! Broca responded disdainfully: 'Monsieur Gratiolet has almost appealed to our patriotic sentiments. But it will be easy for me to show him that he can grant some value to the size of the brain without ceasing, for that, to be a good Frenchman'.[39]

Broca then worked his way systematically through the data. First of all, Gratiolet's figure of 100 grams came from unsupported claims of the German scientist E. Huschke. When Broca collated all the actual data he could find, the difference in size between German and French brains fell from 100 to 48 grams. Broca then applied a series of corrections for non-intellectual factors that also affect brain size. He argued, quite correctly, that brain size increases with body size, decreases with age, and decreases during long periods of poor health (thus explaining why executed criminals often have larger brains than honest folk who die of degenerative diseases in hospitals). Broca noted a mean French age of fifty-six and a half years in his sample, while the Germans averaged only fifty-one. He estimated that this difference would account for 16 grams of the disparity between French and Germans, cutting the German advantage to 32 grams. He then removed from the German sample all individuals who had died by violence or execution. The mean brain weight of twenty Germans, dead from natural causes, now stood at 1,320 grams, already *below* the French average of 1,333 grams. And Broca had not even yet corrected for the larger average body size of Germans. *Vive la France.*

Broca's colleague de Jouvencel, speaking on his behalf against the unfortunate Gratiolet, argued that greater German brawn accounted for all the apparent difference in brain and then some. Of the average German, he wrote:

> He ingests a quantity of solid food and drink far greater than that which satisfies us. This, joined with his consumption of beer, which is pervasive even in areas where wine is made, makes the German much more fleshy [*charnu*] than the Frenchman – so much so that their relation of brain size to total mass, far from being superior to ours, appears to me, on the contrary, to be inferior.[40]

I do not challenge Broca's use of corrections but I do note his skill in wielding them when his own position was threatened. Bear this in mind when I discuss how deftly he avoided them when they might have challenged a congenial conclusion – the small brains of women.

SMALL-BRAINED MEN OF EMINENCE

The American anatomist E. A. Spitzka urged men of eminence to donate their brains to science after their death. 'To me the thought of an autopsy is certainly less repugnant than I imagine the process of cadaveric decomposition in the grave to be'.[41] The dissection of dead colleagues became something of a cottage industry among nineteenth-century craniometricians. Brains exerted their customary fascination, and lists were proudly touted, accompanied by the usual invidious comparisons. (The leading American anthropologists J. W. Powell and W. J. McGee even made a wager over who carried the larger brain. As Ko-Ko told Nanki-Poo about the fireworks that would follow his execution, 'You won't see them, but they'll be there all the same.')

Some men of genius did very well indeed. Against a European average of 1,300 to 1,400 grams, the great Cuvier stood out with his top-heavy, 1,830 grams. Cuvier headed the charts until Turgenev finally broke the 2,000 gram barrier in 1883. (Other potential occupants of this stratosphere, Cromwell and Swift, lay in limbo for insufficiency of record.)

The other end was a bit more confusing and embarrassing. Walt Whitman managed to hear America singing with only 1,282 grams. As a crowning indignity, Franz Josef Gall, one of the two founders of phrenology – the original 'science' of judging various mental capacities by the size of localized brain areas – weighed in at a meager 1,198 grams. (His colleague J. K. Spurzheim yielded a quite respectable 1,559 grams.) And, though Broca didn't know it, his own brain weighed only 1,424 grams, a bit above average to be sure, but nothing to crow about. Anatole France extended the range of famous authors to more than 1,000 grams when, in 1924, he opted for the other end of Turgenev's fame and clocked in at a mere 1,017 grams.

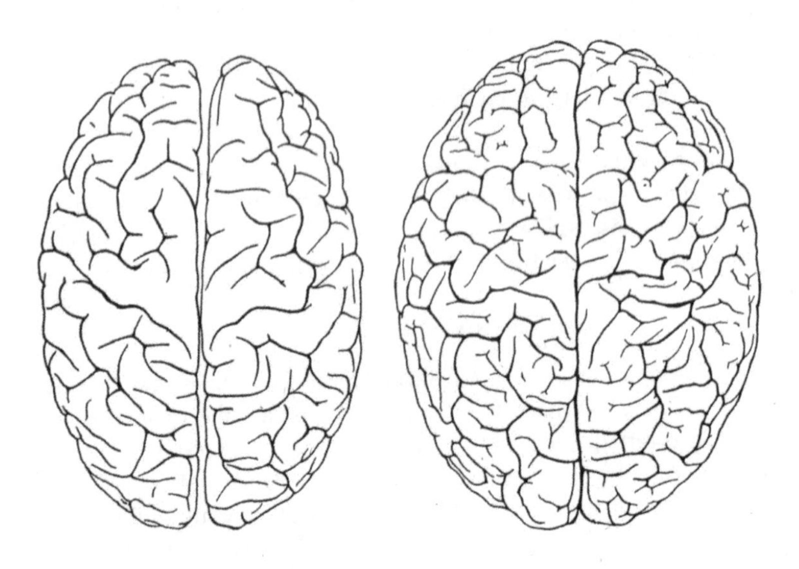

Figure 20. The brain of the great mathematician K. F. Gauss (*right*) proved to be something of an embarrassment since, at 1,492 grams, it was only slightly larger than average. But other criteria came to the rescue. Here, E. A. Spitzka demonstrates that Gauss's brain is much more richly convoluted than that of a Papuan (*left*).

The small brains were troublesome, but Broca, undaunted, managed to account for all of them. Their possessors either died very old, were very short and slightly built, or had suffered poor preservation. Broca's reaction to a study by his German colleague Rudolf Wagner was typical. Wagner had obtained a real prize in 1855, the brain of the great mathematician Karl Friedrich Gauss. It weighed a modestly overaverage 1,492 grams, but was more richly convoluted than any brain previously dissected (Fig. 20). Encouraged, Wagner went on to weigh the brains of all dead and willing professors at Göttingen, in an attempt to plot the distribution of brain size among men of eminence. By the time Broca was battling with Gratiolet in 1861, Wagner had four more measurements. None posed any challenge to Cuvier, and two were distinctly puzzling – Hermann, the professor of philosophy at 1,368 grams, and Hausmann, the professor of mineralogy, at 1,226 grams. Broca corrected Hermann's brain for his age and raised it by 16 grams to 1.19 percent above average – 'not much for a professor of linguistics,' Broca admitted, 'but still something'.[42] No correction could raise Hausmann to the mean of ordinary folks, but considering his venerable seventy-seven years, Broca speculated that his brain may have undergone more than the usual amount of senile degeneration: 'The degree of decadence that old age can impose upon a brain is very variable and cannot be calculated.'

But Broca was still bothered. He could get around the low values, but he couldn't raise them to unusual weights. Consequently, to clinch an unbeatable conclusion, he suggested with a touch of irony that Wagner's post-Gaussian subjects may not have been so eminent after all:

> It is not very probable that 5 men of genius should have died within five years at the University of Göttingen. . . . A professorial robe is not necessarily a certificate of genius; there may be, even at Göttingen, some chairs occupied by not very remarkable men.[43]

At this point, Broca desisted: 'The subject is delicate,' he wrote, 'and I must not insist upon it any longer.'[44]

LARGE-BRAINED CRIMINALS

The large size of many criminal brains was a constant source of bother to craniometricians and criminal anthropologists. Broca tended to dismiss it with his claim that sudden death by execution precluded the diminution that long bouts of disease produced in many honest men. In addition, death by hanging tended to engorge the brain and lead to spuriously high weights.

In the year of Broca's death, T. Bischoff published his study on the brains of 119 assassins, murderers, and thieves. Their average exceeded the mean of honest men by 11 grams, while fourteen of them topped 1,500 grams, and five exceeded 1,600 grams. By contrast, only three men of genius could boast more than 1,600 grams, while the assassin Le Pelley, at 1,809 grams, must have given pause to the shade of Cuvier. The largest female brain ever weighed (1,565 grams) belonged to a woman who had killed her husband.

Broca's successor Paul Topinard puzzled over the data and finally decided that too much of a good thing is bad for some people. Truly inspired criminality may require as much upstairs as professorial virtuosity; who shall decide between Moriarty and Holmes? Topinard concluded: 'It seems established that a certain proportion of criminals are pushed to depart from present social rules by an exuberance of cerebral activity and, consequently, by the fact of a large or heavy brain'.[45]

FLAWS IN A PATTERN OF INCREASE THROUGH TIME

Of all Broca's studies, with the exception of his work on differences between men and women, none won more respect or attention than his supposed demonstration of steady increase in brain size as European civilization advanced from medieval to modern times.[46]

This study merits close analysis because it probably represents the best case of hope dictating conclusion that I have ever encountered. Broca viewed himself as a liberal in the sense that he did not condemn groups to permanent inferiority based on their current status. Women's brains had degenerated through time thanks to a socially enforced underusage; they might increase again under different social conditions. Primitive races had not been sufficiently challenged, while European brains grew steadily with the march of civilization.

Broca obtained large samples from each of three Parisian cemeteries, from the twelfth, the eighteenth, and the nineteenth centuries. Their average cranial capacities were, respectively, 1,426, 1,409, and 1,462 cc – not exactly the stuff for a firm conclusion of steady increase through time. (I have not been able to find Broca's raw data for statistical testing, but with a 3.5 percent mean difference between smallest and largest sample, it is likely that no statistically significant differences exist at all among the three samples.)

But how did these limited data – only three sites with no information on ranges of variation at a given time and no clear pattern through time – lead Broca to his hopeful conclusion? Broca himself admitted an initial disappointment: he had expected to find intermediate values in the eighteenth-century site.[47] Social class, he argued, must hold the answer, for successful groups within a culture owe at least part of their status to superior wits. The twelfth-century sample came from a churchyard and must represent gentry. A common grave provided the eighteenth-century skulls. But the nineteenth-century sample was a mixture, ninety skulls from individual graves with a mean of 1,484 cc, and thirty-five from a common grave with an average of 1,403 cc. Broca claimed that if differences in social class do not explain why calculated values fail to meet expectations, then the data are unintelligible. Intelligible, to Broca, meant steadily increasing through time – the proposition that the data were meant to prove, not rest upon. Again, Broca travels in a circle:

> Without this [difference in social class], we would have to believe that the cranial capacity of Parisians has really diminished during centuries following the 12th. Now during this period . . . intellectual and social progress has been considerable, and even if we are not yet certain that the development of civilization makes the brain grow as a consequence, no one, without doubt, would want to consider this cause as capable of making the brain decrease in size.[48]

But Broca's division of the nineteenth-century sample by social class also brought trouble as well as relief – for he now had two samples from common graves and the earlier one had a larger mean capacity, 1,409 for the eighteenth century vs. 1,403 for the nineteenth. But Broca was not to be defeated; he argued that the eighteenth-century common grave included a better class of people. In these prerevolutionary times, a man had to be really rich or noble to rest in a churchyard. The dregs of the poor measured 1,403 in the nineteenth century; the dregs leavened by good stock yielded about the same value one hundred years before.

Each solution brought Broca new trouble. Now that he was committed to a partition by social class within cemeteries, he had to admit that an additional seventeen skulls from the morgue's grave at the nineteenth-century site yielded a higher value than skulls of middle- and upper-class people from individual graves – 1,517 vs. 1,484 cc. How could unclaimed bodies, abandoned to the state, surpass the cream of society? Broca reasoned in a chain of surpassingly weak inference: morgues stood on river borders; they probably housed a large number of drowned people; many drowned are suicides; many suicides are insane; many insane people, like criminals, have surprisingly large brains. With a bit of imagination, nothing can be truly anomalous.

Front and back

> Tell me about this new young surgeon, Mr Lydgate. I am told he is wonderfully clever; he certainly looks it – a fine brow indeed.
>
> George Eliot, *Middlemarch* (1872)

Size of the whole, however useful and decisive in general terms, did not begin to exhaust the content of craniometry. Ever since the heyday of phrenology, specific parts of the brain and skull had been assigned definite status, thus providing a set of subsidiary criteria for the ranking of groups. (Broca, in his other career as a medical man, made his most important discovery in this area. In 1861 he developed the concept of cortical localization of function when he discovered that an aphasic patient had a lesion in the left inferior frontal gyrus, now called Broca's convolution.)

Most of these subsidiary criteria can be reduced to a single formula: front is better. Broca and his colleagues believed that higher mental functions were localized in anterior regions of the cortex, and that posterior areas busied themselves with the more mundane, though crucial, roles of involuntary movement, sensation, and emotion. Superior people should have more in front, less behind. We have already seen how Bean followed this assumption in generating his spurious data on front and back parts of the corpus callosum in whites and blacks.

Broca often used the distinction of front and back, particularly to extract himself from uncomfortable situations imposed by his data. He accepted Gratiolet's classification of human groups into '*races frontales*' (whites with anterior and frontal lobes most highly developed), '*races pariétales*' (Mongolians with parietal or mid lobes most prominent), and '*races occipitales*' (blacks with most in the back). He often unleashed the double whammy against inferior groups – small size and posterior prominence: 'Negroes, and especially Hottentots, have a simpler brain than ours, and the relative poverty of their convolutions can be found primarily on their frontal lobes'.[49] As more direct evidence, he argued that Tahitians artificially deformed the frontal areas of certain male children in order to make the back portions bulge. These men became courageous warriors, but could never match white heroes for style: 'Frontal deformation produced blind passions, ferocious instincts, and animal courage, all of which I would willingly call occipital courage. We must not confound it with true courage, frontal courage, which we may call Caucasian courage'[50]

Broca also went beyond size to assess the quality of frontal vs. occipital regions in various races. Here, and not only to placate his adversary, he accepted Gratiolet's favorite argument that the sutures between skull bones close earlier in inferior races, thus trapping the brain within a rigid vault and limiting the effectiveness of further education. Not only do white sutures close later; they close in a different order – guess how? In blacks and other inferior people, the front sutures close first, the back sutures later; in whites, the front sutures close last. Extensive modern studies of cranial closure show no difference of timing or pattern among races.[51]

Broca used this argument to extricate himself from a serious problem. He had described a sample of skulls from the earliest populations of *Homo sapiens* (Cro-Magnon type) and found that they exceeded modern Frenchmen in cranial capacity. Fortunately, however, their anterior sutures closed first and these progenitors must have been inferior after all: 'These are signs of inferiority. We find them in all races in which the material life draws all cerebral activity to it. As intellectual life develops among a people, the anterior sutures become more complicated and stay open for a longer time'.[52]

The argument of front and back,* so flexible and far-ranging, served as a powerful tool for rationalizing prejudice in the face of apparently contradictory fact. Consider the following two examples.

* Broca did not confine his arguments on the relative worth of brain parts to the distinction between front and back. Virtually any measured difference between peoples could be given a value in terms of prior conviction about relative worth. Broca once claimed, for example,[53] that blacks probably had larger cranial nerves than whites, hence a larger nonintellectual portion of the brain.

THE CRANIAL INDEX

Beyond brain size itself, the two most hoary and misused measures of craniometry were surely the facial angle (jutting forward of face and jaws – the less the better), and the cranial index. The cranial index never had much going for it beyond ease of measurement. It was calculated as the ratio of maximum width to maximum length of the skull. Relatively long skulls (ratio of .75 or less) were called dolichocephalic; relatively short skulls (over .8), brachycephalic. Anders Retzius, the Swedish scientist who popularized the cranial index, constructed a theory of civilization upon it. He believed that Stone Age peoples of Europe were brachycephalic, and that progressive Bronze Age elements (Indo-European, or Aryan dolichocephalics) later invaded and replaced the original and more primitive inhabitants. Some original brachycephalic stocks survive among such benighted people as Basques, Finns, and Lapps.

Broca disproved this popular tale conclusively by discovering dolichocephalics both among Stone Age skulls and within modern remnants of 'primitive' stocks. Indeed, Broca had good reason to be suspicious of attempts by Nordic and Teutonic scientists to enshrine dolichocephaly as a mark of higher capability. Most Frenchmen, including Broca himself,[54] were brachycephalic. In a passage that recalls his dismissal of Tiedemann's claims for equality between black and white brains, Broca labeled Retzius's doctrine as self-serving gratification rather than empirical truth. Did he ever consider the possibility that he might fall prey to similar motivations?

> Since the work of Mr. Retzius, scientists have generally held, without sufficient study, that dolichocephaly is a mark of superiority. Perhaps so; but we must also not forget that the characters of dolichocephaly and brachycephaly were studied first in Sweden, then in England, the United States and Germany – and that in all these countries, particularly in Sweden, the dolichocephalic type clearly predominates. It is a natural tendency of men, even among those most free of prejudice, to attach an idea of superiority to the dominant characteristics of their race.[55]

Obviously, Broca declined to equate brachycephaly with inherent stupidity. Still, the prestige of dolichocephaly was so great that Broca felt more than a little uncomfortable when clearly inferior people turned up longheaded – uncomfortable enough to invent one of his most striking, unbeatable arguments. The cranial index had run into a stunning difficulty: not only were African blacks and Australian aborigines dolichocephalic, but they turned out to be the world's most longheaded peoples. Adding insult to this injury, the fossil Cro-Magnon skulls were not only larger than those of modern Frenchmen; they were more dolichocephalic as well.

Dolichocephaly, Broca reasoned, could be attained in several ways. The longheadedness that served as a mark of Teutonic genius obviously arose by frontal elongation. Dolichocephalics among people known to be inferior must have evolved by lengthening the back – occipital dolichocephaly in Broca's terms. With one sweep, Broca encompassed both the superior cranial capacity and the dolichocephaly of his Cro-Magnon fossils: 'It is by the greater development of their posterior cranium that their general cranial capacity is rendered greater than ours'.[56] As for blacks, they had acquired both a posterior elongation and a diminution in frontal width, thus giving them both a smaller brain in general and a longheadedness (not to be confused with the Teutonic style) exceeded by no human group. As to the brachycephaly of Frenchmen, it is no failure of frontal elongation (as the Teutonic supremacists claimed), but an addition of width to a skull already admirable.

THE CASE OF THE FORAMEN MAGNUM

The foramen magnum is the hole in the base of our skull. The spinal cord passes through it and the vertebral column articulates to the bone around its edge (the occipital condyle). In the embryology of all mammals, the foramen magnum begins under the skull, but migrates back to a position behind the skull at birth. In humans, the foramen magnum migrates only slightly and remains under the skull in adults. The foramen magnum of adult great apes occupies an intermediate position, not so far forward as in humans, not so far back as in other mammals. The functional significance of these orientations is clear. An upright animal like *Homo sapiens* must have its skull mounted *on top* of its vertebral column in order to look forward when standing erect; four-footed animals mount their vertebral column *behind* their skull and look forward in their usual posture.

	Whites	Blacks	Difference in favour of blacks
Anterior	90.736	100.304	+ 9.568
Facial	12.385	27.676	+15.291
Cranial	78.351	72.628	- 5.723
Posterior	100.385	100.857	+ 0.472

Table 4. Broca's measurements on the relative position of the foramen magnum.

These differences provided an irresistible source for invidious comparison. Inferior peoples should have a more posterior foramen magnum, as in apes and lower mammals. In 1862 Broca entered an existing squabble on this issue. Relative egalitarians like James Cowles Pritchard had been arguing that the foramen magnum lies exactly in the center of the skull in both whites and blacks. Racists like J. Virey had discovered graded variation, the higher the race, the more forward the foramen magnum. Neither side, Broca noted, had much in the way of data. With characteristic objectivity, he set out to resolve this vexatious, if minor, issue.

Broca amassed a sample of sixty whites and thirty-five blacks and measured the length of their skulls both before and behind the anterior border of the foramen magnum. Both races had the same amount of skull behind – 100.385 mm for whites; 100.857 mm for blacks (note precision to third decimal place). But whites had much less in front (90.736 vs. 100.304 mm) and their foramen magnum therefore lay in a more anterior position (see Table 4). Broca concluded: 'In orangutans, the posterior projection [the part of the skull behind the foramen magnum] is shorter. It is therefore incontestable . . . that the conformation of the Negro, in this respect as in many others, tends to approach that of the monkey'.[57]

But Broca then began to worry. The standard argument about the foramen magnum referred only to its relative position on the cranium itself, not to the face projecting in front of the cranium. Yet Broca had included the face in his anterior measure. Now everyone knows, he wrote, that blacks have longer faces than whites. This is an apelike sign of inferiority in its own right, but it should not be confused with the relative position of the foramen magnum within the cranium. Thus Broca set out to subtract the facial influence from his measures. He found that blacks did, indeed, have longer faces – white faces accounted for only 12.385 mm of their anterior measure, black faces for 27.676 mm (see Table 4). Subtracting facial length, Broca obtained the following figures for anterior cranium: 78.351 for whites, 72.628 for blacks. In other words, based on the cranium alone, the foramen magnum of blacks lay *farther forward* (the ratio of front to back, calculated from Broca's data, is .781 for whites, and .720 for blacks). Clearly, by criteria explicitly accepted before the study, blacks are superior to whites. Or so it must be, unless the criteria suddenly shift, as they did forthwith.

The venerable argument of front and back appeared to rescue Broca and the threatened people he represented. The more forward position of the foramen magnum in blacks does not record their superiority after all; it only reflects their lack of anterior brain power. Relative to whites, blacks have lost a great deal of brain in front. But they have added some brain behind, thus reducing the front/back ratio of the foramen magnum and providing a spurious appearance of black advantage. But they have not added to these inferior back regions as much as they lost in the anterior realm. Thus blacks have smaller and more poorly proportioned brains than whites:

> The anterior cranial projection of whites . . . surpasses that of Negroes by 4.9 percent. . . . Thus, while the foramen magnum of Negroes is further back with respect to their incisors [Broca's most forward point in his anterior measure that included the face], it is, on the contrary, further forward with respect to the anterior edge of their brain. To change the cranium of a white into that of a Negro, we would have not only to move the jaws forward, but also to reduce the front of the cranium – that is, to make the anterior brain atrophy and to give, as insufficient compensation, part of the material

> we extracted to the posterior cranium. In other words, in Negroes, the facial and occipital regions are developed to the detriment of the frontal region.[58]

This was a small incident in Broca's career, but I can imagine no better illustration of his method – shifting criteria to work through good data toward desired conclusions. Heads I'm superior; tails, you're inferior.

And old arguments never seem to die. Walter Freeman, dean of American lobotomists (he performed or supervised thirty-five hundred lesions of frontal portions of the brain before his retirement in 1970), admitted late in his career:[59]

> What the investigator misses most in the more highly intelligent individuals is their ability to introspect, to speculate, to philosophize, especially in regard to oneself. . . . On the whole, psychosurgery reduces creativity, sometimes to the vanishing point.

Freeman then added that 'women respond better than men, Negroes better than whites.' In other words, people who didn't have as much up front in the first place, don't miss it as badly.

Women's Brains

Of all his comparisons between groups, Broca collected most information on the brains of women vs. men – presumably because it was more accessible, not because he held any special animus toward women. 'Inferior' groups are interchangeable in the general theory of biological determinism. They are continually juxtaposed, and one is made to serve as a surrogate for all – for the general proposition holds that society follows nature, and that social rank reflects innate worth. Thus, E. Huschke, a German anthropologist, wrote in 1854: 'The Negro brain possesses a spinal cord of the type found in children and women and, beyond this, approaches the type of brain found in higher apes'.[60] The celebrated German anatomist Carl Vogt wrote in 1864:

> By its rounded apex and less developed posterior lobe the Negro brain resembles that of our children, and by the protu-

> berance of the parietal lobe, that of our females. . . . The grown-up Negro partakes, as regards his intellectual faculties, of the nature of the child, the female, and the senile white. . . . Some tribes have founded states, possessing a peculiar organization; but, as to the rest, we may boldly assert that the whole race has, neither in the past nor in the present, performed anything tending to the progress of humanity or worthy of preservation.[61]

G. Hervé, a colleague of Broca, wrote in 1881: 'Men of the black races have a brain scarcely heavier than that of white women'.[62] I do not regard as empty rhetoric a claim that the battles of one group are for all of us.

Broca centered his argument about the biological status of modern women upon two sets of data: the larger brains of men in modern societies and a supposed widening through time of the disparity in size between male and female brains. He based his most extensive study upon autopsies he performed in four Parisian hospitals. For 292 male brains, he calculated a mean weight of 1,325 grams; 140 female brains averaged 1,144 grams for a difference of 181 grams, or 14 percent of the male weight. Broca understood, of course, that part of this difference must be attributed to the larger size of males. He had used such a correction to rescue Frenchmen from a claim of German superiority. In that case, he knew how to make the correction in exquisite detail. But now he made no attempt to measure the effect of size alone, and actually stated that he didn't need to do so. Size, after all, cannot account for the entire difference because we know that women are not as intelligent as men.

> We might ask if the small size of the female brain depends exclusively upon the small size of her body. Tiedemann has proposed this explanation. But we must not forget that women are, on the average, a little less intelligent than men, a difference which we should not exaggerate but which is, nonetheless, real. We are therefore permitted to suppose that the relatively small size of the female brain depends in part upon her physical inferiority and in part upon her intellectual inferiority.[63]

To record the supposed widening of the gap through time, Broca measured the cranial capacities of prehistoric skulls from L'Homme

Mort cave. Here he found a difference of only 99.5 cc between males and females, while modern populations range from 129.5 to 220.7 cc. Topinard, Broca's chief disciple, explained the increasing discrepancy through time as a result of differing evolutionary pressures upon dominant men and passive women:

> The man who fights for two or more in the struggle for existence, who has all the responsibility and the cares of tomorrow, who is constantly active in combatting the environment and human rivals, needs more brain than the woman whom he must protect and nourish, than the sedentary woman, lacking any interior occupations, whose role is to raise children, love, and be passive.[64]

In 1879 Gustave Le Bon, chief misogynist of Broca's school, used these data to publish what must be the most vicious attack upon women in modern scientific literature (it will take some doing to beat Aristotle). Le Bon was no marginal hate-monger. He was a founder of social psychology and wrote a study of crowd behavior still cited and respected today (*La psychologie des foules*, 1895). His writings also had a strong influence upon Mussolini. Le Bon concluded:

> In the most intelligent races, as among the Parisians, there are a large number of women whose brains are closer in size to those of gorillas than to the most developed male brains. This inferiority is so obvious that no one can contest it for a moment; only its degree is worth discussion. All psychologists who have studied the intelligence of women, as well as poets and novelists, recognize today that they represent the most inferior forms of human evolution and that they are closer to children and savages than to an adult, civilized man. They excel in fickleness, inconstancy, absence of thought and logic, and incapacity to reason. Without doubt there exist some distinguished women, very superior to the average man, but they are as exceptional as the birth of any monstrosity, as, for example, of a gorilla with two heads; consequently, we may neglect them entirely.[65]

Nor did Le Bon shrink from the social implications of his views. He was horrified by the proposal of some American reformers to grant women higher education on the same basis as men:

> A desire to give them the same education, and, as a consequence, to propose the same goals for them, is a dangerous chimera. . . . The day when, misunderstanding the inferior occupations which nature has given her, women leave the home and take part in our battles; on this day a social revolution will begin, and everything that maintains the sacred ties of the family will disappear.[66]

Sound familiar?*

I have reexamined Broca's data, the basis for all this derivative pronouncement, and I find the numbers sound but Broca's interpretation, to say the least, ill founded. The claim for increasing difference through time is easily dismissed. Broca based this contention on the sample from L'Homme Mort alone. It consists of seven male, and six female, skulls. Never has so much been coaxed from so little!

In 1888 Topinard published Broca's more extensive data on Parisian hospitals. Since Broca recorded height and age as well as brain size, we may use modern statistical procedures to remove their effect. Brain weight decreases with age, and Broca's women were, on average, considerably older than his men at death. Brain weight increases with height, and his average man was almost half a foot taller than his average woman. I used multiple regression, a technique that permits simultaneous assessment of the influence of height and age upon brain size. In an analysis of the data for women, I found that, at average male height and age, a woman's brain would weigh 1,212 grams.† Correction for height and age reduces the 181 gram difference by more than a third to 113 grams.

* Ten years later, America's leading evolutionary biologist, E. D. Cope, dreaded the result if 'a spirit of revolt become general among women.' 'Should the nation have an attack of this kind,' he wrote,[67] 'like a disease, it would leave its traces in many after-generations.' He detected the beginnings of such anarchy in pressures exerted by women 'to prevent men from drinking wine and smoking tobacco in moderation,' and in the carriage of misguided men who supported female suffrage: 'Some of these men are effeminate and long-haired.'

† I calculate, where y is brain size in grams, x_1 age in years, and x_2 body height in cm: $y = 764.5 - 2.55x_1 + 3.47x_2$

It is difficult to assess this remaining difference because Broca's data contain no information about other factors known to influence brain size in a major way. Cause of death has an important effect, as degenerative disease often entails a substantial diminution of brain size. Eugene Schreider,[68] also working with Broca's data, found that men killed in accidents had brains weighing, on average, 60 grams more than men dying of infectious diseases. The best modern data that I can find (from American hospitals) records a full 100 gram difference between death by degenerative heart disease and by accident or violence. Since so many of Broca's subjects were elderly women, we may assume that lengthy degenerative disease was more common among them than among the men.

More importantly, modern students of brain size have still not agreed on a proper measure to eliminate the powerful effect of body size.[69] Height is partly adequate, but men and women of the same height do not share the same body build. Weight is even worse than height, because most of its variation reflects nutrition rather than intrinsic size – and fat vs skinny exerts little influence upon the brain. Léonce Manouvrier took up this subject in the 1880s and argued that muscular mass and force should be used. He tried to measure this elusive property in various ways and found a marked difference in favor of men, even in men and women of the same height. When he corrected for what he called 'sexual mass,' women came out slightly ahead in brain size.

Thus, the corrected 113 gram difference is surely too large; the true figure is probably close to zero and may as well favor women as men. One hundred thirteen grams, by the way, is exactly the average difference between a five-foot four-inch and a six-foot-four-inch male in Broca's data* – and we would not want to ascribe greater intelligence to tall men. In short, Broca's data do not permit any confident claim that men have bigger brains than women.

Maria Montessori did not confine her activities to educational reform for young children. She lectured on anthropology for several years at the University of Rome and wrote an influential book entitled *Pedagogical Anthropology* (English edition, 1913). She was, to say the least, no egalitarian. She supported most of Broca's work and the theory of innate criminality proposed by her compatriot Cesare Lombroso.

* For his largest sample of males, and using the favored power function for bivariate analysis of brain allometry, I calculate, where y is brain weight in grams and x is body height in cm: $y = 121.6x^{0.47}$

She measured the circumference of children's heads in her schools and inferred that the best prospects had bigger brains. But she had no use for Broca's conclusions about women. She discussed Manouvrier's work at length and made much of his tentative claim that women have slightly larger brains when proper corrections are made. Women, she concluded, are intellectually superior to men, but men have prevailed heretofore by dint of physical force. Since technology has abolished force as an instrument of power, the era of women may soon be upon us: 'In such an epoch there will really be superior human beings, there will really be men strong in morality and in sentiment. Perhaps in this way the reign of woman is approaching, when the enigma of her anthropological superiority will be deciphered. Woman was always the custodian of human sentiment, morality and honor'.[70]

Montessori's argument represents one possible antidote to 'scientific' claims for the constitutional inferiority of certain groups. One may affirm the validity of biological distinctions, but argue that the data have been misinterpreted by prejudiced men with a stake in the outcome, and that disadvantaged groups are truly superior. In recent years, Elaine Morgan has followed this strategy in her *Descent of Woman*, a speculative reconstruction of human prehistory from the woman's point of view – and as farcical as more famous tall tales by and for men.

I dedicate this book to a different position. Montessori and Morgan follow Broca's method to reach a more congenial conclusion. I would rather label the whole enterprise of setting a biological value upon groups for what it is: irrelevant, intellectually unsound, and highly injurious.

Postscript

Craniometric arguments lost much of their luster in our century, as determinists switched their allegiance to intelligence testing – a more 'direct' path to the same invalid goal of ranking groups by mental worth – and as scientists exposed the prejudiced nonsense that dominated most literature on form and size of the head. The American anthropologist Franz Boas, for example, made short work of the fabled cranial index by showing that it varied widely both among adults of a single group and within the life of an individual.[71] Moreover, he found significant differences in cranial index between immigrant parents and their American-born children. The immutable obtuseness of the brachycephalic southern European might veer toward the dolichocephalic Nordic norm in a single generation of altered environment.[72]

Yet the supposed intellectual advantage of bigger heads refuses to disappear entirely as an argument for assessing human worth. We still encounter it occasionally at all levels of determinist contention.

1. Variation within the general population: Arthur Jensen[73] supports the value of IQ as a measure of innate intelligence by claiming that the correlation between brain size and IQ is about 0.30. He doesn't doubt that the correlation is meaningful and that 'there has been a direct causal effect, through natural selection in the course of human evolution, between intelligence and brain size.' Undaunted by the low value of the correlation, he proclaims that it would be even higher if so much of the brain were not 'devoted to noncognitive functions.'

On the same page, Jensen cites an average correlation of 0.25 between IQ and physical stature. Although this value is effectively the same as the IQ vs. brain size correlation, Jensen switches ground and holds that 'this correlation almost certainly involves no causal or functional relationship between stature and intelligence.' Both height and intelligence, he argues, are perceived as desirable traits, and people lucky enough to possess more than the average of both are drawn to each other. But is it not more likely that height vs. brain size represents the primary causal correlation for the obvious reason that tall people tend to have large body parts? Brain size would then be an imperfect measure of height, and IQ might correlate with it (at the low value of 0.3) for the primarily environmental reason that poverty and poor nutrition can lead both to reduced stature and poor IQ scores.

2. Variation among social classes and occupational groups: In a book dedicated to putting educators in touch with latest advances in the brain sciences, H. T. Epstein states:

> First we shall ask if there is any indication of a linkage of any kind between brain and intelligence. It is generally stated that there is no such linkage. . . . But the one set of data I have found seems to show clearly that there is a substantial connection. Hooton studied the head circumferences of white Bostonians as part of his massive study of criminals. The following table shows that the ordering of people according to head size yields an entirely plausible ordering according

VOCATIONAL STATUS	N	MEAN (IN MM)	S.D.
Professional	25	569.9	1.9
Semiprofessional	61	566.5	1.5
Clerical	107	566.2	1.1
Trades	194	565.7	0.8
Public service	25	564.1	2.5
Skilled trades	351	562.9	0.6
Personal services	262	562.7	0.7
Laborers	647	560.7	0.3

Table 5. Mean and standard deviation of head circumference for people of varied vocational statuses.[74]

> to vocational status. It is not at all clear how the impression has been spread that there is no such correlation.[75]

Epstein's chart, reproduced as he presents it in Table 5, seems to support the notion that people in more prestigious jobs have larger heads. But a bit of probing and checking in original sources exposes the chart as a shoddy bit of finagling (not by Epstein who, I suspect, copied it from another secondary source that I have not been able to identify).

i) Epstein's reported standard deviations are so low, and therefore imply such a small range of variation within each occupational class, that the differences in mean head size must be significant even though they are so small. But a glance at Hooton's original table[76] reveals that the wrong column (standard errors of the mean) has been copied and called standard deviation. The true standard deviations, given in another column of Hooton's table, run from 14.4 to 18.6 – large enough to render most mean differences between occupational groups statistically insignificant.

ii) The chart arranges occupational groups by mean head size, but does not include Hooton's ranked assessments of vocational status based upon years of education.[77] In fact, since the column is labeled 'vocational status,' we are led to assume that the jobs have been listed in their proper order of prestige and that a perfect correlation therefore exists between status and head size. But the professions are arranged only by head size. Several professions do not fit the pattern; personal services and skilled trades (Hooton's status 5 and 6) rank just above the bottom in head size but at the middle in prestige.

iii) As a much worse, and entirely inexcusable omission, my consultation of Hooton's original chart shows that data for three trades have been expunged without comment in Table 5. Guess why? All three rank at or near the bottom of Hooton's list of status – factory workers at rank 7 (of 11), transportation employees at rank 8, and 'extractive' trades (farming and mining) at the lowest rank 11. All three have mean head circumferences (564.7, 564.9, and 564.7, respectively) *above* the grand average for all professions (563.9)!

I do not know the source of this disgracefully fudged chart. Jensen[78] reproduces it in Epstein's version with the three trades omitted. But he correctly labels the standard error (though he also omits the standard deviation) and properly denotes the professions as 'occupational category' rather than 'vocational status.' Yet Jensen's version includes the same minor numerical error as Epstein's (standard error of 0.3 for laborers, miscopied as the correct value from the omitted line of 'extractive' workers placed just above laborers in Hooton's chart). Since I doubt that the same insignificant error would have been made twice independently, and since Jensen's book and Epstein's article appeared at virtually the same time, I assume that both took the information from an unidentified secondary source (neither cites anyone but Hooton).

iv) Since Epstein and Jensen make so much of Hooton's data, they might have consulted his own opinion about it. Hooton was no do-gooding environmentalist liberal. He was a strong eugenicist and biological determinist who ended his study of American criminals with these chilling words: 'The elimination of crime can be effected only by the extirpation of the physically, mentally, and morally unfit, or by their complete segregation in a socially aseptic environment'.[79] Yet Hooton himself thought that his chart of head sizes and professions had proved nothing.[80] He noted that only one vocational group, laborers, departed significantly from the average of all groups. And he stated explicitly that his sample for the only profession with noticeably larger than average heads – the professionals – was 'wholly inadequate'[81] as a result of its small size.

v) The primary environmental hypothesis for correlations of head size with social class holds that they are artifacts of a causal correlation between body size and status. Large bodies tend to carry large heads, and proper nutrition and freedom from poverty fosters better growth in childhood. Hooton's data provide tentative support for both parts of this argument, though Epstein doesn't mention these data on stature at all. Hooton provides information on both height and weight (both inadequate measures of stature – see p. 518). Most significant deviations from the grand average support the environmental hypothesis. For weight, two groups departed significantly: professionals (status 1) heavier than average, and laborers (status 10) lighter than average. For height, three groups were deficient and none significantly taller than average: laborers (status 10), personal service (status 5), and clerical (status 2 – and contrary to the environmentalist hypothesis). I also computed correlation coefficients for head circumference vs. stature from Hooton's data. I found no correlation for total height, but significant correlations for both sitting height (0.605) and weight (0.741).

3. Variation among races: In its eighteenth edition of 1964, the *Encyclopaedia Britannica* was still listing 'a small brain in relation to their size' along with woolly hair as characteristic of black people.

In 1970 the South African anthropologist P. V. Tobias wrote a courageous article[82] exposing the myth that group differences in brain size bear any relationship to intelligence – indeed, he argued, group differences in brain size, independent of body size and other biasing factors, have never been demonstrated at all.

This conclusion may strike readers as strange, especially since it comes from a famous scientist well acquainted with the reams of published data on brain size. After all, what can be simpler than weighing a brain? – Take it out, and put it on the scale. Not so. Tobias lists fourteen important biasing factors. One set refers to problems of measurement itself: at what level is the brain severed from the spinal cord; are the meninges removed or not (meninges are the brain's covering membranes, and the dura mater, or thick outer covering, weighs 50 to 60 grams); how much time elapsed after death; was the brain preserved in any fluid before weighing and, if so, for how long; at what temperature was the brain preserved after death. Most literature does not specify these factors adequately, and studies made by different scientists usually cannot be compared. Even when we can be sure that the same object has been measured in the same way under the same conditions, a second set of biases intervenes – influences upon brain size with no direct tie to the desired properties of intelligence or racial affiliation: sex, body size, age, nutrition, nonnutritional environment, occupation, and cause of death. Thus, despite thousands of published pages, and tens of thousands of subjects, Tobias concludes that we do not know – as if it mattered at all – whether blacks, on the average, have larger or smaller brains than whites. Yet the larger size of white brains was an unquestioned 'fact' among white scientists until quite recently.

Many investigators have devoted an extraordinary amount of attention to the subject of group differences in human brain size. They have gotten nowhere, not because there are no answers, but because the answers are so difficult to get and because the a priori convictions are so clear and controlling. In the heat of Broca's debate with Gratiolet, one of Broca's defenders, admittedly as a nasty debating point, made a remark that admirably epitomizes the motivations implicit in the entire craniometric tradition: 'I have noticed for a long time,' stated de Jouvencel, 'that, in general, those who deny the intellectual importance of the brain's volume have small heads.'[83] Self-interest, for whatever reason, has been the wellspring of opinion on this heady issue from the start.

Notes

1. G. Stocking, *From Chronology to Ethnology: James Cowles Prichard and British Anthropology 1800–1850*, in facsimile of 1813 edition of J. C. Prichard, *Researches into the Physical History of Man* (Chicago: University of Chicago Press, 1973), p. lxx.
2. F. Galton, *Memories of My Life* (London: Methuen, 1909), pp. 315–16.
3. Ibid., p. 278.
4. Ibid., pp. 314–15.
5. F. Galton, *Hereditary Genius* (New York: D. Appleton, 1884), pp. 338–9.
6. Galton, *Memories*, pp. 249–50.
7. C. Darwin cited in ibid., pp. 290.
8. R. B. Bean, 'Some Racial Peculiarities of the Negro Brain', *American Journal of Anatomy*, 5 (1906): 390.
9. Ibid., 380.
10. Ibid., 409.
11. R. B. Bean cited in A. Chase, *The Legacy of Malthus* (New York: Alfred A. Knopf, 1977), p. 179.

12. F. P. Mall, 'On Several Anatomical Characters of the Human Brain, Said to Vary according to Race and Sex, with Especial Reference to the Weight of the Frontal Lobe', *American Journal of Anatomy*, 9 (1909): 1–32.
13. G. Myrdal, *An American Dilemma: The Negro Problem and Modern Democracy*, 2 vols (New York: Harper and Brothers, 1944); J. S. Haller Jr., *Outcasts from Evolution: Scientific Attitudes of Racial Inferiority, 1859–1900* (Urbana, Ill.: University of Illinois Press, 1971); Chase, op. cit.
14. P. Broca, 'Sur le volume et la forme du cerveau suivant les individus et suivant les races', *Bulletin Société d'Anthropologie Paris*, 2 (1861): 141.
15. Ibid., 139.
16. Ibid., 304.
17. Ibid., 188.
18. P. Broca, 'Anthropologie', in A. Dechambre (ed.), *Dictionnaire encyclopédique des sciences médicales* (Paris: Masson, 1866), p. 280.
19. Ibid., pp. 295–6.
20. Ibid., p. 296.
21. Broca cited in E. W. Count, *This Is Race* (New York: Henry Schuman, 1950), p. 72.
22. Broca, 1855, cited in ibid., p. 73.
23. P. Broca, 'Sur la mensuration de la capacité du crâne', *Memoire Société Anthropologie*, 2nd series, 1 (1873).
24. Ibid, 12.
25. P. Broca, *Mémoire sur les crânes des Basques* (Paris: Mason, 1868), p. 4.
26. Broca, 'Sur le volume et la forme du cerveau', 183.
27. Broca, 'Sur la mensuration'.
28. C. Cuvier cited in P. Topinard, *Anthropology* (London: Chapman and Hall, 1878), pp. 493–4.
29. P. Broca cited in ibid., p. 660.
30. P. Broca, 'Sur les proportions relatives du bras, de l'avant bras et de la clavicule chez les nègres et les européens', *Bulletin Société d'Anthropologie Paris*, 3/2 (1862): 10.
31. Ibid., 11.
32. P. Broca, 'Sur les crânes de la caverne de l'Homme-Mort (Lozère)', *Revue d'Anthropologie*, 2 (1873): 38.
33. Ibid.
34. Topinard, op. cit., p. 221.
35. Ibid.
36. Broca, 'Sur le volume et la forme du cerveau', 176.
37. E. A. Spitzka, 'A Study of the Brain of the Late Major J. W. Powell', *American Anthropology*, 5 (1903): 604.
38. Mall, 'On Several Anatomical Characters of the Human Brain', 24.
39. Broca, 'Sur le volume et la forme du cerveau', 441–2.
40. M. de Jouvencel, 'Discussion sur le cerveau', *Bulletin Société d'Anthropologie Paris*, 2 (1861): 464–74.
41. E. A. Spitzka, 'A Study of the Brains of Six Eminent Scientists and Scholars Belonging to the American Anthropometric Society, Together with a Description of the Skull of Professor E. D. Cope', *Transactions of the American Philosophical Society*, 21 (1907): 235.

42. Broca, 'Sur le volume et la forme du cerveau', 167.
43. Ibid., 165–6.
44 Ibid., 169.
45. P. Topinard, 'Le poids de l'encéphale d'après les registres de Paul Broca', *Mémoires Société d'Anthropologie Paris*, 2nd series, 3 (1888): 15.
46. P. Broca, 'Sur la capacité des crânes parisiens des diverses époques', *Bulletin Société d'Anthropologie Paris*, 3 (1862): 102–16.
47. Ibid., 106.
48. Ibid.
49. Broca, 'Sur les crânes de la caverne', 32.
50. Broca, 'Sur le volume et la forme du cerveau', 202–3.
51. T. W. Todd and D. W. Lyon Jr., 'Endocranial Suture Closure: Its Progress and Age Relationship. Part 1: Adult Males of White Stock', *American Journal of Physical Anthropology*, 7 (1924): 325–84; *idem*, 'Cranial Suture Closure, II: Ectocranial Closure in Adult Males of White Stock', *American Journal of Physical Anthropology*, 8 (1925): 23–40; *idem*, 'Cranial Suture Closure, III: Endocranial Closure in Adult Males of Negro Stock', *American Journal of Physical Anthropology*, 8 (1925): 47–71.
52. Broca, 'Sur les crânes de la caverne', 19.
53. Broca, 'Sur le volume et la forme du cerveau', 187.
54. L. Manouvrier, 'Conclusions générales sur anthropologie des sexes et applications sociales', *Revue de l'École d'Anthropologie*, 13 (1903): 405–23.
55. Broca, 'Sur le volume et la forme du cerveau', 513.
56. Broca, 'Sur les crânes de la caverne', 41.
57. P. Broca, 'Sur les projections de la tête et sur un nouveau procédé de céphalométrie', *Bulletin Société d'Anthropologie Paris*, 3 (1862): 16.
58. Ibid., 18.
59. W. Freeman, cited in S. L. Chorover, *From Genesis to Genocide* (Cambridge, MA: Massachusetts Institute of Technology Press. 1979).
60. E. Huschke, cited in F. P. Mall, op. cit., pp. 1–2.
61. C. Vogt, *Lectures on Man* (London: Longman, Green, Longman and Roberts, 1864), pp. 183–92.
62. G. Hervé, 'Du poids de l'encéphale', *Revue d'Anthropologie*, 2nd series, 4 (1881): 692.
63. Broca, 'Sur le volume et la forme du cerveau', 153.
64. Topinard, 'Le poids de l'encéphale', 22.
65. G. Le Bon, 'Recherches anatomiques et mathématiques sur les lois des variations du volume du cerveau et sur leurs relations avec l'intelligence', *Revue d'Anthropologie*, 2nd series, 2 (1879): 60–1.
66. Ibid., 62.
67. E. D. Cope, 'Two Perils of the Indo-European', *The Open Court*, 3 (1890): 2071.
68. E. Schreider, 'Brain Weight Correlations calculated from Original Results of Paul Broca', *American Journal of Physical Anthropology*, 25 (1966): 153–8.
69. J. J. Jerison, *The Evolution of the Brain and Intelligence* (New York: Academic Press, 1973).
70. M. Montessori, *Pedagogical Anthropology* (New York: F. A. Stokes Company, 1913).

71. F. Boas, 'The Cephalic Index', *American Anthropology*, 1 (1899): 448–61.
72. F. Boas, 'Changes in the Bodily Form of Descendants of Immigrants', Senate Document 208, 61st Congress, 2nd Session, 1911.
73. A. R. Jensen, *Bias in Mental Testing* (New York: Free Press, 1979), pp. 361–2.
74. Source: E. A. Hooton, *The American Criminal*, vol. 1 (Cambridge, MA: Harvard University Press, 1939), Table VIII-17.
75. H. T. Epstein, 'Growth Spurts during Brain Development: Implications for Educational Policy and Practice', in J. S. Chall and A. F. Mirsky (eds), *Education and the Brain*, 77th Yearbook of the National Society for the Study of Education (Chicago: University of Chicago Press, 1978), pp. 349–50.
76. Ibid.
77. Ibid., p. 150.
78. Jensen, op. cit., p. 361.
79. Hooton, op. cit., p. 309.
80. Ibid., 154.
81. Ibid., 153.
82. P. V. Tobias, 'Brain-size, Grey Matter, and Race – Fact or Fiction?', *American Journal of Physical Anthropology*, 32 (1970): 3–26.
83. Jouvencel, op. cit.

THE MOST UNKINDEST CUT OF ALL

CONSIDER this unremarkable description of a perennially pleasant pastime: unwinding after a hard day's work – a smoke and a drink around the fire.

> I remember that at the end of this——conference,——and I sat very cozily near the stove and then I saw——smoke for the first time, and I thought to myself, '——smoking today'; I'd never seen him do that. 'He is drinking brandy'; I hadn't seen him do that for years . . . We all sat together like comrades. Not to talk shop, but to rest after long hours of effort.

Now let's play 'fill in the blanks.' The man who made such an exception to his usual abstemiousness in order to celebrate his pleasure at such a successful outcome? Reinhard Heydrich, head of the Nazi security police and chief deputy to SS director Heinrich Himmler. The location of the cozy stove? The Wannsee Conference of January 20, 1942, held to prepare a plan for the *Endlösung der Judenfrage* – the 'final solution to the Jewish question,' the systematic murder of 11 million human beings (by Heydrich's own reckoning) and the genocide of a people. The man who remembered the closing scene quoted above? Adolf Eichmann, another participant and author of the Wannsee Protocol, the infamous document that summarized the hard day's work.

The maximally chilling content of the Wannsee Protocol is enhanced by its euphemistic and circumlocutory language. Killing and murder are never mentioned directly, and genocide sounds even more evil (if anything can make the ultimately hateful still worse) in its obtuse, but unmistakable, description as a 'final solution.' But Eichmann recalled at his trial that the verbal discussions could not have been more direct: 'What I know is that the gentlemen convened their session, and then in very plain terms – not in the language that I had to use in the minutes, but in absolutely blunt terms – they addressed the issue, with no mincing of words . . . The discussion covered killing, elimination, and annihilation.'

In an evasive first half, the Wannsee Protocol enumerates the Jewish population of Europe at some 11 million and then reviews the first two stages of action, now deemed unsuccessful and insufficient. At first, Hitler and company attempted 'the expulsion of the Jews from every particular sphere of life of the German people' – read *Kristallnacht*, confiscation and terrorism. The second strategy stressed physical removal: 'the expulsion of Jews from the living space (*Lebensraum*) of the German people.' But, Eichmann writes in the Wannsee Protocol, emigration spawned too many obstacles and had not worked fast enough: 'Financial difficulties, such as the demand for increasing sums of money to be presented at the time of the landing on the part of various foreign governments, lack of shipping space, increasing restriction of entry permits, or canceling of such, extraordinarily increased the difficulties of emigration.'

On, then, to a third (and truly ultimate) 'solution' – kill them all. (In some sense, as so many others have noted, the most chilling aspect of the Wannsee Conference and Protocol, held to initiate and implement this third strategy, lies neither in Heydrich's ability to conceptualize evil at such grand scale, nor in Eichmann's propensities for composing plans in euphemistic bureaucratese, but in the painstaking and deliberate construction of detailed logistics for such an extensive undertaking – the careful calculation of railroad cars and their volumes, the siting of death camps at the hubs of transportation lines, complex efforts to mask the true intent by depicting genocide as relocation and forced labor.)

Eichmann then introduces the new plan: 'Another possible solution of the problem has now taken the place of emigration, i.e., the evacuation of the Jews to the East, provided the Führer agrees to the plan' (Hitler had, in fact, already ordered such a strategy). Eichmann continues with his masked description of forced transportation to death as emigration for labor:

> Under proper guidance the Jews are now to be allocated for labor to the East in the course of the final solution. Able-bodied Jews will be taken in large labor columns to these districts for work on roads, separated according to sexes, in the course of which action a great part will undoubtedly be eliminated by natural causes . . . In the course of the practical execution of this final settlement of the problem, Europe will be cleaned up from West to East . . . The evacuated Jews will first be sent, group by group, into so-called transit ghettos from which they will be taken to the East . . . It is intended not to evacuate Jews of more than sixty-five years of age but to send them to an old-age ghetto.

Some camps – most notably Auschwitz – served both as killing places and prisons for forced labor (only a slower form of death). But others were gassing sites, pure and simple – Treblinka, Chelmno, Sobibor, Belzec. Architects of the final solution never intended to implement their 'cover story' of transportation for labor. We sometimes get a false impression of numbers allocated to immediate death vs. starvation and forced labor at the camps – though little preference can be specified in such a grisly 'Sophie's choice.' Several thousand survived Auschwitz, and many have told their stories. But the pure death camps are less well known because virtually no one lived to remember. Two people survived at Belzec, three at Chelmno (M. Gilbert, *The Holocaust*). The final solution was always and only about murder – total and unvarnished.

Clearly, I have nothing to add to this ultimate story of human atrocity. I am neither a poet nor historian, and I was not there. Why take up such a subject in a series of essays on natural history and evolutionary theory? The answer lies in the second half of the Wannsee Protocol – the part of the document that is rarely discussed and almost never quoted. My rationale also rests with Hitler's misuse of genetics and evolutionary biology as a centerpiece of plans explicitly stated right from the beginning, at the publication of *Mein Kampf* in 1925.

Insofar as Hitler's evil scheme should be dignified as a 'theory' at all, he sunk his argument squarely in paranoia about racial purity and its biological necessity for triumphant peoples in a world of natural selection. The Aryan nation had been great, but its strength had been sapped by racial mixing, encouraged by the insidious and liberal propaganda of parasitic Jews, out to rule the world or at least to feed off the higher morality of Aryan virtue. Hitler wrote in *Mein Kampf*:

> The Aryan gave up the purity of his blood and therefore he also lost his place in the Paradise which he had created for himself. He became submerged in the race-mixture, he gradually lost his cultural ability till . . . he began to resemble more the subjected and aborigines than his ancestors . . . Blood-mixing, with the lowering of the racial level caused by it, is the sole cause of the dying-off of old cultures; for the people do not perish by lost wars [a comment on Germany's defeat in World War I], but by the loss of that force of resistance which is contained only in the pure blood. All that is not race in this world is trash.

(I am quoting from the first complete English translation of *Mein Kampf*, published in America by John Chamberlain and others in 1939 as a warning about the enemy we would soon need to confront. My parents bought this book before my father left to join the battle. Throughout my youth, I stared at this volume on my parents' shelves, taking it down now and again – more to experience the frisson of touching evil than from any desire to read. When my father died a few years ago, and my mother offered me his collection of books, I included this familiar volume, with its blood-red jacket, among the few items that I wanted for my own. To hold the book now, and to quote from it for the first time, gives me an eerie feeling of connectivity with my past, and rekindles my dimmest three-year-old impression of World War II as a fight between my daddy and a bad man named Hitler.)

If the first (and oft-quoted) half of the Wannsee Protocol is a cursory and euphemistic account of genocide at worksites and death camps, the highly specific (and usually neglected) second half is a detailed and specific disquisition on genetics and race-mixing – and for an obvious reason, given Hitler's eugenical ideas. From the Führer's standpoint, killing 11 million people will not solve the Jewish problem if many German citizens still carry tainted blood as a result of partial Jewish ancestry through mixed marriages. A truly final solution demands a set of rules and policies for these *Mischlinge* (literally, mixtures). Kill the others and purify yourselves. The two halves of the Protocol could not be more intimately connected through the sick logic of Hitler's racial doctrines. Eichmann's transition sentence, written in optimally opaque bureaucratese, states:

> The implementation of the final solution problem is supposed to a certain extent to be based on the Nuremberg Laws [Nazi legislation on eugenic marriage and sterilization], in which connection also the solution of the problems presented by the mixed marriages and the persons of mixed blood is seen to be conditional to an absolutely final clarification of the question.

These pages on racial mixing reveal a form of madness different from Eichmann's first half, on genocide in the east. The first part of the Protocol wallows in the ultimate evil of euphemized mass murder; this second half follows the steely logic of total craziness reasoned right through, with full coverage of all details. Only once before have I experienced such a feeling while reading an official state document: when I studied the annual proceedings of the South African racial classification board (under the old regime, before Nelson Mandela) as they tortured the logic of continuity to find a discrete pigeonhole for each individual under strict apartheid.

Needless to say, any such effort faces an intractable dilemma at the start: people are complexly interbred at all degrees of mixture, and neither pure solution will work. Is everyone with even the slightest trace of Jewish ancestry a Jew (which would condemn most of the nation), or does salvation arise with any Aryan infusion (which would be far too lenient for zealous madmen)? Heydrich, Eichmann, and company therefore invoke the usual trick of argument for breaking a true continuum lacking a compelling point for separation: choose an arbitrary dividing line and then treat your division as a self-evident fact of nature.

The Protocol basically proclaims that half-breeds are Jews (offspring of one pure Jewish and one pure Aryan parent); quarter-breeds are German (offspring of a half-breed and an Aryan). But this tidy little rule required some nuancing at the borders. At one end, half-breeds could avoid their death warrant if they had children by marriage to a person of German blood (but not if such a union had produced no offspring), or 'if exemption licenses have been issued by the highest Party or State authorities.'

Even this hope for mitigation carried two provisos: first, each exemption must be granted on a case-by-case basis and only for 'personal essential merit of the person of mixed blood.' Second, anyone so spared must make a little gesture in return: 'Any person of mixed blood of the first degree to whom exemption from the evacuation is granted will be sterilized – in order to eliminate the possibility of offspring and to secure a final solution of the problem presented by the persons of mixed blood.' But fear not: 'The sterilization will take place on a voluntary basis.' Consider the alternative, however: 'But it will be conditional to a permission to stay in the Reich.' One participant at the conference made an astute observation on this score: 'SS-Gruppenführer Hofmann advocates the opinion that sterilization must be applied on a large scale; in particular as the person of mixed blood placed before the alternative as whether to be evacuated or to be sterilized would rather submit to sterilization.'

At the other end, the supposedly acceptable quarter-breeds could be 'demoted' to genocide for any of three reasons all designed to ferret out an unacceptable degree of racial taint: (1) 'The person of mixed blood of the second degree is the result of a marriage where both parents are persons of mixed blood' (apparently, a child needs at least one purely Aryan parent to enter the realm of the blessed). (2) 'The general appearance of the person of mixed blood of the second degree is racially particularly objectionable so that he already outwardly must be included among the Jews.' (3) 'The person of mixed blood of the second degree has a particularly bad police and political record sufficient to reveal that he feels and behaves like a Jew.'

In the gray area between, marriages of two people with mixed blood sealed everyone's doom, parents and children alike. Even a normally exempt quarter-breed must die (with all children) if such a person marries a half-breed: 'Marriage between persons of mixed blood of the first degree and persons of mixed blood of the second degree: Both partners will be evacuated, regardless of whether or not they have children . . . Since as a rule these children will racially reveal the admixture of Jewish blood more strongly than persons of mixed blood of the second degree.' What can be more insane than madness that constructs its own byzantine taxonomy – or are we just witnessing the orderly mind of the petty bureaucrat applied to human lives rather than office files?

The Protocol's most stunning misuse of evolutionary biology, however, appears not in this tedious taxonomy of genetic nonsense, but right in the heart of the document's chief operational paragraph. I quoted the words above, but inserted ellipses to signify an omission that I now wish to restore. Eichmann speaks of deportation to the east, and of hard labor on the roads, leading to the death of most evacuees. He then continues (to flesh out my ellipses): 'The possible final remnant will, as it must undoubtedly consist of the toughest, have to be treated accordingly . . .' Let me now switch to the original German: '. . . *da dieser, eine natürliche Auslese darstellend, bei Freilassung als Keimzelle eines neuen jüdischen Aufbaues anzusprechen ist.*' Or: '. . . as it is the *product of natural selection*, and would, if liberated, act as a bud cell of a Jewish reconstruction.'

Perhaps you do not see the special horror of this line (embedded, as it is, in such maximal evil). But what can be more wrenching than the violation of one's own child, or the perversion for vicious purpose of the most noble item in a person's world? I am an evolutionary biologist by original training and a quarter-century of practice. Charles Darwin is the resident hero in my realm – and few professions can name a man so brilliant, so admirable, and so genial as both founder and continuing inspiration. Darwin, of course, gave a distinctive name to his theory of evolutionary change: natural selection. This theory has a history of misuse almost as long as its proper pedigree. Claptrap and bogus Darwinian formulations have been used to justify every form of social exploitation – rich over poor, technologically complex over traditional, imperialist over aborigine, conqueror over defeated in war. Every evolutionist knows this history only too well, and we bear some measure of collective responsibility for the uncritical fascination that many of us have shown for such unjustified extensions. But most false expropriations of our chief phrase have been undertaken without our knowledge and against our will.

I have known this story all my professional life. I can rattle off lists of such misuses, collectively called 'social Darwinism.' (See my book *The Mismeasure of Man*.) But until the fiftieth anniversary of the Wannsee Conference piqued my curiosity and led me to read Eichmann's Protocol for the first time, I had not known about the absolute ultimate in all conceivable misappropriation – and the discovery hit me as a sudden, visceral haymaker, especially since I had steeled myself to supposed unshockability before reading the document. *Natürliche Auslese* is the standard German translation of Darwin's 'natural selection.' To think that the key phrase of my professional world lies so perversely violated in the very heart of the chief operative paragraph in the most evil document ever written! What symbol of misuse could possibly be more powerful? Surely this is the literary equivalent of imagining one's daughter shackled in a dungeon operated by sadistic rapists.

It scarcely matters that the phrase is so ludicrously inverted in its misapplication – for we dare hardly dignify Eichmann's argument with space for refutation. Natural selection is nature's process of differential reproductive success, however that advantage be attained (sometimes by dominating one's fellows, to be sure, but also by cooperation for mutual benefit). What could be more unnatural, more irrelevant to Darwin's process, than the intricately planned murder and starvation of several million people by human technology?

I could simply end this essay here, washing my profession's hands of all conceivable responsibility. After all, the Wannsee Protocol would have proclaimed and implemented its horrors even if Eichmann had not employed Darwin's phrase to justify and embellish his major recommendation. And Eichmann's use, in any case, ranks as a perverse misinterpretation of Darwin on both key grounds – the false application of a natural principle to human moral conduct, and the distortion of a statement about differential reproductive success into a bogus validation of mass murder as natural. Of the second mistake, Darwin wrote in a key passage in *The Origin of Species*:

> I should presume that I use the term Struggle for Existence in a large and metaphorical sense, including dependence of one being on another, and including (which is more important) not only the life of the individual, but success in leaving progeny. Two canine animals in a time of dearth, may be truly said to struggle with each other which shall get food and live. But a plant on the edge of a desert is said to struggle for life against the drought.

But resolutions are never so clean or simple. Science, as a profession, does have a little something to answer for, or at least something to think about. Darwin may have been explicit in labeling the struggle for existence as metaphorical, but most nineteenth-century versions (including Darwin's own illustrations, most of the time) stressed overt competition and victory by death – surely a more congenial image than peaceful cooperation for an age of aggressive expansion and conquest, both ethnographic and industrial.

Hitler didn't invent the mistaken translation to human affairs. Claptrap Darwinism had served as an official rationale for German military conquest in World War I (while our side often used the same argument, though less zealously and systematically). In fact, William Jennings Bryan (see essay 28 in my book *Bully for Brontosaurus*) first decided to oppose evolution when he mistook Darwin's actual formulation for the egregious German misuse that so deeply disturbed him.

Many scientists consistently opposed this misapplication, but others, probably the majority, remained silent (many enjoying the prestige, even if falsely won), while a few actively abetted the cooptation of their field for a variety of motives, including misplaced patriotism and immediate personal reward. Several English and American eugenicists offered initial praise for Hitler's laws on restriction of marriage and enforced sterilization – before they realized what the Führer really intended. The text of the German legislation borrowed heavily from eugenical sterilization statutes then on the books of several American states, and upheld by the Supreme Court in 1927. German evolutionists did not raise a chorus of protest against Hitler's misuse of natural selection, dating to *Mein Kampf* in 1925. Wannsee is the logical extension of the following fulmination in *Mein Kampf*, with its final explicit misanalogy to nature, and its call for elimination and sterilization of the supposedly unfit:

> The fight for daily bread [in Nature] makes all those succumb who are weak, sickly and less determined . . . The fight is always a means for the promotion of the species' health and force of resistance, and thus a cause for its development to a higher level. If it were different, every further development towards higher levels would stop, and rather the contrary would happen. For, since according to numbers, the inferior element always outweighs the superior element, under the same preservation of life and under the same propagating possibilities, the inferior element would increase so much more rapidly that finally the best element would be forced to step into the background, if no correction of this condition were carried out. But just this is done by Nature, by subjecting the weaker part to such difficult living conditions that even by this the number is restricted, and finally by

> preventing the remainder, without choice, from increasing . . . [Did Eichmann have this passage in mind when he advocated the same two-step process for the final solution: starve to a remnant by 'difficult living conditions'; then kill the rest.] Man, by trying to resist this iron logic of Nature, becomes entangled in a fight against the principles to which alone he, too, owes his existence as a human being. Thus his attack is bound to lead to his own doom.

A scientist's best defense against such misappropriation lies in a combination that may seem to mix two disparate traits, vigilance and humility: vigilance in combating misuses that threaten effectiveness (you don't have to refute every kook who writes a letter to the local newspaper – time doesn't permit – but how can we distinguish the young Hitler from 'just another nut'?); humility in recognizing that science does not, and cannot in principle, find answers to moral questions. A popular linguistic construction of late would have us believe that morality may be measured as a kind of 'fiber' – as though one might pour ethics out of a cereal box prominently labeled (by exacting laboratory standards) with precise content and contribution to minimal daily requirements!

Science can supply information as input to a moral decision, but the ethical realm of 'oughts' cannot be logically specified by the factual 'is' of the natural world – the only aspect of reality that science can adjudicate. As a scientist, I can refute the stated genetic rationale for Nazi evil and nonsense. But when I stand against Nazi policy, I must do so as everyman – as a human being. For I win my right to engage moral issues by my membership in *Homo sapiens* – a right vested in absolutely every human being who has ever graced this earth, and a responsibility for all who are able.

If we ever grasped this deepest sense of a truly universal community – the equal worth of all as members of a single entity, the species *Homo sapiens*, whatever our individual misfortunes or disabilities – then Isaiah's vision could be realized, and our human wolves would dwell in peace with lambs, for 'they shall not hurt nor destroy in all my holy mountain.' We are freighted by heritage, both biological and cultural, granting us capacity both for infinite sweetness and unspeakable evil. What is morality but the struggle to harness the first and suppress the second? Darwin's soulmate, the great American born on the very same day, said much the same in the famous words of his first inaugural address, in

March 1861, when he still hoped to spare the nation from the horrors of civil war. Lincoln asked us to remember the former unity of North and South, and to avoid destruction by applying our better nature to this memory. Let us extend his hope to all bodies of the single human species:

> The mystic chords of memory, stretching from every battle-field and patriot grave to every living heart and hearthstone all over this broad land, will yet swell the chorus of the Union when again touched, as surely they will be, by the better angels of our nature.

Epilogue

I doubt that I have ever written a more thoroughly serious essay, without even an attempt at passing humor. What else can one do with this most totally tragic of all conceivable subjects? Still, an old literary tradition holds that some lightness must be introduced – for the relief and variety that our emotional natures crave – especially in such situations of sustained sadness. Thus, Hamlet jests with the gravedigger about poor Yorick, while Puccini introduces three courtiers, Ping, Pang, and Pong, as comic relief between Turandot's beheading of several sequential suitors.

Therefore, with such a historically distinguished excuse, I have decided to cite in full a remarkable letter received by the editor of *Natural History* in response to the original version of this essay:

> As a long-time faculty member concerned about correct usage of the English language, I was disturbed by the title of Stephen Jay Gould's latest article. Then I decided something in the article would explain why the double superlative was used. When that supposition proved groundless, I could only conclude that someone intended it as a joke or as a means to provoke correspondence from readers like myself. Am I correct? I will continue to enjoy your periodical, but hope that you will not continue to provide shocks of this kind.

I replied that I grasped both her point and her distress, and that, while I hate to pass the buck and believe that a man must bear the consequences of his own actions, she really would have to air her complaint with Mr Shakespeare!

> If you have tears, prepare to shed them now . . .
> Look! in this place ran Cassius' dagger through:
> See what a rent the envious Casca made:
> Through this the well-beloved Brutus stabb'd . . .
> Judge O you gods, how dearly Caesar loved him!
> This was the most unkindest cut of all.
>
> Mark Antony in *Julius Caesar*

A TALE OF TWO WORK SITES

CHRISTOPHER Wren, the leading architect of London's reconstruction after the great fire of 1666, lies buried beneath the floor of his most famous building, St Paul's cathedral. No elaborate sarcophagus adorns the site. Instead, we find only the famous epitaph written by his son and now inscribed into the floor: '*si monumentum requiris, circumspice*' – if you are searching for his monument, look around. A tad grandiose perhaps, but I have never read a finer testimony to the central importance – one might even say sacredness – of actual places, rather than replicas, symbols, or other forms of vicarious resemblance.

An odd coincidence of professional life turned my thoughts to this most celebrated epitaph when, for the second time, I received an office in a spot laden with history, a place still redolent of ghosts of past events both central to our common culture and especially meaningful for my own life and choices.

In 1971, I spent an academic term as a visiting researcher at Oxford University. I received a cranny of office space on the upper floor of the University Museum. As I set up my books, fossil snails, and microscope, I noticed a metal plaque affixed to the wall, informing me that this reconfigured space of shelves and cubicles had been, originally, the site of the most famous public confrontation in the early history of Darwinism. On this very spot, in 1860, just a few months after Darwin published *The Origin of Species*, T. H. Huxley had drawn his rhetorical sword, and soundly skewered the slick but superficial champion of creationism, Bishop 'Soapy Sam' Wilberforce.

(As with most legends, the official version ranks as mere cardboard before a much more complicated and multifaceted truth. Wilberforce and Huxley did put on a splendid, and largely spontaneous, show – but no clear victor emerged from the scuffle, and Joseph Hooker, Darwin's other champion, made a much more effective reply to the bishop, however forgotten by history. See my essay on this debate, entitled 'Knight Takes Bishop?' and published in *Bully for Brontosaurus*.)

I can't claim that the lingering presence of these Victorian giants increased my resolve or improved my work, but I loved the sense of continuity vouchsafed to me by this happy circumstance. I even treasured the etymology – for *circumstance* means 'standing around' (as Wren's *circumspice* means 'looking around'), and here I stood, perhaps in the very spot where Huxley had said, at least according to legend, that he preferred an honest ape for an ancestor to a bishop who would distort a known truth for rhetorical advantage.

Not so long ago, I received a part-time appointment as visiting research professor of biology at New York University. I was given an office on the tenth floor of the Brown building on Washington Place, a nondescript early-twentieth-century structure now filled with laboratories and other academic spaces. As the dean took me on a casual tour of my new digs, he made a passing remark, intended as little more than 'tour-guide patter,' but producing an electric effect upon his new tenant. Did I know, he asked, that this building had been the site of the infamous Triangle Shirtwaist fire of 1911, and that my office occupied a corner location on one of the affected floors – in fact, as I later discovered, right near the escape route used by many workers to safety on the roof above. The dean also told me that, each year on the March 25 anniversary of the fire, the International Ladies' Garment Workers Union still holds a ceremony at the site and lays wreaths to memorialize the 146 workers killed in the blaze.

If the debate between Huxley and Wilberforce defines a primary legend of my chosen profession, the Triangle Shirtwaist fire occupies an even more central place in my larger view of life. I grew up in a family of Jewish immigrant garment workers, and this holocaust (in the literal meaning of a thorough sacrifice by burning) had set their views and helped to define their futures.

The shirtwaist – a collared blouse designed on the model of a man's shirt and worn above a separate skirt – had become the fashionable symbol of more independent women. The Triangle Shirtwaist Company, New York City's largest manufacturer of shirtwaists, occupied three floors (eighth through tenth) of the Asch Building (later bought by New

York University and rechristened Brown, partly to blot out the infamy of association with the fire). The company employed some five hundred workers, nearly all young women who had recently arrived either as Jewish immigrants from eastern Europe or as Catholics from Italy. Exits from the building, in addition to elevators, included only two small stairways and one absurdly inadequate fire escape. But the owners had violated no codes, both because general standards of regulation were then so weak, and because the structure was supposedly fireproof – as the framework proved to be (for the building, with my office, still stands), though inflammable walls and ceilings could not prevent an internal blaze on floors crammed full of garments and cuttings. The Triangle company was, in fact, a deathtrap – for fire hoses of the day could not pump above the sixth floor, while nets and blankets could not sustain the force of a human body jumping from greater heights.

The fire broke out at quitting time. Most workers managed to escape by the elevators, down one staircase (we shall come to the other staircase later), or by running up to the roof. But the flames trapped 146 employees, nearly all young women. About fifty workers met a hideous, if dramatic, end by jumping in terror from the ninth-floor windows, as a wall of fire advanced from behind. Firemen and bystanders begged them not to jump, and then tried to hold improvised nets of sheets and blankets. But these professionals and good Samaritans could not hold the nets against the force of fall, and many bodies plunged right through the flimsy fabrics onto the pavement below, or even through the 'hollow sidewalks' made of opaque glass circles designed to transmit daylight to basements below, and still a major (and attractive) feature of my SoHo neighborhood. (These sidewalks carry prominent signs warning delivery trucks not to back in.) Not a single jumper survived, and the memory of these forced leaps to death remains the most searing image of America's prototypical sweatshop tragedy.

All defining events of history develop simplified legends as official versions – primarily, I suppose, because we commandeer such events for shorthand moral instruction, and the complex messiness of actual truth always blurs the clarity of a pithy epigram. Thus, Huxley, representing the righteousness of scientific objectivity, must slay the dragon of ancient and unthinking dogma. The equally oversimplified legend of the Triangle fire holds that workers became trapped because management had locked all the exit doors to prevent pilfering, unscheduled breaks, or access to union organizers – leaving only the fire escape as a mode of exit. All five of my guidebooks to New York architecture tell this 'official' version. My favorite book, for example, states: 'Although the building was equipped with fire exits, the terrified workers discovered to their horror that the ninth-floor doors had been locked by supervisors. A single fire-escape was wholly inadequate for the crush of panic-stricken employees.'

These traditional (indeed, virtually 'official') legends may exaggerate for moral punch, but such interpretations emerge from a factual basis of greater ambiguity – and this reality, as we shall see in the Triangle case, often embodies a deeper and more important lesson. Huxley did argue with Wilberforce, after all, even if he secured no decisive victory, and Huxley did represent the side of the angels – the true angels of light and justice. And although many Triangle workers escaped by elevators and one staircase, another staircase (that might have saved nearly everyone else) was almost surely locked.

If Wilberforce and his minions had won, I might be a laborer, a linguist, or a lawyer today. But the Triangle fire might have blotted me out entirely. My grandmother arrived in America in 1910. On that fatal March day in 1911, she was working as a sixteen-year-old seamstress in a sweatshop – but, thank God, not for the Triangle Shirtwaist Company. My grandfather, at the same moment, was cutting cloth in yet another nearby factory.

These two utterly disparate stories – half a century and an ocean apart, and with maximal contrast between an industrial tragedy and an academic debate – might seem to embody the most unrelatable of items: the apples and oranges, or chalk and cheese (the British version), of our mottoes. Yet I feel that an intimate bond ties these two stories together in illustrating opposite poles of a central issue in the history of evolutionary theory: the application of Darwinian thought to the life and times of our own troubled species. I claim nothing beyond personal meaning – and certainly no rationale for boring anyone else – in the accidental location of my two offices in such sacred spots of history. But the emotion of a personal prod often dislodges a general theme well worth sharing.

The application of evolutionary theory to *Homo sapiens* has always troubled Western culture deeply – not for any reason that might be called scientific (for humans are biological objects, and must therefore take their place with all other living creatures on the genealogical tree of life), but only as a consequence of ancient prejudices about human distinctiveness and unbridgeable superiority. Even Darwin tiptoed lightly across this subject when he wrote *The Origin of Species* in 1859 (though he plunged in later, in 1871, by publishing *The Descent of Man*). The first edition of the *Origin* says little about *Homo sapiens* beyond a cryptic promise that 'light will be thrown on the origin of man and his history.' (Darwin became a bit bolder in later editions and ventured the following emendation: 'Much light will be thrown . . .')

Troubling issues of this sort often find their unsurprising resolution in a bit of wisdom that has permeated our traditions from such sublime sources as Aristotle's *aurea mediocritas* (or golden mean) to the vernacular sensibility of Goldilocks's decisions to split the difference between two extremes, and find a solution 'just right' in the middle. Similarly, one can ask either too little or too much of Darwinism in trying to understand 'the origin of man and his history.' As usual, a proper solution lies in the intermediary position of 'a great deal, but not everything.' Soapy Sam Wilberforce and the Triangle Shirtwaist fire gain their odd but sensible conjunction as illustrations of the two extremes that must be avoided – for Wilberforce denied evolution altogether and absolutely, while the major social theory that hindered industrial reform (and permitted conditions that led to such disasters as the Triangle Shirtwaist fire) followed the most overextended application of biological evolution to patterns of human history – the theory of 'social Darwinism.' By understanding the fallacies of Wilberforce's denial and social Darwinism's uncritical and total embrace, we may find the proper balance between.

They didn't call him Soapy Sam for nothing. The orotund bishop of Oxford saved his finest invective for Darwin's attempt to apply his heresies to human origins. In his review of *The Origin of Species* (published in the *Quarterly Review*, England's leading literary journal, in 1860), Wilberforce complained above all: 'First, then, he not obscurely declares that he applies his scheme of the action of the principle of natural selection to Man himself, as well as to the animals around him.' Wilberforce then uncorked a passionate argument for a human uniqueness that could only have been divinely ordained:

> Man's derived supremacy over the earth; man's power of articulate speech; man's gift of reason; man's free-will and responsibility; man's fall and man's redemption; the incarnation of the Eternal Son; the indwelling of the Eternal Spirit, – all are equally and utterly irreconcilable with the degrading

> notion of the brute origin of him who was created in the image of God, and redeemed by the Eternal Son.

But the tide of history quickly engulfed the good bishop. When Wilberforce died in 1873, from a head injury after a fall from his horse, Huxley acerbically remarked that, for once, the bishop's brains had come into contact with reality – and the result had been fatal. Darwinism became the reigning intellectual novelty of the late nineteenth century. The potential domain of natural selection, Darwin's chief explanatory principle, seemed nearly endless to his devotees (though not, interestingly, to the master himself, as Darwin remained cautious about extensions beyond the realm of biological evolution). If a 'struggle for existence' regulated the evolution of organisms, wouldn't a similar principle also explain the history of just about anything – from the cosmology of the universe, to the languages, economics, technologies, and cultural histories of human groups?

Even the greatest of truths can be overextended by zealous and uncritical acolytes. Natural selection may be one of the most powerful ideas ever developed in science, but only certain kinds of systems can be regulated by such a process, and Darwin's principle cannot explain all natural sequences that develop historically. For example, we may talk about the 'evolution' of a star through a predictable series of phases over many billion years from birth to explosion, but natural selection – a process driven by the differential survival and reproductive success of some individuals in a variable population – cannot be the cause of stellar development. We must look, instead, to the inherent physics and chemistry of light elements in such large masses.

Similarly, although Darwinism surely explains many universal features of human form and behavior, we cannot invoke natural selection as the controlling cause of our cultural changes since the dawn of agriculture – if only because such a limited time of some ten thousand years provides so little scope for any general biological evolution at all. Moreover, and most importantly, human cultural change operates in a manner that precludes a controlling role for natural selection. To mention the two most obvious differences: first, biological evolution proceeds by continuous division of species into independent lineages that must remain forever separated on the branching tree of life. Human cultural change works by the opposite process of borrowing and amalgamation. One good look at another culture's wheel or alphabet may alter the course of a civilization forever. If we wish to identify a biological analog for cultural change, I suspect that infection will work much better than evolution.

Second, human cultural change runs by the powerful mechanism of Lamarckian inheritance of acquired characters. Anything useful (or alas, destructive) that our generation invents can be passed directly to our offspring by direct education. Change in this rapid Lamarckian mode easily overwhelms the much slower process of Darwinian natural selection, which requires a Mendelian form of inheritance based on small-scale and undirected variation that can then be sifted and sorted through a struggle for existence. Genetic variation is Mendelian, so Darwinism rules biological evolution. But cultural variation is largely Lamarckian, and natural selection cannot determine the recent history of our technological societies.

Nonetheless, the first blush of high Victorian enthusiasm for Darwinism inspired a rush of attempted extensions to other fields, at least by analogy. Some efforts proved fruitful, including the decision of James Murray, editor of *The Oxford English Dictionary* (first volume published in 1884, but under way for twenty years before then), to work strictly by historical principles and to treat the changing definitions of words not by current preferences in use (as in a truly normative dictionary), but by the chronology and branching evolution of recorded meanings (making the text more an encyclopedia about the history of words than a true dictionary).

But other extensions proved both invalid in theory, and also (or so most of us would judge by modern moral sensibilities) harmful, if not tragic, in application. As the chief offender in this category, we must cite a highly influential theory that acquired the inappropriate name of 'social Darwinism.' (As many historians have noted, this theory should really be called 'social Spencerism,' since Herbert Spencer, chief Victorian pundit of nearly everything, laid out all the basic postulates in his *Social Statics* of 1850, nearly a decade before Darwin published *The Origin of Species*. Darwinism did add the mechanism of natural selection as a harsher version of the struggle for existence, long recognized by Spencer. Moreover, Darwin himself maintained a highly ambivalent relationship to this movement that borrowed his name. He felt the pride of any creator toward useful extensions of his theory – and he did hope for an evolutionary account of human origins and historical patterns. But he also understood only too well why the mechanism of natural selection applied poorly to the causes of social change in humans.)

Social Darwinism often serves as a blanket term for any genetic or biological claim made about the inevitability (or at least the 'naturalness') of social inequalities among classes and sexes, or military conquests of one group by another. But such a broad definition distorts the history of this important subject – although pseudo-Darwinian arguments have long been advanced, prominently and forcefully, to cover all these sins. Classical social Darwinism operated as a more specific theory about the nature and origin of social classes in the modern industrial world. The *Encyclopaedia Britannica* article on this subject correctly emphasizes this restriction by first citing the broadest range of potential meaning, and then properly narrowing the scope of actual usage:

> *Social Darwinism:* the theory that persons, groups, and races are subject to the same laws of natural selection as Charles Darwin had perceived in plants and animals in nature. . . . The theory was used to support laissez-faire capitalism and political conservatism. Class stratification was justified on the basis of 'natural' inequalities among individuals, for the control of property was said to be a correlate of superior and inherent moral attributes such as industriousness, temperance, and frugality. Attempts to reform society through state intervention or other means would, therefore, interfere with natural processes; unrestricted competition and defense of the status quo were in accord with biological selection. The poor were the 'unfit' and should not be aided; in the struggle for existence, wealth was a sign of success.

Spencer believed that we must permit and welcome such harshness to unleash the progressive development that all 'evolutionary' systems undergo if allowed to follow their natural course in an unimpeded manner. As a central principle of his system, Spencer believed that progress – defined by him as movement from a simple undifferentiated homogeneity, as in a bacterium or a 'primitive' human society without social classes, to complex and structured heterogeneity, as in 'advanced' organisms or industrial societies – did not arise as an inevitable property of matter in motion, but only through interaction between evolving systems and their environments. These interactions must therefore not be obstructed.

The relationship of Spencer's general vision to Darwin's particular theory has often been misconstrued or overemphasized. As stated above, Spencer had published the outline (and most of the details) of his system nearly ten years before Darwin presented his evolutionary theory. Spencer certainly did welcome the principle of natural selection as an even more ruthless and efficient mechanism for driving evolution forward. (Ironically, the word *evolution*, as a description for the genealogical history of life, entered our language through Spencer's urgings, not from Darwin. Spencer favored the term for its vernacular English meaning of 'progress,' in the original Latin sense of *evolutio*, or 'unfolding.' At first, Darwin resisted the term – he originally called his process 'descent with modification' – because his theory included no mechanism or rationale for general progress in the history of life. But Spencer prevailed, largely because no society has ever been more committed to progress as a central notion or goal than Victorian Britain at the height of its colonial and industrial expansion.)

Spencer certainly used Darwin's mechanism of natural selection to buttress his system. Few people recognize the following historical irony: Spencer, not Darwin, coined the term 'survival of the fittest,' now our conventional catch-phrase for Darwin's mechanism. Darwin himself paid proper tribute in a statement added to later editions of *The Origin of Species*: 'I have called this principle, by which each slight variation, if useful, is preserved, by the term Natural Selection . . . But the expression often used by Mr Herbert Spencer of the Survival of the Fittest is more accurate, and is sometimes equally convenient.'

As a mechanism for driving his universal 'evolution' (of stars, species, languages, economics, technologies, and nearly anything else) toward progress, Spencer preferred the direct and mechanistic 'root, hog, or die' of natural selection (as William Graham Sumner, the leading American social Darwinian, epitomized the process), to the vaguer and largely Lamarckian drive toward organic self-improvement that Spencer had originally favored as a primary cause. (In this colorful image, Sumner cited a quintessential American metaphor of self-sufficiency that my dictionary of catchphrases traces to a speech by Davy Crockett in 1834.) In a post-Darwinian edition of his *Social Statics*, Spencer wrote:

> The lapse of a third of a century since these passages were published, has brought me no reason for retreating from the position taken up in them. Contrariwise, it has brought a

> vast amount of evidence strengthening that position. The beneficial results of the survival of the fittest, prove to be immeasurably greater than [I formerly recognized]. The process of 'natural selection,' as Mr Darwin called it . . . has shown to be a chief cause . . . of that evolution through which all living things, beginning with the lower, and diverging and re-diverging as they evolved, have reached their present degrees of organization and adaptation to their modes of life.

But putting aside the question of Darwin's particular influence, the more important, underlying point remains firm: the theory of Social Darwinism (or social Spencerism) rests upon a set of analogies between the causes of change and stability in biological and social systems – and on the supposedly direct applicability of these biological principles to the social realm. In his founding document, the *Social Statics* of 1850, Spencer rests his case upon two elaborate analogies to biological systems.

1. The struggle for existence as purification in biology and society. Darwin recognized the 'struggle for existence' as metaphorical shorthand for any strategy that promotes increased reproductive success, whether by outright battle, cooperation, or just simple prowess in copulation under the old principle of 'early and often.' But many contemporaries, including Spencer, read 'survival of the fittest' only as overt struggle to the death – what T. H. Huxley later dismissed as the 'gladiatorial' school, or the incarnation of Hobbes's *bellum omnium contra omnes* (the war of all against all). Spencer presented this stark and limited view of nature in his *Social Statics*:

> Pervading all Nature we may see at work a stern discipline which is a little cruel that it may be very kind. That state of universal warfare maintained throughout the lower creation, to the great perplexity of many worthy people, is at bottom the most merciful provision which the circumstances admit of. . . . Note that carnivorous enemies, not only remove from herbivorous herds individuals past their prime, but also weed out the sickly, the malformed, and the least fleet or powerful. By the aid of which purifying process . . . all vitiation of the race through the multiplication of its inferior samples is

> prevented; and the maintenance of a constitution completely adapted to surrounding conditions, and therefore most productive of happiness, is ensured.

Spencer then compounds this error by applying the same argument to human social history, without ever questioning the validity of such analogical transfer. Railing against all governmental programs for social amelioration – Spencer opposed state-supported education, postal services, regulation of housing conditions, and even public construction of sanitary systems – he castigated such efforts as born of good intentions but doomed to dire consequences by enhancing the survival of social dregs who should be allowed to perish for the good of all. (Spencer insisted, however, that he did not oppose private charity, primarily for the salutary effect of such giving upon the moral development of donors. Does this discourse remind you of arguments now advanced as reformatory and spanking-new by our 'modern' ultraconservatives? Shall we not profit from Santayana's famous dictum that those ignorant of history must be condemned to repeat it?) In his chapter on poor laws (which he, of course, opposed), Spencer wrote in the *Social Statics*:

> We must call those spurious philanthropists who, to prevent present misery, would entail greater misery on future generations. That rigorous necessity which, when allowed to operate, becomes so sharp a spur to the lazy and so strong a bridle to the random, these paupers' friends would repeal, because of the wailings it here and there produces. Blind to the fact that under the natural order of things society is constantly excreting its unhealthy, imbecile, slow, vacillating, faithless members, these unthinking, though well-meaning, men advocate an interference which not only stops the purifying process, but even increases the vitiation – absolutely encouraging the multiplication of the reckless and incompetent by offering them an unfailing provision. . . . Thus, in their eagerness to prevent the salutary sufferings that surround us, these sigh-wise and groan-foolish people bequeath to posterity a continually increasing curse.

2. The stable body and the stable society. In the universal and progressive 'evolution' of all systems, organization becomes increasingly more

complex by division of labor among the growing number of differentiating parts. All parts must 'know their place' and play their appointed role, lest the entire system collapse. A primitive hydra, constructed of simple 'all purpose' modules, can regrow any lost part, but nature gives a man only one head, and one chance. Spencer recognized the basic inconsistency in validating social stability by analogy to the integrated needs of a single organic body – for he recognized the contrary rationales of the two systems: the parts of a body serve the totality, but the social totality (the state) supposedly exists only to serve the parts (individual people). But Spencer never allowed himself to be fazed by logical or empirical difficulties when pursuing such a lovely generality. (Huxley was speaking about Spencer's penchant for building grandiose systems when he made his famous remark about 'a beautiful theory, killed by a nasty, ugly little fact.') So Spencer barged right through the numerous absurdities of such a comparison, and even claimed that he had found a virtue in the differences. In his famous 1860 article, 'The Social Organism,' Spencer described the comparison between a human body and a human society: 'Such, then, are the points of analogy and the points of difference. May we not say that the points of difference serve but to bring into clearer light the points of analogy.'

Spencer's article then lists the supposed points of valid comparison, including such far-fetched analogies as the historical origin of a middle class to the development, in complex animals, of the mesoderm, or third body layer between the original ectoderm and endoderm; the likening of the ectoderm itself to the upper classes, for sensory organs that direct an animal arise in ectoderm, while organs of production, for such activities as digesting food, emerge from the lower layer, or endoderm; the comparison of blood and money; the parallel courses of nerve and blood vessels in higher animals with the side-by-side construction of railways and telegraph wires; and finally, in a comparison that even Spencer regarded as forced, the likening of a primitive all-powerful monarchy with a simple brain, and an advanced parliamentary system with a complex brain composed of several lobes. Spencer wrote: 'Strange as this assertion will be thought, our Houses of Parliament discharge in the social economy, functions that are in sundry respects comparable to those discharged by the cerebral masses in a vertebrate animal.'

Spencer surely forced his analogies, but his social intent could not have been more clear: a stable society requires that all roles be filled and well executed – and government must not interfere with a natural process of sorting out and allocation of appropriate rewards. A humble worker must toil, and may remain indigent forever, but the industrious poor, as an organ of the social body, must always be with us:

> Let the factory hands be put on short time, and immediately the colonial produce markets of London and Liverpool are depressed. The shopkeeper is busy or otherwise, according to the amount of the wheat crop. And a potato-blight may ruin dealers in consols. . . . This union of many men into one community – this increasing mutual dependence of units which were originally independent – this gradual segregation of citizens into separate bodies with reciprocally subservient functions – this formation of a whole consisting of unlike parts – this growth of an organism, of which one portion cannot be injured without the rest feeling it – may all be generalized under the law of individuation.

Social Darwinism grew into a major movement, with political, academic, and journalistic advocates for a wide array of particular causes. But as historian Richard Hofstadter stated in the most famous book ever written on this subject – *Social Darwinism in American Thought*, first published in 1944, in press ever since, and still full of insight despite some inevitable archaisms – the primary impact of this doctrine lay in its buttressing of conservative political philosophies, particularly through the central (and highly effective) argument against state support of social services and governmental regulation of industry and housing:

> One might, like William Graham Sumner, take a pessimistic view of the import of Darwinism, and conclude that Darwinism could serve only to cause men to face up to the inherent hardship of the battle of life; or one might, like Herbert Spencer, promise that, whatever the immediate hardships for a large portion of mankind, evolution meant progress and thus assured that the whole process of life was tending toward some very remote but altogether glorious consummation. But in either case the conclusions to which Darwinism

> was at first put were conservative conclusions. They suggested that all attempts to reform social processes were efforts to remedy the irremediable, that they interfered with the wisdom of nature, that they could lead only to degeneration.

The industrial magnates of America's gilded age ('robber barons,' in a terminology favored by many people) adored and promoted this argument against regulation, evidently for self-serving reasons, and however frequently they mixed their lines about nature's cruel inevitability with standard Christian piety. John D. Rockefeller stated in a Sunday school address:

> The growth of a large business is merely a survival of the fittest. . . . The American Beauty rose can be produced in the splendor and fragrance which bring cheer to its beholder only by sacrificing the early buds which grow up around it. This is not an evil tendency in business. It is merely the working-out of a law of nature and a law of God.

And Andrew Carnegie, who had been sorely distressed by the apparent failure of Christian values, found his solution in Spencer's writings, and then sought out the English philosopher for friendship and substantial favors. Carnegie wrote about his discovery of Spencer's work: 'I remember that light came as in a flood and all was clear. Not only had I got rid of theology and the supernatural, but I had found the truth of evolution. "All is well since all grows better" became my motto, and true source of comfort.' Carnegie's philanthropy, primarily to libraries and universities, ranks as one of the great charitable acts of American history, but we should not forget his ruthlessness and resistance to reforms for his own workers (particularly his violent breakup of the Homestead strike of 1892) in building his empire of steel – a harshness that he defended with the usual Spencerian line that any state regulation must derail an inexorable natural process eventually leading to progress for all. In his most famous essay (entitled 'Wealth,' and published in *North American Review* for 1889), Carnegie stated:

> While the law may be sometimes hard for the individual, it is best for the race, because it insures the survival of the fittest in every department. We accept and welcome, therefore, as

> conditions to which we must accommodate ourselves, great inequality of environment, the concentration of wealth, business, industrial and commercial, in the hands of a few, and the law of competition between these, as being not only beneficial, but essential for the future progress of the race.

I don't want to advocate a foolishly grandiose view about the social and political influence of academic arguments – and I also wish to avoid the common fallacy of inferring a causal connection from a correlation. Of course I do not believe that the claims of social Darwinism directly caused the ills of unrestrained industrial capitalism and the suppression of workers' rights. I know that most of these Spencerian lines functioned as mere window dressing for social forces well in place, and largely unmovable by any academic argument.

On the other hand, academic arguments should not be regarded as entirely impotent either – for why else would people in power invoke such claims so forcefully? The general thrust of social change unfolded in its own complex manner without much impact from purely intellectual rationales, but many particular issues – especially the actual rates and styles of changes that would have eventually occurred in any case – could be substantially affected by academic discourse. Millions of people suffered when a given reform experienced years of legislative delay, and then became vitiated in legal battles and compromises. The social Darwinian argument of the super-rich and the highly conservative did stem, weaken, and slow the tides of amelioration, particularly for workers' rights.

Most historians would agree that the single most effective doctrine of social Darwinism lay in Spencer's own centerpiece – the argument against state-enforced standards for industry, education, medicine, housing, public sanitation, and so on. Few Americans, even the robber barons, would go so far, but Spencerian dogma did become a powerful bludgeon against the regulation of industry to ensure better working conditions for laborers. On this particular point – the central recommendation of Spencer's system from the beginning – we may argue for a substantial effect of academic writing upon the actual path of history.

Armed with this perspective, we may return to the Triangle Shirtwaist fire, the deaths of 146 young workers, and the palpable influence of a doctrine that applied too much of the wrong version of Darwinism to human history. The battle for increased safety of workplaces, and healthier environments for workers, had been waged with intensity for several decades. The trade union movement put substantial priority upon these issues, and management had often reacted with intransigence, or even violence, citing their Spencerian rationale for the perpetuation of apparent cruelty. Government regulation of industry had become a major struggle of American political life – and the cause of benevolent state oversight had advanced from the Sherman Anti-Trust Act of 1890 to the numerous and crusading reforms of Theodore Roosevelt's presidency (1901–9). When the Triangle fire broke out in 1911, regulations for the health and safety of workers were so weak, and so unenforceable by tiny and underpaid staffs, that the company's managers – cynically and technically 'up to code' in their firetrap building – could pretty much impose whatever the weak and nascent labor union movement couldn't prevent.

If the standard legend were true – and the Triangle workers died because all the doors had been locked by cruel owners – then this heart-wrenching story might convey no moral beyond the personal guilt of management. But the loss of 146 lives occurred for much more complicated reasons, all united by the pathetic weakness of legal regulations for the health and safety of workers. And I do not doubt that the central thrust of social Darwinism – the argument that governmental regulation can only forestall a necessary and natural process – exerted a major impact in slowing the passage of laws that almost everyone today, even our archconservatives, regard as beneficial and humane. I accept that these regulations would eventually have been instituted even if Spencer had never been born – but life or death for the Triangle workers rode upon the 'detail' that forces of pure laissez-faire, buttressed by their Spencerian centerpiece, managed to delay some implementations to the 1920s, rather than acceding to the just demands of unions and social reformers in 1910.

One of the two Triangle stairways almost surely had been locked on that fateful day – although lawyers for company owners won acquittal of their clients on this issue, largely by using legal legerdemain to confuse, intimidate, and draw inconsistencies from young witnesses with poor command of English. Two years earlier, an important strike had begun at the Triangle company, and had spread to shirtwaist manufacturers throughout the city. The union won in most factories but not, ironically, at Triangle – where management held out, and compelled the return of workers without anything gained. Tensions remained high at

Triangle in 1911, and management had become particularly suspicious, even paranoid, about thefts. Therefore, at quitting time (when the fire erupted, and against weakly enforced laws for maintaining multiple active exits), managers had locked one of the doors to force all the women to exit by the Greene Street stairwell, where a supervisor could inspect every handbag to guard against thefts of shirtwaists.

But if the bosses broke a weak and unenforceable law in this instance, all other causes of death can be traced to managerial compliance with absurdly inadequate standards, largely kept so weak by active political resistance to legal regulation of work sites, buttressed by the argument of social Darwinism. Fire hoses could not pump above the sixth floor, but no law prevented the massing of workers into crowded floors above. No statute required fire drills or other forms of training for safety. In other cases, weak regulations were risibly inadequate, easy to flaunt, and basically unenforced in any case. For example, by law, each worker required 250 cubic feet of air space – a good rule to prevent crowding. But companies had managed to circumvent the intent of this law, and maintain their traditional (and dangerous) density of workers, by moving into large loft buildings with high ceilings and substantial irrelevant space that could be included in calculating the 250-cubic-foot minimum.

When the Asch Building opened in 1900, an inspector for the Buildings Department informed the architect that a third staircase should be provided. But the architect sought and received a variance, arguing that the single fire escape could count as the missing staircase required by law for structures with more than ten thousand square feet per floor. Moreover, the single fire escape – which buckled and fell during the fire, as a result of poor maintenance and the weight of too many workers trying to escape – led only to a glass skylight in a closed courtyard. The building inspector had also complained about this arrangement, and the architect had promised to make the necessary alterations. But no changes had been made, and the falling fire escape plunged right through the skylight, greatly increasing the death toll.

Two final quotations highlight the case for inadequate legal protection as a primary cause for the unconscionable death toll in the Triangle Shirtwaist fire (Leon Stein's excellent book, *The Triangle Fire*, J. B. Lippincott Company, 1962, served as my chief source for information about this event). Rose Safran, a survivor of the fire and supporter of the 1909 strike, said: 'If the union had won we would have been safe.

Two of our demands were for adequate fire escapes and for open doors from the factories to the street. But the bosses defeated us and we didn't get the open doors or the better fire escapes. So our friends are dead.' A building inspector who had actually written to the Triangle management just a few months before, asking for an appointment to discuss the initiation of fire drills, commented after the blaze: 'There are only two or three factories in the city where fire drills are in use. In some of them where I have installed the system myself, the owners have discontinued it. The neglect of factory owners in the matter of safety of their employees is absolutely criminal. One man whom I advised to install a fire drill replied to me: "Let 'em burn. They're a lot of cattle, anyway."'

The Triangle fire galvanized the workers' reform movement as never before. An empowered force, now irresistible, of labor organizers, social reformers, and liberal legislators pressed for stronger regulation under the theme of 'never again.' Hundreds of laws passed as a direct result of this belated agitation. But nothing could wash the blood of 146 workers from the sidewalks of New York.

This tale of two work sites – of a desk situated where Huxley debated Wilberforce, and an office built on a floor that burned during the Triangle Shirtwaist fire – has no end, for the story illustrates a theme of human intellectual life that must always be with us, however imbued with an obvious and uncontroversial solution. Extremes must usually be regarded as untenable, even dangerous places on complex and subtle continua. For the application of Darwinian theory to human history, Wilberforce's 'none' marks an error of equal magnitude with the 'all' of an extreme social Darwinism. In a larger sense, the evolution of a species like *Homo sapiens* should fill us with notions of glory for our odd mental uniqueness, and of deep humility for our status as a tiny and accidental twig on such a sturdy and luxuriantly branching tree of life. Glory *and* humility! Since we can't abandon either feeling for a unitary stance in the middle, we had best make sure that both attitudes *always* walk together, hand in hand, and secure in the wisdom of Ruth's promise to Naomi: 'Whither thou goest, I will go; and where thou lodgest, I will lodge.'

CARRIE BUCK'S DAUGHTER

THE Lord really put it on the line in his preface to that prototype of all prescription, the Ten Commandments:

> . . . for I, the Lord thy God, am a jealous God, visiting the iniquity of the fathers upon the children unto the third and fourth generation of them that hate me (Exodus. 20:5).

The terror of this statement lies in its patent unfairness – its promise to punish guiltless offspring for the misdeeds of their distant forebears.

A different form of guilt by genealogical association attempts to remove this stigma of injustice by denying a cherished premise of Western thought – human free will. If offspring are tainted not simply by the deeds of their parents but by a material form of evil transferred directly by biological inheritance, then 'the iniquity of the fathers' becomes a signal or warning for probable misbehavior of their sons. Thus Plato, while denying that children should suffer directly for the crimes of their parents, nonetheless defended the banishment of a personally guiltless man whose father, grandfather, and great-grandfather had all been condemned to death.

It is, perhaps, merely coincidental that both Jehovah and Plato chose three generations as their criterion for establishing different forms of guilt by association. Yet we maintain a strong folk, or vernacular, tradition for viewing triple occurrences as minimal evidence of regularity. Bad things, we are told, come in threes. Two may represent an accidental association; three is a pattern. Perhaps, then, we should not wonder that our own century's most famous pronouncement of blood guilt employed the same criterion – Oliver Wendell Holmes's defense of compulsory sterilization in Virginia (Supreme Court decision of 1927 in *Buck* v. *Bell*): 'three generations of imbeciles are enough.'

Restrictions upon immigration, with national quotas set to discriminate against those deemed mentally unfit by early versions of IQ testing, marked the greatest triumph of the American eugenics movement – the flawed hereditarian doctrine, so popular earlier in our century and by no means extinct today, that attempted to 'improve' our human stock by preventing the propagation of those deemed biologically unfit and encouraging procreation among the supposedly worthy. But the movement to enact and enforce laws for compulsory 'eugenic' sterilization had an impact and success scarcely less pronounced. If we could debar the shiftless and the stupid from our shores, we might also prevent the propagation of those similarly afflicted but already here.

The movement for compulsory sterilization began in earnest during the 1890s, abetted by two major factors – the rise of eugenics as an influential political movement and the perfection of safe and simple operations (vasectomy for men and salpingectomy, the cutting and tying of Fallopian tubes, for women) to replace castration and other socially unacceptable forms of mutilation. Indiana passed the first sterilization act based on eugenic principles in 1907 (a few states had previously mandated castration as a punitive measure for certain sexual crimes, although such laws were rarely enforced and usually overturned by judicial review). Like so many others to follow, it provided for sterilization of afflicted people residing in the state's 'care,' either as inmates of mental hospitals and homes for the feeble-minded or as inhabitants of prisons. Sterilization could be imposed upon those judged insane, idiotic, imbecilic, or moronic, and upon convicted rapists or criminals when recommended by a board of experts.

By the 1930s, more than thirty states had passed similar laws, often with an expanded list of so-called hereditary defects, including alcoholism and drug addiction in some states, and even blindness and deafness in others. These laws were continually challenged and rarely enforced in most states; only California and Virginia applied them zealously. By January 1935, some 20,000 forced 'eugenic' sterilizations had been performed in the United States, nearly half in California.

No organization crusaded more vociferously and successfully for these laws than the Eugenics Record Office, the semiofficial arm and repository of data for the eugenics movement in America. Harry Laughlin, superintendent of the Eugenics Record Office, dedicated most of his career to a tireless campaign of writing and lobbying for eugenic sterilization. He hoped, thereby, to eliminate in two generations the genes of what he called the 'submerged tenth' – 'the most worthless one-tenth of our present population.' He proposed a 'model sterilization law' in 1922, designed

> to prevent the procreation of persons socially inadequate from defective inheritance, by authorizing and providing for eugenical sterilization of certain potential parents carrying degenerate hereditary qualities.

This model bill became the prototype for most laws passed in America, although few states cast their net as widely as Laughlin advised. (Laughlin's categories encompassed 'blind, including those with seriously impaired vision; deaf, including those with seriously impaired hearing; and dependent, including orphans, ne'er-do-wells, the homeless, tramps, and paupers.') Laughlin's suggestions were better heeded in Nazi Germany, where his model act inspired the infamous and stringently enforced *Erbgesundheitsrecht*, leading by the eve of World War II to the sterilization of some 375,000 people, most for 'congenital feeblemindedness,' but including nearly 4,000 for blindness and deafness.

The campaign for forced eugenic sterilization in America reached its climax and height of respectability in 1927, when the Supreme Court, by an 8–1 vote, upheld the Virginia sterilization bill in *Buck* v. *Bell*. Oliver Wendell Holmes, then in his mid-eighties and the most celebrated jurist in America, wrote the majority opinion with his customary verve and power of style. It included the notorious paragraph, with its chilling tag line, cited ever since as the quintessential statement of eugenic principles. Remembering with pride his own distant experiences as an infantryman in the Civil War, Holmes wrote:

> We have seen more than once that the public welfare may call upon the best citizens for their lives. It would be strange if it could not call upon those who already sap the strength of the state for these lesser sacrifices. . . . It is better for all the world, if instead of waiting to execute degenerate offspring

> for crime, or to let them starve for their imbecility, society can prevent those who are manifestly unfit from continuing their kind. The principle that sustains compulsory vaccination is broad enough to cover cutting the Fallopian tubes. Three generations of imbeciles are enough.

Who, then, were the famous 'three generations of imbeciles,' and why should they still compel our interest?

When the state of Virginia passed its compulsory sterilization law in 1924, Carrie Buck, an eighteen-year-old white woman, lived as an involuntary resident at the State Colony for Epileptics and Feeble-Minded. As the first person selected for sterilization under the new act, Carrie Buck became the focus for a constitutional challenge launched, in part, by conservative Virginia Christians who held, according to eugenical 'modernists,' antiquated views about individual preferences and 'benevolent' state power. (Simplistic political labels do not apply in this case, and rarely in general for that matter. We usually regard eugenics as a conservative movement and its most vocal critics as members of the left. This alignment has generally held in our own decade. But eugenics, touted in its day as the latest in scientific modernism, attracted many liberals and numbered among its most vociferous critics groups often labeled as reactionary and antiscientific. If any political lesson emerges from these shifting allegiances, we might consider the true inalienability of certain human rights.)

But why was Carrie Buck in the State Colony and why was she selected? Oliver Wendell Holmes upheld her choice as judicious in the opening lines of his 1927 opinion:

> Carrie Buck is a feeble-minded white woman who was committed to the State Colony. . . . She is the daughter of a feeble-minded mother in the same institution, and the mother of an illegitimate feeble-minded child.

In short, inheritance stood as the crucial issue (indeed as the driving force behind all eugenics). For if measured mental deficiency arose from malnourishment, either of body or mind, and not from tainted genes, then how could sterilization be justified? If decent food, upbringing, medical care, and education might make a worthy citizen of Carrie Buck's daughter, how could the State of Virginia justify the severing of

Carrie's Fallopian tubes against her will? (Some forms of mental deficiency are passed by inheritance in family lines, but most are not – a scarcely surprising conclusion when we consider the thousand shocks that beset us all during our lives, from abnormalities in embryonic growth to traumas of birth, malnourishment, rejection, and poverty. In any case, no fair-minded person today would credit Laughlin's social criteria for the identification of hereditary deficiency – ne'er-do-wells, the homeless, tramps, and paupers – although we shall soon see that Carrie Buck was committed on these grounds.)

When Carrie Buck's case emerged as the crucial test of Virginia's law, the chief honchos of eugenics understood that the time had come to put up or shut up on the crucial issue of inheritance. Thus, the Eugenics Record Office sent Arthur H. Estabrook, their crack field-worker, to Virginia for a 'scientific' study of the case. Harry Laughlin himself provided a deposition, and his brief for inheritance was presented at the local trial that affirmed Virginia's law and later worked its way to the Supreme Court as *Buck* v. *Bell.*

Laughlin made two major points to the court. First, that Carrie Buck and her mother, Emma Buck, were feebleminded by the Stanford-Binet test of IQ, then in its own infancy. Carrie scored a mental age of nine years, Emma of seven years and eleven months. (These figures ranked them technically as 'imbeciles' by definitions of the day, hence Holmes's later choice of words – though his infamous line is often misquoted as 'three generations of idiots.' Imbeciles displayed a mental age of six to nine years; idiots performed worse, morons better, to round out the old nomenclature of mental deficiency.) Second, that most feeblemindedness resides ineluctably in the genes, and that Carrie Buck surely belonged with this majority. Laughlin reported:

> Generally feeble-mindedness is caused by the inheritance of degenerate qualities; but sometimes it might be caused by environmental factors which are not hereditary. In the case given, the evidence points strongly toward the feeble-mindedness and moral delinquency of Carrie Buck being due, primarily, to inheritance and not to environment.

Carrie Buck's daughter was then, and has always been, the pivotal figure of this painful case. I noted in beginning this essay that we tend (often at our peril) to regard two as potential accident and three as an established pattern. The supposed imbecility of Emma and Carrie might have been an unfortunate coincidence, but the diagnosis of similar deficiency for Vivian Buck (made by a social worker, as we shall see, when Vivian was but six months old) tipped the balance in Laughlin's favor and led Holmes to declare the Buck lineage inherently corrupt by deficient heredity. Vivian sealed the pattern – *three* generations of imbeciles are enough. Besides, had Carrie not given illegitimate birth to Vivian, the issue (in both senses) would never have emerged.

Oliver Wendell Holmes viewed his work with pride. The man so renowned for his principle of judicial restraint, who had proclaimed that freedom must not be curtailed without 'clear and present danger' – without the equivalent of falsely yelling 'fire' in a crowded theater – wrote of his judgment in *Buck* v. *Bell:* 'I felt that I was getting near the first principle of real reform.'

And so *Buck* v. *Bell* remained for fifty years, a footnote to a moment of American history perhaps best forgotten. Then, in 1980, it reemerged to prick our collective conscience, when Dr K. Ray Nelson, then director of the Lynchburg Hospital where Carrie Buck had been sterilized, researched the records of his institution and discovered that more than 4,000 sterilizations had been performed, the last as late as 1972. He also found Carrie Buck, alive and well near Charlottesville, and her sister Doris, covertly sterilized under the same law (she was told that her operation was for appendicitis), and now, with fierce dignity, dejected and bitter because she had wanted a child more than anything else in her life and had finally, in her old age, learned why she had never conceived.

As scholars and reporters visited Carrie Buck and her sister, what a few experts had known all along became abundantly clear to everyone. Carrie Buck was a woman of obviously normal intelligence. For example, Paul A. Lombardo of the School of Law at the University of Virginia, and a leading scholar of *Buck* v. *Bell,* wrote in a letter to me:

> As for Carrie, when I met her she was reading newspapers daily and joining a more literate friend to assist at regular bouts with the crossword puzzles. She was not a sophisticated woman, and lacked social graces, but mental health professionals who examined her in later life confirmed my impressions that she was neither mentally ill nor retarded.

On what evidence, then, was Carrie Buck consigned to the State Colony for Epileptics and Feeble-Minded on January 23, 1924? I have seen the text of her commitment hearing; it is, to say the least, cursory and contradictory. Beyond the bald and undocumented say-so of her foster parents, and her own brief appearance before a commission of two doctors and a justice of the peace, no evidence was presented. Even the crude and early Stanford-Binet test, so fatally flawed as a measure of innate worth (see my book *The Mismeasure of Man*, although the evidence of Carrie's own case suffices) but at least clothed with the aura of quantitative respectability, had not yet been applied.

When we understand why Carrie Buck was committed in January 1924, we can finally comprehend the hidden meaning of her case and its message for us today. The silent key, again as from the first, is her daughter Vivian, born on March 28, 1924, and then but an evident bump on her belly. Carrie Buck was one of several illegitimate children borne by her mother, Emma. She grew up with foster parents, J. T. and Alice Dobbs, and continued to live with them as an adult, helping out with chores around the house. She was raped by a relative of her foster parents, then blamed for the resulting pregnancy. Almost surely, she was (as they used to say) committed to hide her shame (and her rapist's identity), not because enlightened science had just discovered her true mental status. In short, she was sent away to have her baby. Her case never was about mental deficiency; Carrie Buck was persecuted for supposed sexual immorality and social deviance. The annals of her trial and hearing reek with the contempt of the well-off and well-bred for poor people of 'loose morals.' Who really cared whether Vivian was a baby of normal intelligence; she was the illegitimate child of an illegitimate woman. Two generations of bastards are enough. Harry Laughlin began his 'family history' of the Bucks by writing: 'These people belong to the shiftless, ignorant and worthless class of anti-social whites of the South.'

We know little of Emma Buck and her life, but we have no more reason to suspect her than her daughter Carrie of true mental deficiency. Their supposed deviance was social and sexual; the charge of imbecility was a cover-up, Mr Justice Holmes notwithstanding.

We come then to the crux of the case, Carrie's daughter, Vivian. What evidence was ever adduced for her mental deficiency? This and only this: At the original trial in late 1924, when Vivian Buck was seven months old, a Miss Wilhelm, social worker for the Red Cross, appeared before the court. She began by stating honestly the true reason for Carrie Buck's commitment:

> Mr. Dobbs, who had charge of the girl, had taken her when a small child, had reported to Miss Duke [the temporary secretary of Public Welfare for Albemarle County] that the girl was pregnant and that he wanted to have her committed somewhere – to have her sent to some institution.

Miss Wilhelm then rendered her judgment of Vivian Buck by comparing her with the normal granddaughter of Mrs Dobbs, born just three days earlier:

> It is difficult to judge probabilities of a child as young as that, but it seems to me not quite a normal baby. In its appearance – I should say that perhaps my knowledge of the mother may prejudice me in that regard, but I saw the child at the same time as Mrs Dobbs' daughter's baby, which is only three days older than this one, and there is a very decided difference in the development of the babies. That was about two weeks ago. There is a look about it that is not quite normal, but just what it is, I can't tell.

This short testimony, and nothing else, formed all the evidence for the crucial third generation of imbeciles. Cross-examination revealed that neither Vivian nor the Dobbs grandchild could walk or talk, and that 'Mrs Dobbs' daughter's baby is a very responsive baby. When you play with it or try to attract its attention – it is a baby that you can play with. The other baby is not. It seems very apathetic and not responsive.' Miss Wilhelm then urged Carrie Buck's sterilization: 'I think,' she said, 'it would at least prevent the propagation of her kind.' Several years later, Miss Wilhelm denied that she had ever examined Vivian or deemed the child feebleminded.

Unfortunately, Vivian died at age eight of 'enteric colitis' (as recorded on her death certificate), an ambiguous diagnosis that could mean many things but may well indicate that she fell victim to one of the preventable childhood diseases of poverty (a grim reminder of the real subject in *Buck* v. *Bell*). She is therefore mute as a witness in our reassessment of her famous case.

When *Buck* v. *Bell* resurfaced in 1980, it immediately struck me that Vivian's case was crucial and that evidence for the mental status of a child who died at age eight might best be found in report cards. I have therefore been trying to track down Vivian Buck's school records for the past four years and have finally succeeded. (They were supplied to me by Dr Paul A. Lombardo, who also sent other documents, including Miss Wilhelm's testimony, and spent several hours answering my questions by mail and Lord knows how much time playing successful detective re Vivian's school records. I have never met Dr Lombardo; he did all this work for kindness, collegiality, and love of the game of knowledge, not for expected reward or even requested acknowledgment. In a profession – academics – so often marred by pettiness and silly squabbling over meaningless priorities, this generosity must be recorded and celebrated as a sign of how things can and should be.)

Vivian Buck was adopted by the Dobbs family, who had raised (but later sent away) her mother, Carrie. As Vivian Alice Elaine Dobbs, she attended the Venable Public Elementary School of Charlottesville for four terms, from September 1930 until May 1932, a month before her death. She was a perfectly normal, quite average student, neither particularly outstanding nor much troubled. In those days before grade inflation, when C meant 'good, 81–87' (as defined on her report card) rather than barely scraping by, Vivian Dobbs received As and Bs for deportment and C's for all academic subjects but mathematics (which was always difficult for her, and where she scored D) during her first term in Grade 1A, from September 1930 to January 1931. She improved during her second term in 1B, meriting an A in deportment, C in mathematics, and B in all other academic subjects; she was placed on the honor roll in April 1931. Promoted to 2A, she had trouble during the fall term of 1931, failing mathematics and spelling but receiving A in deportment, B in reading, and C in writing and English. She was 'retained in 2A' for the next term – or 'left back' as we used to say, and scarcely a sign of imbecility as I remember all my buddies who suffered a similar fate. In any case, she again did well in her final term, with B in deportment, reading, and spelling, and C in writing, English, and mathematics during her last month in school. This daughter of 'lewd and immoral' women excelled in deportment and performed adequately, although not brilliantly, in her academic subjects.

In short, we can only agree with the conclusion that Dr Lombardo has reached in his research on *Buck* v. *Bell* – there were no imbeciles, not a one, among the three generations of Bucks. I don't know that such correction of cruel but forgotten errors of history counts for much, but I find it both symbolic and satisfying to learn that forced eugenic sterilization, a procedure of such dubious morality, earned its official justification (and won its most quoted line of rhetoric) on a patent falsehood.

Carrie Buck died last year. By a quirk of fate, and not by memory or design, she was buried just a few steps from her only daughter's grave. In the umpteenth and ultimate verse of a favorite old ballad, a rose and a brier – the sweet and the bitter – emerge from the tombs of Barbara Allen and her lover, twining about each other in the union of death. May Carrie and Vivian, victims in different ways and in the flower of youth, rest together in peace.

JUST IN THE MIDDLE

THE case for organic integrity was stated most forcefully by a poet, not a biologist. In his romantic paean *The Tables Turned*, William Wordsworth wrote:

> Sweet is the lore which nature brings;
> Our meddling intellect
> Misshapes the beauteous form of things:
> We murder to dissect.

The whiff of anti-intellectualism that pervades this poem has always disturbed me, much as I appreciate its defense of nature's unity. For it implies that any attempt at analysis, any striving to understand by breaking a complex system into constituent parts, is not only useless but even immoral.

Yet caricature and dismissal from the other side have been just as intense, if not usually stated with such felicity. Those scientists who study biological systems by breaking them down into ever smaller parts, until they reach the chemistry of molecules, often deride biologists who insist upon treating organisms as irreducible wholes. The two sides of this oversimplified dichotomy even have names, often invoked in a derogatory way by their opponents. The dissectors are 'mechanists' who believe that life is nothing more than the physics and chemistry of its component parts. The integrationists are 'vitalists' who hold that life and life alone has that 'special something,' forever beyond the reach of chemistry and physics and even incompatible with 'basic' science. In this reading you are, according to your adversaries, either a heartless mechanist or a mystical vitalist.

I have often been amused by our vulgar tendency to take complex issues, with solutions at neither extreme of a continuum of possibilities, and break them into dichotomies, assigning one group to one pole and the other to an opposite end, with no acknowledgment of subtleties and intermediate positions – and nearly always with moral opprobrium attached to opponents. As the wise Private Willis sings in Gilbert and Sullivan's *Iolanthe*:

> I often think it's comical
> How nature always does contrive
> That every boy and every gal
> That's born into the world alive
> Is either a little Liberal
> Or else a little Conservative!
> Fal la la!

The categories have changed today, but we are still either rightists or leftists, advocates of nuclear power or solar heating, pro choice or against the murder of fetuses. We are simply not allowed the subtlety of an intermediate view on intricate issues (although I suspect that the only truly important and complex debate with no possible stance in between is whether you are for or against the designated hitter rule – and I'm agin it).

Thus, the impression persists that biologists are either mechanists or vitalists, either advocates of an ultimate reduction to physics and chemistry (with no appreciation for the integrity of organisms) or supporters of a special force that gives life meaning (and modern mystics who would deny the potential unity of science). For example, a popular article on research at the Marine Biological Laboratory of Woods Hole (in the September–October 1983 issue of *Harvard Magazine*) discusses the work of a scientist with a physicist's approach to neurological problems:

> In the parlance of philosophers of science, [he] could be considered a 'reductionist' or 'mechanist.' He believes that fundamental laws of mechanics and electromagnetism suffice to account for all phenomena at this level. Vitalists, in contrast, maintain that some vital principle, some spark of life, separates living from nonliving matter. Thomas Hunt

> Morgan, a confirmed vitalist, once remarked acidly that scientists who compared living organisms to machines were like 'wild Indians who derailed trains and looked for the horses inside the locomotive.' Most mechanists, in turn, regard their opponents' vital principle as so much black magic.

But this dichotomy is an absurd caricature of the opinions held by most biologists. Although I have known a few mechanists, as defined in this article, I don't think that I have ever met a vitalist (although the argument did enjoy some popularity during the nineteenth century). The vast majority of biologists, including the great geneticist T. H. Morgan (who was as antivitalist as any scientist of our century), advocate a middle position. The extremes may make good copy, and a convenient (if simplistic) theme for discussion, but they are occupied by few, if any, practicing scientists. If we can understand this middle position, and grasp why it has been so persistently popular, perhaps we can begin to criticize our lamentable tendency to dichotomize complex issues in the first place. I therefore devote this essay to defining and supporting this middle way by showing how a fine American biologist, Ernest Everett Just, developed and defended it in the course of his own biological research.

The middle position holds that life, as a result of its structural and functional complexity, cannot be taken apart into chemical constituents and explained in its entirety by physical and chemical laws working at the molecular level. But the middle way denies just as strenuously that this failure of reductionism records any mystical property of life, any special 'spark' that inheres in life alone. Life acquires its own principles from the hierarchical structure of nature. As levels of complexity mount along the hierarchy of atom, molecule, gene, cell, tissue, organism, and population, new properties arise as results of interactions and interconnections emerging at each new level. A higher level cannot be fully explained by taking it apart into component elements and rendering their properties in the absence of these interactions. Thus, we need new, or 'emergent,' principles to encompass life's complexity; these principles are additional to, and consistent with, the physics and chemistry of atoms and molecules.

This middle way may be designated 'organizational,' or 'holistic'; it represents the stance adopted by most biologists and even by most physical scientists who have thought hard about biology and directly experienced its complexity. It was, for example, espoused in what may be our century's most famous book on 'what is life?' – the short masterpiece of the same title written in 1944 by Erwin Schrödinger, the great quantum physicist who turned to biological problems at the end of his career. Schrödinger wrote:

> From all we have learnt about the structure of living matter, we must be prepared to find it working in a manner that cannot be reduced to the ordinary laws of physics. And that not on the ground that there is any 'new force' or what not, directing the behavior of the single atoms within a living organism, but because the construction is different from anything we have yet tested in the physical laboratory.

Schrödinger then presents a striking analogy. Compare the ordinary physicist to an engineer familiar only with the operation of steam engines. When this engineer encounters, for the first time, a more complicated electric motor, he will not assume that it works by intrinsically mysterious laws just because he cannot understand it with the principles appropriate to steam engines: 'He will not suspect that an electric motor is driven by a ghost because it is set spinning by the turn of a switch, without boiler and steam.'

Ernest Everett Just, a thoughtful embryologist who developed a similar holistic attitude as a direct consequence of his own research, was born 100 years ago in Charleston, South Carolina.* He graduated as valedictorian of Dartmouth in 1907 and did most of his research at the Marine Biological Laboratory of Woods Hole during the 1920s. He continued his work at various European biological laboratories during the 1930s, and was briefly interned by the Nazis when France fell in 1940. Repatriated to the United States, and broken in spirit, he died of pancreatic cancer in 1941 at age fifty-eight.

Just began as an experimentalist, studying problems of fertilization at the cellular level, and in the great tradition of careful, descriptive research so characteristic of the 'Woods Hole school.' As this work developed, and particularly after he left for Europe, his career entered a new phase: he became fascinated with the biology of cell surfaces. This shift emerged directly from his interest in fertilization and his

* I wrote this essay in 1983, for the centenary of Just's birth.

particular concern with an old problem: How does the sperm penetrate an egg's outer membrane, and how does the egg's surface then react in physical and chemical terms? At the same time, Just's work took on a more philosophical tone (although he never abandoned his experiments), and he slowly developed a holistic, or organizational, perspective midway between the caricatured extremes of classical mechanism and vitalism. Just expounded this biological philosophy, a direct result of his growing concern with the properties of cell surfaces considered as wholes, in *The Biology of the Cell Surface*, published in 1939.

Just's early work on fertilization was a harbinger of things to come. He was not particularly interested in how the genetic material of egg and sperm fuse and then direct the subsequent architecture of development – a classical theme of the reductionist tradition (an attempt to explain the properties of embryology in terms of genes housed in a controlling nucleus). He was more concerned with the effects that fertilization imposes upon the entire cell, particularly its surface, and on the interaction of nucleus and cytoplasm in subsequent cell division and differentiation of the embryo.

Just had an uncanny knack for devising simple and elegant experiments that spoke to the primary theoretical issues of his day. In his very first paper, for example, he showed that, for some species of marine invertebrates at least, the sperm's point of entry determines the plane of first cleavage (the initial division of the fertilized egg into two cells). He also proved that the egg's surface is 'equipotential' – that is, the sperm has an equal probability of entering at any point. At this time, biologists were pursuing a vigorous debate (here we go with dichotomies again) between preformationists who held that an embryo's differentiation into specialized parts and organs was already prefigured in the structure of an unfertilized egg, and epigeneticists who argued that differentiation arose during development from an egg initially able to form any subsequent structure from any of its regions.

By showing that the direction of cleavage followed the happenstance of a sperm's penetration (and that a sperm could enter anywhere on the egg's surface), Just supported the epigenetic alternative. This first paper already contains the basis for Just's later and explicit holism – his concern with properties of entire organisms (the egg's complete surface) and with interactions of organism and environment (the epigenetic character of development contrasted with the preformationist view that pathways of later development lie within the egg's structure).

I believe that Just's mature holism had two primary sources in his earlier experimental work on fertilization. First, Just distinguished himself at Woods Hole as the great 'green thumb' of his generation. He was a stickler for proper procedure and cleanliness in the laboratory. He had an uncanny rapport with the various species of marine invertebrates that inhabit the waters about Woods Hole. He knew where to find them and he understood their habits intimately. He could extract eggs and keep them normal and healthy under laboratory conditions. He became the chief source of technical advice for hotshot young researchers who had mastered all the latest techniques of experimentation but knew little natural history.

Just therefore understood better than anyone else the importance of healthy normality in eggs used for experiments on fertilization – the integrity of whole cells in their ordinary conditions of life could not be compromised. Over and over again, he showed how many famous experiments by eminent scientists had no validity because they used moribund or abnormal cells and their results could be traced to these 'unlifelike' conditions, not to the experimental intervention itself. For example, Just refuted an important set of experiments on abnormalities of development produced when eggs are fertilized by sperm of another species. He proved that the peculiar patterns of embryology must be traced, not to the foreign sperm itself, but to the moribund state of eggs produced by environmental conditions (of temperature and water chemistry) necessary to induce the abnormal fertilization, but uncongenial for the eggs' good health.

Just derided the lack of concern for natural history shown by so many experimenters who knew all the latest about fancy physics and chemistry, but ever so little about organisms. They referred to their eggs and sperm as 'material' (I have the same reaction to modern reductionists who call the living cells and organs of their experiments a 'preparation') and accepted their experimental objects in any condition because they couldn't distinguish normality from abnormality: 'If the condition of the eggs is not taken into account,' Just wrote, 'the results obtained by the use of sub-normal eggs in experiments may be due wholly or in part to the poor physiological condition of the eggs.'

Second, and more important, Just's twenty years of research on fertilization led directly, almost inexorably, to his interest in the cell's surface and to his holistic philosophy. Since his work, as previously mentioned, centered upon the changes that cell surfaces undergo during fertilization,

Just soon realized that the cell's surface was no simple, passive boundary, but a complex and essential part of cellular organization:

> The surface-cytoplasm cannot be thought of as inert or apart from the living cell-substance. The ectoplasm [Just's name for the surface material] is more than a barrier to stem the rising tide within the active cell-substance; it is more than a dam against the outside world. It is a living mobile part of the cell.

Later, pursuing a common concern of holistic biology, Just emphasized that the cell surface, as the domain of communication between organism and environment, embodies the theme of interaction – an organizational complexity that cannot be reduced to chemical parts:

> It is keyed to the outside world as no other part of the cell. It stands guard over the peculiar form of the living substance, is buffer against the attacks of the surroundings and the means of communication with it.

Moreover, as his major experimental contribution, Just showed that the cell surface responded to fertilization as a continuous and indivisible entity, even though the sperm only entered at a single point. If the surface has such integrity, and if it regulates so many cellular processes, how can we meaningfully interpret the functions of cells by breaking them apart into molecular components?

> Under the impact of a spermatozoon the egg-surface first gives way and then rebounds; the egg-membrane moves in and out beneath the actively moving spermatozoon for a second or two. Then suddenly the spermatozoon becomes motionless with its tip buried in a slight indentation of the egg-surface, at which point the ectoplasm develops a cloudy appearance. The turbidity spreads from here so that at twenty seconds after insemination – the mixing of eggs and spermatozoa – the whole ectoplasm is cloudy. Now like a flash, beginning at the point of sperm-attachment, a wave sweeps over the surface of the egg, clearing up the ectoplasm as it passes.

As his work progressed, Just claimed more and more importance for the cellular surface, eventually going too far. He wisely denied the reductionistic premise that all cellular features are passive products of directing genes in the nucleus, but his alternative view of ectoplasmic control over nuclear motions cannot be supported either. Moreover, his argument that the history of life records an increasing dominance of ectoplasm, since nerve cells are most richly endowed with surface material, and since brain size increases continually in evolution, reflects the common misconception that evolution inevitably yields progress measured by mental advance as a primary criterion. The following passage may reflect Just's literary skill, but it stands as confusing metaphor, not enlightening biology:

> Our minds encompass planetary movements, mark out geological eras, resolve matter into its constituent electrons, because our mentality is the transcendental expression of the age-old integration between ectoplasm and nonliving world.

Finally, Just's work also suffered because he had the misfortune to pursue his research and publish his book just before the invention of the electron microscope. The cell surface is too thin for light microscopy to resolve, and Just could never fathom its structure. He was forced to work from inferences based upon transient changes of the cell surface during fertilization – and he succeeded brilliantly in the face of these limitations. But within a decade of his death, much of his painstaking work had been rendered obsolete.

Thus, Just fell into obscurity partly because he claimed too much and alienated his colleagues, and partly because he knew too little as a consequence of limited techniques available to him. Yet the current invisibility of Just's biology seems unfair for two reasons. First, he was basically right about the integrity and importance of cell surfaces. With electron microscopy, we have now resolved the membrane's structure – a complex and fascinating story worth an essay in its own right. Moreover, we accept Just's premise that the surface is no mere passive barrier, but an active and essential component of cellular structure. The most popular college text in biology (Keeton's *Biological Science*) proclaims:

> The cell membrane not only serves as an envelope that gives mechanical strength and shape and some protection to the cell. It is also an active component of the living cell, preventing

> some substances from entering it and others from leaking out. It regulates the traffic in materials between the precisely ordered interior of the cell and the essentially unfavorable and potentially disruptive outer environment. All substances moving between the cell's environment and the cellular interior in either direction must pass through a membrane barrier.

Second, and more important for this essay, whatever the factual status of Just's views on cell surfaces, he used his ideas to develop a holistic philosophy that represents a sensible middle way between extremes of mechanism and vitalism – a wise philosophy that may continue to guide us today.

We may epitomize Just's holism, and identify it as a genuine solution to the mechanist-vitalist debate, by summarizing its three major premises. First, nothing in biology contradicts the laws of physics and chemistry; any adequate biology must conform with the 'basic' sciences. Just began his book with these words:

> Living things have material composition, are made up finally of units, molecules, atoms, and electrons, as surely as any nonliving matter. Like all forms in nature they have chemical structure and physical properties, are physico-chemical systems. As such they obey the laws of physics and chemistry. Would one deny this fact, one would thereby deny the possibility of any scientific investigation of living things.

Second, the principles of physics and chemistry are not sufficient to explain complex biological objects because new properties emerge as a result of organization and interaction. These properties can only be understood by the direct study of whole, living systems in their normal state. Just wrote in a 1933 article:

> We have often striven to prove life as wholly mechanistic, starting with the hypothesis that organisms are machines! Thus we overlook the organo-dynamics of protoplasm – its power to organize itself. Living substance is such because it possesses this organization – something more than the sum of its minutest parts. . . . It is . . . the organization of protoplasm, which is its predominant characteristic and which

> places biology in a category quite apart from physics and chemistry. . . . Nor is it barren vitalism to say that there is something remaining in the behavior of protoplasm which our physico-chemical studies leave unexplained. This 'something' is the peculiar organization of protoplasm.

In striking metaphor, Just illustrates the inadequacy of mechanistic studies:

> The living thing disappears and only a mere agglomerate of parts remains. The better this analysis proceeds and the greater its yield, the more completely does life vanish from the investigated living matter. The state of being alive is like a snowflake on a windowpane which disappears under the warm touch of an inquisitive child. . . . Few investigators, nowadays, I think, subscribe to the naïve but seriously meant comparison once made by an eminent authority in biology, namely that the experimenter on an egg seeks to know its development by wrecking it, as one wrecks a train for understanding its mechanism. . . . The days of experimental embryology as a punitive expedition against the egg, let us hope, have passed.

Third, the insufficiency of physics and chemistry to encompass life records no mystical addition, no contradiction to the basic sciences, but only reflects the hierarchy of natural objects and the principle of emergent properties at higher levels of organization:

> The direct analysis of the state of being alive must never go below the order of organization which characterizes life; it must confine itself to the combination of compounds in the life-unit, never descending to single compounds and, therefore, certainly never below these. . . . The physicist aims at the least, the indivisible, particle of matter. The study of the state of being alive is confined to that organization which is peculiar to it.

Finally, I must emphasize once again that Just's arguments are not unique or even unusual. They represent the standard opinion of most practicing biologists and, as such, refute the dichotomous scheme that sees biology as a war between vitalists and mechanists. The middle way is both eminently sensible and popular. I chose Just as an illustration because his career exemplifies how a thoughtful biologist can be driven to such a position by his own investigation of complex phenomena. In addition, Just said it all so well and so forcefully; he qualifies as an exemplar of the middle way under our most venerable criterion – 'what oft was thought, but ne'er so well expressed.'

This essay should end here. In a world of decency and simple justice it would. But it cannot. E. E. Just struggled all his life for judgment by the intrinsic merit of his biological research alone – something I have tried so uselessly (and posthumously) to grant him here. He never achieved this recognition, never came close, for one intrinsically biological reason that should not matter, but always has in America. E. E. Just was black.

Today, a black valedictorian at a major Ivy League school would be inundated with opportunity. Just secured no mobility at all in 1907. As his biographer, M.I.T. historian of science Kenneth R. Manning, writes: 'An educated black had two options, both limited: he could either teach or preach – and only among blacks.' (Manning's biography, *Black Apollo of Science: The Life of Ernest Everett Just*, was published in 1983 by Oxford University Press. It is a superbly written and documented book, the finest biography I have read in years. Manning's book is an institutional history of Just's life. It discusses his endless struggle for funding and his complex relationships with institutions of teaching and research, but says relatively little about his biological work per se – a gap that I have tried, in some respects, to fill with this essay.)

So Just went to Howard and remained there all his life. Howard was a prestigious school, but it maintained no graduate program, and crushing demands for teaching and administration left Just neither time nor opportunity for the research career that he so ardently desired. But Just would not be beaten. By assiduous and tireless self-promotion, he sought support from every philanthropy and fund that might sponsor a black biologist – and he succeeded relatively well. He garnered enough support to spend long summers at Woods Hole, and managed to publish more than seventy papers and two books in what could never be more than a part-time research career studded with innumerable obstacles, both overt and psychological.

But eventually, the explicit racism of his detractors and, even worse, the persistent paternalism of his supporters, wore Just down. He dared not even hope for a permanent job at any white institution that might foster research, and the accumulation of slights and slurs at Woods Hole eventually made life intolerable for a proud man like Just. If he had fit the mold of an acceptable black scientist, he might have survived in the hypocritical world of white liberalism in his time. A man like George Washington Carver, who upheld Booker T. Washington's doctrine of slow and humble self-help for blacks, who dressed in his agricultural work clothes, and who spent his life in the practical task of helping black farmers find more uses for peanuts, was paraded as a paragon of proper black science. But Just preferred fancy suits, good wines, classical music, and women of all colors. He wished to pursue theoretical research at the highest levels of abstraction, and he succeeded with distinction. If his work disagreed with the theories of eminent white scientists, he said so, and with force (although his general demeanor tended toward modesty).

The one thing that Just so desperately wanted above all else – to be judged on the merit of his research alone – he could never have. His strongest supporters treated him with what, in retrospect, can only be labeled a crushing paternalism. Forget your research, de-emphasize it, go slower, they all said. Go back to Howard and be a 'model for your race'; give up personal goals and devote your life to training black doctors. Would such an issue ever have arisen for a white man of Just's evident talent?

Eventually, like many other black intellectuals, Just exiled himself to Europe. There, in the 1930s he finally found what he had sought – simple acceptance for his excellence as a scientist. But his joy and productivity were short-lived, as the specter of Nazism soon turned to reality and sent him back home to Howard and an early death.

Just was a brilliant man, and his life embodied strong elements of tragedy, but we must not depict him as a cardboard hero. He was far too fascinating, complex, and ambiguous a man for such simplistic misconstruction. Deeply conservative and more than a little elitist in character, Just never identified his suffering with the lot of blacks in general and considered each rebuff as a personal slight. His anger became so deep, and his joy at European acceptance so great, that he completely misunderstood Italian politics of the 1930s and became a supporter of Mussolini. He even sought research funds directly from Il Duce.

Yet how can we dare to judge a man so thwarted in the land of his birth? Yes, Just fared far better than most blacks. He had a good job and reasonable economic security. But, truly, we do not live by bread alone. Just was robbed of an intellectual's birthright – the desire to be taken seriously for his ideas and accomplishments. I know, in the most direct and personal way, the joy and the need for research. No fire burns more deeply within me, and no scientist of merit and accomplishment feels any differently. (One of my most eminent colleagues once told me that he regarded research as the greatest joy of all, for it was like continual orgasm.) Just's suffering may have been subtle compared with the brutalization of so many black lives in America, but it was deep, pervasive, and soul destroying. The man who understood holism so well in biology was not allowed to live a complete life. We may at least mark his centenary by considering the ideas that he struggled to develop and presented so well.

PART VIII

RELIGION

THE reasons for Steve's increasing interest in religion, and specifically in Christianity, may lie in aspects of his personal and psychological development, or a dissatisfaction with the explanatory power of science, to which we have no privileged access. But a more direct explanation could lie in the peculiarly American fights over the incompatibility of evolutionary theory with creationism, and its place in the school curriculum, which so astonishes most European observers. The battle with fundamentalist Protestant creationists engaged and still engages many prominent biologists in the United States – a fight in which Steve played a full part (see 'Trouble in Our Own House'). The clear disengagement of the Catholic Church from this battle, indicated in successive papal encyclicals, seems to have encouraged Steve to attempt a more conciliatory approach in his turn, with his doctrine of NOMA – non-overlapping magisteria – with which we begin this section. However, this approach to reconciliation was never to be allowed to override his knowledge, especially as a Jew, of the darker history of the Church, as portrayed in 'The Diet of Worms'.

But his primary focus, inevitably, because it challenged his core values and scientific principles was on the problem of creationism. It could be treated with contempt, as in 'Darwin and the Munchkins of Kansas', or by reasoned argument, as in 'Hooking Leviathan by Its Past,' our final choice in this edited collection. The struggle for rationality is one of Gould's legacies to his successors, and in the context of creationism it is a fitting tribute to him that after his death so many hundreds of Stevens, Stephens, Steves, Stefans and Stephanies should have signed up to a statement defending the fact of evolution and the theory of natural selection against the forces of obscurantism.

NON-OVERLAPPING MAGISTERIA

INCONGRUOUS places often inspire anomalous stories. In early 1984, I spent several nights at the Vatican housed in a hotel built for itinerant priests. While pondering over such puzzling issues as the intended function of the bidet in each bathroom, and hungering for something more than plum jam on my breakfast rolls (why did the basket only contain hundreds of identical plum packets and not one of, say, strawberry?), I encountered yet another among the innumerable issues of contrasting cultures that can make life so expansive and interesting. Our crowd (present in Rome to attend a meeting on nuclear winter, sponsored by the Pontifical Academy of Sciences) shared the hotel with a group of French and Italian Jesuit priests who were also professional scientists. One day at lunch, the priests called me over to their table to pose a problem that had been troubling them. What, they wanted to know, was going on in America with all this talk about 'scientific creationism'? One of the priests asked me: 'Is evolution really in some kind of trouble; and, if so, what could such trouble be? I have always been taught that no doctrinal conflict exists between evolution and Catholic faith, and the evidence for evolution seems both utterly satisfying and entirely overwhelming. Have I missed something?'

A lively pastiche of French, Italian, and English conversation then ensued for half an hour or so, but the priests all seemed reassured by my general answer – 'Evolution has encountered no intellectual trouble; no new arguments have been offered. Creationism is a home-grown phenomenon of American sociocultural history – a splinter movement (unfortunately rather more of a beam these days) of Protestant fundamentalists who believe that every word of the Bible must be literally true, whatever such a claim might mean.' We all left satisfied, but I certainly felt bemused by the anomaly of my role as a Jewish agnostic, trying to reassure a group of priests that evolution remained both true and entirely consistent with religious belief.

Another story in the same mold: I am often asked whether I ever encounter creationism as a live issue among my Harvard undergraduate students. I reply that only once, in thirty years of teaching, did I experience such an incident. A very sincere and serious freshman student came to my office with a question that had clearly been troubling him deeply. He said to me, 'I am a devout Christian and have never had any reason to doubt evolution, an idea that seems both exciting and well documented. But my roommate, a proselytizing evangelical, has been insisting with enormous vigor that I cannot be both a real Christian and an evolutionist. So tell me, can a person believe both in God and in evolution?' Again, I gulped hard, did my intellectual duty, and reassured him that evolution was both true and entirely compatible with Christian belief – a position that I hold sincerely, but still an odd situation for a Jewish agnostic.

These two stories illustrate a cardinal point, frequently unrecognized but absolutely central to any understanding of the status and impact of the politically potent, fundamentalist doctrine known by its self-proclaimed oxymoron as 'scientific creationism' – the claim that the Bible is literally true, that all organisms were created during six days of twenty-four hours, that the earth is only a few thousand years old, and that evolution must therefore be false. Creationism does not pit science against religion (as my opening stories indicate), for no such conflict exists. Creationism does not raise any unsettled intellectual issues about the nature of biology or the history of life. Creationism is a local and parochial movement, powerful only in the United States among Western nations, and prevalent only among the few sectors of American Protestantism that choose to read the Bible as an inerrant document, literally true in every jot and tittle.

I do not doubt that one could find an occasional nun who would prefer to teach creationism in her parochial school biology class, or an occasional rabbi who does the same in his yeshiva, but creationism based on biblical literalism makes little sense either to Catholics or Jews, for neither religion maintains any extensive tradition for reading the Bible as literal truth, rather than illuminating literature based partly on metaphor and allegory (essential components of all good writing), and demanding interpretation for proper understanding. Most Protestant groups, of course, take the same position – the fundamentalist fringe notwithstanding.

The argument that I have just outlined by personal stories and general statements represents the standard attitude of all major Western religions (and of Western science) today. (I cannot, through ignorance, speak of Eastern religions, though I suspect that the same position would prevail in most cases.) The *lack of conflict* between science and religion arises from a *lack of overlap* between their respective domains of professional expertise – science in the empirical constitution of the universe, and religion in the search for proper ethical values and the spiritual meaning of our lives. The attainment of wisdom in a full life requires extensive attention to both domains – for a great book tells us both that the truth can make us free, and that we will live in optimal harmony with our fellows when we learn to do justly, love mercy, and walk humbly.

In the context of this 'standard' position, I was enormously puzzled by a statement issued by Pope John Paul II on October 22, 1996, to the Pontifical Academy of Sciences, the same body that had sponsored my earlier trip to the Vatican. In this document, titled 'Truth Cannot Contradict Truth,' the Pope defended both the evidence for evolution and the consistency of the theory with Catholic religious doctrine. Newspapers throughout the world responded with front-page headlines, as in the *New York Times* for October 25: 'Pope Bolsters Church's Support for Scientific View of Evolution.'

Now I know about 'slow news days,' and I do allow that nothing else was strongly competing for headlines at that particular moment. Still, I couldn't help feeling immensely puzzled by all the attention paid to the Pope's statement (while being wryly pleased, of course, for we need all the good press we can get, especially from respected outside sources). The Catholic Church does not oppose evolution, and has no reason to do so. Why had the Pope issued such a statement at all? And why had the press responded with an orgy of worldwide front-page coverage?

I could only conclude at first, and wrongly as I soon learned, that journalists throughout the world must deeply misunderstand the relationship between science and religion, and must therefore be elevating a minor papal comment to unwarranted notice. Perhaps most people really do think that a war exists between science and religion, and that evolution cannot be squared with a belief in God. In such a context, a papal admission of evolution's legitimate status might be regarded as major news indeed – a sort of modern equivalent for a story that never happened, but would have made the biggest journalistic splash of 1640: Pope Urban VIII releases his most famous prisoner from house arrest and humbly apologizes: 'Sorry, Signor Galileo . . . the sun, er, is central.'

But I then discovered that such prominent coverage of papal satisfaction with evolution had not been an error of non-Catholic anglophone journalists. The Vatican itself had issued the statement as a major news release. And Italian newspapers had featured, if anything, even bigger headlines and longer stories. The conservative *Il Giornale*, for example, shouted from its masthead: 'Pope Says We May Descend from Monkeys.'

Clearly, I was out to lunch; something novel or surprising must lurk within the papal statement, but what could be causing all the fuss? – especially given the accuracy of my primary impression (as I later verified) that the Catholic Church values scientific study, views science as no threat to religion in general or Catholic doctrine in particular, and has long accepted both the legitimacy of evolution as a field of study and the potential harmony of evolutionary conclusions with Catholic faith.

As a former constituent of Tip O'Neill, I certainly know that 'all politics is local' – and that the Vatican undoubtedly has its own internal reasons, quite opaque to me, for announcing papal support of evolution in a major statement. Still, I reasoned that I must be missing some important key, and I felt quite frustrated. I then remembered the primary rule of intellectual life: When puzzled, it never hurts to read the primary documents – a rather simple and self-evident principle that has, nonetheless, completely disappeared from large sectors of the American experience.

I knew that Pope Pius XII (not one of my favorite figures in twentieth-century history, to say the least) had made the primary statement in a 1950 encyclical entitled *Humani Generis*. I knew the main thrust of his message: Catholics could believe whatever science determined about the evolution of the human body, so long as they accepted that, at some time of his choosing, God had infused the soul into such a creature. I also knew that I had no problem with this argument – for, whatever my private beliefs about souls, science cannot touch such a subject and therefore cannot be threatened by any theological position on such a legitimately and intrinsically religious issue. Pope Pius XII, in other words, had properly acknowledged and respected the separate domains of science and theology. Thus, I found myself in total agreement with *Humani Generis* – but I had never read the document in full (not much of an impediment to stating an opinion these days).

I quickly got the relevant writings from, of all places, the Internet. (The Pope is prominently online, but a Luddite like me is not. So I got a cyberwise associate to dredge up the documents. I do love the fracture of stereotypes implied by finding religion so hep and a scientist so square.) Having now read in full both Pope Pius's *Humani Generis* of 1950 and Pope John Paul's proclamation of October 1996, I finally understand why the recent statement seems so new, revealing, and worthy of all those headlines. And the message could not be more welcome for evolutionists, and friends of both science and religion.

The text of *Humani Generis* focuses on the *Magisterium* (or Teaching Authority) of the Church – a word derived not from any concept of majesty or unquestionable awe, but from the different notion of teaching, for *magister* means 'teacher' in Latin. We may, I think, adopt this word and concept to express the central point of this essay and the principled resolution of supposed 'conflict' or 'warfare' between science and religion. No such conflict should exist because each subject has a legitimate magisterium, or domain of teaching authority – and these magisteria do not overlap (the principle that I would like to designate as NOMA, or 'non-overlapping magisteria'). The net of science covers the empirical realm: what is the universe made of (fact) and why does it work this way (theory). The net of religion extends over questions of moral meaning and value. These two magisteria do not overlap, nor do they encompass all inquiry (consider, for starters, the magisterium of art and the meaning of beauty). To cite the usual clichés, we get the age of rocks, and religion retains the rock of ages; we study how the heavens go, and they determine how to go to heaven.

This resolution might remain entirely neat and clean if the non-overlapping magisteria of science and religion stood far apart, separated by an extensive no-man's-land. But, in fact, the two magisteria bump right up against each other, interdigitating in wondrously complex ways along their joint border. Many of our deepest questions call upon aspects of both magisteria for different parts of a full answer – and the sorting of legitimate domains can become quite complex and difficult. To cite just two broad questions involving both evolutionary facts and moral arguments: Since evolution made us the only earthly creatures with advanced consciousness, what responsibilities are so entailed for our relations with other species? What do our genealogical ties with other organisms imply about the meaning of human life?

Pius XII's *Humani Generis* (1950), a highly traditionalist document written by a deeply conservative man, faces all the 'isms' and cynicisms that rode the wake of World War II and informed the struggle to rebuild human decency from the ashes of the Holocaust. The encyclical bears the subtitle 'concerning some false opinions which threaten to undermine the foundations of Catholic doctrine,' and begins with a statement of embattlement:

> Disagreement and error among men on moral and religious matters have always been a cause of profound sorrow to all good men, but above all to the true and loyal sons of the Church, especially today, when we see the principles of Christian culture being attacked on all sides.

Pius lashes out, in turn, at various external enemies of the Church: pantheism, existentialism, dialectical materialism, historicism, and, of course and preeminently, communism. He then notes with sadness that some well-meaning folks within the Church have fallen into a dangerous relativism – 'a theological pacifism and egalitarianism, in which all points of view become equally valid' – in order to include those who yearn for the embrace of Christian religion, but do not wish to accept the particularly Catholic magisterium.

Speaking as a conservative's conservative, Pius laments:

> Novelties of this kind have already borne their deadly fruit in almost all branches of theology . . . Some question whether angels are personal beings, and whether matter and spirit differ essentially . . . Some even say that the doctrine of Transubstantiation, based on an antiquated philosophic notion of substance, should be so modified that the Real Presence of Christ in the Holy Eucharist be reduced to a kind of symbolism.

Pius first mentions evolution to decry a misuse by overextension among zealous supporters of the anathematized 'isms':

> Some imprudently and indiscreetly hold that evolution . . . explains the origin of all things . . . Communists gladly subscribe to this opinion so that, when the souls of men have been deprived of every idea of a personal God, they may the more efficaciously defend and propagate their dialectical materialism.

Pius presents his major statement on evolution near the end of the encyclical, in paragraphs 35 through 37. He accepts the standard model of non-overlapping magisteria (NOMA) and begins by acknowledging that evolution lies in a difficult area where the domains press hard against each other. 'It remains for Us now to speak about those questions which, although they pertain to the positive sciences, are nevertheless more or less connected with the truths of the Christian faith.'*

Pius then writes the well-known words that permit Catholics to entertain the evolution of the human body (a factual issue under the magisterium of science), so long as they accept the divine creation and infusion of the soul (a theological notion under the magisterium of religion).

> The Teaching Authority of the Church does not forbid that, in conformity with the present state of human sciences and sacred theology, research and discussions, on the part of men experienced in both fields, take place with regard to the doctrine of evolution, in as far as it inquires into the origin of the human body as coming from pre-existent and living

* Interestingly, the main thrust of these paragraphs does not address evolution in general, but lies in refuting a doctrine that Pius calls 'polygenism,' or the notion of human ancestry from multiple parents – for he regards such an idea as incompatible with the doctrine of original sin 'which proceeds from a sin actually committed by an individual Adam and which, through generation, is passed on to all and is in everyone as his own.' In this one instance, Pius may be transgressing the NOMA principle – but I cannot judge, for I do not understand the details of Catholic theology and therefore do not know how symbolically such a statement may be read. If Pius is arguing that we cannot entertain a theory about derivation of all modern humans from an ancestral population rather than through an ancestral individual (a potential fact) because such an idea would question the doctrine of original sin (a theological construct), then I would declare him out of line for letting the magisterium of religion dictate a conclusion within the magisterium of science.

> matter – for the Catholic faith obliges us to hold that souls are immediately created by God.

I had, up to here, found nothing surprising in *Humani Generis*, and nothing to relieve my puzzlement about the novelty of Pope John Paul's recent statement. But I read further and realized that Pius had said more about evolution, something I had never seen quoted, and something that made John Paul's statement most interesting indeed. In short, Pius forcefully proclaimed that while evolution may be legitimate in principle, the theory, in fact, had not been proven and might well be entirely wrong. One gets the strong impression, moreover, that Pius was rooting pretty hard for a verdict of falsity.

Continuing directly from the last quotation, Pius advises us about the proper study of evolution:

> However, this must be done in such a way that the reasons for both opinions, that is, those favorable and those unfavorable to evolution, be weighed and judged with the necessary seriousness, moderation and measure . . . Some however, rashly transgress this liberty of discussion, when they act as if the origin of the human body from preexisting and living matter were already completely certain and proved by the facts which have been discovered up to now and by reasoning on those facts, and as if there were nothing in the sources of divine revelation which demands the greatest moderation and caution in this question.

To summarize, Pius generally accepts the NOMA principle of non-overlapping magisteria in permitting Catholics to entertain the hypothesis of evolution for the human body so long as they accept the divine infusion of the soul. But he then offers some (holy) fatherly advice to scientists about the status of evolution as a scientific concept: the idea is not yet proven, and you all need to be especially cautious because evolution raises many troubling issues right on the border of my magisterium. One may read this second theme in two rather different ways: either as a gratuitous incursion into a different magisterium, or as a helpful perspective from an intelligent and concerned outsider. As a man of goodwill, and in the interest of conciliation, I am content to embrace the latter reading.

In any case, this rarely quoted second claim (that evolution remains both unproven and a bit dangerous) – and not the familiar first argument for the NOMA principle (that Catholics may accept the evolution of the body so long as they embrace the creation of the soul) – defines the novelty and the interest of John Paul's recent statement.

John Paul begins by summarizing Pius's older encyclical of 1950, and particularly by reaffirming the NOMA principle – nothing new here, and no cause for extended publicity:

> In his encyclical 'Humani Generis' (1950), my predecessor Pius XII had already stated that there was no opposition between evolution and the doctrine of the faith about man and his vocation.

To emphasize the power of NOMA, John Paul poses a potential problem and a sound resolution: How can we possibly reconcile science's claim for physical continuity in human evolution with Catholicism's insistence that the soul must enter at a moment of divine infusion?

> With man, then, we find ourselves in the presence of an ontological difference; an ontological leap, one could say. However, does not the posing of such ontological discontinuity run counter to that physical continuity which seems to be the main thread of research into evolution in the field of physics and chemistry? Consideration of the method used in the various branches of knowledge makes it possible to reconcile two points of view which would seem irreconcilable. The sciences of observation describe and measure the multiple manifestations of life with increasing precision and correlate them with the time line. The moment of transition to the spiritual cannot be the object of this kind of observation.

The novelty and news value of John Paul's statement lies, rather, in his profound revision of Pius's second and rarely quoted claim that evolution, while conceivable in principle and reconcilable with religion, can cite little persuasive evidence in support, and may well be false. John Paul states – and I can only say amen, and thanks for noticing – that the half century between Pius surveying the ruins of World War II and his own pontificate heralding the dawn of a new millennium has witnessed such a growth of data, and such a refinement of theory, that evolution can no longer be doubted by people of goodwill and keen intellect:

> Pius XII added . . . that this opinion [evolution] should not be adopted as though it were a certain, proven doctrine . . . Today, almost half a century after the publication of the encyclical, new knowledge has led to the recognition of the theory of evolution as more than a hypothesis.* It is indeed remarkable that this theory has been progressively accepted by researchers, following a series of discoveries in various fields of knowledge. The convergence, neither sought nor fabricated, of the results of work that was conducted independently is in itself a significant argument in favor of the theory.

In conclusion, Pius had grudgingly admitted evolution as a legitimate hypothesis that he regarded as only tentatively supported and potentially (as he clearly hoped) untrue. John Paul, nearly fifty years later, reaffirms the legitimacy of evolution under the NOMA principle – no news here – but then adds that additional data and theory have placed the factuality of evolution beyond reasonable doubt. Sincere Christians must now accept evolution not merely as a plausible possibility, but also as an effectively proven fact. In other words, official Catholic opinion on evolution has moved from 'say it ain't so, but we can deal with it if we have to' (Pius's grudging view of 1950) to John Paul's entirely welcoming 'it has

* This passage, here correctly translated, provides a fascinating example of the subtleties and inherent ambiguities in rendering one language into another. Translation may be the most difficult of all arts, and meanings have been reversed (and wars fought) for perfectly understandable reasons. The Pope originally issued his statement in French, where this phrase read '. . . *de nouvelles connaissances conduisent à reconnaitre dans la théorie de l'évolution plus qu'une hypothèse.' L'Osservatore Romano*, the official Vatican newspaper, translated this passage as: 'new knowledge has led to the recognition of more than one hypothesis in the theory of evolution.' This version (obviously, given the official Vatican source) then appeared in all English commentaries, including the original version of this essay.

I included this original translation, but I was profoundly puzzled. Why should the Pope be speaking of several hypotheses within the framework of evolutionary theory? But I had no means to resolve my confusion, so I assumed that the Pope had probably fallen under the false impression (a fairly common misconception) that, although evolution had been documented beyond reasonable doubt, natural selection had fallen under suspicion as a primary mechanism, while other alternatives had risen to prominence. [contd.]

been proven true; we always celebrate nature's factuality, and we look forward to interesting discussions of theological implications.' I happily endorse this turn of events as gospel – literally good news. I may represent the magisterium of science, but I welcome the support of a primary leader from the other major magisterium of our complex lives. And I recall the wisdom of King Solomon: 'As cold waters to a thirsty soul, so is good news from a far country' (Proverbs 25:25).

Just as religion must bear the cross of its hardliners, I have some scientific colleagues, including a few in prominent enough positions to wield influence by their writings, who view this rapprochement of the separate magisteria with dismay. To colleagues like me – agnostic scientists who welcome and celebrate the rapprochement, especially the Pope's latest statement – they say, 'C'mon, be honest; you know that religion is addlepated, superstitious, old-fashioned BS. You're only making those welcoming noises because religion is so powerful, and we need to be diplomatic in order to buy public support for science.' I do not think that many scientists hold this view, but such a position fills me with dismay – and I therefore end this essay with a personal statement about religion, as a testimony to what I regard as a virtual consensus among thoughtful scientists (who support the NOMA principle as firmly as the Pope does).

I am not, personally, a believer or a religious man in any sense of

---

Other theologians and scientists were equally puzzled, leading to inquiries and a resolution of the problem as an error in translation (as many of us would have realized right away if we had seen the original French, or even known that the document had been issued in French). The problem lies with ambiguity in double meaning for the indefinite article in French – where *un* (feminine *une*) can mean either 'a' or 'one.' Clearly, the Pope had meant that the theory of evolution had now become strong enough to rank as 'more than *a* hypothesis' (*plus qu'une hypothèse*), but the Vatican originally read *une* as 'one' and gave the almost opposite rendition: 'more than *one* hypothesis.' *Caveat emptor.*

I thank about a dozen correspondents for pointing out this error, and the Vatican's acknowledgment, to me. I am especially grateful to Boyce Rensberger, one of America's most astute journalists on evolutionary subjects, and David M. Byers, executive director of the National Conference of Catholic Bishops' Committee on Science and Human Values. Byers affirms the NOMA principle by writing to me: 'Thank you for your recent article . . . It admirably captures the relationship between science and religion that the Catholic Bishops' Committee works to promote and to realize. The text of the October 1996 papal statement from which you were working contains a mistranslation of a key phrase; the correct translation supports your thesis with even greater force.'

institutional commitment or practice. But I have great respect for religion, and the subject has always fascinated me, beyond almost all others (with a few exceptions, like evolution and paleontology). Much of this fascination lies in the stunning historical paradox that organized religion has fostered, throughout Western history, both the most unspeakable horrors and the most heartrending examples of human goodness in the face of personal danger. (The evil, I believe, lies in an occasional confluence of religion with secular power. The Catholic Church has sponsored its share of horrors, from Inquisitions to liquidations – but only because this institution held great secular power during so much of Western history. When my folks held such sway, more briefly and in Old Testament times, we committed similar atrocities with the same rationales.)

I believe, with all my heart, in a respectful, even loving, concordat between our magisteria – the NOMA concept. NOMA represents a principled position on moral and intellectual grounds, not a merely diplomatic solution. NOMA also cuts both ways. If religion can no longer dictate the nature of factual conclusions residing properly within the magisterium of science, then scientists cannot claim higher insight into moral truth from any superior knowledge of the world's empirical constitution. This mutual humility leads to important practical consequences in a world of such diverse passions.

Religion is too important for too many people to permit any dismissal or denigration of the comfort still sought by many folks from theology. I may, for example, privately suspect that papal insistence on divine infusion of the soul represents a sop to our fears, a device for maintaining a belief in human superiority within an evolutionary world offering no privileged position to any creature. But I also know that the subject of souls lies outside the magisterium of science. My world cannot prove or disprove such a notion, and the concept of souls cannot threaten or impact my domain. Moreover, while I cannot personally accept the Catholic view of souls, I surely honor the metaphorical value of such a concept both for grounding moral discussion, and for expressing what we most value about human potentiality: our decency, our care, and all the ethical and intellectual struggles that the evolution of consciousness imposed upon us.

As a moral position (and therefore not as a deduction from my knowledge of nature's factuality), I prefer the 'cold bath' theory that nature can be truly 'cruel' and 'indifferent' – in the utterly inappropriate terms of our ethical discourse – because nature does not exist for us, didn't know we were coming (we are, after all, interlopers of the latest geological moment), and doesn't give a damn about us (speaking metaphorically). I regard such a position as liberating, not depressing, because we then gain the capacity to conduct moral discourse – and nothing could be more important – in our own terms, free from the delusion that we might read moral truth passively from nature's factuality.

But I recognize that such a position frightens many people, and that a more spiritual view of nature retains broad appeal (acknowledging the factuality of evolution, but still seeking some intrinsic meaning in human terms, and from the magisterium of religion). I do appreciate, for example, the struggles of a man who wrote to the *New York Times* on November 3, 1996, to declare both his pain and his endorsement of John Paul's statement:

> Pope John Paul II's acceptance of evolution touches the doubt in my heart. The problem of pain and suffering in a world created by a God who is all love and light is hard enough to bear, even if one is a creationist. But at least a creationist can say that the original creation, coming from the hand of God, was good, harmonious, innocent and gentle. What can one say about evolution, even a spiritual theory of evolution? Pain and suffering, mindless cruelty and terror are its means of creation. Evolution's engine is the grinding of predatory teeth upon the screaming, living flesh and bones of prey . . . If evolution be true, my faith has rougher seas to sail.

I don't agree with this man, but we could have a terrific argument. I would push the 'cold bath' theory; he would (presumably) advocate the theme of inherent spiritual meaning in nature, however opaque the signal. But we would both be enlightened and filled with better understanding of these deep and ultimately unanswerable issues. Here, I believe, lies the greatest strength and necessity of NOMA, the non-overlapping magisteria of science and religion. NOMA permits – indeed enjoins – the prospect of respectful discourse, of constant input from both magisteria toward the common goal of wisdom. If human beings can lay claim to anything special, we evolved as the only creatures that must ponder and talk. Pope John Paul II would surely point out to me that his magisterium has always recognized this uniqueness, for John's gospel begins by stating *in principio erat verbum* – in the beginning was the word.

THE DIET OF WORMS AND THE DEFENESTRATION OF PRAGUE

I once ate an ant (chocolate-covered) on a dare. I have no awful memories of the experience, but I harbor no burning desire for a repeat performance. I therefore feel poor Martin Luther's pain when, at the crux of his career, in April 1521, he devoted ten days to the Diet of Worms (washed down with a good deal of wine, or so I read).

I am a collector by nature, and mental drawers have more room for phrases and facts than physical cabinets maintain for specimens. I therefore reserve one cranial shelf for the best funny or euphonious phrases of history. 'The Diet of Worms' remains my prize specimen, but I award second place to another *D*-phrase of European history: 'the Defenestration of Prague' in 1618 – the 'official' trigger of the Thirty Years War, one of the most extended, horrendous, and senseless conflicts in Western culture.

I do not believe in vicarious experience and will go to great, even absurd, lengths to stand on the true spot, or place a hand on the very wall. I could have written *Wonderful Life* without a visit to the Burgess Shale, but what a sacrilege! Walcott's fossil quarry is holy ground, and only a four-mile trek from the main road.

I therefore accepted a recent invitation for a lecture in Heidelberg on the stipulation that my hosts drive me to nearby Worms, site of the Diet. (I had, three years earlier, stood on the square in Prague where those bodies once landed after ejection from an upper-story window.) Now, with pilgrimages completed to the sources of both phrases that most caught my fancy in Mrs Ponti's fifth-grade European history class, I can muse more formally upon the sadly common theme behind the two *D*'s – our cursed tribal tendency to factionalize, fight, and then, so often in our righteous certainty, to define our opponents as vermin and try to expunge either their doctrines (by censorship and fire) or their very being (by genocide). The Diet of Worms and the Defenestration of Prague mark two cardinal events in the sad chronology of hatred and bloodshed surrounding a central theme of Western history, one filled with aspects of grandeur as well – the schism of 'universal' Christianity into Catholic and Protestant portions.

The Diet, or governing body, of the Holy Roman Empire met at the great medieval Rhineland city of Worms in 1521, partly to demand the recantation of Martin Luther. (German sources call the Diet a *Reichstag*. Moreover, German fishbait is spelled with a 'u,' not an 'o' as in English. Thus, the *Reichstag zu Worms* packs no culinary punch in the original vernacular.)

In school, I learned the heroic version of Luther before the Diet of Worms. This account (so far as I know, and have just affirmed by reading several recent biographies) reports factual material in an accurate manner – and is therefore 'true' in one crucial sense, yet frightfully partial and therefore misleadingly incomplete in other equally important ways. Luther, excommunicated by Pope Leo X in January 1521, arrived in Worms under an imperial guarantee of safe conduct to justify or recant his apostasies before the militantly Catholic and newly elected Holy Roman Emperor, Charles V, heir to the Hapsburg dynasty of central Europe and Spain, and twenty-one-year-old grandson of Ferdinand and Isabella, monarchs of Spain, and patrons of Christopher Columbus.

Luther, with substantial support from local people of all classes, including his most powerful protector, Frederick the Wise, Elector of Saxony, appeared before Charles and the Imperial Diet on April 17. Asked if he would retract the contents of his books, Luther begged some time for consideration (and, no doubt, for preparation of a rip-roaring speech). The emperor granted a one-day recess, and Luther returned on April 18 to make his most famous statement.

Speaking first in German and then in Latin, Luther argued that he could not disavow his work unless he could be proved wrong either by the Scriptures or by logic. He may or may not have ended his speech (reports vary) with one of the most famous statements in Western history: *Hier stehe ich; ich kann nicht anders; Gott helfe mir; Amen* – Here I stand; I cannot do otherwise; God help me; Amen.

Faced with Luther's intransigence, the Emperor and a rump session of the Diet issued the Edict of Worms on May 8. But that document, banning Luther's work and enjoining his detention, could not be enforced, given the strength of Luther's local support. Instead, under Frederick's protection, Luther 'escaped' to the castle of Wartburg, where he translated the New Testament into German.

A stirring story, invoking some of the finest themes in Western liberal and intellectual traditions: freedom of thought, personal bravery against authority, the power of one man with a grand idea before the crumbling weight of centuries. But dig just a little deeper, below the overt level of hagiography and school-day moralisms, and you enter a quagmire of intolerance and mayhem on all sides. Scratch the surface of soaring notions like 'justification by faith,' and you encounter a world where any major idea becomes a political instrument in a quest for social order, or a tool in the struggle for power between distant popes and local princes. Consider the operative paragraph of the Edict of Worms, complete with a closing metaphor about diets in the modern culinary sense:

> We want all of Luther's books to be universally prohibited and forbidden, and we also want them to be burned . . . We follow the very praiseworthy ordinance and custom of the good Christians of old who had the books of heretics like the Arians, Priscillians, Nestorians, Eutychians, and others burned and annihilated, even everything that was contained in these books, whether good or bad. This is well done, since if we are not allowed to eat meat containing just one drop of poison because of the danger of bodily infection, then we surely should leave out every doctrine (even if it is good) which has in it the poison of heresy and error, which infects and corrupts and destroys under the cover of charity everything that is good.

These words may be chilling enough when confined to the destruction of documents. But annihilation often extended to the inventors of unorthodoxies, and to the genocide of followers. Of the early heretics mentioned above, Priscillian, bishop of Avila in Spain, was convicted of sorcery and immorality, and executed by the Roman emperor Maximus in 385. The later Albigensians fared far worse. These ascetic communitarians of southern France frightened papal and other authorities with their views on the corruption of clergy and secular rulers. In 1209, Pope Innocent III urged a crusade against them – just one among so many examples of Christians annihilating other Christians – and the resulting war effectively destroyed the Provençal civilization of southern France. The Inquisition mopped up during the next several decades, thus completing the extirpation of an unpopular view by genocide. Grisly, but effective. The *Encyclopaedia Britannica* simply states: 'It is exceedingly difficult to form any very precise idea of the Albigensian doctrines because present knowledge of them is derived from their opponents.'

If Luther and other reformers had promoted their new versions of Christianity in the name of love, toleration, and respect, then I might accept the heroic version of history as progress inspired by rare individuals of broader vision. But Luther could be just as dogmatic, just as unforgiving, and just as bloodthirsty as his opponents – and when his folks took the reins of power, the old tactics of banning, book burning, and doctrinal murder continued. For example, Luther had originally held little animus toward Jews, for he hoped that his reforms, by eliminating papal abuses, might lead to their conversion. But when his hopes withered, Luther turned on his vitriol and, in a 1543 pamphlet titled *On the Jews and Their Lies*, recommended either forced deportation to Palestine, or the burning of all synagogues and Jewish books (including the Bible), and the restriction of Jews to agrarian pursuits.

In his most horrific recommendation (and on the eve of supposed personal happiness in his marriage to Katherine von Bora), Luther advocated the wholesale slaughter of German peasants, whose rebellion had recently been so brutally suppressed. Luther had his reasons and frustrations, to be sure. He had never supported uprising against secular authority, although some of the more moderate peasant groups had used his teachings as justifications. Moreover, the militant faction of peasants had been led by his bitter theological enemy, Thomas Müntzer. Political conservatives like Luther always take a dim view (if only to save their own skins) of insurrections by large and poorly disciplined groups of disenfranchised people, but Luther's recommendations for virtual genocide, as presented in his tract of 1525 *Against the Murderous and Thieving Hordes of Peasants*, makes my skin crawl, especially as a recommendation (however secular) from a supposed man of God:

> If the peasant is in open rebellion, then he is outside the law of God . . . Rebellion brings with it a land full of murders and bloodshed, makes widows and orphans, and turns everything upside down like a great disaster. *Therefore, let everyone who can, smite, slay, and stab, secretly or openly*, remembering that nothing can be more poisonous, hurtful, or devilish than a rebel. It is just as when one must kill a mad dog; if you don't strike him, he will strike you, and the whole land with you [my italics].

The victorious nobility followed Luther's recommendations, and estimates of the death toll (mostly inflicted upon rebels who had already surrendered and therefore posed no immediate threat) range to 100,000 people.

Sad tales of mass murder perpetrated by differing factions of a supposedly united cause haunt human history. I don't think that Christians are worse than other folks in this regard; we just know these stories better as defining incidents of a culture shared by most readers of this book. I am not speaking of isolated executions, but of wholesale slaughters, however unknown to us today for two eerie reasons. First, Hitlers of the past didn't possess the technology (though they probably had the will) to kill six million in a few years, so their depredations, though thorough, were more local. Second, obliterated cultures of bygone times featured fewer people, living in limited areas, and publishing little or no documentation. An older style of genocide could therefore be devastatingly complete and effective, truly wiping out all memory of a vibrant people.

I have already mentioned the Albigensian crusade. In 1204, the Fourth Crusade, having failed to reach Palestine through Egypt to conquer the Holy Land, sacked the Byzantine Christian capital of Constantinople instead, imposing more mayhem upon people and art than the 'infidel' Ottomans exacted when the city finally fell from Christian rule in 1453. The rupture of Europe into Protestant and Catholic parts provided more opportunity for such divisive destruction – and Luther's legacy surely includes as much darkness as light. Which brings me to the Thirty Years War and the Defenestration of Prague.

Throwing people out of windows has a long legacy in this beautiful city – a true Scandal in Bohemia, so to speak. In each major incident but the last, rebelling Protestants (or proto-Protestants) tossed entrenched Catholics out of their strongholds. The 'official' Defenestration (with a capital D) occurred in 1618. Local Protestants, justifiably enraged when the very Catholic King Ferdinand II reneged on promises of religious freedom, stormed Hradčany Castle and threw three Catholic councilors out of the window and into the moat. (Legend states that they walked away, embarrassed but unharmed, thanks either to good fortune or to the good aim of their adversaries – for they landed in a large and soft dunghill.)

The rebels of 1618 had consciously reenacted a past incident that they wished to claim as part of a proud and continuous history. (A Latin window, by the way, is a *fenestra* – so *defenestration* is just a fancy word for throwing something out such an opening.) The memory of Bohemian religious reformer Jan Hus, burned for heresy in 1415 and claimed by later Protestants as a precursor, inspired the initial defenestration of Prague in 1419. A Hussite army (if you like them) or a rabble (if you don't) stormed the New Town Hall and threw three Catholic consuls and seven citizens out the window (some to their death, for no cushioning dunghill broke these falls), and Bohemia did pass to Hussite rule for a time. Yet another, but lesser known, defenestration occurred in 1483. King Vladislav had restored Catholic dominion, so another dissident band of Hussites threw the Catholic mayor out the window.

The tragic epilogue to this sequence will be remembered by readers just a few years older than me. Jan Masaryk, son of Thomas Masaryk, the founder of Czechoslovakia, continued to serve as the only non-communist minister of the postwar puppet government. On March 10, 1948, his body was found in the courtyard of Czernin Palace. He had fallen to his death from a window forty-five feet above. Had he jumped as a suicide (with ironic consciousness of his nation's history), or had he been pushed in a murder? The case has never been solved.

The Protestant triumph after the official defenestration lasted only two years and ended in yet another splurge of murder and destruction. With powerful Hapsburg support, Catholics regrouped, and decisively defeated the Protestants in the Battle of the White Mountain on November 8, 1620. Weeks of plunder and pillage followed in Prague. A few months later, twenty-seven nobles and other citizens were tortured and executed in the Old Town Square. The victors hung twelve heads, impaled on iron hooks, from the Bridge Tower as a warning.

The resulting Thirty Years War cannot be reduced to a dichotomous struggle between Catholics and Protestants – but this essential division did define much of the temper and zealotry of the controversy (for we seem able to kill 'apostates' far more easily than merely errant compatriots). Much of central Europe lay in ruins, as the mercenary armies of various potentates ravaged their way through the countryside, burning, raping, and pillaging as they went. Nor did the battles of Protestants against Catholics end with the Treaty of Westphalia in 1648. The ruined castle of Heidelberg, beautifully lit at night, turns the entire city into a romantic stage set for *The Student Prince*. But Heidelberg retains no medieval buildings, and the castle lies in ruins, as the consequence of yet another disastrous internecine war among Christians – when the Protestant Elector of the Rhineland Palatinate (with Heidelberg as a center) died without heir, and Catholic France claimed the territory because the Elector's sister had married Philip of Orléans, brother of Louis XIV.

A somewhat cynical, but sadly accurate, principle of human history states that when things look bad, they can still get far worse. If Christians could slaughter each other with such gusto and ferocity, what could true outsiders expect – for non-Christians could be defined even more easily as beyond human worth and therefore ripe for elimination. For this final chapter in man's inhumanity, we turn to the obvious test: the fate of Jewish communities in medieval and Renaissance Europe at the time of diets and defenestrations.

Jewish settlements persisted for one thousand years in the Rhineland. Every city that I visited – Worms, Speyer, Rothenburg – maintains memorials to the persecution and elimination of these communities, while tourist outlets sell long and informative pamphlets on their history, lest we forget. One almost feels a concerted and commendable attempt to expiate what cannot be undone – a tale of prolonged intolerance capped by such recent memory of the brutally and entirely effective last curtain (at least locally) of the 'final solution.'

Gershom ben Judah, known as the 'light of the exile,' headed the rabbinic academy of Mainz at the end of the tenth century, before the first millennium of our era's common time had passed. His most celebrated disciple, the great Talmudist Rashi, studied in Worms around 1060. Rashi's most noted follower, Meir Ben Baruch (known to pious Jews by his acronym as the Maharam), headed the Jewish community in Rothenburg, now the most perfectly preserved, entirely walled, and touristically flooded of all medieval towns. In 1286, the Emperor Rudolph I abrogated the political freedom of Jews and imposed special taxes to make these despised people *servi camerae* (or serfs of the treasury). Rabbi Meir tried to lead a group of Jews to Palestine, but he was arrested and confined in an Alsatian fortress. His people raised an enormous ransom, but Meir refused (and died in prison) because he knew that a purchased freedom would only encourage the emperor to capture other rabbis for revenue. Fourteen years later, a Jewish merchant in Worms ransomed the great rabbi's body. His tomb, and that of the merchant, occupy adjacent places in the Jewish cemetery of Worms. Following an ancient custom, Jewish visitors and residents (mostly Russian émigrés) still write their prayers and requests on scraps of paper and place them, weighted down by small stones, atop the Maharam's tomb.

After Meir's exile, the Jews of Rothenburg were expelled to a ghetto beyond the city walls, and then, in 1520, banished entirely and forevermore. Only the small dance hall remains (because it became a poorhouse for Christians), with a few tombstones in Hebrew, mounted on the garden wall.

The larger Jewish community of Luther's Worms survived longer, but just as precariously. In 1096, soldiers of the First Crusade passed through Worms and ravaged the Jewish quarter. In 1349, nearly all the Jews of Worms were murdered on the false accusation that they had brought the plague by poisoning the wells. In 1938, on the infamous *Kristallnacht*, the Jewish synagogue burned to the ground. More than one thousand Jews of Worms perished in the Holocaust. The reconstructed synagogue now serves as the centerpiece of a Jewish museum and memorial, but not as a place of worship, for no active Jewish community now exists in Worms.

Two plaques on the synagogue wall tell a tale of hope and despair. The first, mounted shortly after the end of World War II, contains the names of Jewish citizens presumed dead in the Holocaust. Happily, some of these people had survived (in refugee camps, unknown to makers of the plaque). Their raised bronze names have been filed off, leaving blank spaces of victory. But further records of the Holocaust then documented more deaths, and these names adorn the second plaque – greatly exceeding in number the names happily erased from the first memorial.

Ironically, only the Jewish cemetery survived intact, thanks to a ruse (according to local tradition) of the town archivist, a sincere Christian with great respect for Jewish traditions. Himmler had expressed a passing interest in the cemetery during a prewar visit. When local Nazis later ordered the destruction of the cemetery (located on the other side of town, beyond the walls), the archivist exaggerated Himmler's casual comment into an explicit order for preservation. Cautious local authorities never checked with Berlin – and a place of death remains as the only unscathed survivor of a millennium's existence for one of Europe's most illustrious Jewish communities.

If you have been wondering why I recount these tales from the dark side of human history in an essay on evolutionary biology, I do intend to segue toward an ending on both a positive and a Darwinian note. Humans are capable of such glory – and such horror: the pogroms of Worms, and Luther's stirring speech at the Diet of Worms; the numerous defenestrations of Prague, and the magnificent baroque architecture of Prague. We bask in the glory with simple pleasure; but we contemplate the horror with anguish and puzzlement – and with a burning urge to explain how creatures capable of such decency can promote such iniquity of their own free will (and with apparent moral calm and intensity of supposed purpose).

But do we perpetrate the darkness 'of our own free will'? Perhaps the most popular of all explanations for our genocidal capacity cites evolutionary biology as an unfortunate source – and as an ultimate escape from full moral responsibility. Perhaps we evolved these capacities as active adaptations now gone awry in the modern world. Current genocide may be a sad legacy of behaviors that originated for Darwinian benefit during our ancestral construction as small bands of hunters and gatherers on the savannahs of Africa. Darwin's mechanism, after all, encourages only the reproductive success of individuals, not the moral dream of human fellowship across an entire species. Perhaps the traits that lead to modern genocide – xenophobia, tribalism, anathematization of outsiders as subhuman and therefore subject to annihilation – rose to prominence during our early evolution because they enhanced survival in tiny, nontechnological societies based on kinship and living in a world of limited resources under a law of kill-or-be-killed.

A group devoid of xenophobia and unschooled in murder might invariably succumb to others replete with genes to encode a propensity for such categorization and destruction. Chimpanzees, our closest relatives, will band together and systematically kill the members of adjacent groups. Perhaps we are programmed to act in such a manner as well. These grisly propensities once promoted the survival of groups armed with nothing more destructive than teeth and stones. In a world of nuclear bombs, such unchanged (and perhaps unchangeable) inheritances may now spell our undoing (or at least propagate our tragedies) – but we cannot be blamed for these moral failings. Our accursed genes have made us creatures of the night.

This superficially attractive balm to our collective conscience is nothing but a cop-out based on deep fallacies of reasoning. (Perhaps the tendency to think by such fallacies represents our real evolutionary legacy – but this is another speculation for another time.) I am quite happy to acknowledge that we have a biologically based capacity to categorize humans as insiders or outsiders, and then to view outsiders as beyond fellowship and ripe for slaughter. But where can such an argument take us in terms of modern moral discourse, or even social observation? For this claim is entirely empty and devoid of explanatory power. We gain nothing by speculating that a capacity for genocide lies within our evolutionary heritage. We already know that we have such a capacity because human history provides so many examples of actualization.

An evolutionary speculation, to be useful, must suggest something we don't know already – if, for example, we learned that genocide has been biologically enjoined by certain genes, or even that a positive propensity, rather than a mere capacity, regulated our murderous potentiality. But the observational facts of human history speak against determination, and only for potentiality. Each case of genocide can be matched with numerous incidents of social benevolence; each murderous band can be paired with a pacific clan. Genocide gains greater prominence only for superior 'news value' – and for devastating effectiveness (as the pacific clan disappears and murderers control the resulting media). But if both darkness and light lie within our capacities, and if both tendencies operate at high frequency in human history, then we learn nothing by speculating that either (or probably both) lie within our evolutionary, adaptive, Darwinian heritage. At the very most, biology might help us to delimit the environmental circumstances that tend to elicit one behavior rather than the other.

To cite the example most under current discussion in the 'pop science' press, numerous books and articles preach to us that a new science of evolutionary psychology has discovered the biological basis of behavioral differences between sexes. Women produce only a few large eggs and must spend years of their lives growing embryos within their bodies and then nurturing the resulting babies. Men, on the other hand, produce millions of tiny sperm each time, and need invest nothing more in a potential offspring than the effort of an ejaculation. Therefore, the argument continues, in the great Darwinian quest for passing more genes to future generations, women will behave in ways that encourage male investment after impregnation (protection, feeding, economic wealth, and subsequent child care), whereas men would rather wander right off in search of other mates in a never-ending quest for maximal genetic spread. From this basic dichotomy of evolutionary purpose, all else in the lexicon of pop psychology follows. We now know why men rape, lust for power, dominate politics, have affairs, and abandon families with young children – and why women act coy, love to nurture children, and preferentially enter the caring professions.

Perhaps I have caricatured this position – but I don't think so, having read so many articles of support. In fact, I don't even think that the basic argument is wrong. Such differences in behavioral strategy do make Darwinian sense in the light of structural disparity between male and female reproduction. But the attributions could not be more deeply erroneous for the same reasons noted above in discussing the fallacy of biological explanations for genocide. Men are not programmed by genes to maximize matings, or women devoted to monogamy on the same basis. We can only speak of capacities, not of requirements or even determining propensities. Therefore, our biology does not make us do it. Moreover, what we share in common genetics can easily overwhelm what men and women might tend to do differently. Any man who has fiercely loved his little child – including most fathers, I trust (and I happen to be writing this essay on Father's Day) – knows that no siren song from distinctive genes or hormones can overcome this drive for nurturing behavior shared with the child's mother.

Finally, when we note the crucial differences in fundamental pattern and causation between biological evolution and cultural change – and when we recognize that everything distinctive about the cultural style enjoins flexibility rather than determination – we can understand even more generally why a cultural phenomenon like genocide (despite any underlying biological capacity for such action) cannot be explained in evolutionary terms. As the fundamental difference in pattern, biological evolution is a topological tree – a process of separation and divergence. A new species, arising as an independent lineage, acquires genetic distinction from all other lineages forever, and must evolve on its own path. Cultural change, on the other hand, is virtually defined by possibilities of amalgamation among different traditions – as Marco Polo brings pasta from China and I speak English as a 'native' tongue. Our distinctive flexibilities arise from this constant interweaving.

As the fundamental difference in causation, biological evolution is Mendelian. Organisms can only pass their genes, not the heritage of their efforts, as physical contributions to future generations. But cultural change is Lamarckian, as we transmit the fruits of our acquired wisdom and inventiveness directly to future generations in the form of books, instruction manuals, tools, and buildings. Again, this Lamarckian style grants to cultural change a speed, a lability, and a flexibility that Darwinian evolution cannot muster.

In 1525, thousands of German peasants were slaughtered (with Luther's approbation), and Michelangelo worked on the Medici Chapel. In 1618, the upper windows of Prague disgorged a few men, and Rubens painted some mighty canvases. The Cathedral of Canterbury is both the site of Becket's murder and the finest Gothic building in England. Both sides of this dichotomy represent our common, evolved humanity; which, ultimately, shall we choose? As for the potential path of genocide and destruction, let us take this stand. It need not be. We can do otherwise.

DARWIN AND THE MUNCHKINS OF KANSAS

IN 1999 the Kansas Board of Education voted six to four to remove evolution, and the big bang theory as well, from the state's science curriculum. In so doing, the board transported its jurisdiction to a never-never land where a Dorothy of a new millennium might exclaim, 'They still call it Kansas, but I don't think we're in the real world anymore.' The new standards do not forbid the teaching of evolution, but the subject will no longer be included in statewide tests for evaluating students – a virtual guarantee, given the realities of education, that this central concept of biology will be diluted or eliminated, thus reducing biology courses to something like chemistry without the periodic table, or American history without Lincoln.

The Kansas skirmish marks the latest episode of a long struggle by religious fundamentalists and their allies to restrict or eliminate the teaching of evolution in public schools – a misguided effort that our courts have quashed at each stage, and that saddens both scientists and the vast majority of theologians as well. No scientific theory, including evolution, can pose any threat to religion – for these two great tools of human understanding operate in complementary (not contrary) fashion in their totally separate realms: science as an inquiry about the factual state of the natural world, religion as a search for spiritual meaning and ethical values.

In the early 1920s, several states simply forbade the teaching of evolution outright, opening an epoch that inspired the infamous 1925 Scopes trial (leading to the conviction of a Tennessee high school teacher), and that ended only in 1968, when the Supreme Court declared such laws unconstitutional on First Amendment grounds. In a second round in the late 1970s, Arkansas and Louisiana required that if evolution be taught, equal time must be given to Genesis literalism, masquerading as oxymoronic 'creation science.' The Supreme Court likewise rejected those laws in 1987.

The Kansas decision represents creationism's first – and surely temporary* – success with a third strategy for subverting a constitutional imperative: that by simply deleting, but not formally banning, evolution, and by not demanding instruction in a biblical literalist 'alternative,' their narrowly partisan religious motivations might not derail their goals by legal defeat.

Given this protracted struggle, Americans of goodwill might be excused for supposing that some genuine scientific or philosophical dispute motivates this issue: Is evolution speculative and ill-founded? Does evolution threaten our ethical values or our sense of life's meaning? As a paleontologist by training, and with abiding respect for religious traditions, I would raise three points to alleviate these worries:

First, no other Western nation has endured any similar movement, with any political clout, against evolution – a subject taught as fundamental, and without dispute, in all other countries that share our major sociocultural traditions.

Second, evolution is as well documented as any phenomenon in science, as firmly supported as the earth's revolution around the sun rather than vice versa. In this sense, we can call evolution a 'fact.' (Science does not deal in certainty, so 'fact' can only mean a proposition affirmed to such a high degree that it would be perverse to withhold one's provisional assent.)

The major argument advanced by the school board – that large-scale evolution must be dubious because the process has not been directly observed – smacks of absurdity and only reveals ignorance about the nature of science. Good science integrates observation with inference. No process that unfolds over such long stretches of time (mostly, in this case, before humans evolved), or beneath our powers of direct visualization (subatomic particles, for example), can be seen directly. If

* This 'viewpoint' appeared in *Time* magazine on August 23, 1999. Supporters of good scientific education defeated the creationists in the next school board election in 2000. This newly elected board immediately restored evolution to the biology curriculum.

justification required eyewitness testimony, we would have no sciences of deep time – no geology, no ancient human history, either. (Should I believe Julius Caesar ever existed? The hard, bony evidence for human evolution surely exceeds our reliable documentation of Caesar's life.)

Third, no factual discovery of science (statements about how nature 'is') can, in principle, lead us to ethical conclusions (how we 'ought' to behave), or to convictions about intrinsic meaning (the 'purpose' of our lives). These last two questions – and what more important inquiries could we make? – lie firmly in the domains of religion, philosophy, and humanistic study. Science and religion should operate as equal, mutually respecting partners, each the master of its own domain, and with each domain vital to human life in a different way.

Why get excited over this latest episode in the long, sad history of American anti-intellectualism? Let me suggest that, as patriotic Americans, we should cringe in embarrassment that, at the dawn of a new, technological millennium, a jurisdiction in our heartland has opted to suppress one of the greatest triumphs of human discovery. Evolution cannot be dismissed as a peripheral subject, for Darwin's concept operates as the central organizing principle of all biological science. No one who has not read the Bible or the Bard can be considered educated in Western traditions; similarly, no one ignorant of evolution can understand science.

Dorothy followed her yellow brick road as the path spiraled outward toward redemption and homecoming (to the true Kansas of our dreams and possibilities). The road of the newly adopted Kansas curriculum can only spiral inward toward restriction and ignorance.

HOOKING LEVIATHAN BY ITS PAST

THE landscape of every career contains a few crevasses, and usually a more extensive valley or two – for every Ruth's bat a Buckner's legs; for every lopsided victory at Agincourt, a bloodbath at Antietam. Darwin's *The Origin of Species* contains some wonderful insights and magnificent lines, but this masterpiece also includes a few notable clunkers. Darwin experienced most embarrassment from the following passage, curtailed and largely expunged from later editions of his book:

> In North America the black bear was seen by Hearne swimming for hours with widely open mouth, thus catching, like a whale, insects in the water. Even in so extreme a case as this, if the supply of insects were constant, and if better adapted competitors did not already exist in the country, I can see no difficulty in a race of bears being rendered, by natural selection, more aquatic in their structure and habits, with larger and larger mouths, till a creature was produced as monstrous as a whale.

Why did Darwin become so chagrined about this passage? His hypothetical tale may be pure speculation and conjecture, but the scenario is not entirely absurd. Darwin's discomfort arose, I think, from his failure to follow a scientific norm of a more sociocultural nature. Scientific conclusions supposedly rest upon facts and information. Speculation is not entirely taboo, and may sometimes be necessary *faute de mieux*. But when scientists propose truly novel and comprehensive theories – as Darwin tried to do in advancing natural selection as the primary mechanism of evolution – they need particularly good support, and invented hypothetical cases just don't supply sufficient confidence for crucial conclusions.

Natural selection (or the human analogue of differential breeding) clearly worked at small scale – in the production of dog breeds and strains of wheat, for example. But could such a process account for the transitions of greater scope that set our concept of evolution in the fullness of time – the passage of reptilian lineages to birds and mammals; the origin of humans from an ancestral stock of apes? For these larger changes, Darwin could provide little direct evidence, for a set of well-known and much-lamented reasons based on the extreme spottiness of the fossil record.

Some splendid cases began to accumulate in years following *The Origin of Species*, most notably the discovery of *Archaeopteryx*, an initial bird chock-full of reptilian features, in 1861; and the first findings of human fossils late in the nineteenth century. But Darwin had little to present in his first edition of 1859, and he tried to fill this factual gap with hypothetical fables about swimming bears eventually turning into whales – a fancy that yielded far more trouble in easy ridicule than aid in useful illustration. Just two years after penning his bear-to-whale tale, Darwin lamented to a friend (letter to James Lamont, February 25, 1861), 'It is laughable how often I have been attacked and misrepresented about this bear.'

The supposed lack of intermediary forms in the fossil record remains the fundamental canard of current anti-evolutionism. Such transitional forms are sparse, to be sure, and for two sets of good reasons – geological (the gappiness of the fossil record) and biological (the episodic nature of evolutionary change, including patterns of punctuated equilibrium, and transition within small populations of limited geographic extent). But paleontologists have discovered several superb examples of intermediary forms and sequences, more than enough to convince any fair-minded skeptic about the reality of life's physical genealogy.

The first 'terrestrial' vertebrates retained six to eight digits on each limb (more like a fish paddle than a hand), a persistent tailfin, and a lateral-line system for sensing sound vibrations underwater. The anatomical transition from reptiles to mammals is particularly well documented in the key anatomical change of jaw articulation to hearing bones. Only one bone, called the dentary, builds the mammalian jaw, while reptiles retain several small bones in the rear portion of the jaw. We can trace, through a lovely sequence of intermediates, the reduction of these small reptilian bones, and their eventual disappearance or exclusion from the jaw, including the remarkable passage of the reptilian articulation bones into the mammalian middle ear (where they became our malleus and incus, or hammer and anvil). We have even found the transitional form that creationists often proclaim inconceivable in theory – for how can jawbones become ear bones if intermediaries must live with an unhinged jaw before the new joint forms? The transitional species maintains a double jaw joint, with both the old articulation of reptiles (quadrate to articular bones) and the new connection of mammals (squamosal to dentary) already in place! Thus, one joint could be lost, with passage of its bones into the ear, while the other articulation continued to guarantee a properly hinged jaw.

Still, our creationist incubi, who would never let facts spoil a favorite argument, refuse to yield, and continue to assert the absence of *all* transitional forms by ignoring those that have been found, and continuing to taunt us with admittedly frequent examples of absence. Darwin's old case for the origin of whales remains a perennial favorite, for if Darwin had to invent a fanciful swimming bear, and if paleontologists haven't come to the rescue by discovering an intermediary form with functional legs and potential motion on land, then Jonah's scourge may gobble up the evolutionary heathens as well. God's taunt to Job might be sounded again: 'Canst thou draw out leviathan with an hook?' (The biblical Leviathan is usually interpreted as a crocodile, but many alternate readings favor whales.)

Every creationist book on my shelf cites the actual absence and inherent inconceivability of transitional forms between terrestrial mammals and whales. Alan Haywood, for example, writes in his *Creation and Evolution*:

> Darwinists rarely mention the whale because it presents them with one of their most insoluble problems. They believe that somehow a whale must have evolved from an ordinary land-dwelling animal, which took to the sea and lost its legs . . . A land mammal that was in process of becoming a whale would fall between two stools – it would not be fitted for life on land or at sea, and would have no hope of survival.

Duane Gish, creationism's most ardent debater, makes the same argument in his more colorful style (*Evolution: The Challenge of the Fossil Record*):

> There simply are no transitional forms in the fossil record between the marine mammals and their supposed land mammal ancestors . . . It is quite entertaining, starting with cows, pigs, or buffaloes, to attempt to visualize what the intermediates may have looked like. Starting with a cow, one could even imagine one line of descent which prematurely became extinct, due to what might be called an 'udder failure.'

The most 'sophisticated' (I should really say 'glossy') of creationist texts, *Of Pandas and People* by P. Davis, D. H. Kenyon, and C. B. Thaxton says much the same, but more in the lingo of academese:

> The absence of unambiguous transitional fossils is strikingly illustrated by the fossil record of whales . . . If whales did have land mammal ancestors, we should expect to find some transitional fossils. Why? Because the anatomical differences between whales and terrestrial mammals are so great that innumerable in-between stages must have paddled and swam the ancient seas before a whale as we know it appeared. So far these transitional forms have not been found.

Three major groups of mammals have returned to the ways of distant ancestors in their seafaring modes of life (while smaller lineages within several other mammalian orders have become at least semi-aquatic, often to a remarkable degree, as in river and sea otters): the suborder Pinnepedia (seals, sea lions, and walruses) within the order Carnivora (dogs, cats, and Darwin's bears among others); and two entire orders – the Sirenia (dugongs and manatees) and Cetacea (whales and dolphins). I confess that I have never quite grasped the creationists' point about inconceivability of transition – for a good structural (though admittedly not a phylogenetic) series of intermediate anatomies may be extracted from these groups. Otters have remarkable aquatic abilities, but retain fully functional limbs for land. Sea lions are clearly adapted for water, but can still flop about on land with sufficient dexterity to negotiate ice floes, breeding grounds, and circus rings.

But I admit, of course, that the transition to manatees and whales represents no trivial extension, for these fully aquatic mammals propel themselves by powerful, horizontal tail flukes and have no visible hind limbs at all – and how can a lineage both develop a flat propulsive tail from the standard mammalian length of rope, and then forfeit the usual equipment of back feet so completely? (Sirenians have lost every vestige of back legs; whales often retain tiny, splintlike pelvic and leg bones, but no foot or finger bones, embedded in musculature of the body wall, but with no visible expression in external anatomy.)

The loss of back legs, and the development of flukes, fins, and flippers by whales, therefore stands as a classic case of a supposed cardinal problem in evolutionary theory – the failure to find intermediary fossils for major anatomical transitions, or even to imagine how such a bridging form might look or work. Darwin acknowledged the issue by constructing a much-criticized fable about swimming bears, instead of presenting any direct evidence at all, when he tried to conceptualize the evolution of whales. Modern creationists continue to use this example and stress the absence of intermediary forms in this supposed (they would say impossible) transition from land to sea.

Goethe told us to 'love those who yearn for the impossible.' But Pliny the Elder, before dying of curiosity by straying too close to Mount Vesuvius at the worst of all possible moments, urged us to treat impossibility as a relative claim: 'How many things, too, are looked upon as quite impossible until they have been actually effected.' Armed with such wisdom of human ages, I am absolutely delighted to report that our usually recalcitrant fossil record has come through in exemplary fashion. During the past fifteen years, new discoveries in Africa and Pakistan have greatly added to our paleontological knowledge of the earliest history of whales. The embarrassment of past absence has been replaced by a bounty of new evidence – and by the sweetest series of transitional fossils an evolutionist could ever hope to find. Truly, we have met the enemy and he is now ours. Moreover, to add blessed insult to the creationists' injury, these discoveries have arrived in a gradual and sequential fashion – a little bit at a time, step by step, from a tentative hint fifteen years ago to a remarkable smoking gun early in 1994. Intellectual history has matched life's genealogy by spanning the gaps in sequential steps. Consider the four main events in chronological order.

Case One. *Discovery of the oldest whale.* Paleontologists have been fairly confident, since Leigh Van Valen's demonstration in 1966, that whales descended from mesonychids, an early group of primarily carnivorous running mammals that spanned a great range of sizes and habits from eating fishes at river edges to crushing bones of carrion. Whales must have evolved during the Eocene epoch, some 50 million years ago, because Late Eocene and Oligocene rocks already contain fully marine cetaceans, well past any point of intermediacy.

In 1983, my colleague Phil Gingerich from the University of Michigan, along with N. A. Wells, D. E. Russell, and S. M. Ibrahim Shah, reported their discovery of the oldest whale, named *Pakicetus* to honor its country of present residence, from Middle Eocene sediments some 52 million years old in Pakistan. In terms of intermediacy, one could hardly have hoped for more from the limited material available, for only the skull of *Pakicetus* has been found. The teeth strongly resemble those of terrestrial mesonychids, as anticipated, but the skull, in feature after feature, clearly belongs to the developing lineage of whales.

Both the anatomy of the skull, particularly in the ear region, and the inferred habitat of the animal in life, testify to transitional status. The ears of modern whales contain modified bones and passageways that permit directional hearing in the dense medium of water. Modern whales have also evolved enlarged sinuses that can be filled with blood to maintain pressure during diving. The skull of *Pakicetus* lacks both these features, and this first whale could neither dive deeply nor hear directionally with any efficiency in water.

In 1993, J. G. M. Thewissen and S. T. Hussain affirmed these conclusions and added more details on the intermediacy of skull architecture in *Pakicetus*. Modern whales achieve much of their hearing through their jaws, as sound vibrations pass through the jaw to a 'fat pad' (the technical literature, for once, invents no jargon and employs the good old English vernacular in naming this structure), and thence to the middle ear. Terrestrial mammals, by contrast, detect most sound through the ear hole (called the 'external auditory meatus,' which means the same thing in more refined language). Since *Pakicetus* lacked the enlarged jaw hole that holds the fat pad, this first whale probably continued to hear through the pathways of its terrestrial ancestors. Gingerich concluded that 'the auditory mechanism of *Pakicetus* appears more similar to that of land mammals than it is to any group of extant marine mammals.'

As for place of discovery, Gingerich and colleagues found *Pakicetus* in river sediments bordering an ancient sea – an ideal habitat for the first stages of such an evolutionary transition (and a good explanation for lack of diving specializations if *Pakicetus* inhabited the mouths of rivers and adjacent shallow seas). My colleagues judged *Pakicetus* as 'an amphibious stage in the gradual evolutionary transition of primitive whales from land to sea . . . *Pakicetus* was well equipped to feed on fishes in the surface waters of shallow seas, but it lacked auditory adaptations necessary for a fully marine existence.'

Verdict: In terms of intermediacy, one could hardly hope for more from the limited material of skull bones alone. But the limit remains severe, and the results therefore inconclusive. We know nothing of the limbs, tail, or body form of *Pakicetus*, and therefore cannot judge transitional status in these key features of anyone's ordinary conception of a whale.

Case Two. *Discovery of the first complete hind limb in a fossil whale.* In the most famous mistake of early American paleontology, Thomas Jefferson, while not engaged in other pursuits usually judged more important, misidentified the claw of a fossil ground sloth as a lion. My prize for second worst error must go to R. Harlan, who, in 1834, named a marine fossil vertebrate *Basilosaurus* in the *Transactions of the American Philosophical Society. Basilosaurus* means 'king lizard,' but Harlan's creature is an early whale. Richard Owen, England's greatest anatomist, corrected Mr Harlan before the decade's end, but the name sticks – and must be retained by the official rules of zoological nomenclature. (The Linnaean naming system is a device for information retrieval, not a guarantor of appropriateness. The rules require that each species have a distinctive name, so that data can be associated unambiguously with a stable tag. Often, and inevitably, the names originally given become literally inappropriate for the unsurprising reason that scientists make frequent mistakes, and that new discoveries modify old conceptions. If we had to change names every time our ideas about a species altered, taxonomy would devolve into chaos. So *Basilosaurus* will always be *Basilosaurus* because Harlan followed the rules when he gave the name. And we do not change ourselves to *Homo horribilis* after Auschwitz, or to *Homo ridiculosis* after Tonya Harding – but remain, however dubiously, *Homo sapiens*, now and into whatever forever we allow ourselves.)

Basilosaurus, represented by two species, one from the United States and the other from Egypt, is the 'standard' and best-known early whale.

A few fragments of pelvic and leg bones had been found before, but not enough to know whether *Basilosaurus* bore working hind legs – the crucial feature for our usual concept of a satisfying intermediate form in both anatomical and functional senses.

In 1990, Phil Gingerich, B. H. Smith, and E. L. Simons reported their excavation and study of several hundred partial skeletons of the Egyptian species *Basilosaurus isis*, which lived some 5 to 10 million years after *Pakicetus*. In an exciting discovery, they reported the first complete hind limb skeleton found in any whale – a lovely and elegant structure (put together from several partial specimens), including all pelvic bones, all leg bones (femur, tibia, fibula, and even the patella, or kneecap), and nearly all foot and finger bones, right down to the phalanges (finger bones) of the three preserved digits.

This remarkable find might seem to clinch our proof of intermediacy, but for one small problem. The limbs are elegant but tiny (see the accompanying illustration), a mere 3 percent of the animal's total length. They are anatomically complete, and they did project from the body wall (unlike the truly vestigial hind limbs of modern whales), but these miniature legs could not have made any important contribution to locomotion – the real functional test of intermediacy. Gingerich et al. write: 'Hind limbs of *Basilosaurus* appear to have been too small relative to body size to have assisted in swimming, and they could not possibly have supported the body on land.' The authors strive bravely to invent some potential function for these minuscule limbs, and end up speculating that they may have served as 'guides during copulation, which may otherwise have been difficult in a serpentine aquatic mammal.' (I regard such guesswork as unnecessary, if not ill-conceived. We need not justify the existence of a structure by inventing some putative Darwinian function. All bodies contain vestigial features of little, if any, utility. Structures of lost usefulness in genealogical transitions do not disappear in an evolutionary overnight.)

Verdict: Terrific and exciting, but no cigar, and no bag-packer for creationists. The limbs, though complete, are too small to work as true intermediates must (if these particular limbs worked at all) – that is, for locomotion on both land and sea. I intend no criticism of *Basilosaurus*, but merely point out that this creature had already crossed the bridge (while retaining a most informative remnant of the other side). We must search for an earlier inhabitant of the bridge itself.

Case Three. *Hind limb bones of appropriate size. Indocetus ramani* is an early whale, found in shallow-water marine deposits of India and Pakistan, and intermediate in age between the *Pakicetus* skull and the *Basilosaurus* hind legs (cases one and two above). In 1993, P. D. Gingerich, S. M. Raza, M. Arif, M. Anwar, and X. Zhou reported the discovery of leg bones of substantial size from this species.

Gingerich and colleagues found pelvic bones and the ends of both femur and tibia, but no foot bones, and insufficient evidence for reconstructing the full limb and its articulations. The leg bones are large and presumably functional on both land and sea (the tibia, in particular differs a little in size and complexity from the same bone in the related and fully terrestrial mesonychid *Pachyaena ossifraga*). The authors conclude: 'The pelvis has a large and deep acetabulum [the socket for articulation of the femur, or thighbone], the proximal femur is robust, the tibia is long . . . All these features, taken together, indicate the *Indocetus* was probably able to support its weight on land, and it was almost certainly amphibious, as early Eocene *Pakicetus* is interpreted to have been . . . We speculate that *Indocetus*, like *Pakicetus*, entered the sea to feed on fish, but returned to land to rest and to birth and raise its young.'

Verdict: Almost there, but not quite. We need better material. All the right features are now in place – primarily leg bones of sufficient size and complexity – but we need more and better-preserved fossils.

Case Four: *Large, complete, and functional hind legs for land and sea – finding the smoking gun.* The first three cases, all discovered within ten years, surely indicate an increasingly successful paleontological assault upon an old and classic problem. Once you know where to look, and once high interest spurs great attention, full satisfaction often follows in short order. I was therefore delighted to read, in the January 14, 1994, issue of *Science*, an article by J. G. M. Thewissen, S. T. Hussain, and M. Arif, titled 'Fossil evidence for the origin of aquatic locomotion in archaeocete whales.'

In Pakistan, in sediments 120 meters above the beds that yielded *Pakicetus* (and therefore a bit younger in age), Thewissen and colleagues collected a remarkable skeleton of a new whale – not complete, but far better preserved than anything previously found of this age, and with crucial parts in place to illustrate a truly transitional status between land and sea. The chosen name, *Ambulocetus natans* (literally, the swimming walking-whale) advertises the excitement of this discovery.

Ambulocetus natans weighed some 650 pounds, the size of a hefty sea lion. The preserved tail vertebra is elongated, indicating that *Ambulocetus* still retained the long, thin mammalian tail, and had not yet transmuted this structure to a locomotory blade (as modern whales do in shortening the tail and evolving a prominent horizontal fluke as the animal's major means of propulsion). Unfortunately, no pelvic bones have been found but most elements of a large and powerful hind leg were recovered – including a complete femur, parts of the tibia and fibula, an astragalus (ankle bone), three metatarsals (foot bones), and several phalanges (finger bones). To quote the authors: 'The feet are enormous.' The fourth metatarsal, for example, is nearly six inches long, and the associated toe almost seven inches in length. Interestingly, the last phalanx of each toe ends in a small hoof, as in terrestrial mesonychid ancestors.

Moreover, this new bounty of information allows us to infer not only the form of this transitional whale, but also, with good confidence, an intermediary style of locomotion and mode of life (an impossibility with the first three cases, for *Pakicetus* is only a skull, *Basilosaurus* had already crossed the bridge, and *Indocetus* is too fragmentary). The forelimbs were smaller than the hind, and limited in motion; these front legs were, to quote the authors, 'probably used in maneuvering and steering while swimming, as in extant cetaceans ['modern whales' in ordinary language], and they lacked a major propulsive force in water.'

Modern whales move through the water by powerful beats of their horizontal tail flukes – a motion made possible by strong undulation of a flexible rear spinal column. *Ambulocetus* had not yet evolved a tail fluke, but the spine had requisite flexibility. Thewissen et al. write: '*Ambulocetus* swam by means of dorsoventral [back-to-belly] undulations of its vertebral column, as evidenced by the shape of the lumbar [lower back] vertebra.' These undulations then functioned with (and powered) the paddling of *Ambulocetus*'s large feet – and these feet provided the major propulsive force for swimming. Thewissen et al. conclude their article by writing: 'Like modern cetaceans – it swam by moving its spine up and down, but like seals, the main propulsive surface was provided by its feet. As such, *Ambulocetus* represents a critical intermediate between land mammals and marine cetaceans.'

Ambulocetus was no ballet dancer on land, but we have no reason to judge this creature as any less efficient than modern sea lions, which do manage, however inelegantly. Forelimbs may have extended out to the sides, largely for stability, with forward motion mostly supplied by extension of the back and consequent flexing of the hind limbs – again, rather like sea lions.

Verdict: Greedy paleontologists, used to working with fragments in reconstructing wholes, always want more (some pelvic bones would be nice, for starters), but if you had given me both a blank sheet of paper and a blank check, I could not have drawn you a theoretical intermediate any better or more convincing than *Ambulocetus*. Those dogmatists who can make white black, and black white, by verbal trickery will never be convinced by anything, but *Ambulocetus* is the very animal that creationists proclaimed impossible in theory.

Some discoveries in science are exciting because they revise or reverse previous expectations, others because they affirm with elegance something well suspected, but previously undocumented. Our four-case story, culminating in *Ambulocetus*, falls into this second category. This sequential discovery of picture-perfect intermediacy in the evolution of whales stands as a triumph in the history of paleontology. I cannot imagine a better tale for popular presentation of science, or a more satisfying, and intellectually based, political victory over lingering creationist opposition. As such, I present the story in this series of essays with both delight and relish.

Still, I must confess that this part of the tale does not intrigue me most as a scientist and evolutionary biologist. I don't mean to sound jaded or dogmatic, but *Ambulocetus* is so close to our expectation for a transitional form that its discovery could not provide a professional paleontologist with the greatest of all pleasures in science – surprise. As a public illustration and sociopolitical victory, transitional whales may provide the story of the decade, but paleontologists didn't doubt their existence or feel that a central theory would collapse if their absence continued. We love to place flesh upon our expectations (or put bones under them, to be more precise), but this kind of delight takes second place to the intellectual jolting of surprise.

I therefore find myself far more intrigued by another aspect of *Ambulocetus* that has not received much attention, either in technical or popular reports. For the anatomy of this transitional form illustrates a vital principle in evolutionary theory – one rarely discussed, or even explicitly formulated, but central to any understanding of nature's fascinating historical complexity.

In our Darwinian traditions, we focus too narrowly on the adaptive nature of organic form, and too little on the quirks and oddities encoded into every animal by history. We are so overwhelmed – as well we should be – by the intricacy of aerodynamic optimality of a bird's wing, or by the uncannily precise mimicry of a dead leaf by a butterfly. We do not ask often enough why natural selection had homed in upon this *particular* optimum – and not another among a set of unrealized alternatives. In other words, we are dazzled by good design and therefore stop our inquiry too soon when we have answered, 'How does this feature work so well?' – when we should also be asking the historian's questions: 'Why *this* and not *that?*' or 'Why *this* over here, and *that* in a related creature living elsewhere?'

To give the cardinal example from seagoing mammals: The two fully marine orders, Sirenia and Cetacea, both swim by beating horizontal tail flukes up and down. Since these two orders arose separately from terrestrial ancestors, the horizontal tail fluke evolved twice independently. Many hydrodynamic studies have documented both the mode and the excellence of such underwater locomotion, but researchers too often stop at an expression of engineering wonder, and do not ask the equally intriguing historian's question. Fishes swim in a truly opposite manner – also by propulsion from the rear, but with vertical tail flukes that beat from side to side (seals also hold their rear feet vertically and move them from side to side while swimming).

Both systems work equally well; both may be 'optimal.' But why should ancestral fishes favor one system, and returning mammals the orthogonal alternative? We do not wish to throw up our hands, and simply say 'six of one, half a dozen of the other.' Either way will do, and the manner chosen by evolution is effectively random in any individual case. 'Random' is a deep and profound concept of great positive utility and value, but some vernacular meanings amount to pure cop-out, as in this case. It may not matter in the 'grand scheme of things' whether optimality be achieved vertically or horizontally, but one or the other solution occurs for a reason in any particular case. The reasons may be unique to an individual lineage, and historically bound – that is, not related to any grand concept of pattern or predictability in the overall history of life – but local reasons do exist and should be ascertainable.

This subject, when discussed at all in evolutionary theory, goes by the name of 'multiple adaptive peaks.' We have developed some standard examples, but few with any real documentation; most are hypothetical, with no paleontological backup. (For example, my colleague

Dick Lewontin loves to present the following case in our joint introductory course in evolutionary biology: some rhinoceros species have two horns, others one horn. The two alternatives may work equally well for whatever rhinos do with their horns, and the pathway chosen may not matter. Two and one may be comparable solutions, or multiple adaptive peaks. Lewontin then points out that a reason must exist for two or one in any case, but that the explanation probably resides in happenstances of history, rather than in abstract predictions based on universal optimality. So far, so good. History's quirkiness, by populating the earth with a *variety* of *unpredictable* but sensible and well-working anatomical designs, does constitute the main fascination of evolution as a subject. But we can go no further with rhinos, for we have no data for understanding the particular pathway chosen in any individual case.)

I love the story of *Ambulocetus* because this transitional whale has provided hard data on reasons for a chosen pathway in one of our best examples of multiple adaptive peaks. Why did both orders of fully marine mammals choose the solution of horizontal tail flukes? Previous discussions have made the plausible argument that particular legacies of terrestrial mammalian ancestry established an anatomical predisposition. In particular, many mammals (but not other terrestrial vertebrates), especially among agile and fast-moving carnivores, run by flexing the spinal column up and down (conjure up a running tiger in your mind, and picture the undulating back). Mammals that are not particularly comfortable in water – dogs dog-paddling, for example – may keep their backs rigid and move only by flailing their legs. But semiaquatic mammals that swim for a living – notably the river otter (*Lutra*) and the sea otter (*Enhydra*) – move in water by powerful vertical bending of the spinal column in the rear part of the body. This vertical bending propels the body forward both by itself (and by driving the tail up and down), and by sweeping the hind limbs back and forth in paddling as the body undulates.

Thus, horizontal tail flukes may evolve in fully marine mammals because inherited spinal flexibility for movement up and down (rather than side to side) directed this pathway from a terrestrial past. This scenario has only been a good story up to now, with limited symbolic support from living otters, but no direct evidence at all from the ancestry of whales or sirenians. *Ambulocetus* provides this direct evidence in a most elegant manner – for all pieces of the puzzle lie within the recovered fossil skeleton.

We may infer from a tail vertebra that *Ambulocetus* retained a long and thin mammalian tail, and had not yet evolved the horizontal fluke. We know from the spinal column that this transitional whale retained its mammalian signature of flexibility for up and down movement – and from the large hind legs that undulation of the back must have supplied propulsion to powerful paddling feet, as in modern otters.

Thewissen and colleagues draw the proper evolutionary conclusion from these facts, thus supplying beautiful evidence to nail down a classic case of multiple adaptive peaks with paleontological data: '*Ambulocetus* shows that spinal undulation evolved before the tail fluke . . . Cetaceans have gone through a stage that combined hindlimb paddling and spinal undulation, resembling the aquatic locomotion of fast swimming otters.' The horizontal tail fluke, in other words, evolved because whales carried their terrestrial system of spinal motion to the water.

History channels a pathway among numerous theoretical alternatives. In his last play, Shakespeare noted that 'what's past is prologue; what to come, in yours and my discharge.' But present moments build no such wall of separation between a past that molds us and a future under our control. The hand of the past reaches forward right through us and into an uncertain future that we cannot fully specify.

Epilogue

I wrote this essay in a flush of excitement during the week that Thewissen and colleagues published their discovery of the definitive intermediate whale *Ambulocetus*, in January 1994. With my lead time of three months from composition to the first publication of these essays in *Natural History* magazine, 'Hooking Leviathan by Its Past' appeared in April 1994 – complete with central theme of a chronologically developing story in four stages.

I think of the old spiritual: 'Sometimes I get discouraged, and think my work's in vain. But then the Holy Spirit revives my soul again.' I'm actually a fairly cheerful soul, but we all need replenishment now and then. If 'there is a balm in Gilead' (the song's title) for scientists, that elixir, that infusion of the holy spirit, takes the form of new discoveries. On the very week of my essay's publication, Phil Gingerich and colleagues published their description of yet another intermediate fossil whale, a fifth tale for this gorgeous sequence of evolutionary and paleontological affirmation. (I did feel a bit funny about the superannuation of my essay on the day of its birth, but all exciting science must be obsolescent from inception – and I knew I could write this epilogue for my next book!)

Gingerich and colleagues discovered and named a new fossil Eocene whale from Pakistan, *Rodhocetus kasrani* (*Rodho* for the local name of the region, *kasrani* for the group of Baluchi people living in the area). *Rodhocetus*, estimated at some ten feet in length, lived about 46.5 million years ago. This new whale is thus about 3 million years younger than the 'smoking gun' *Ambulocetus* (Case Four and the key story in the main essay), and about the same age as *Indocetus* (stage three in the main essay). No forelimb bones have been found, and the spinal column lacks tail vertebrae, but much of the skull has been recovered with, perhaps more important, a nearly complete vertebral column from the neck all the way back to the beginning of the tail. Most of the pelvis has also been found and, crucial to evidence about intermediacy, a complete femur (but no other elements of the hind limb).

We may summarize the importance of *Rodhocetus*, and its gratifying extension of our story about 'hard' evidence for intermediacy in the evolution of whales from terrestrial ancestors, by summarizing evidence in the three great categories of paleontological data: form (anatomy), habitat (environment), and function.

Form. I was most struck by two features of *Rodhocetus*'s anatomy. First, the excellent preservation of the vertebral column provides good evidence of intermediacy in a mixture of features retained from a terrestrial past with others newly acquired for an aquatic present. The high neural spines (upward projections) of the anterior thoracic vertebrae (just behind the neck) support muscles that help to hold up the head in terrestrial animals (not a functional necessity in the buoyancy of marine environments; whales evolved from a terrestrial group, the mesonychids, with particularly large heads). Direct articulation of the pelvis with the sacrum (the adjacent region of the vertebral column) also characterizes both *Rodhocetus* and terrestrial mammals (where gravity requires this extra strength), but does not occur in modern whales. Gingerich and colleagues conclude: 'These are primitive characteristics of mammals that support their weight on land, and both suggest that *Rodhocetus* or an immediate predecessor was still partly terrestrial.'

But other features of the spinal column indicate adaptation for swimming: short cervical (neck) vertebrae, implying rigidity for the front end of the body (good for cutting through the water as the rear parts of the animal provide propulsion); and, especially, the seamless flexibility of posterior vertebrae (sacral vertebrae are fused together in most large terrestrial mammals, but unfused in both modern whales and *Rodhocetus*), an important configuration for providing forward thrust in swimming. Gingerich and colleagues conclude: 'These are derived characteristics of later archaeocetes [ancient whales] and modern whales associated with aquatic locomotion.'

Second, and even more striking for this essay's case of graded intermediacy, sequentially discovered during the past twenty years, *Rodhocetus* is about 3 million years younger than the 'smoking gun' *Ambulocetus* (a marine whale with limbs large enough for movement on land as well), and a good deal older than later whales that had already crossed the bridge to fully marine life (*Basilosaurus*, my Case Two, with well-formed but tiny hind limbs that could not have functioned on land, and probably didn't do much in water either). In the most exciting discovery of this new Case Five, the femur of *Rodhocetus* is about two thirds as long as the same bone in the older *Ambulocetus* – still functional on land (probably), but already further reduced after 3 million additional years of evolution.

Habitat. *Rodhocetus* is the oldest whale from fully and fairly deep marine waters. The oldest of all whales, *Pakicetus* of Case One, lived around the mouths of rivers; *Ambulocetus* and *Indocetus* of Cases Three and Four inhabited very shallow marine waters. Interestingly, the more fully marine habitat of *Rodhocetus* correlates with greater reduction of the hind limb, for *Indocetus* is a contemporary of *Rodhocetus*, yet grew a larger femur comparable in length with the earlier *Ambulocetus*. (All three creatures had about the same body size). Thus, admittedly on limited evidence, limbs decreased in size over time and became smaller faster in whales from more fully marine environments. (Perhaps *Rodhocetus* had already ceased making excursions on land, while the earlier *Ambulocetus*, with a larger femur, almost surely inhabited both land and water.) In any case, the contemporaneity of *Rodhocetus* (shorter femur and deeper water) and *Indocetus* (longer femur with life in shallower water) illustrates the diversity that already existed in cetacean evolution. Evolution, as I always say, no doubt to the point of reader's boredom, is a copiously branching bush, not a ladder.

Function. *Rodhocetus* lacks tail vertebrae, so we can't tell for sure whether or not this whale had yet evolved a tail fluke. But evidence of the beautifully preserved spinal column – particularly the unfused sacral vertebrae, 'making,' in the words of Gingerich et al., 'the lumbocaudal [back-to-tail] column seamlessly flexible' – indicates strong dorsoventral (back-to-belly) flexion at the rear end of the body – the prerequisites for swimming in the style of modern whales (with propulsion provided by a horizontal tail fluke, driven up and down by bending the vertebral column). I was particularly pleased by this result, since I closed my essay with a mini-disquisition on multiple adaptive peaks and the importance of historical legacies, as illustrated by vertical tail fins in fishes vs. horizontal flukes in whales – both solutions working equally well, but with whales limited to this less familiar alternative because they evolved from terrestrial ancestors with backs that flexed dorsoventrally in running. Gingerich and colleagues conclude: 'This indicates that the characteristic cetacean mode of swimming by dorsoventral oscillation of a heavily muscled tail evolved within the first three million years or so of the appearance of the archaeocetes.'

A tangential comment in closing: The sociology of science includes much that I do not like, but let us praise what we do well. Science at its best is happily and vigorously international – and I can only take great pleasure in the following list of authors for research done in an American lab based on fieldwork in Asia, supported by the Geological Survey of Pakistan: Philip D. Gingerich, S. Mahmood Raza, Muhammad Arif, Mohammad Anwar, and Xiaoyuan Zhou. Bravo to you all. I also couldn't help noting the paper's first sentence: 'The early evolution of whales is illustrated by partial skulls and skeletons of five archaeocetes of Ypresian (Early Eocene) . . . age.' The geological time scale is just as international, for our fossil record is a global scheme for correlating the ages of rocks. So a layer of sediments in Pakistan may be identified as representing a time named for a place that later became the bloodiest European battle site of World War I – the dreaded Ypres (or 'Wipers' as British soldiers named and pronounced their hecatomb).

But so much for lugubrious and sentimental thoughts. Let's just end in the main essay's format for our new case of *Rodhocetus*:

Case Five. Open and shut.

Verdict: sustained in spades, wine and roses.

SOURCES AND ACKNOWLEDGEMENTS

'I Have Landed', from *I Have Landed: Splashes and Reflections in Natural History* (London: Jonathan Cape and New York: Ballantine Books, 2002), pp. 13–25. Copyright © Turbo Inc. 2002. Reprinted by permission of The Random House Group Ltd and Random House Inc., USA.

'The Median Isn't the Message', from *Bully for Brontosaurus: Reflections in Natural History* (London: Hutchinson Radius, 1991), pp. 473–8. Copyright © Stephen Jay Gould 1991. Reprinted by permission of The Random House Group Ltd and Random House Inc., USA.

'The Streak of Streaks', from *Bully for Brontosaurus: Reflections in Natural History* (London: Hutchinson Radius, 1991), pp. 463–72. Copyright © Stephen Jay Gould 1991. Reprinted by permission of The Random House Group Ltd and Random House Inc., USA.

'Seven Inning Stretch: Baseball, Father and Me', from *Triumph and Tragedy in Mudsville: A Lifelong Passion for Baseball* (London: Jonathan Cape, 2004), pp. 25–34. Copyright © Turbo Inc. 2003. Reprinted by permission of The Random House Group Ltd.

'Trouble in Our Own House: A Brief Legal Survey from Scopes to Scalia', from *Rocks of Ages: Science and Religion in the Fullness of Life* (New York: The Library of Contemporary Thought/The Ballantine Publishing Group, 1999 and London: Jonathan Cape, 2001), pp. 133–50. Copyright © Stephen Jay Gould 1999. Reprinted by permission of The Random House Group Ltd and Random House Inc., USA.

'Of Two Minds and One Nature', co-written with Rhonda Shearer, reprinted by permission of the American Association for the Advancement of Science from *Science*, 286, 5442 (1999): 1093–4. Copyright © AAAS 1999.

'Thomas Burnet's Battleground of Time', reprinted by permission of the publisher from *Time's Arrow, Time's Cycle: Myth and Metaphor in the Discovery of Geological Time* (Cambridge, Mass. and London: Harvard

University Press, 1987), pp. 21–41. Copyright © 1987 by the President and Fellows of Harvard College.

'The Lying Stones of Marrakech', from *The Lying Stones of Marrakech: Penultimate Reflections in Natural History* (London: Jonathan Cape and New York: Ballantine Books, 2000), pp. 9–26. Copyright © Turbo Inc. 2000. Reprinted by permission of The Random House Group Ltd and Random House Inc., USA

'The Stinkstones of Oeningen', from *Hen's Teeth and Horse's Toes: Further Reflections on Natural History* (New York and London: W. W. Norton, 1983), pp. 94–106. Copyright © Stephen Jay Gould 1983. Reprinted by permission of W. W. Norton & Company.

'The Razumovsky Duet', from *Dinosaur in a Haystack: Reflections in Natural History* (New York: Ballantine Books, 1995 and London: Jonathan Cape, 1996), pp. 260–71. Copyright © Stephen Jay Gould 1995. Reprinted by permission of The Random House Group Ltd and Random House Inc., USA.

'The Power of Narrative', from *An Urchin in the Storm: Essays about Books and Ideas* (London: Harvill, 1988), pp. 75–92. Copyright © Stephen Jay Gould 1987. Reprinted by permission of The Random House Group Ltd.

'Not Necessarily a Wing', from *Bully for Brontosaurus: Reflections in Natural History* (London: Hutchinson Radius, 1991), pp. 199–51. Copyright © Stephen Jay Gould 1991. Reprinted by permission of The Random House Group Ltd and Random House Inc., USA.

'Worm for a Century, and All Seasons', from *Hen's Teeth and Horse's Toes: Further Reflections on Natural History* (New York and London: W. W. Norton, 1983), pp. 120–133. Copyright © Stephen Jay Gould 1983. Reprinted by permission of W. W. Norton & Company.

'The Darwinian Gentleman at Marx's Funeral: Resolving Evolution's Oddest Coupling', from *I Have Landed: Splashes and Reflections in Natural History* (London: Jonathan Cape and New York: Ballantine Books, 2002), pp. 113–29. Copyright © Turbo Inc. 2002. Reprinted by permission of The Random House Group Ltd and Random House Inc., USA.

'The Piltdown Conspiracy', from *Hen's Teeth and Horse's Toes: Further Reflections on Natural History* (New York and London: W. W. Norton, 1983), pp. 201–26. Copyright © Stephen Jay Gould 1983. Reprinted by permission of W. W. Norton & Company.

'The Evolution of Life on the Earth', *Scientific American* (October 1994): 63–9. Reprinted by kind permission of *Scientific American*.

'Challenges to Neo-Darwinism and Their Meaning for a Revised View of Human Consciousness', delivered as a Tanner Lecture in Human Values at Clare Hall, Cambridge University, 30 April and 1 May 1984. Printed with the permission of the Tanner Lectures on Human Values, a Corporation, University of Utah, Salt Lake City, Utah.

'The Structure of Evolutionary Theory: Revising the Three Central Features of Darwinian Logic', reprinted by permission of the publisher from *The Structure of Evoluionary Theory* (Cambridge, Mass.: The Belknap Press of Harvard University Press, 2002), pp. 12–15; 75–89. Copyright © 2002 by the President and Fellows of Harvard College.

'The Episodic Nature of Evolutionary Change', from *The Panda's Thumb: More Reflections in Natural History* (New York and London: W. W. Norton, 1980), pp. 149–54. Copyright © Stephen Jay Gould 1980. Reprinted by permission of W. W. Norton & Company.

'Betting on Chance – and No Fair Peeking', from *Eight Little Piggies: Reflections in Natural History* (London: Jonathan Cape, 1993), pp. 396–406. Copyright © Stephen Jay Gould 1993. Reprinted by permission of The Random House Group Ltd and Random House Inc., USA.

'The Power of the Modal Bacter, or Why the Tail Can't Wag the Dog', from *Life's Grandeur: The Spread of Excellence from Plato to Darwin* (London: Jonathan Cape, 1996; published in the USA as *Full House: The Spread of Excellence from Plato to Darwin*, New York: Ballantine Books, 1996), pp. 167–75. Copyright © Stephen Jay Gould 1996. Reprinted by permission of The Random House Group Ltd and Random House Inc., USA.

'The Great Dying', from *Ever since Darwin: Reflections in Natural History* (London: Andre Deutsch, 1977), pp. 134–8. Copyright © Stephen Jay Gould 1977. Reprinted by permission of Andre Deutsch.

'The Validation of Continental Drift', from *Ever since Darwin: Reflections in Natural History* (London: Andre Deutsch, 1977), pp. 160–7. Copyright © Stephen Jay Gould 1977. Reprinted by permission of Andre Deutsch.

'Phyletic Size Decrease in Hershey Bars', from *Hen's Teeth and Horse's Toes: Further Reflections on Natural History* (New York and London: W. W. Norton, 1983), pp. 313–19. Copyright © Stephen Jay Gould 1983. Reprinted by permission of W. W. Norton & Company.

'Opus 100', from *The Flamingo's Smile: Reflections in Natural History* (London: Penguin Books, 1987; revised edition 1995), pp. 167–84. Copyright © Stephen Jay Gould 1985, 1995. Reprinted by permission of Penguin Books Ltd.

'Size and Shape', from *Ever since Darwin: Reflections in Natural History* (London: Andre Deutsch, 1977), pp. 171–8. Copyright © Stephen Jay Gould 1977. Reprinted by permission of Andre Deutsch.

'How the Zebra Gets Its Stripes', from *Hen's Teeth and Horse's Toes: Further Reflections on Natural History* (New York and London: W. W.

Norton, 1983), pp. 366–75. Copyright © Stephen Jay Gould 1983. Reprinted by permission of W. W. Norton & Company.
'Size and Scaling in Human Evolution', co-written with David Pilbeam, reprinted by permission of the American Association for the Advancement of Science from *Science* 186 (1974): 892–901. Copyright © AAAS 1974.
'The Ladder and the Cone: Iconographies of Progress', from *Wonderful Life: The Burgess Shale and the Nature of History* (London: Hutchinson Radius, 1990), pp. 27–45. Copyright © Stephen Jay Gould 1989. Reprinted by permission of The Random House Group Ltd and Random House Inc., USA.
'Up against a Wall', from *Leonardo's Mountain of Clams and the Diet of Worms: Essays on Natural History* (London: Jonathan Cape and New York: Ballantine Books, 1998). Copyright © Turbo Inc. 1998. Reprinted by permission of The Random House Group Ltd and Random House Inc., USA.
'Ontogeny and Phylogeny', reprinted by permission of the publisher from *Ontogeny and Phylogeny* (Cambridge, Mass. and London: Harvard University Press, 1977), pp. 115–135. Copyright © 1977 by the President and Fellows of Harvard College.
'The Spandrels of San Marco and the Panglossian Paradigm: A Critique of the Adaptationist Programme', co-written with Richard Lewontin, *Proceedings of the Royal Society of London. Series B: Biological Sciences*, 205 (1979): 581–98. Reproduced by permission of the Royal Society.
'More Things in Heaven and Earth', from Hilary Rose and Steve Rose (eds), *Alas, Poor Darwin: Arguments against Evolutionary Psychology* (London: Jonathan Cape and New York: Ballantine Books, 2000), pp. 85–105. Copyright © Stephen Jay Gould 2000. Reprinted by permission of The Random House Group Ltd and Random House Inc., USA.
'Posture Maketh the Man', from *Ever since Darwin: Reflections in Natural History* (London: Andre Deutsch, 1977), pp. 207–13. Copyright © Stephen Jay Gould 1977. Reprinted by permission of Andre Deutsch.
'Freud's Evolutionary Fantasy', from *I Have Landed: Splashes and Reflections in Natural History* (London: Jonathan Cape and New York: Ballantine Books, 2002), pp. 147–58. Copyright © Turbo Inc. 2002. Reprinted by permission of The Random House Group Ltd and Random House Inc., USA.
'Measuring Heads: Paul Broca and Heyday of Craniology', from *The Mismeasure of Man* (London: Penguin Books, 1984), pp. 73–112. Copyright © Stephen Jay Gould 1984. Reprinted by permission of Penguin Books Ltd.
'The Most Unkindest Cut of All', from *Dinosaur in a Haystack: Reflections*

INDEX